Dairy Product Technology

— Recent Advances —

THE EDITORS

Dr. Subrota Hati is working as an Assistant Professor in Department of Dairy Microbiology, SMC College of Dairy Science, Anand Agricultural University, Anand, Gujarat. He graduated from West Bengal University of Animal and Fishery Sciences in Dairy Technology in 2006. Then he did his M.Sc. in Dairy Microbiology from National Dairy Research Institute, Karnal. He also obtained Doctoral degree in Dairy Microbiology from National Dairy Research Institute, Karnal in 2012. Now he is working presently on Isolation and characterization of milk derived bioactive peptides and its biofunctional properties. He also served Dairy Industry as Quality Assurance Executive in Mother Dairy, New Delhi. He is also handling externally funded projects by ICAR, DST and DBT as PI or Co-PI. He was the recipient of various awards: URMILABALA GOLD MEDAL by Indian Dairy Association, New Delhi; PROF. SUKUMER DE GOLD MEDAL as a Topper by WBUAFS, Kolkata; Two Best Paper awards by IDA, New Delhi; 03 Best Poster awards; Silver Medal by All Bengal Teachers' Asso, Govt. West Bengal, Young Scientist Award by Bioved, Allhabad; Best Young Scientist award by SASNET-FF, AAU, Anand and Lund University, Sweden and Best Oral Presentation Award by IIFANS, New Delhi etc. He is the recipient of Junior and Senior Research fellowship during his Master and Doctoral programmes in National Dairy Research Institute, Karnal, Haryana. He is member of various Scientific Society: Life member of Dairy Technology Society of India, Karnal; Life membership of SASNET-Fermented Foods, Anand; Indian Dairy Associations, New Delhi. He is also acting as Editorial Board member of American Journal of Food and Nutrition, Journal of Dairy and Food Technology and also reviewers of National and International Journals. He is recently awarded the Young scientist Project by DST, Govt. India, New Delhi on ACE inhibitory Bioactive Peptides derived from fermented milks. He has published 18 research papers, 05 Review Articles, 22 Technical articles in various National and International peer reviewed Journals and also published 16 Book Chapters and presented 35 abstracts in various National and International Seminars/Conferences.

Dr. Surajit Mandal is working as Scientist in National Dairy Research Institute, Karnal, Haryana. He did his B.Tech. (Dairy Technology) from West Bengal University of Animal and Fishery Sciences, Kolkata; M.Sc. in Dairy Microbiology and Ph.D. in Dairy Microbiology from National Dairy Research Institute, Karnal, India. He has eight years teaching and research experience so far. Currently he is doing research in the area of functional foods and dairy products incorporating live probiotics, microencapsulation of probiotics, technology of functional lactic cultures for fermented and non-fermented dairy products as well as actively engaged with different external in house research projects. Regularly teach to undergraduate and post graduate students of Dairy Technology and Dairy Microbiology and also guided 11 Master scholars and guiding 04 Ph.D. scholars. He developed several teaching laboratory manuals for undergraduate and postgraduate microbiology courses. He published several research, review and popular articles in national and international Journals. Also published several book chapters and teaching reviews in various training programmes. Attended various national and international conferences, seminars, symposiums, workshops and presented papers and posters. He received several awards for best papers and posters/presentations. He attended various advanced training programme and organized training programme for industry personnel and academicians. He is heading membership of several scientific organizations and societies. He involves in regular paper setter/moderator and evaluator for various microbiology courses at undergraduate and postgraduate degrees for different universities/institutions, reviewer for national and international Journals.

Dr. Birendra Kumar Mishra is working as an Assistant Professor in the Department of Rural Development and Agricultural Production at North-Eastern Hill University, Tura campus, Meghalaya. He received a Ph.D. degree in Animal Husbandry and Dairying (Dairy Technology) in 2003 from Institute of Agricultural Sciences, BHU, Varanasi. He completed his M.Sc. (Ag) in Dairy Science from C.S.A.Agriculture University, Kanpur in1995 with gold medal for obtaining first class and position. He obtained his B.Sc. (Ag) in 1991 with first class and first position from Gorakhpur University, Gorakhpur. Dr.Mishra has published more than two dozen research papers and articles in referred journals and also published six edited book titled-Livestock Production and Rural Development in India, Sustainability and Economic Development in Hill Agriculture, Livestock Production and Management, Advances in Livestock Production and Management and Dairy and Food Processing Industry-Recent Trends in two volumes. He served private dairy industry in various capacities as Quality Control Manager and Technical Officer etc. during 1995 to 1999.He is also a member and life member of various professional bodies and societies at national and international level. He is the recipients of Young Scientist Award 2010 from BIOVED Research Society, Allahabad, U.P, Founder Fellow Award 2011, from Indian Society of Hill Agriculture for his outstanding contribution in the field of Rural Development and Livestock Production and BIOVED Fellowship award 2012 from BIOVED research society, Allahabad, U.P.

Dairy Product Technology

— *Recent Advances* —

Editors

Subrota Hati

Dairy Microbiology Department
SMC College of Dairy Science
Anand Agricultural University, Anand
Gujarat (India)

Surajit Mandal

Dairy Microbiology Division
National Dairy Research Institute, Karnal
Haryana (India)

Birendra Kumar Mishra

Department of Rural Development and Agricultural Production
North-Eastern Hill University
Tura Campus-794002
Meghalaya (India)

2015

Daya Publishing House®

A Division of

Astral International Pvt. Ltd.

New Delhi – 110 002

Cataloging in Publication Data--DK
Courtesy: D.K. Agencies (P) Ltd. <docinfo@dkagencies.com>

Dairy product technology : recent advances / editors, Subrota Hati, Surajit Mandal, Birendra Kumar Mishra.
pages ; 23 cm
Includes bibliographical references and index.
ISBN 9789351306337 (International Edition)

1. Dairy processing--India. 2. Dairy products industry--India. 3. Milk--Microbiology--India. 4. Cheese--Microbiology--India. I. Hati, Subrota, editor. II. Mandal, Surajit, editor. III. Mishra, Birendra Kumar, editor.

DDC 637.0954 23

Published by : **Daya Publishing House®**
A Division of
Astral International Pvt. Ltd.
– ISO 9001:2008 Certified Company –
4760-61/23, Ansari Road, Darya Ganj
New Delhi-110 002
Ph. 011-43549197, 23278134
E-mail: info@astralint.com
Website: www.astralint.com

Laser Typesetting : **Classic Computer Services**, Delhi - 110 035

Printed at : **Thomson Press India Limited**

PRINTED IN INDIA

Dedicated to My Respected Teacher

Mr. Shivshankar Pathak

and

Mr. Mihir Barman

SMC COLLEGE OF DAIRY SCIENCE
ANAND AGRICULTURAL UNIVERSITY
ANAND-388 110 (GUJARAT)

Phone: +91 (02692) 261030
Email: principaldsc@yahoo.com

Fax: (02692) 261314
Date: 28.08.2014

Foreword

Milk is a food for all. Milk and milk products are the daily diet in the food basket. Advancement of Science and Technology changes the processing infrastructure in Dairy Plants for the complete automation of the whole processing system. Scientists are working to develop functional fermented or non-fermented value added dairy products to meet the demand of the health conscious consumers. Consumers are now focusing on the medicinal foods, which are manufactured to target the life style disorder diseases like hypertension, diabetes, cancer and many more. Dairy Science and Technology is helping the Dairy Industries to provide the cost effective technology by modifying the traditional methods, replacing the traditional starter cultures with functional ones, addition of dietary fibers and probiotics, incorporation of fat replacers, bioactive components such as bioactive lipids, peptides etc.

It is a great pleasure that this book entitled *"Dairy Product Technology: Recent Advances"* has been edited by Subrota Hati, Surajit Mandal and BK Mishra. These editors are fully involved in teaching, guiding the research scholars and handling various edge-cutting research projects in the emerging field of Dairy Science and Technology. This book has been well written by the eminent Professors/Scientists from different State Agricultural Universities and National Institute covering all the chapters like scope and emerging need of value chain in Indian dairying, effect of production and processing on quality of milk and dairy products, bioactive fermented dairy products - value addition through probiotics, biotechnological approaches, value addition to indigenous dairy products and infant food formulations, milk derived bio-active peptides, functional lactic acid bacteria and fermented dairy products, their health benefits, clinical studies for fermented foods, packaging strategies for indigenous and functional dairy products, status and management of dairy waste in India, recent biotechnological approaches in Dairy and Food industry. This book will enrich the knowledge of the undergraduate and post graduate students in Dairy Technology and Food Technology discipline as well as it will also serve the purpose of personnel working in the Dairy and Food Plants. Infact, it is the need of the hour to compile maximum possible information useful for Dairy and Food Technology to enhance the research and development as well as to find the new horizon for the expansion of the growth of the Dairy Industries. I congratulate the Editors for compiling the fruitful information contributed by the eminent Professor/Scientists and published as an essential Book to upgrade the knowledge of Dairy Science and Technology.

28/8/13

Dr. B.P. Shah

Principal and Dean

Preface

India is the largest milk producing country in the globe (approx. 133 million tones in 2014). Dairy industry is of enormous importance for development of our country because of the central linkages and synergies that it promotes between the two pillars of our economy *i.e.,* industry and agriculture. More than 2445 million people are economically involved in agriculture worldwide; probably two-third or even more are entirely or partly dependent on livestock farming. Fast growth in the food processing sector and progressive improvement in value addition sequences are also of great significance for achieving favorable terms of the trade for Indian agriculture both in the domestic and overseas markets. The total food production in India has been projected likely to double in the next ten years and there is a golden opportunity for large investments in food and food processing technologies, skills and equipment, especially in the areas of canning, dairy and food processing, specialty processing, packaging, frozen food/refrigeration and thermo-processing. Fruits and vegetables, fisheries, milk and milk products, meat and poultry, packaged/convenience foods, alcoholic beverages and soft drinks etc are important sub-sectors of the food processing industry. Health food and health food supplements are rapidly growing segment of this industry, which is gaining vast popularity amongst the health conscious populace.

This book is mainly focused on effect of production and processing on microbiological quality of milk, bioactive fermented dairy products with special reference to cheese: scope and challenges, cheese: a novel milieu of probiotics, genetically modified cheese: a novel biotechnological development, value addition to indigenous dairy products: scope and future strategies, health benefits of milk derived bio-active peptides, Indian dairy value chain: opportunities and challenges, structural changes of Indian dairy farming- smallholder prospects, clinical studies for novel fermented foods, *Lactobacillus reuteri*: a multifaceted lactic acid bacterium for fermented foods, aspects of infant formulations: an overview, nutritional significance of milk, fermented dairy products, recent status of dairy waste in India, health benefits of fermented milk products, LAB: Potential application on dairy and food industry, traditional packaging strategies for Indigenous dairy products, recent biotechnological approaches in Dairy and Food industry. This Book has covered all the integral segments of dairy industry as well as food industry. This book would be helpful towards the Food Technological Colleges, Dairy Science Colleges, Dairy Industry, Food Industry as a upgraded information related to their work fields.

Subrota Hati

Surajit Mandal

Birendra Kumar Mishra

Contents

List of Contributors

Ahuja, Kunal
Assistant Professor, Department Dairy Technology, Dairy Science College, Amreli, Gujarat

Behare, Pradip V.
Scientist, Dairy Microbiology Division, National Dairy Research Institute, Karnal – 132 001, Haryana

Bera, S.
Ph.D. Scholar, Department Dairy Technology, WBUAFS, Kolkata, West Bengal

Charan, Rohit
M.V.Sc Research Scholar, College of Veterinary Science and Animal Husbandry, SDAU, S.K. Nagar, Gujarat

Chavda, Viral Ray Sinh
Assistant Professor, Dairy Science College, Kamdhenu University, Amreli

Dahale, Pravin
ITC Ltd., R&D Center, Bangalore

Das, Anamika
Assistant Professor, Department of Dairy Chemistry, SHIATS, Allahabad

Datta, K.K.
Dairy Economics Statistics and Management, National Dairy Research Institute, Karnal – 132 001, Haryana

Goyal, G.K.
National Dairy Research Institute, Karnal – 132 001, Haryana

Goyal, Sumit
National Dairy Research Institute, Karnal – 132 001, Haryana

Hati, Subrota
Assistant Professor, Department of Dairy Microbiology, Anand Agricultural University, Anand – 388 110, Gujarat

Jagbir, Rehal
Assistant Professor (Food Science and Technology), PAU, Punjab

Jain, Amit Kumar
Assistant Professor, Department of Dairy Chemistry, Anand Agricultural University, Anand – 388 110, Gujarat

Kaur, Gagan Jyot
Assistant Professor (Agri. Engineering), PAU, Punjab

Kiran Bala
National Dairy Research Institute, Karnal – 132 001, Haryana

Kumar, Amit
Associate Professor, Shri G.N. Patel Dairy Science and Food Technology College, Sardarkrushinagar Dantiwada Agricultural University, Sardarkrushinagar – 385 506, Gujarat

Lule, Vaibhao
Ph.D. Scholar, Dairy Microbiology Division, National Dairy Research Institute, Karnal – 132 001, Haryana

Makhal, S.
Research Scientist, ITC Ltd., R&D Center, Bangalore

Malik, R.K.
Principal Scientist, Dairy Microbiology Division, National Dairy Research Institute, Karnal – 132 001, Haryana, India

Mandal, Surajit
Scientist, Dairy Microbiology Division, National Dairy Research Institute, Karnal – 132 001, Haryana

Mann, B.
Principal Scientist, Dairy Chemistry Division, National Dairy Research Institute, Karnal – 132 001, Haryana

Mishra, K.K.
Associate Professor, Department of Medicine, College of Veterinary Science and A.H. (NDVSU), Rewa, M.P.

Mishra, Santosh Kumar
Assistant Professor, College of Dairy Science and Technology, GADVASU, Ludhiana, Punjab

Mogha, Kanchan
Ph.D. Scholar, Department of Dairy Microbiology, Anand Agricultural University, Anand – 388 110, Gujarat

Mudgal, Sreeja
Assistant Professor, Department of Dairy Microbiology, Anand Agricultural University, Anand – 388 110, Gujarat

Narwader, B.M.
Assistant Professor, Department of Animal Husbandry and Dairy Science, College of Agriculture, Parbhani, Vasantrao Naik Marathwada Krishi Vidyapeeth, Parbhani, Maharastra

Padghan, P.V.
Assistant Professor, Department of Animal Husbandry and Dairy Science, College of Agriculture, Latur, Vasantrao Naik Marathwada Krishi Vidyapeeth, Parbhani, Maharastra

Palit, Dipanjan
Deputy Manager (Production), Glaxo Smithkline Consumer Healthcare Limited, Haryana

Parmar, Satish
Assistant Professor, Department of Dairy Chemistry, Anand Agricultural University, Anand – 388 110, Gujarat

Pawar, N.S.
Assistant Professor, Department of Dairy Chemistry, Udgir, Maharashtra

Roy, S.K.
Professor, Shri G.N. Patel Dairy Science and Food Technology College, Sardarkrushinagar Dantiwada Agricultural University, Sardarkrushinagar – 385 506, Gujarat

Salma
M.Tech., Research Scholar, Department Dairy Microbiology, Anand Agricultural University, Anand – 388 110, Gujarat

Sarkate, Santosh
Scientist, ITC Ltd., R&D Center, Bangalore

Shaikh, Ahesanvarish
Assistant Professor, Department of Dairy Chemistry, Anand Agricultural University, Anand – 388 110, Gujarat

Shendurse, A.M.

Assistant Professor, Shri G.N. Patel Dairy Science and Food Technology College, Sardarkrushinagar Dantiwada Agricultural University, Sardarkrushinagar – 385 506, Gujarat

Singh, Shiv Raj

Assistant Professor, Department of Dairy and Food Business Management, SDAU, Banaskantha – 385 506, Gujarat

Talwar, Gopika

Assistant Professor, College of Dairy Science and Technology, GADVASU, Ludhiana, Punjab

Tomar, S.K.

Principal Scientist, Dairy Microbiology Division, National Dairy Research Institute, Karnal – 132 001, Haryana

2015, Dairy Product Technology: Recent Advances ***Pages 1–12***
Editors: **Subrota Hati, Surajit Mandal and Birendra Kumar Mishra**
Published by: **DAYA PUBLISHING HOUSE, NEW DELHI**

Chapter 1

Effect of Production and Processing on Microbiological Quality of Milk

Surajit Mandal, Subrota Hati and Pradip V. Behare

Introduction

Milk is one of the perishable food articles. The quality and safety of milk can be affected if proper care is not taken during production, procurement, transportation, processing, storage and distribution, and consumption. Various types of milk and milk products like Ghee, Powder, Butter, Paneer, Curd, Buttermilk, sweets, etc. are manufactured from raw milk. Thus, the production and processing of milk ultimately determine the quality (shelf-life and safety) of the finished products. The production of clean and wholesome milk and processing under proper conditions are important for better quality milk and milk products. Cleanliness and health of milch animals, cattle care, and personal hygiene of milking person, cleanliness of milking place and cattle-shed, milking vessel and utensils of milk collection at the society are key parameters for clean and wholesome raw milk production and procurements. The processing of milk at the organized industry under good manufacturing practices determines the quality milk and milk products to customers.

Milk and dairy products are highly nutritious media. Microorganisms can multiply milk under ambient conditions and cause spoilages and health hazards. The levels and types of microorganisms in milk and milk products depends on the microbial quality of the raw milk, conditions under which the products are produced, temperature and duration of storage, etc. The common spoilage microorganisms in milk and milk products are gram negative rod-shaped bacteria (*Pseudomonas* spp., Colifroms), Gram positive, spore-froming bacteria (*Bacillus* spp., *Clostridium* spp.),

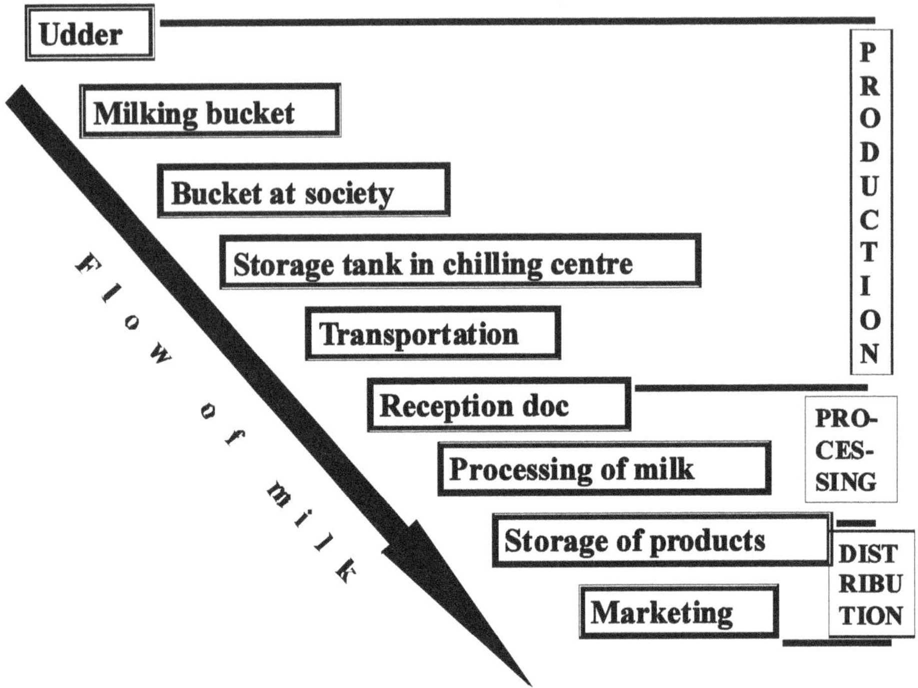

Figure 1.1: Flow of Milk from Production to Consumption.

lactic acid bacteria (*Strepotococcus* spp., *Lactococcus* spp., *Lactobacillus* spp.) and yeasts and moulds. Milk and milk products are also to a limited extent, associated with food-borne illness. Disease is mainly due to consumption of un-pasteurized milk containing pathogenic microorganisms (*Salmonella* spp., *Listeria monocytogenes*, *Campylobacter jejuni*, etc.).

Effect of Production on Microbiological Quality of Raw Milk

Quality milk production is an important part of any dairy operation. The milk quality is determined by the composition and hygiene. The compositional quality of milk is mainly influenced by feeding, management systems, genetics, breed, etc. Hygienic parameters are decisive for food safety. However, these may influence the composition of the milk as in case of mastitis or elevated numbers of somatic cells. The hygienic requirement to be met by raw milk and milk products is vary between categorical postulates for protecting human health. The main criteria for milk and milk products of high hygienic value are: (*i*) low in saprophytic microorganisms, (*ii*) absence or very low number of pathogenic microorganisms including mastitis pathogens, (*iii*) avoidance of residues due to the measures of mastitis prophylaxis and control and (*iv*) reduction or minimization of contaminants. This can be achieved by adopting the followings:

- ☆ Clean milking practices - key for quality milk production.

- ✰ Quality milk production - key ingredient to the future dairy industry.
- ✰ Consistent milking practices - by all people milking cows - clean, dry and well stimulated teat.
- ✰ Keeping cows clean, dry, and comfortable are essential ways to produce quality milk.

Bacteriological Contamination of Raw Milk

Most the microorganisms are undesirable in milk because they may be pathogenic or produce undesirable bio-transformation of milk components. The pathogenic microorganisms in milk can affect humans or animals.

- ✰ Contamination of the raw milk in an order of magnitude of 10^2 to 10^3 microorganisms/ml is practically unavoidable. This contamination is due to microorganisms present in the interior of the udder or in teat canal.
- ✰ Under condition of careful milking, 10^4 microorganisms/ml could be expected. The main contribution of this order of magnitude comes from the microorganisms present at the surface of the teat and in the milking systems.
- ✰ Under appropriate conditions of the storage with cooling to 4°C, the number of microorganisms can be kept in the range of 10^4 to 10^5/ml.
- ✰ Sensorial detectable differences occur at total counts between 10^6 to 10^7 cfu/ml and are depends on the species and activity of the respective microorganisms.

Most of the bacteria present in raw milk are contaminants of the outside and gain entrance into the milk from various sources including soil, bedding, manure, feed, milking equipments, etc.

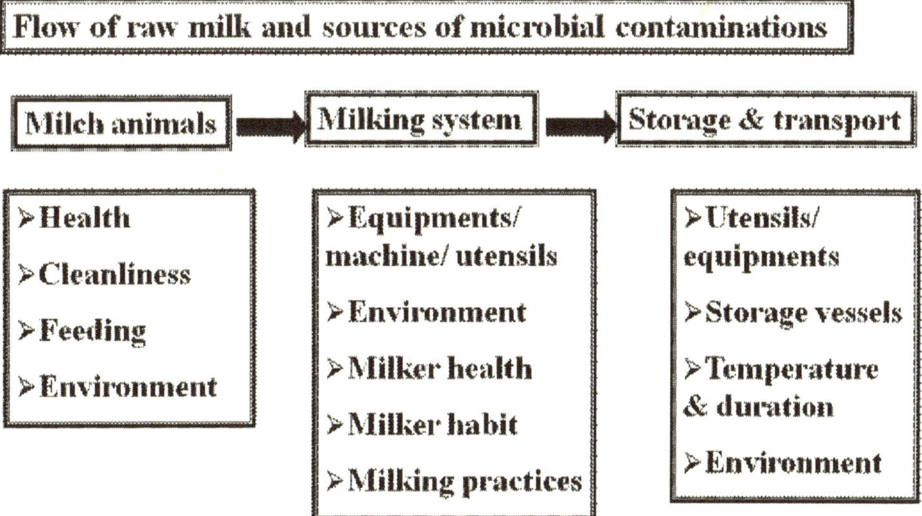

Figure 1.2: Sources of Contamination during Production and Transport of Raw Milk.

Table 1.1: Sources of Microbial Contamination in Milk at the Dairy Farm and Control Measures

Sl.No.	*Sources*	*Microbial contaminants*	*Control measures*
1.	Inside udder	*Streptococcus, Micrococcus, Corynebacterium*	Appropriate housing and care of cows to promote clean udders.
2.	Outside udder and teats	*Micrococcus, Staphylococcus, Enterococcus Bacillus*	Dry treatment *i.e.* removal of loose dirt. Cleaning of dirty udder before milking.
3.	Feed	*Clostridium, Listeria, Bacillus,* Lactic acid bacteria	Avoid feeding during milking process.
4.	Soil	*Clostridium, Bacillus, Pseudomonas, Mycobacterium,* Yeast and molds	Cleaning of milking parlor. Restfulness of the cows during milking.
5.	Bedding	*Clostridium, Bacillus, Klebsiella*	
6.	Faeces	*Escherichia coli, Staphylococcus, Listeria, Mycobacterium, Salmonella*	
7.	Air	*Streptococcus, Micrococcus,* Corynforms, *Bacillus,* Yeast and molds	Free from suspended dirt and dusts. Proper ventilated air supply design. Closed milking system.
8.	Water	Coliforms, *Listeria, Bacillus, Alcaligenes, Pseudomonas*	Use of good microbiological quality of water used for different purposes. Regular checking of water quality. Sanitation and disinfection in water supply.
9.	Human	Coliforms, *Salmonella, Enterococcus, Staphylococcus*	Through washing and disinfection of hands of the milking personnel. Good habit during milking Clean milking time dresses.
10.	Milking machine and utensils	*Micrococcus, Streptococcus, Bacillus*, Coliforms	Proper cleaning and sanitation of milking machine and utensils.

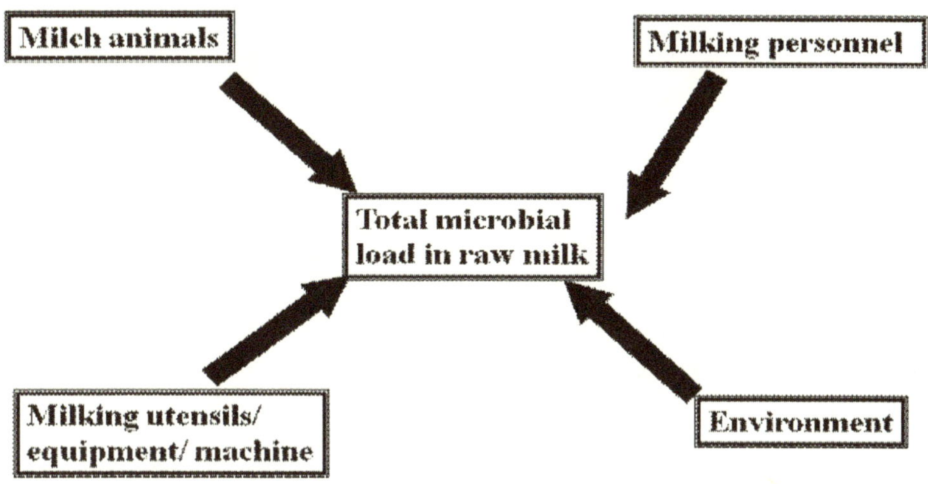

Figure 1.3: Microbial Contamination of Raw Milk at Farm Level.

Microorganisms in Raw Milk and Significances

The presence and multiplication of saprophytic and pathogenic bacteria in raw milk might change the milk composition and produce toxins, and influence the quality and safety of the milk and milk products. Moreover, flavour of the raw milk may be adversely influenced and heat-stable bacterial enzymes may continue to act in products particular during long storage and adversely affect the stability and the flavour of cream and UHT milk. The pathogenic bacteria include "classical" microorganisms and "emerging pathogens". At present Salmonella, pathogenic *Escherichia coli* strains, *Yersinia enterocolitica, Staphylococcus aureus, Listeria monocytogenes, Campylobacter jejuni* are the most important.

Saprophytic Bacteria in Milk and Significances

According to the main points of attack on the major milk constituents, the saprophytic bacteria are subdivided as follows:

a) Microorganisms degrading milk carbohydrate (lactose) are classified as glycolates, *e.g. Streptococci, Lactobacilli, Coliforms*

b) Microorganisms degrading proteins are classified as proteolytes, *e.g. pseudomonas, enterobacteriaceae,* aerobic spore-formers.

c) Microorganisms degrading lipids are classified as lipolytes, *e.g. pseudomonas, micrococci, aeromonas, corynebacteria.*

The effect of growth of saprophytic bacteria in milk may be important in three ways as follows:

a) The change in milk composition may interfere with manufacture, if a fermentation is involved in the manufacture process, and this may affect the yield and quality of the product, *e.g.* cheese.

Table 1.2: Saprophytic Bacteria in Milk – Significance and Control Measures

Sl.No.	*Group of Microorganism*	*Representative Organisms*	*Significances*	*Control Measures*
1.	Lactic acid bacteria	*Lactococcus* spp. *Lactobacillus* spp. *Strpotococcus thermophilus*	At room temperature ferment lactose to lactic acid and milk gets sour and become unfit for processing due to the loss of heat stability	Pasteurization of milk to kill mesophilic lactic acid bacteria. Storage of milk at low temperature (4°C)
2.	Coliforms	*Escherichia coli* *Enterobacter aerogenes* *Klebsiella* spp. *Citrobacter* spp.	Ferment lactose using hetero-lactic pathway Degraded proteins Produce gas and cause "unclean flavour" in milk Indicator of post processing contamination and hygienic operations	Pasteurization kill all the coliforms Hygienic processing of raw milk Clean milk production protocol reduces the initial number in raw milk
3.	Psychro-trophs	*Pseudomonas, Achromobacter, Flavobacterium, Alcaligenes*	Grow at refrigerated temperature Produce heat stable proteases and lipases which break down the milk proteins and fats Cause off flavour and taste, off colour	Thermization of raw milk before low temperature storage for processing Pasteurization of raw milk Hygienic processing of raw milk Clean milk production protocol reduces the initial number in raw milk
4.	Heat resistant bacteria	*Microbacterium lacticum*, *Micrococcus* spp. Thermophilic streptococci, sproe-froming bacillus: *Bacillus* spp. (*B. cereus, B. subtilis*), *Clostridium* spp. (*Cl. tyrobutyricum*)	Spoil heat treated milk by sweet curdling, off flavour, clumping of fat globules	Clean milk production protocol reduces the initial number in raw milk UHT sterilization of milk Use of bactofugation and microfiltration in processing operation

b) The flavour of the raw milk may be adversely influenced (*e.g.* rancidity) and this may directly affect the flavour of the product *e.g.* pasteurized milk or cream.

c) Heat-stable bacterial enzymes may continue to act in the product, particularly during long storage, and adversely affect the stability and/or flavour of cream and UHT milk.

Pathogenic Bacteria in Milk and Significance

The human pathogens transmitted through milk are classified into food infection and food poisoning groups. In food infection milk act as a carrier of the microorganisms, this enters in human body through milk. It takes time to a person to

Table 1.3: Pathogenic Bacteria in Raw Milk – Significances and Control

Pathogens	*Disease*	*Control Measures*
Mycobacterium bovis/ M. tuberculosis	Tuberculosis	Clean milk production Low temperature storage Pasteurization Hygienic processing
Brucella abortus/ B. melitensis	Brucellosis	Clean milk production Thermization before low temperature storage for processing Pasteurization Hygienic processing
Coxiella burnetii	Q fever	Clean milk production Low temperature storage Pasteurization Hygienic processing
Staphylococcus aureus	Enterotoxin	Clean milk production Thermization before low temperature storage for processing Pasteurization Hygienic processing
Escherichia coli	Some serotypes pathogenic for men, faecal contamination	Clean milk production Thermization before low temperature storage for processing Pasteurization Hygienic processing
Listeria monocytogenes	Listeriosis	Clean milk production protocol reduces the initial number in raw milk Pasteurization
Bacillus cereus	Enterotoxin	Clean milk production UHT sterilization of milk Use of bactofugation and microfiltration in processing operation
Clostridium perfringens	Gas gangrene/survive pasteurization	Clean milk production UHT sterilization of milk Use of bactofugation and microfiltration in processing operation

become ill and fairly small numbers of microorganisms may suffice to cause illness. In food poisoning, preformed toxins in milk are responsible. Consumers rapidly fall ill. Large numbers of the pathogenic microorganisms are usually needed to cause food poisoning.

Effect of Processing on Microbiological Quality of Milk

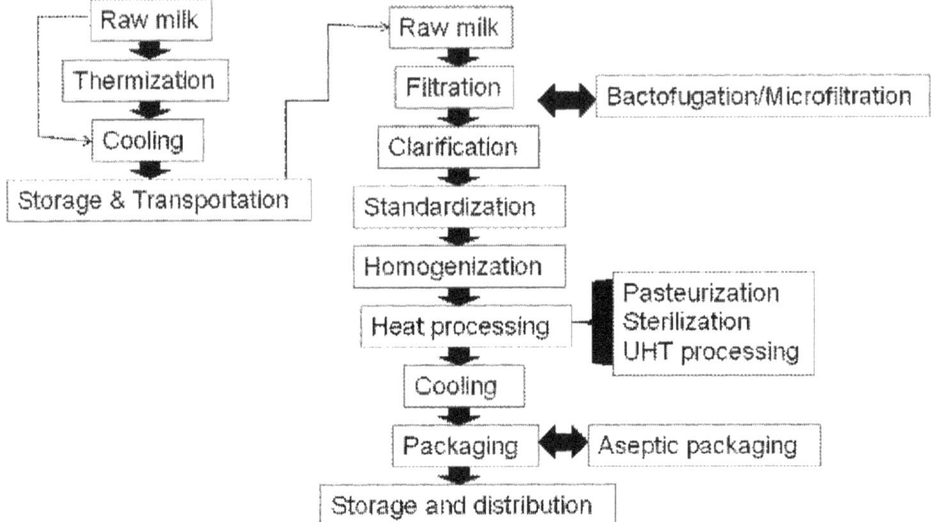

Figure 1.4: General Flow Chart of Processing of Raw Milk.

Pasteurization of Milk

Pasteurization is applied to destroy heat sensitive spoilage and pathogenic bacteria present in raw milk. However, pasteurization is not sufficient to inactivate bacterial spores and heat resistant bacteria. Pasteurization is a process applied with the aim of avoiding public health hazards arising from pathogenic microorganisms associated with milk by heat treatment (63°C for 30 min in holder process or 72°C for 15 s in HTST process) which is consistent with minimal chemical, physical and organoleptic changes in the product. Pasteurized products should last for upto 24 hours without refrigeration and several days at refrigerated condition. Recently, it has been required that pasteurized milk should show a positive lactoperoxidase activity to prevent the milk being over processed. Freshly pasteurized milk should deemed to pass the coliform test and the palte count tests if its colifrom count is < 1/ml and its plate count is < 30,000/ml. Pasteurization is a mild form of heat treatment, causing minimum whey protein denaturation, little loss of heat-sensitive vitamins, and no change in its colour, flavour and texture.

Factors Affecting the Effectiveness of Pasteurization

The main control points for ensuring good quality pasteurized products are:

- ☆ Raw milk quality
- ☆ Processing condition

- ☆ Post-processing contamination
- ☆ Storage temperature

Raw milk contains both saprophytic and pathogenic microorganisms. Most of the vegetative forms of microorganisms are destroyed during pasteurization process. However, thermoduric, thermophilic, spore-formers can survive during pasteurization and causes problems in the final products during storage. During processing, each and every particulate of milk should be heated to suitable pasteurization temperature for at least specified duration and immediately cooling to below 5°C are important to obtain good quality pasteurized products. Problems are arising from the build up to thermophilic bacteria in heating and cooling section associated with long operating times in continuous heat exchangers. Good keeping quality pasteurized milk can be achieved by eliminating post processing contamination and can be further enhanced by using low storage temperature. Post-pasteurization contamination (PPC) is very important determinant of keeping quality of the final product. These are recontamination in down-stream of the end of the holding tube, in the regeneration section or cooling section, storage tank, final packaging of the product and poor hygienic practices.

The spoilage of pasteurized milk product is caused by:

- ☆ Growth and enzyme production by psychrotrophs before pasteurization
- ☆ Activity of thermo-resistant enzymes
- ☆ Growth of thermo-resistant psychrotrophs
- ☆ Post pasteurization contaminants

Thus, the shelf life of pasteurized milk is depends on the following:

- ☆ Microbiological quality of raw milk.
- ☆ Time and temperature of pasteurization
- ☆ Presence and activity of post pasteurization contaminants.
- ☆ Types and activity of pasteurization-resistant microorganisms
- ☆ Storage temperature of milk

Sterilization of Milk

Sterilization improves the keeping quality of milk for long at ambient temperature by destroying all microorganisms, heat resistant spores and enzymes in raw milk. Sterilization of milk can be done by three processes, such as a) complete in bottle system, b) Two-stage process and c) UHT process. In complete in-bottle system milk is filled in glass bottles and capping followed by heating (110-120°C for few minutes). Strong cooked off-flavour and browning maillard reaction defects are observed in the final product. In two stage process, milk is sterilized at 130-150°C for 1-20 sec in bulk in first stage followed by bottling and capping and again heating at 100-110°C for 5-10 minutes in second stage. This is milder than complete in bottle process and the product taste and colour is good. In UHT sterilization milk is heated at 130-150°C for few seconds in continuous sterilizer followed aseptic packaging in suitable container.

The main aim of in-container method of sterilization is to inactivate heat resistant spores, thereby producing a product which is "commercially sterile" with an extended shelf life. The legal requirement initially was that it would be expected to remain fit fro human consumption for at least 7 days, though as a general rule it would keep sweet for several weeks at ordinary temperature. The turbidity test is useful for ensuring proper sterilization of milk. The inactivation of most heat-resistant pathogenic spores (*Clostridium botulinum*) is achieved through sterilization. Since, milk is a low-acid food; the main aim of sterilization is to achieve 12D (D – Decimal reduction) reduction of *Cl. botulinum* spores (121°C for 3 min.). The microbial severity of a process is traditionally expressed in terms of its F_O value. This takes into account the contribution of the heating, holding and cooling periods to the total lethality and is expressed in terms of minutes at 121°C. The minimum F_O value for any low acid food should be 3 and thus minimum botulinum cook will produce a safe product but not necessarily commercially sterile.

UHT processes are continuous process of sterilization, which involve in excess of 135°C for time of greater than 1 sec, followed by aseptic packaging with much less chemical change as compared sterilize milk in terms of colour development, thiamine inactivation, lactulose formation and whey proteins denaturation. Both sterilized and UHT milks have a shelf life upto 6 months, but chemical reactions and physical changes take place during storage, which alters the sensory characteristics. To ensure a long shelf-life the sterile product is packed into pre-sterilized containers in sterile containers in a sterile environment and an airtight seal is formed. Higher temperature and shorter time process results in less chemical damage to important nutrient and functional ingredients within foods, thereby leading to an improvement in product quality. Chemical reactions are less temperature sensitive, so the use of higher temperatures, combined more rapid heating and cooling rates, helps to reduce the amount of chemical reaction. Two other parameters introduced for UHT processing of dairy products are the B* and C* values. B* is a microbial parameter used to measure the total integrated lethal effect of a process. A process, given a B* value of 1 would be sufficient to produce 9 D reduction of mesophilic spores and would equivalent to 10.1 sec at 135°C. C* parameter measures the amount of chemical damage taking place during the process. A process given the C* value of 1 would cause 3 per cent destruction of thiamine and is equivalent to 30.5 sec at 135°C. The chemical damage could be further reduced by using temperature > 145°C and very short period of processing.

Good Hygienic Practices in Milk Production

The presence of bacteria in milk is what causes the milk to become unfit for human food. If there were no germs in milk, it would keep sweet and wholesome indefinitely. The problem of producing clean milk is therefore one of keeping bacteria out of the milk.

The following rules are comparatively simple and inexpensive to follow, and at the same time they will do much to help the dairyman produce clean milk:

- Keep the cow clean.
- Clip the hair about the flank and udder at least twice each year.

☆ Wipe the udder with a damp cloth just before milking.

☆ Do not brush or feed the cow just before milking.

☆ Do not sweep the floor within three-quarters of an hour before milking.

☆ Use a small-top or covered milk-pail.

☆ Milk with clean hands and clean suits.

☆ Rinse all of the milk utensils with cold water, and then wash them thoroughly with a brush and hot water in which washing powder has been dissolved. Then scald everything in boiling water.

☆ Have the barns well lighted and ventilated. Bacteria do not thrive in sunlight. Have not less than four square feet of glass per cow.

☆ Keep the milk utensils in a place free from dust.

☆ In purchasing dairy apparatus, insist that all seams be filled with solder. Cracks and seams make an ideal place in which germs grow.

☆ Keep the milk cold (at least 50° F.) after milking.

Table 1.4: Microbiological Hazards and Control Points during Production and Storage of Raw Milk

Sl.No.	*Source(s)*	*Hazards*	*Control Point*
1.	Cow	Milk is obtained from unhealthy animals	Checking of milch animal health regularly (farmers and the supervising veterinarian)
2.	Milking	Milking routines may damage tissues leading to infection of the udder, contaminate the milk by the environment, milking equipment, etc.	Cleaning the udder before and after milking
3.	Storage and milk cooler	High storage temperature and time may create growth of pathogens	Temperature and time
4.	Collection and transportation	High transportation temperature and time as well as unhygienic routines may create growth and contamination of pathogens	Temperature and time
5.	Storage in silo tanks	High storage temperature and time may create growth of pathogens.	Temperature and time

Monitoring and Corrective Actions in Milk Processing

At all stages in processing, good hygiene of manufacturing plant is essential to ensure that the product stream is not contaminated after heat treatment of raw milk. Sources of post pasteurization contaminations are equipment, packaging materials, air, aerosols, water, lubricants, etc. Pasteurization equipment should be properly designed, installed, maintained and operated to ensure that the milk is heated to at least the specified temperature for at least the specified time. Maintenance of the plant hygiene includes the cleaning and sanitation program, cleaning and disinfection by CIP system once in a day. Formation of biofilms on the surface of milk processing equipment threatens the quality and safety of dairy products. Dead ends, corners, cracks, crevices, gaskets, valves and joints are vulnerable points for biofilm formation. The development of biofilm depends on the type and number of microorganisms in milk, type of product, operating conditions of plant (temperature, length of production run), type of surfaces, etc. The formation of biofilm can be controlled by hygienic design of processing equipment, effective cleaning and sanitation process.

Conclusions

The various stages in the milk processing chain from milking the cow to consumption have to be under control in order to assure the quality and safety of milk and milk products.

- ☆ Adherence of basic good manufacturing practices is one of the first steps to achieve this.
- ☆ HACCP can be applied as a tool to assess hazards and establish control system that focus on preventive measure rather than relying mainly on end product testing.
- ☆ Ensuring that the raw materials are of best quality.
- ☆ Elimination of spoilage and pathogenic bacteria from milk and other raw materials by heat treatments.
- ☆ Prevention of contamination, and growth limitation of undesirable microorganisms during storage.
- ☆ Training programmes to people involved in processing, distribution and handling of milk and dairy products on principles of personal hygiene, milk spoilage and need of cooling chain.
- ☆ Education to consumers about the importance of keeping milk cold.
- ☆ Label information on packages may help to achieve improved quality control of milk and dairy products.

Suggested Readings

Boor, J.K. and Murphy, S.C. 2002. Microbiology of market milk. In: *Dairy Microbiology Handbook: The Microbiology of Milk and Milk products.* Robinson, R. K. (Eds.), 3rd Edn., 2002, pp. 91-112.

Chambers, J. V. 2002. The microbiology of raw milk. In: *Dairy Microbiology Handbook: The Microbiology of Milk and Milk products.* Robinson, R. K. (Eds.), 3rd Edn., 2002, pp. 39-89.

Heeschen, W. H. 1996. Bacteriological quality of raw milk: legal requirements and payment system. In: *Symposium on Bacteriological Quality of Raw Milk",* International Dairy Federation, Wolfpassing, Austria, March 13-15, 1996, pp. 1-18.

Slaghuis, B.A. 1996. Sources and significances of contaminants on different of raw milk production. In: *Symposium on Bacteriological Quality of Raw Milk",* International Dairy Federation, Wolfpassing, Austria, March 13-15, 1996, pp. 19-27.

2015, Dairy Product Technology: Recent Advances ***Pages 13–37***
Editors: **Subrota Hati, Surajit Mandal and Birendra Kumar Mishra**
Published by: **DAYA PUBLISHING HOUSE, NEW DELHI**

Chapter 2

Bioactive Fermented Dairy Products with Special Reference to Cheese: Scope and Challenges

S. Makhal, S. Hati, S. Bera and Santosh Sarkate

Introduction

Living longer and better is in the minds of most people; especially in wealthy countries and diet is one of the most important facets of a healthy life. Consequently, diet-health link is now an integral part of healthy life style. The role of diet and specific foods for the prevention and treatment of diseases and improvement of body functions is being investigated around the world. Present day consumers prefer foods that promote good health and prevent disease. Such foods need to fit into current life styles providing convenience of use, good taste and acceptable price-value ratio. Therefore, it seems very likely that the concept of functional foods, which enables the consumers to exercise a level of "self-health maintenance", significantly influences the food and drink manufacturers worldwide.

Since nowadays, consumers are more diverted towards the health driven foods, hence the recent advances in research on bioactive peptides show much promise in new product development using these biomolecules to derive multifarious health benefits. It is now evident that during fermentation of milk with certain dairy starters, peptides with various bioactivities are formed and are detected in an active form even in the final products, such as fermented milks and cheeses (Meisel and Bockelmann, 1999). These peptides are likely to be produced industrially in the future in the form of hydrolysates or peptide mixtures in foods, which can be used as ingredients for various dietary or pharmaceutical products or can be consumed as functional foods. Bioactive foods enriched with health giving bioactive peptide holds a very promising prospect because researchers are realizing the importance of the polyfunctional roles of bioactive peptides. The *in situ* production of bioactive peptides in fermented dairy products, such as *Dahi*, yoghurt, cheese, etc has now been conceptualized as a novel approach to improve the health value of the products. While cheese has long been associated with a high quality nutritional image, more recently research efforts devoted to the development of bioactive cheese with demonstrated health promoting properties of bioactive peptides have been made to push cheese into the "functional foods" category.

Bioactive Foods: Looking Beyond their Nutrition

Functional foods, in addition to their basic nutritive value and natural being, will contain the proper balance of ingredients, which help us to function better and more effectively in many aspects of our lives, including helping directly in the prevention and treatment of illness and disease ultimately leading to a sound and healthy life. Functional foods serve to promote health or help to prevent disease, and in general the term is used to indicate a food that contains some health-promoting components beyond the traditional nutrients (Berner and O'Donnell, 1998). Functional foods have been variously termed as *nutraceuticals, designed foods, medicinal foods, bioactive foods, therapeutic foods, superfoods, foodiceuticals, medifoods,* etc. In most cases, the term refers to a food that has been modified in some way to become 'functional'. Modifications can be achieved by incorporation of phytochemicals, bioactive peptide, þ-3 PUFA and probiotics and/or prebiotics. Milk naturally enriched with bioactive peptides, protective enzymes and other biologically active curative components, is argued to be a "functional food" (Chandan, 2001). It is clear that the market for functional foods is at presents, blooming and is predicted to expand substantially in the coming decade.

Bioactive food is an important category of functional foods, which contains health promoting bioactive peptides, serving to regulate a particular body process, such as enhancement of the biological defense mechanisms, prevention and recovery of a specific disease. Milk proteins have been identified as an important source of bioactive peptides, which can be released during hydrolysis induced by digestive or microbial enzymes. The peptides derived from milk proteins have been shown to have a variety of different functions *in vitro* and *in vivo*, for example, antihypertensive, antimicrobial, antioxidative, antithrombotic, immunomodulatory, mineral-carrying and opioid. There is increasing commercial interest in the production of bioactive peptides with the purpose of using them as active ingredients for bioactive foods or in the development of fermented milk products with elevated level of these bioactive peptides by the manipulation of processing conditions.

Development of Bioactive Cheese: A Holistic Biotechnological Approach

Dietary patterns of today's consumers have been changed and they are going for foods having polyfunctional health benefits. Hence, an important appeal of the modern health conscious consumers is to consume functional foods to derive health-giving and curative functions, naturally imparted to foods. In an endeavour to expand the range of bioactive dairy products, a small number of researchers and companies have attempted to manufacture bioactive cheeses, which would contain a considerable amount of these wonder molecules *viz.* bioactive peptides. Cheese will have a number of advantages over the non-fermented products as a delivery system for bioactive peptides. Given the increasingly competitive scenery of the European market in food products, the cheese industries in Ireland and throughout the Europe are trying to derive benefit from a marketing advantage, such as added-value probiotics and bioactive peptides containing cheese, which would afford a competitive edge over existing products. The development of bioactive cheese could thus lead to a major economic advantage.

While cheese has long been recognized as a highly nutritive food, more recently research efforts devoted to the development of bioactive cheese with demonstrated health benefits of bioactive peptides have been made to make cheese more promising as functional foods. Generally, incorporation of probiotic bacteria into milk-based food systems, including cheese still remains at the Biblical age in terms of maintaining viability and probiotic functionality during manufacture and shelf life. Therefore, with the biotechnological approach to manipulate dairy starter cultures, there is a good potential to incorporate some health attributes in cheese by *in situ* synthesis of bioactive peptides at an elevated level to derive some health benefits. This approach of enriching cheese with bioactive peptides seems to be more attractive in terms of its stability and maintaining functionality. Genetic engineering, the use of recombinant DNA technology to produce genetically modified cheese starters, is of the most important scientific advances of the 20th century, which can be judiciously exploited to modulate gene expression yielding strains with optimal properties (Henriksen *et al.,* 1999), for example specific bioactive peptide production.

Bioactive Peptides: The Wonder Biomolecules

Milk proteins possess unique biological and physico-chemical properties. For example, milk is known to contain a wide range of proteins, which either provide protection against enteropathogens or are essential for the manufacture and characteristics of certain dairy products. Dietary proteins of milk are certainly among the most potent sources of physiologically active molecules. Their breakdown by proteolytic enzymes during the digestion process releases short or medium size peptides whose structures may be identical or very close to that of animal or human hormones. Bioactive peptide can interact with target sites at the luminal side of the intestine tract. Furthermore, they can be absorbed and then reach peripheral organs (Meisel and Bockelmann, 1999).

Researches carried out during the last 15 years have demonstrated that major milk protein groups, caseins and whey proteins, are the important source of biologically active peptides, termed as functional or bioactive peptides, which are described as defined sequences of amino acids, which are inactive within the native protein, but which display specific properties once they have been released by enzymatic activity (Nayak *et al.,* 1999). Bioactive peptides usually contain 3 to 20 amino acid residues per molecule, except for glycomacropeptides (GMPs), which consists of 64 amino acid residues. The sequence of the amino acids in these small molecules is a crucial factor in their activity. The amino acid located in the C-terminal or N-terminal position is often significant (Nayak *et al.*, 1999).

Milk-derived peptides have been shown to exert beneficial physiological effects. They serve to modulate metabolic processes like digestion, circulation, immunological responsiveness, cell growth and repair, and nutrient intake. These functional peptides display partial resistance to hydrolysis and can exert their effects either locally in the digestive tract or elsewhere in the body. The multifunctionality of physiologically active peptides derived from milk and their wide distribution among mammals could confer on them a role of messenger molecules. Their contribution to the health of the newborn would be three-fold: an easily assimilated source of organic nitrogen, a good source of essential amino acids and a potential source of bioactive molecules. Several casein-derived peptides may play a significant role in the stimulation of the immune system. They have been found to exert a protective effect against microbial infections and to enhance some functions of the immune system (Migliore-Sammour *et al.*, 1989).

The biological action of the peptides is intimately related to their amino acids sequence. The presence of some amino acids, such as proline, arginine or the branched amino acids strongly affects the relation between structure and activity of the peptides. An active site within an immunostimulating peptide could consist of proline and branch-chain amino acids, such as leucine, isoleucine, and valine (Suetsuna *et al.*, 1991). Arginine present on the C-terminal residue of many antihypertensive peptides and some immunodulatory peptides contribute to the bioactivity of the peptides since the positively charged arginine enhances the interaction with the different receptors (Meisel, 1993). Peptides possessing an increased content of amino acids with basic side groups are able to interact with T-lymphocyte membranes and exhibit immunomodulating activities.

Milk: An Important Source of Bioactive Peptides

The formation and properties of milk protein-derived peptides have been reviewed in many recent articles. Peptides with biological activity can be produced from milk proteins in three ways (Korhonen and Pihlanto-Leppala, 2001): (i) Enzymatic hydrolysis with digestive enzymes, (ii) Fermentation of milk with proteolytic starter cultures and (iii) through the action of enzymes derived from proteolytic microorganisms. The enzymes involved occur naturally in foods or are derived from exogenous or microbial sources. The bioactivities of the peptides encrypted in major milk proteins are dormant until released and activated by enzymatic proteolysis, *e.g.* during gastrointestinal digestion or food processing like during fermentation and ripening. The proteolytic system of lactic acid bacteria (LAB) contribute to the liberation of bioactive peptides. *In vitro*, the purified cell wall proteinase of *Lactococcus lactis* was shown to liberate oligopeptides from β- and α- caseins, which contain amino acid sequences present in casomorphins, casokinins and immunopeptides. Further degradation of these peptides by endopeptidases and exopeptidases of LAB could lead to the liberation of bioactive peptides in fermented milk products.

Hydrolysis of β-casein has been well investigated to liberate a number of peptides. More than 40 per cent of the peptide bonds were cleaved resulting in the formation of more than 100 different oligopeptides (Juillard *et al.*, 1995). Many of the peptides had chain lengths of less than 10 amino acids and thus were substrates for the oligopeptide transport system. A further intracellular breakdown is catalyzed by a number of peptidases (Mierau *et al.*, 1997). Casein-derived peptides with chain lengths of more than 10 amino acids are abundant and contain various amino acid sequences representing bioactive peptides within a longer peptide (Juillard *et al.*, 1995). It is now evident that during the fermentation of milk with certain dairy starters, peptides with various bioactivities are formed, which are detected in an active form even in the final products, such as fermented milks and cheese. After ingestion along with foods, they are absorbed and then reach peripheral organs. These bioactive peptides can be produced by conditioned fermentation in several fermented milk products including cheese.

Multifarious Health Attributes of Bioactive Peptides

In contrast to endogenous bioactive peptides, many milk-derived peptides have multifunctional properties. Regions in the primary structure of caseins contain overlapping peptide sequences, which exert different biological effects. These regions, considered to be strategic zones, are partially protected from proteolytic breakdown. For example, most β-casomorphins and casokinins are both ACE-inhibitory and immunostimulatory and α- and β-lactorphin contain sequences with both opioid and ACE-inhibitory activities (Meisel, 1998). Mann (1993) has reported that physiologically active peptides also have the ability to promote the growth of *Bifidobacteria* in the GIT, thus increasing the number of these beneficial organisms. This is attributed particularly to k-casein, which can affect bacterial adhesion as a receptor analogue (Lonnerdal, 1998). Many milk-derived peptides reveal multifunctional properties, that is, specific peptide sequences may exert two or more different biological activities. It is probable that part of the beneficial effects attributed to the probiotics containing foods is associated with bioactive peptides formed by enzymes produced by these bacteria. Table 2.1 exhibits different bioactive milk derived peptides and their biological activities:

Opiod Activity

Opioid peptides derived from milk proteins appear to have physiological significance in the female organism (liberation of casomorphins in the mammary gland) and in neonates, and they appear to participate in the control of gastrointestinal functions in adults (Teschemacher *et al.*, 1997). Opioid receptors are located in the nervous endocrine and immune systems as well as in the intestinal tract of the mammalian and can interact with their endogenous ligands as well as with exogenous opioids and opioid antagonists (Teschemacher and Brantl, 1994). Orally given opioid peptides derived from milk proteins are able to modulate absorption processes in the gut. The enhancement of net water and electrolyte absorption by β-casomorphins in the small and large intestine is a major component of their antidiarrhoeal action (Daniel *et al.*, 1990a, b).

Table 2.1: Bioactive Peptides Derived from Milk Proteins

Bioactive Peptides	*Protein Precursor*	*Bioactivity*
α-Casomorphins	α-Casein	Opioid agonist
β-Casomorphins	β-Casein	Opioid agonist
α-Lactorphin	α-Lactalbumin	Opioid agonist
β-Lactorphin	β-Lactoglobulin	Opioid agonist
Lactoferroxins	Lactoferrin	Opioid antagonists
Casoxins	X-Casein	Opioid antagonists
Casokinins	α-, β-Casein	ACE-inhibitory
Casoplatelins	X-Casein, Transferrin	Antithrombotic
Lactoferriein	Lactoferrin	Antimicrobial
Immunopeptides	α-, β-Casein	Immunomodulatory
Phosphopeptides	α-, β-Casein	Mineral carriers

Source: Meisel and Bockelmann (1999).

Opioid antagonists have been identified in peptide sequences in bovine and human κ-casein (casoxins) and in α_{s1}-casein (Yoshikawa *et al.*, 1994). The casoxins are opioid receptor ligands of the μ-type but they are of relatively low potency compared with the opioid antagonist, naloxone (Meisel, 1997). Casomorphins may produce analgesia, modulate social behaviour, influence postprandial metabolism by motivating the secretion of insulin and somatostatin, and may influence gastrointestinal absorption of nutrients by prolonging the gastrointestinal transit time and exerting an antidiarrhoeal activity (Meisel and Schlimme, 1990). Physiological effects arise when casomorphins are absorbed as long-chain precursors and hydrolyzed into smaller bioactive fragments in the intestinal tissue. After crossing the intestinal mucosa, the fragments then react with μ- and δ-type receptors located all through the intestinal tract and in the brain. β-Casomorphin-7 with amino acid sequence of β-casein f60-66 having an opioid effect also inhibits ACE activity and LAB endo- and aminopeptidases (Meisel and Schlimme 1994 and Stepaniak *et al.*, 1995).

Antihypertensive Activity

Angiotensin, a blood polypeptide, exists in two forms *i.e.*, the physiologically inactive Angiotensin I and the active Angiotensin II. Angiotensin I Converting Enzyme (ACE) is a multifunctional enzyme located in different tissues *e.g.* plasma, lung, kidney, heart, skeletal muscle, pancreas, brain, mammary arteries, testes, uterus, and intestine. ACE is associated with the Renin Angiotensin System (RAS), which regulates peripheral blood pressure (Meisel and Bockelmann, 1999). In the RAS, the aspartic proteinase rennin hydrolyses Angiotensinogen releasing the decapeptide Angiotensin I. Its inhibition results in an antihypertensive effect and may influence different regulatory systems of the host organism involved in modulating blood pressure, immune defense and nervous activity (Meisel, 1997). ACE hydrolyses Angiotensin I to octapeptide Angiotensin II, a vasoconstrictor, which increases blood pressure. Angiotensin II also inactivates bradykinin, a peptide, which acts as a vasodilator and increases the production of aldosterone, which decreases renal excretion of fluid and salts, increasing water retention and the volume of extracellular fluids (Fiat *et al.*, 1993 and Tirelli *et al.*, 1997). The inactive form *i.e.* Angiotensin I is converted into the active one *i.e.* Angiotensin II by the ACE, a peptidyl dipeptidase, which cleaves dipeptides from the carboxyl terminal (Schlimme and Meisel, 1995). Since Angiotensin-II elevates blood pressure by constricting blood vessels, inhibition of ACE causes lowering of blood pressure. The functional peptides have found to inhibit ACE by blocking its active site. The ACE inhibitory peptides having pro residues, mainly at the C-terminal end withstand degradation by digestive enzymes. This ACE inhibitory activity of milk derived peptides would therefore make dairy foods a natural functional food for controlling hypertension. The ACE-inhibiting effect of bioactive peptides has also been observed *in vivo* in rats that ingested peptides from milk fermented by *L. helveticus* and *S. cerevisiae* (Masuda *et al.*, 1996). A blood pressure-lowering effect of peptides derived from milk fermented by *L. helveticus* has also been observed in rats (Yamamoto *et al.*, 1994). Another study has demonstrated a strong antihypertensive effect of a dipeptide (Tyr-pro) formed upon fermentation with *L. helveticus* CPN4 strain (Yamamato *et al.*, 1999).

Antimicrobial Activity

Bactericidal peptides may support in defending against microbial challenge, especially in the neonatal intestinal tract, and thus support the nonimmune defense of the gut (Schanbacher *et al.*, 1997). Antimicrobial peptides have been observed to restrain *in vitro* the growth of many pathogenic and non-pathogenic microbes (Clare *et al.*, 2000). In particular, lactoferricin, a peptide derived from lactoferrin by pepsin digestion, has been demonstrated to display antimicrobial activity *in vitro* against both Gram-positive and Gram-negative microorganisms, including *Bacillus, E. coli, Klebsiella, Listeria, Proteus, Pseudomonas, Salmonella, Streptococcus* and *Candida* (Bellamy *et al.*, 1992 and 1994 and Dionysius and Milne, 1997). Lactoferricin, which causes disruption of bacterial cell membranes, is more effective than undigested lactoferrin. It is an iron-binding glycoprotein present in most mammalian body fluids as a component of the host defense against microbial infection (Meisel and Schlimme, 1996). Lactoferricin is reported to have 100 to 1000 times antimicrobial activity of intact lactoferrin (Schanbacher *et al.*, 1998). Lactoferricin is shown to have potent broad-spectrum antimicrobial properties and its effect is lethal, causing a rapid loss of colony-forming capability. This peptide is now produced commercially in Japan with several expected applications as a potent antimicrobial (Regester *et al.*, 1997).

Casocidin, a chymosin digest of casein *in vitro* has exhibited antimicrobial activity against *Staphylococci, Sarcina, Bacillus subtilis, Diplococcus pneumoniae* and *Streptococcus pyogenes*. Casocidin-I, a αs_{1}- casein fragment has been shown to inhibit the growth of *E. coli* and *Staphylococcus carnosus* (Meisel, 1998). Isracidin, a αs_1- casein fragment of f (1-23), has both therapeutic and prophylactic effect. Injection of isracidin into the udder has been shown to give protection against mastitis in sheep and cows (Lahov and Regelson, 1996). Similarly, fragments of human β-casein have a protective effect against *Klebsiella pneumoniae* in mice (Migliore-Samour *et al.,* 1989). Peptides derived from human β-casein have been demonstrated to confer a protection against *K. pneumoniae* infections in mice. These results have been obtained by parenteral administration of peptides (Migliore-Samour0 *et al.,* 1989). Among other bioactivities related to milk fermented by *L. helveticus* and somewhat correlated to the degree of protein hydrolysis, there is also the protective effect on mice against infection with *Klebsiella pneumonia* resulting from administration of the fermented milk for 8 days (Moineau and Goulet, 1991).

Anticarcinogenic Activity

There are plenty of evidences that consumption of fermented milk products may help prevent mucosal cancers. A large number of information stating that fermented milks and yoghurt have effectively prevented, inhibited or cured cancer, tumours in experimental animals have been published in the scientific literature. However, the nature of the milk components involved in cancer prevention remains to be identified. According to MacDonald *et al.* (1994), peptides may reduce the risk of colon cancer by altering the intestinal kinetics. Kim *et al.* (1995) exhibited the anticarcinogenicity of hydrophobic peptide fractions isolated from cheese slurries.

Antimutagenic Activity

A good deal of work has suggested that fermented dairy products exhibit antimutagenic and antitumor activities. Fermented dairy products have been implicated to decrease the incidence of certain cancers. Some workers have exhibited the antimutagenic activity of milk derived peptides. The antimutagenic activity, against 4-nitroquinoline N′-oxide of a similar fermented milk was also reported by Matar *et al.* (1997). Antimutagenic compounds are produced in milk during fermentation by *L. helveticus* and the liberation of peptides is one possible contributing mechanisms (Matar *et al.,* 1997). Casein degradation may increase the accessibility of the mutagens to specific binding sites of the peptides. Peptides resulting from the degradation of casein by the combined action of rennet and LAB during cheese have been released in cheese and have been responsible for this antimutagenic activity of the cheese.

The mechanism of antimutagenicity may be due either to physical binding by entrapping the mutagens by casein micelle or adsorption of the mutagens on the protein molecules, thus, preventing their interactions with the target cells. Some workers have suggested that a quenching reaction may take place between the mutagens and the peptides that were derived from protein hydrolysis or a chemical binding by a scavenging mechanisms or a chemical antagonism. The binding of the mutagens within the protein may depend on specific amino acid composition and the number of nucleophilic groups involved in the reaction.

Immunostimulatory Activity

Much evidence now exists that fermented milks are more competent at stimulating the immune response than are non-fermented milks. They seem to give a better immunostimulation than bacteria suspensions (Goulet *et al.,* 1989). Some peptides that simulate phagocytosis have been isolated from α_{s1}-casein. Phagocytosis is the process of ingestion and digestion of microparasites by amoeboid WBCs called phagocytes. The immunomodulatory outcome of many peptides has been demonstrated *in vitro* but their *in vivo* effects are not well identified. The tyr-Gly and Tyr-Gly-Gly peptides, potentially derived from k-casein and α-lactalbumin, have shown to modulate the lymphokine production *in vitro* (Kayser and Meisel, 1996). These peptides have been employed for immunotherapy of human immunodeficiency virus infection. Several casein-derived peptides may play a significant role in the stimulation of the immune system. They have been found to exert a protective effect against microbial infections and to enhance some functions of the immune system (Migliore-Sammour *et al.,* 1989).

Milk protein-derived peptides, which have been obtained by hydrolysis with digestive and/or microbial enzymes, affect the human immune system regulated by peripheral blood lymphocytes. Sutas *et al.* (1996a) established that digestion of casein fractions by both pepsin and trypsin produced peptides, which *in vitro* had either immunostimulatory or immunosuppressive effect on human blood lymphocytes. Peptides derived from total casein and α_{s1}-casein was found to be mainly suppressive, while those derived from β- and k-casein were principally stimulatory. When the caseins were subjected to hydrolysis by enzymes isolated from a probiotic *Lactobacillus GG* var. *casei* strain prior to pepsin-trypsin treatment, all hydrolysate fractions were immunosuppressive and the highest activity was again found in α_{s1}-casein. This modulation may be beneficial in the down regulation of hypersensitivity reactions to ingest proteins in patients with food protein allergy. Further studies by Sutas *et al.* (1996b) suggested that casein hydrolysates modulate the *in vitro* production of cytokines by human blood T lymphocytes. Cytokines are connected in the regulation of allergic reactions. Definitely, promising results can be obtained in the management of atopic reactions of infants by oral bacteriotherapy with the probiotic *Lactobacillus* GG strain. Laffieneur *et al.* (1996) also supported this result, which showed that β-casein hydrolysed by LAB, for example, *L. helveticus,* has immunomodulatory action, which could be interrelated to interaction with monocyte-macrophage and T-helper cells.

Certain functional peptides from casein components are also involved in the production of the immunoglobulins. With regard to *in vitro* effects, isracidin develops long-term immunity to sheep against reinfection. Immunopeptides obtained from α_{s1}- and β-casein and α-lactalbumin stimulates phagocytic activities of murine and human macrophages, and defends against *Klebsiella pneumoniae* contagion in mice (Parker *et al.,* 1984; Migliore-Samour *et al.,* 1989 and Meisel and Schlimme 1990). Glycomacropeptide (GMP) down-regulates the immune systems of neonates by suppressing the proliferative responses of both B- and T-lymphocytes. The antihypertensive peptides from C-terminal α_{s1} casein and the peptide from β-casein fragment f (177-183) may also show an immunostimulation activity (Maruyama *et al.,* 1987 and Migliore-Sammour *et al.,* 1989). Perdigon *et al.* (1995) attributed the increase of the anti-SRBC in yoghurt stored for 20 days to the stimulation of the lymphoid cells associated with the intestinal mucosa by the peptides released during the storage period.

Feeding mice with milk fermented by *B. longum, Lactobacillus casei* or *Lactobacillus helveticus* has shown a noteworthy stimulation of phagocytosis by pulmonary macrophages (Moineau and Goulet, 1991). It was reported that the increased number of cells secreting IgA (but not IgG) in the large intestine of mice fed with yoghurt could contribute to limit the inflammatory immune response (Perdigon *et al.,* 1998). Since IgA is considered to be an immune barrier in colonic neoplasia. The modulation of mucosal inflammation by IgA is vital to prevent the tissue damaging consequences of a permanent inflammatory response, which occurs during the development of tumors and neoplasia. This suggests that the mechanisms by which fermented milk or yoghurt cause immune stimulation and/or inhibits tumour development, which could be attributed to the role of bioactive peptides on the immunostimulation and the regression of mucosal tumours. An interesting property associated with CPPs is their potential to enhance mucosal immunity. This idea is supported by a recent study, which showed that oral administration of a commercial caseinophosphopeptide preparation enhanced intestinal IgA levels in piglets.

Anticariogenic Activity

Most minerals from food are dissociated at a low pH in the stomach, and subsequently transferred to the duodenum. These mineral ions may gradually become insoluble as the pH increases. It has been observed that CPPs exhibit a potent ability to form soluble complexes with Ca^{2+}, preventing Ca^{2+} from precipitation as Ca-phosphate in the intestine (Korhonen and Pihlanto-Leppala, 2001).

The highly anionic character of the phosphopeptides renders them resistant to further proteolytic attack and allows them to form soluble complexes with calcium, which obstruct calcium phosphate precipitation (Tirelli *et al.,* 1997) and enhance intestinal absorption of calcium and its retention in the body (Meisel and Schlimme 1990). CCP added to toothpaste help prevent enamel demineralization and exert an anticariogenic effect (Tirelli *et al.,* 1997). The negatively-charged side chains of CCPs represent the binding sites for minerals. It has been claimed that these peptides may function as carrier for different minerals, especially Ca and trace elements, like Fe, Mn, Cu and Se (West, 1986 and Kitts and Yuan, 1992) and thereby enhancing mineral solubility at an intestinal pH. Furthermore, these phosphopeptides can have an anticariogenic effect, based on their ability to localize amorphous phosphate in dental plaque (Reynolds, 1994), by inhibiting caries lesions through recalcification of the dental enamel.

Bio-Peptides derived from tryptic hydrolysates of casein, known as CCPs possess physico-chemical properties that enable the chelation of various bi- and trivalent minerals to be carried out, thereby enhancing mineral solubility in the lower small intestine (Kitts and Yuan, 1992). It has been suggested that moderate and exchangeable binding of calcium to CPP is responsible for the high absorbability of calcium from milk (Yamauchi, 1992). The Japanese market has introduced products containing CPPs that promote calcium absorption in the intestine. Calcium that is removed from teeth as a result of sugar consumption can be replaced by anticariogenic CPPs (FitzGerald, 1998). Researches at the University of Melbourne, School of Dental Studies have proved that there is a significant decline in the tooth decay in laboratory animals fed with anticariogenic CPPs (Anon, 1996). It has been reported that consumption of cheese at the end of meal prevents dental caries. There is a possibility that this may be attributed to the presence of anticariogenic CPP in ripened cheese.

Antithrombic Activity

Antithrombotic peptides are present in milk. Biologically active peptides, isolated from both casein and lactotransferrin, inhibited platelet function. The peptide, called casopiastrin, a k-casein fragment f (106-110), produced by trypsin hydrolyzation, shows antithrombotic activity through inhibition of fibrinogen binding to platelets. The peptides also combines with the receptor sites, thus preventing the binding of human fibrinogen γ-chain to a specific receptor site on the platelet surface, hence expressing its antithrombotic activity (Fiat *et al.,* 1989). A second segment of the trypsin treated k-casein fragment f (103-111), inhibits platelet aggregation but did not affect fibrinogen binding to the platelet receptor. The C-terminal of dodecapeptide of human fibrinogen γ-chain and the undecapeptide from bovine k-casein are structurally and functionally somewhat similar. This casein-derived peptide sequence termed cseoplatelin influences the activity of platelet and restrains both the aggregation of ADP-activated platelets and the binding of human fibrinogen λ-chain to its receptor region on the platelets' surface. Bovine k-caseinoglycopeptide, the C-terminal end of k-casein fragment f (106-169), also inhibits von Willebrand factor-dependent platelet aggregation. The C-terminal residues f (106-171) of sheep k-casein or k-caseinoglycopeptide, decreased thrombin- and collagen-induced platelet aggregation in a dose dependent manner. Lastly, thrombin induced platelet aggregation was inhibited with pepsin digest of sheep and human lactoferrin (Clare and Swaisgood, 2000).

Antigastric Activity

It has been reported that certain bioactive peptides of milk origin possesses antigastric activity. Casopiastrin, a k-casein fragment with sequence f (106-169) and the so-called macropeptide fragment, inhibits gastric secretion and exerts antigastric activity (Meisel *et al.,* 1989).

Fermented Dairy Products: A Milieu of Bioactive Peptides

Fermented dairy products have been consumed through many centuries by millions of human beings long before the existence of microorganisms was scientifically demonstrated. Their nutritive value and therapeutic properties have been and still are widely recognized in all parts of the world and allow them to be classified as nutraceuticals. More and more evidence is now suggesting that substrate derived molecules may play a momentous role in the reported beneficial effects of some of these products. Several peptides derived from milk proteins, mainly caseins and whey proteins, have already found interesting applications in pharmaceutical preparations or in dietary supplements. Bioactive peptides, isolated in fermented dairy products, have braced the evidence that "tertiary metabolites" resulting from the enzymatic alterations of native or denatured proteins, might be responsible for many of the so called probiotic effects of traditional and newly developed fermented dairy products.

The formation of bioactive peptides by LAB in fermented milk products seems to be a rare event. Various long oligopeptides are liberated by degradation of casein, which are the precursors of peptides with biological activity when cleaved by other enzymes. In fermented milk products intracellular peptidases of LAB most likely contribute to further degradation after cell lysis (Mierau *et al.,* 1997). LAB are indirectly implicated to the modulation of the immune response of the host by contributing, during the fermentation process, to the release of peptides bearing hormone-like activities. During the fermentation of milk by LAB, casein undergoes a slight proteolysis capable of generating potentially bioactive peptides (Matar *et al.,* 1996). The enzymes involved occur naturally in foods or derives from exogenous or microbial sources (Smacchi and Gobbetti, 2000). The type of starter used is one of the main factors, which influence the synthesis of bioactive peptides in fermented milks.

A number of studies have demonstrated the antihypertensive effect of the Japanese *Calpis* sour milk, a yoghurt like product. It contains two ACE inhibitory tripeptides (Val-Pro-Pro and lle-Pro-Pro), which are formed from β-casein and k-casein during fermentation of milk with *Lactobacillus helveticus* and *Saccharomyces cerevisiae* (Nakamura *et al.,* 1995a, b). Oral administration of 5 ml of sour milk/kg of body weight (BW) significantly decreased systolic blood pressure and the peptides showed a dose-dependent activity upto a dosage of 5 mg/kg BW. A placebo controlled study showed that the blood pressure of hypertensive human subjects decreased considerably between four and eight weeks after daily ingestion of 95 ml of sour milk (Hata *et al.,* 1996). The ingested dose of ACE inhibitory peptides ranged from 1.2 to 1.6 mg. The formation of calcium-binding phosphopeptides in fermented milks and various types of cheese has been observed in many studies (Korhonen and Pihlanto-Leppala, 2001). Rokka *et al.* (1997) reported on the release of variety of bioactive peptides by enzyme proteolysis of UHT milk fermented with a probiotic *Lactobacillus casei* ssp. *rhamnosus* strain. Upon fermentation, the product was treated with pepsin and trypsin with the intention to simulate gastrointestinal conditions. In the hydrolysate, many bioactive peptide, which corresponded fragments of α_{s1}-and β-casein, and of α-lactalbumin, having different degrees of immunostimulating, opioid and ACE-inhibitory activities were identified. The production of these bioactive peptides could partially explain the probiotic properties attributed to the *Lactobacillus* GG strain. A similar suggestion was made by Moineau and Goulet (1991), who observed increased pulmonary macrophage activity in mice fed with *Lactobacillus helveticus* fermented milk.

Yamamoto *et al.* (1994) reported that β-casein (mainly casokinins) hydrolysed by the cell wall-associated serine-type proteinase from *L. helveticus* CP790 showed antihypertensive activity in Spontaneous Hypertensive Rats (SHR). Yoghurt contains peptides in the range of 500-10,000 Dalton, which *in vitro* reduce the risk of colon cancer (Ganjam *et al.,* 1997). Yamamoto *et al.* (1999) observed a strong antihypertensive effect in SHR after oral administration of whey from a yoghurt-like product where a dipeptide (Tyr-Pro) was released upon fermentation with *L. helveticus* CPN4 strain. They also showed that the dipeptide had an ACE inhibitory activity. The concentration of Tyr-Pro peptide increased during fermentation and reached about 8.1 μg/ml of whey in the yoghurt-like product. Fermented milk containing ACE-inhibitory peptides and other biologically active peptides can be useful ingredients in functional foods to prevent hypertension.

Belem *et al.* (1999) fermented whey with *Kluyveromyces marxianus* var. *marxianus* and identified in the hydrolysate a tetrapeptide with a sequence of β-lactorphin (Tyr-Leu-Leu-Phe). It was suggested that this peptide may have antihypertensive properties. Pihlanto-Leppala *et al.* (2000) studied the potential formation of ACE inhibitory peptides from cheese whey and caseins during fermentation with different commercial lactic acid starters used in the manufacture of yoghurt, ropy milk and sour milk. Matar *et al.* (1996) isolated immunostimulatory peptides in milk fermented with a *L. helveticus* strain. Matar and Goulet (1996) also studied that bioactive peptides, especially β-casomorphins could be released during the fermentation of milk by *L. helveticus.* Fermented milk enriched with the opioid β-casomorphin 1-4 f (6—63) was produced using a mutant strain of *Lb. helveticus* (Matar and Goulet, 1996). ACE inhibitory peptides are released both from αs_1– and β-caseins and also from the major whey proteins α-lactalbumin and β-lactoglobulin (Pihlanto-Leppala *et al.*, 2000). When ACE-inhibitory peptides were orally given to rats, blood pressure was reduced in a dose-dependent manner. ACE-inhibitory peptides having Proline residues, especially at the C-terminal end, are highly resistant to the degradation by digestive enzymes.

A recent study has revealed that when adults ingest 500 ml of milk or yoghurt, certain functional peptides could be identified in their stomach, duodenum and blood, notably caseinoglycopeptides with an antithrombotic sequence (Chabance *et al.*, 1998). It is interesting to note that the level of these peptides is higher after ingesting yoghurt rather than milk, which suggests the role of LAB in the formation of functional peptides. b-casomorphins have been identified in several fermented dairy products. Table 2.2 shows a number of bioactive peptides with their functional activities derived from different fermented dairy products.

Table 2.2: Bioactive Peptides Identified in Fermented Milk Products

Product	*Bioactivity*	*References*
Sour milk	Phosphopeptides	Kahala *et al.* (1993)
Sour milk	ACE inhibitory	Nakamura *et al.* (1995a,b)
Sour milk	β-casomorphin-4	Matar and Goulet (1996)
Fermented milk	ACE inhibitory	Matar *et al.* (1996)
(treated with pepsin and trypsin)	Immunomodulatory and Opioid	Rokka *et al.*(1997)
Yoghurt	ACE Inhibitory (weak)	Meisel *et al.* (1997)
Yoghurt	Immunomodulatory, Antihypertensive, Antiamnesic Microbicidal, Antithrombotic	Dionysius *et al.* (2000)

Studies confirmed that the antimutagenic and anticarcinogenic activities were greater in fermented milk and casein hydrolysates than in unfermented milk and native casein. This may be considered as more indirect evidence that peptides resulting from the proteolytic action of LAB on milk proteins may act in a prophylactic manner. Bioactive peptides produced during bacterial fermentation may alter the risk of colon cancer via modification of cell proliferation in the colon. Many studies have reported on the formation of various bioactive peptides in fermented milk products. Further studies are needed on the importance in fermented milk products of added starters and non-starter proteolytic bacteria for the formation and transformation of bioactive peptides.

Cheese: A Novel Milieu of Bioactive Peptides

LAB, used as starter microorganisms in cheese manufacture or endogenous to milk microflora, have a proteolytic system, which include cell bound proteinases and intracellular peptidases (Kunji *et al.,* 1996). Cell bound proteinases release many different oligopeptides into shorter fragments and amino acids, which contribute directly or as precursors to flavour (Stepaniak and Fox, 1995). Some of these peptides are bioactive. In particular, antihypertensive and mineral binding peptides have been identified in fermented milk, whey and ripened cheese. A few of these peptides have proved effective also in humans after ingestion of fermented milk products. LAB used in cheese manufacture may be indirectly involved in the modulation of different physiological response of the host by contributing, during the fermentation process, to the release of several bioactive peptides. During the fermentation and ripening of cheese by LAB, casein undergoes a slight proteolysis capable of generating a number of potentially bioactive peptides.

Bioactive peptides can also be produced in cheese from milk proteins through fermentation by starters employed in manufacture of cheese. In particular, antihypertensive peptides have also been identified in ripened cheese. A number of peptides with the sequence of β-casein f193-209, f94-209, f69-97, f141-163 and f69-84 have been produced by chymosin, trypsin, and chymotrypsin and by cell bound proteinases from *Lc. lactis* subsp. *cremoris* (Fox *et al.,* 1995; Monnet *et al.,* 1989; Reid *et al.,* 1991; Visser, 1993; Stepaniak *et al.,* 1995 and 1996). After synthesis, these peptides are not hydrolyzed or are hydrolyzed very slowly by chymosin and cell bound proteinases, and therefore these peptides probably accumulate during cheese ripening (Fox *et al.,* 1995). In addition to the possible liberation of bioactive peptides during intestinal proteolysis, such peptides may already be generated during manufacture of several milk products like cheese and thus be ingested as food components.

Cheese contains phosphopeptides as natural constituents and secondary proteolysis during cheese ripening leads to formation of various ACE-inhibitory peptides (Meisel *et al.,* 1997). Several bioactive peptides are liberated in cheeses as the consequence of intense proteolysis during cheese ripening. Meisel *et al.* (1997) analyzed a variety of dairy products including several cheese, sports nutrients and infant formulas for the presence of ACE inhibitory peptides. Low ACE inhibitory activity was observed in samples having a low degree of proteolysis, for example, yoghurt, fresh cheese, and Quarg and sports nutrition-related products. In ripened cheese types, the inhibitory activity increased with developing proteolysis but started decreasing when cheese maturation exceeded a certain level, as measured by the free peptide-bound amino acids ratio. These results were supported by the findings about the occurrence of ACE inhibitory peptides in Finnish (Ryhanen *et al.,* 2001) and Italian cheese varieties, such as Crescenza, Italico and Gorgonzola (Smacchi and Gobbetti, 1998). Similar observations have been made on various bioactive peptides identified in Australian yoghurt and cheese samples. This peptide is reported to have immunomodulatory properties. Also antihypertensive, antiamnesic, antithrombotic, opioid agonist and microbicidal peptides were identified (Dionysius *et al.,* 2000).

In a recent study, Saito *et al.* (2000) identified water-soluble ACE inhibitory peptides from several ripened cheese like Gouda, Emmental, Edam, Havarti, Blue and Camembert types of ripened cheeses and assessed their activity by *in vitro* and *in vivo* experiments. The highest activity was detected in the peptides isolated from an eight month aged Gouda cheese. These peptides were derived both from α_{s1}- and β-caseins and comprised nine amino acid residues, which required further digestion by intestinal protease or peptidase before absorption. The strong antihypertensive activity observed after oral administration of each peptide sample are supposed to be generated after digestion. Advance studies are required on the importance in cheese of added starters and non-starter proteolytic bacteria for the formation and transformation of bioactive peptides. In this context, it is of considerable interest to note the recent findings of Smacchi and Gobbetti (2000) that peptides isolated from Italian cheeses showed inhibitory activity to amino- and endopeptidases derived from yoghurt and cheese starters and also to ACE.

In another study, the Festivo cheese, manufactured by using cheese starter cultures in combination with *Lactobacillus acidophilus* and bifidobacteria, was found to contain bioactive peptides with potential antihypertensive effects (Ryhanen *et al.,* 2001). Their results showed that the ACE-inhibitory activity increased during 'Festivo' cheese ripening, and decreased when proteolysis exceeded a certain level during the storage period. The ACE inhibitory activity in medium aged Gouda was also observed to be about double that of the long ripened Gouda (Meisel *et al.,* 1997). A αs_1-casein derived antihypertensive peptide, isolated from 6 month-ripened Parmesan cheese (Addeo *et al.,* 1992), was also reported not to be detectable after 15 months of ripening, which indicates that the bioactive peptides liberated by proteolytic enzymes from LAB during cheese ripening are degraded further to inactive fragments as a result of further proteolysis (Meisel *et al.,* 1997). These results would suggest that ACE-inhibitory peptides and probably other biologically active peptides as well, are naturally formed in cheese, and remain active for a limited period before splitting into other peptides and amino acids as ripening proceeds. This called for the development of a totally new type of bioactive cheese using defined starter culture and controlled ripening. The research group developed this new 'Festivo' cheese based on an innovative concept in the production of healthy foods. Probably, this was the first time that a probiotic cheese coupled with the health benefits of bioactive peptides was produced on an industrial scale. 'Festivo' is now manufactured commercially in Finland and has attracted growing interest among health-aware consumers.

Many industrially used dairy starter cultures are highly proteolytic. This property is traditionally exploited by the cheese industry, as the peptides and amino acids degraded from milk proteins during fermentation contribute to the typical flavour, aroma and texture of the products. The proteolytic system of LAB, such as *Lactococcus lactis, Lactobacillus helveticus* and *Lactobacillus delbrueckii* var. *bulgaricus,* is already well known. Bioactive peptides may be liberated during manufacture of cultured milk products, such as yoghurt, *Dahi*, cheese, etc. Bacterial starter cultures contain several proteolytic enzymes that are responsible for the breakdown of protein into peptides and amino acids during cheese maturing. The GMP released from k-casein as result of rennin action during cheese making may be involved in regulating digestion as well as in modulating platelet function and thrombosis in a beneficial way. It is reported to suppress appetite by stimulating CCK hormone. Accordingly, it may be an important ingredient of satiety diets designed for weight reduction. Furthermore, this peptide may inhibit toxin binding in the gastrointestinal tract (Chandan, 2001). In fermented milk products and cheese, intracellular peptidases of LAB will most likely contribute to further degradation after cell lysis. Relatively high amounts of bioactive peptides could potentially be produced during proteolysis of 1 g of each of the major casein and whey protein components (Meisel, 1998).

Cheese contains phosphopeptides as natural constituents and extensive proteolysis during cheese ripening leads to the formation of other bioactive peptides. β-casein f58-72 containing the β-casomorphin-7 sequence having antihypertensive activity has been isolated from Crescenza cheese (Smacchi and Gobbetti, 1998) and Cheddar cheeses (Stepaniak *et al.,* 1995). Proteolytic enzymes of LAB produce CPPs during ripening of cooked curd cheeses like Comte (Roudot-Algaron *et al.,* 1994) or Grana Padano (Pellegrino *et al.,* 1998). Peptides derived from a cheese slurry prepared using *Lactococcus lactis* subsp. *lactis* as starter culture have been observed to possess anticarcinogenic effect (Kim *et al.,* 1995). Sabikhi (1999) identified β-casomorphin 3 in Edam cheese after 3 months of ripening. The peptide was present at the level of 34.5 mg/100 gm of cheese. The Table 2.3 enlists different bioactive peptides isolated in a number of cheeses by different workers.

Numerous bioactive peptides have been found in cheeses and may be due to the intense proteolysis during cheese ripening. Ripening conditions and the type of starters used affect bioactive peptides synthesis in cheese, which are to be considered in developing bioactive cheese with elevated level of bioactive peptides.

Development of Bioactive Cheese: Challenges Ahead

Bioactive peptides or related sequences are released in several fermented dairy products by the action of proteolytic enzymes from LAB, but, once liberated, bioactive peptides also influence the biochemical activity of the microbial communities in cheese. This role is probably been underestimated in dairy processing, which seems to be a challenge to the development of bioactive cheese. The type of starter used is one of the key factors, which influence the synthesis of bioactive peptides in cheese. In the selection of microbial starters, the sensitivity to peptides should be carefully considered, because after liberation upon cheese ripening, these peptides may accumulate that would retard or block ripening due to enzyme inhibition, which interfere the natural ripening process.

Table 2.3: Bioactive Peptides Identified in Several Cheese

Product	*Bioactivity*	*References*
Quarg	ACE inhibitory	Meisel *et al.* (1997)
Parmesan Reggiano	β-casomorphin precursors	Addeo *et al.* (1992)
Comte'	Phosphopeptides	Roudot-Algaron *et al.* (1994)
Cheddar	Phosphopeptides	Singh *et al.* (1997)
Edam, Emmental, Gouda, Roquefort, Tilsit	ACE inhibitory	Meisel *et al.* (1997)
Mozzarella, Italico, Crescenza, Gorgonzola (Italian varieties)	ACE inhibitory	Smacchi and Gobbetti (1998)
Edam, Emmental, Turunmaa, Cheddar, Festivo (Finnish varieties)	ACE inhibitory	Ryhanen *et al.* (2000)
Cheddar, Edam, Swiss, Feta, Camembert, Blue vein (Australian varieties)	Immunomodulatory	Dionysius *et al.* (2001)
Gouda, Havarti, Emmental	Antihypertensive	
	Antiamnesic, Opioid agonist	
	ACE inhibitory	Saito *et al.* (2000)

Bioactive peptides or related sequences produced by exogenous or LAB enzymes selectively inhibit microbial proteolysis during ripening. For instance, bioactive peptides isolated from Cheddar and Jarlsberg cheese (Gobbetti *et al.,* 1995 and Stepaniak *et al.,* 1995) and from several Italian cheeses, such as Parmesan, Pecorino Romano, Crescenza, Mozzarella (Smacchi and Gobbetti, 1998) have been found to inhibit the intracellular peptidases of LAB. It is also observed that β-casein f58-72 containing the β-casomorphin-7 sequence released during Cheddar and Crescenza ripening and selectively inhibits endopeptidase, aminopeptidase and X-prolyl-dipeptidyl-aminopeptidase of LAB (Gobbetti *et al.,* 1995; Stepaniak *et al.,* 1995 and Smacchi and Gobbetti, 1998). Inhibition of these intracellular peptidase of LAB by the bioactive peptides pose a crucial challenge in developing cheese with high content of bioactive peptides, because of the fact that inhibition of these peptides would hampers normal ripening process responsible for the proper body and texture and flavour characteristics of final cheese. Another technological challenge concerned with successful manufacture of bioactive cheese is careful control over the ripening progress. It has been well documented that overripening further degrades the physiologically active peptides liberated, thus making them inactive.

Bioactive Cheese: The Pragmatic Vision

Microbial starters, such as LAB used in cheese making contribute to the synthesis of bioactive peptides. An increasing number of *in vitro* and *in vivo* studies divulge that biologically active peptides are released from bovine milk proteins upon microbial fermentation and hydrolysis by digestive enzymes. Peptides with different bioactivities can be released in fermented milks and cheese varieties, but the specificity and amount of peptides formed need to be regulated by starter cultures used for cheese ripening and the rate of proteolysis, which needs further extensive studies on the controlled maturation of cheese. Use of highly proteolytic starter cultures, for example, *Lactobacillus helveticus*, would seem to produce short peptides with various bioactivities in several cheeses.

Food researchers are presently considering bioactive peptides as health-enhancing nutraceuticals for use in functional foods. Besides supplementation of food, production of desirable bioactive peptides during food processing, *e.g.* by use of specific enzymes or genetically transformed microorganisms is also a target area that seems to hold the interest of these workers. Presently, the development of bioactive cheese with the *in situ* production of bioactive peptides gets immense scientific interest. The liberation of bioactive peptides during cheese ripening, for example by use of specific bacterial enzymes or genetically modified microorganisms (GMMs), is of interest for future research work. May be in the forthcoming days, cheese with the enriched level of bioactive peptides would occupy a place in market shelves. It is in this context that cheeses, the most distinct category of fermented milk product, are the natural sources of peptides with their physiologically significance and have a potential for future food research.

Processing conditions should be carefully controlled in order to synthesize specific bioactive peptide during cheese production by selected or genetically modified micro-enzymatic processes. Given the advantages of modern biotechnology and genetic engineering, recombinant DNA approaches for expression of relevant genes responsible for the synthesis of enzymes involved in liberation of specific bioactive peptide in cheese may have a potential application. Trypsin, thermolisin, a-chymotrypsin, papain and subtilism have been used for the preparative synthesis of biologically active peptides, such as Angiotensin, caerulein, enkephalin, oxytocin and dynorphin. However, the restricted availability of suitable biocatalysts is still problem. With the modified stability, activity, and specificity via site-directed mutagenesis of well-known proteinases, proteolytic enzymes of LAB, due to their high regiospecificity and proven efficiency, may be assayed for *in situ* enzymatic synthesis of biopeptides in cheese. The use of specific peptide inhibitors may be helpful in order to keep the different proteolytic enzymes active in generating the desired flavour as well as body and texture of cheese (Smacchi and Gobbetti, 1998).

Conclusion

Fermented food products have been consumed through many centuries by millions of human beings long before the existence of microorganism was scientifically demonstrated. Their nutritive value and therapeutic properties have been and still are widely recognized in all parts of the world and allow them to be classified as nutraceuticals. More and more evidence is now suggesting that substrate derived molecules may play a significant role in the reported beneficial effects of some of these products. Several peptides derived from milk proteins, mainly caseins and whey proteins, have already found interesting applications in pharmaceutical preparations or dietary supplements. Bioactive peptides, isolated from fermented dairy products mainly from cheese, have strengthened the interest to develop bioactive cheese with controlled maturation using defined cheese starters. Optimization of fermentation and maturation on the one hand, as well as molecular cloning and gene manipulation of the starter cultures employed cheese making and immunity *via* gene cloning and amplification, site-directed mutagenesis, and hybrid gene formation, etc on the other hand may further improve the production of these biologically active molecules.

With the triumphal achievement in recombinant DNA technology to produce GMMs, one might foresee very promising future for the bioactive cheese with the health promoting effects of bioactive peptides in the field of functional foods and nutraceuticals particularly with regard to the prevention or attenuation of several symptoms related to physiological or infectious diseases. The combinations of special strains of bacteria, specific substrates and optimal fermentation/growth conditions are likely to yield synergies that would not be observed in conventional cheese ripening process. Molecular biology, biochemistry, microbiology, chemistry and many other fields of knowledge will certainly contribute to a better understanding of what can be considered as a new level of interaction between foods, microorganisms and the digestive system of most animal species including man. Development of bioactive cheese might have been successful in prevention of several physiological disorders of human beings. To decipher such a new reservoir of biological messages that is well encrypted in our food is quite a challenge. Fortunately, we can count on probiotics to help us better understand this new language and continue to improve our quality of life.

References

Addeo, F., Chianes, L., Salzano, A., Sacchi, R., Cappuccio, U., Ferranti, P. and Malorni, A. (1992). Characterization of the 12 per cent tricholoacetic acid-insoluble oligopeptides of Parmigiano-Reggiano cheese. *J. Dairy Res.*, **59**: 401-411.

Anon. (1996). Milk derivative combats decay. *Dairy Ind. Int.*, **61** (4): 6.

Bellamy, W.R., Takase, M., Yamauchi, K., Kawase, K., Shimamura, S. and Tomita, M. (1992). Antibacterial spectrum of lactoferricin B, a potent bactericidal peptide derived form the N-terminal region of bovine lactoferrin. *Biochem. Biophys. Acta*, **1121**: 130-136.

Bellamy, W.R., Yamauchi, K., Wakabayashi, H., Takase, M., Shimamura, S. and Tomita, M. (1994). Antifungal properties of lactoferricin, a peptide derived from the N-terminal region of bovine lactoferrin. *Let. Appl. Microbiol.*, **18**: 230-233.

Berner, L.A. and O' Dannell, J.A. (1998). Functional foods and health claims legislation: Application to dairy foods. *Int. Dairy J.*, **8**: 355-362.

Bleme, M.A.F., Gibbs, B.F. and Lee, B.H. (1999). Proposing sequences for peptides derived from whey fermentation with potential bioactive sites. *J. Dairy Sci.*, **82**: 486-493.

Chabance, B., Marteau, P., Rambaud, J.C., Migliore-Samour, D., Boynard, M., Perrotin, P., Guillet, R., Jolles, P. and Fiat, A.M. (1998). Casein peptide release and passage to the blood in humans during digestion of milk or yoghurt. *Biochemie*, **80**: 155-165.

Chandan, R.C. (2001). Functional foods and bioactive dairy ingredients. *Indian Dairyman*, **83**(9): 43-50.

Clare, D.A. and Swaisgood, H.E. (2000). Bioactive milk peptides: A prospectus. *J. Dairy Sci.,* 83: 1187-1195.

Daniel, H., Vohwinkel, M. and Rehner, G. (1990a). Effect of casein and β-casomorphins on gastrointestinal motility in rats. *J. Nutr.,* **120**: 252-257.

Daniel, H., Wessendorf, A., Vohwinkel, M. and Brantl, V. (1990b). Effect of D-Ala ^{2,4}Tyr^{5}-β-casomorphins-5-amide on gastrointestinal functions. In: *β-casomorphins and Related Peptides* (Eds. Nyberg, F. and Brantl, V.), Fyris-Tryck AB, Uppsala, pp. 95-104.

Dionysius, D.A. and Milne, J.M. (1997). Antibacterial peptides of bovine lactoferrin: Purification and characterization. *J. Dairy Sci.*, **80**: 667-674.

Dionysius, D.A., Marschke, R.J., Wood, A.J., Milne, J., Beattie, T.R., Jiang, H., Treloar, T., Alewood, P.F. and Grieve, P.A. (2000). Identification of physiologically functional peptides in dairy products. *Aust. J. Dairy Technol.,* **55**: 103.

Fiat, A.M., Levy-Toledano, S., Caen, J.P. and Jolles, P. (1989). Biologically active peptides of casein and lactoferrin implicated in platelet function. *J. Dairy Res.*, **56** (3): 351-355.

Fiat, A.M., Migliore-Samour, Jolles, P., Drouet, L., Sollier, C.B.D. and Caen, J. (1993). Biologically active peptides from milk proteins with emphasis two examples concerning antithrombotic and immunomodulating activities. *J. Dairy. Sci.,* **76**: 301-310.

FitzGerald, R.J. (1998). Potential uses of caseinophosphopeptides. *Int. Dairy J.*, **8**: 451-457.

Fox, P.F., Singh, T. and Mcsweeney, P.L.H. (1995). Proteolysis in cheese during ripening. In: *Biochemistry of milk products* (Eds. A.T. Andrew and J. Varbey), London, The Royal Society of Chemistry, pp. 1-31.

Ganjam, L.S., Thornton, W.H. Marshall, R.T. and Macdonald, R.S. (1997). Antiproliferative effects of yoghurt fractions obtained by membrane dialysis on cultured mammalian intestinal cells. *J. Dairy Sci.,* **80**: 2325-2329.

Gobbetti, M., Stepaniak, L., Fox, P.F., Sorhaug, T. and Tobiaseen, R. (1995). Inhibition of endo- and amino-peptidase activities in cytoplasmic fractions of *Lactococcus, Lactobacillus* and *Propionibaterium* by peptides from different cheeses. *Milchwissenschaft,* **50**: 565-570.

Goulet, J., Saucier, L. and Moineau, S. (1989). In: *Yoghurt-Nutritional and Health Properties,* Chand, R.C. (Eds.), USA.

Hata, Y., Yanamoto, B., Ohni, M., Nakajima, K., Nakanura, Y. and Takano, T.C. (1996). A placebo-controlled study of the effect of sour milk on blood pressure in hypertensive subtects. *Am. J.Clin. Nutr.*, **64**: 767-771.

Henriksen, C. M., Nilsson, D., Hansen, S. and Johansen, E. (1999). Industrial applications of genetically modified microorganisms: gene technology at Chr. Hansen A/S. *Intl. Dairy J.,* **9**: 17-23.

Juillard, V., Laan, H., Kunji, E.R.S., Jeronimus-Stratingh, C.M., Bruins, A.P. and Konings, W.N. (1995). The extracellular PI-type proteinase of *Lactococcus lactis* hydrolyzes β-casein into more than one hundred different oligopeptides. *J. Bacteriol.*, **177**: 3472-3478.

Kahala, M., Pahkala, E. and Pihlanto-Leppala, A. (1993). Peptides in fermented Finnish milk products. *Agric. Sci. Finl.*, **2**: 379-386.

Kayser, H. and Meisel, H. (1996). Stimulation of human peripheral blood lymphocytes by bioactive peptides derived from bovine milk proteins. *FEBS letters*, **383**: 18-20.

Kim, H.D., Lee, H.J., Shin, Z.I. Nam, H.S. and Wood, H.J. (1995). Anticancer effects of hydrophobic peptides derived from cheese slurry. *Food Biotech.*, **4**: 268-272.

Kitts, D.D. and Yuan, Y.V. (1992). Caseinophosphopeptides and calcium bioavailability. *Trends Food Sci. Technol.*, **3**(2): 31-35.

Korhonen, H. and Philanto-Leppala, A. (2001). Milk protein derived bioactive peptides- novel opportunistic for health promotion. *Bull. Int. Dairy Fed.*, Nos 363, Pp: 17-26.

Kunji, E.R.S., Mierau, I., Hating, A., Poolman, B. and Konings, N. (1996). The proteolytic system of lactic acid bacteria. *Antonie von Leeuwenhock*, **70**: 187-221.

Laffineur, E., Genetet, N. and Leonil, J. (1996). Immunomodulatory activity of β-permeate medium fermented by lactic acid bacteria. *J. Dairy Sci.*, **79**: 2112-2120.

Lahov, E. and Regelson, W. (1996). Antibacterial and immunostimulating casein-derived substances from milk: Casecidin, isracidin peptides. *Food Chem. Toxicol.*, **34** (1), 131-145.

Lonnerdal, B. (1998). Milk proteins and gut microflora of infants. Pages 5-6. In: Abstracts and Poster Presentations. 25th International Dairy Congress. Aarthus, Denmark. Sep. 21-24.

Maan, E.J. (1993). Milk products in special diets. *Dairy Ind. Int.*, **58**(10): 18-19.

Macdonald, R.S., Thornton, W.H. and Marshall, R.T. (1994). A cell culture model to identify biologically active peptides generated by bacterial hydrolysis of casein. *J. Dairy Sci.*, **77**: 1167-1175.

Maruyama, S., Mitachi, H., Awaja, J., Kurono, M., Tomizaka, N. and Suzuki, H. (1987). Angiotensin 1-converting enzyme inhibitory activity of C-terminal hexapeptide of α_{s1}-casein. *Agri. Biol. Chem.*, **51**: 2557-2561.

Masuda, O., Nakamura, Y. and Takano, T. (1996).Antihypertensive peptides are present in aorta after oral administration of sour milk containing these peptides to spontaneously hypertensive rats. *J. Nutr.*, **126**: 3063-3068.

Matar, C. and Goulet, J. (1996). β-casomorphin-4 from milk fermented by a mutant of *Lactobacillus helveticus. Int. Dairy J.*, **6**: 383-397.

Matar, C., Amiot, J., Savoie, L. and Goulet, J. (1996). The effect of milk fermentation by *Lactobacillus helveticus* on the release of peptides during *in vitro* digestion. *J. Dairy Sci.*, **79**: 971-979.

Matar, C., Nadathur, S., Bakalinsky, A. and Goulet, J. (1997). Antimutagenic effects of milk fermented by *Lactobacillus helveticus* L89 and aprotease-deficient derivative, *J. Dairy Sci.*, **80**: 1965-70.

Meisel, H. (1993). Casokinins as inhibitors of Angiotensin Converting-Enzyme. In: *New Perspectives in Infant Nutrition* (Eds. G. Sawazki and B. Renner), Thieme, Stuttgart, pp. 153-159.

Meisel, H. (1997). Biochemical properties of bioactive peptides derived from milk proteins: potential nutraceuticals for food and pharmacological applications. *Livestock Prod. Sci.*, **50**: 125-138.

Meisel, H. (1998). Overview on milk protein-derived peptides. *Int. Dairy J.*, **8**: 363-373.

Meisel, H. and Bockelmann, W. (1999). Bioactive peptides encrypted in milk proteins: proteolytic activation and thropho-functional properties. *Antonie van Leeuwenhoek*, **76**: 207-215.

Meisel, H. and Schlimme, E. (1990). Milk proteins: Precursors of bioactive peptides. *Trends Food Sci. Technol.*, **1**: 41-43.

Meisel, H. and Schlimme, E. (1994). Inhibitors of Angiotensin converting enzyme derived from bovine casein (casokinins). In: *β-casomorphins and Related Peptides: Recent Developments* (Eds. V. Brantl and H. Teschemacher), VCH, NY, Tokyo, pp. 27-33.

Meisel, H. and Schlimme, E. (1996).Bioactive peptides derived from milk proteins: Ingredients for functional foods. *Kieler Milchwirtsch. Forschungsber*, **48**: 343-357.

Meisel, H., Goepfert, A., and Gunther S. (1997). Occurrence of ACE inhibitory peptides in milk products. *Milchwissenschaft*, **52**: 307-311.

Mierau I., Kunji, E.R.S., Venema, G. and Kok, J. (1997). Casein and peptide degradation in lactic acid bacteria. *Biotech. Genetic Engg. Rev.*, **14**: 279-301.

Miesel, H., Frister, H. and Schlimme, E. (1989). Biologically active peptides in milk proteins. *Z. Ernahrung*, **28**: 267-278.

Migliore-Samour, D., Floc'h, F. and Jolles, P. (1989). Biologically active peptides implicated in immunomodulation. *J. Dairy Res.*, **56**: 357.

Moineau, S. and Goulet, J. (1991). Effect of feeding fermented milks on the pulmonary macrophage activity in mice. *Milchwissenschaft*, **46**: 551-554.

Monnet, V., Bockelmann, W., Gripon, J.C. and Teuber, M. (1989). Comparison of cell-wall proteinases form *Lactococcus lactis* subsp. *Cremoris* ACI and *Lc. Lactis* subsp. *Lactis* NCDO 763. I. Specificity towards β–casein. *Appl. Microbiol. Biotechnol.*, **31**: 112-118.

Nakamura, Y., Yamamoto, N., Sakai, K., Okubo, A. Yamazaki, S. and Takano, T. (1995a). Purification and characterization of Angiotensin I-converting enzyme inhibitors from a sour milk. *J. Dairy Sci.*, **78**: 777-83.

Nakamura, Y., Yamamoto, N., Sakai, K. and Takano, T. (1995b). Antihypertensive effect of sour milk and peptides isolated from it that are inhibitors to Angiotensin I-converting enzyme. *J. Dairy Sci.*, 78: 1253-1257.

Nayak, S.K., Pattnaik, P., Arora, S. and Sindhu, J.S. (1999). Functional peptides from milk proteins. *Indian Dairyman*, **51** (8): 29-34.

Parker, F., Migliore-Samour, D., Flotch. F., Zerial, A., Werner, G.H. Jolles, J. Casaretto, M., Zahn, H. and Jolles, P. (1984). Immunostimulating hexapeptide from human casein: amino acid sequence, synthesis and biological properties. *Euro. J. Biochem.*, **145**: 677-682.

Pellegrino, L., Battelli, G., Resmini, P., Ferranti, P., Barone, F. And Addeo, F. (1997). Alkaline phosphatase inactivation during Grana Padano cheese-making and related effects on cheese characterization and ripening. *Lait*, **77**: 217-220.

Perdigon, G., Alverez, S., Medici, M., Vintini, E., De Giori, G. De Kairuz, M. and Holgado de Ruiz, A.P. (1995). Effect of yoghurt with different storage period on the immune system in mice. *Milchwissenschaft*, **50** (7): 367-371.

Perdigon, G., Valdez, J.C. and Rachid, M. (1998). Antitumor activity of yogurt study on possible immune mechanisms. *J. Dairy Res.*, **65**: 129-138.

Pihlanto-Leppala, A., Kosskinen, P., Piilola, K. and Korhonen, H. (2000). Angiotensin I converting enzyme inhibitory properties of whey protein digest: concentration and characterization of active peptides. *Dairy Res.*, **67**: 53-64.

Regester, G.O., Smithers, G.W., Mitchell, I.R., McIntosh, G.H. and Dionysius, D.A. (1997). In: *Milk Composition, Production and Biotechnology* (Eds. R.A.S. Welch, D.J.W. Burns, S.R. Davis, A.I. Popay and C.G. Prosser.), CAB International, New York, pp. 119-132.

Reid, R.R., Ng. K.H., Moore, C.H., Coolbear,T. andPritchard, G.G. (1991). Comparison of bovine β-casein hydrylysis by P_I and P_{III} proteinases from *Lactococcus lactis* supsp. *cremoris. Appl. Microbiol. Biotechnol.*, **36**: 344-351.

Reynolds, E.C. (1994). Anticariogenic casein phosphopeptides. In: Proceedings 24th International Dairy Congress. Melbourne Australia International Dairy Federation, Brussels, Belgium No 7698-7796/10379.

Rokka, T., Syvaoja, E.L., Tuominen, J. and Korhonen, H. (1997). Release of bioactive peptides by enzymatic proteolysis of *Lactobacillus* GG fermented UHT-milk. *Milchwissenschaft*, **52**: 675-678.

Roudot-Algaron, F., LeBars, D., Kerhoas, L., Einhorn, J., and Gripon, J.C. (1994). Phosphopeptides from Comte cheese: Nature and origin. *J. Food Sci.*, **59**: 544-547.

Ryhanen, E., Leppala, A.P. and Pahkala, E. (2001). A new type of ripened, low fat cheese with bioactive properties. *Int. Dairy J.*, **11**: 441-447.

Sabikhi, L. (1999). *Biotechnological Studies on The Enhancement of Probiotic Attributes Through Bifidobacterium Bifidum in Edam Cheese. Ph.D. Thesis,* National Dairy Research Institute (Deemed University), Karnal, India.

Saito, T., Nakamura, T., Kitazawa, H., Kawai, Y. and Itoh, T. (2000). Isolation and structural analysis of antihypertensive peptides that exist naturally in Gouda cheese. *J. Dairy Sci.,* **83**: 1434-1440.

Schanbacher, F.L., Talhouk, R.S. and Murray, F.A. (1997). Biology and origin of bioactive peptides in milk. *Livestock Prod. Sci.,* **50**: 105-123.

Schanbacher, F.L., Talhouk, R.S., Murray, F.A., Gherman, L.I. and Willet, L.B. (1998). Milk-borne bioactive peptides. *Int. Dairy J.,* **8**: 393-403.

Schlimme, E. and Meisel, H. (1995). *Nahrung,* **39**: 1-20.

Singh, T.K., Fox P.F. and Healy, A. (1997). Isolation and identification of further peptides in the diafiltration retentate of the water-soluble fraction of Cheddar cheese. *J. Dairy Res.,* **64**: 433-443.

Smacchi, E. and Gobbetti, M. (1998). Peptides from several Italian cheeses inhibitory to proteolytic enzymes of lactic acid bacteria, *Pseudomonas fluorescens* ATCC 948 and to the Angiotensin I-converting enzyme. *Enz. Microbial Technol.,* **22**: 687-694.

Smacchi, E. and Gobbetti, M. (2000). Bioactive peptides in dairy products: synthesis and interaction with proteolytic enzymes. *Food Microbiol.,* **17** (2): 129-141.

Stepaniak, L. and Fox, P.F. (1995). Characterization of the principal intracellular endopeptidase from *Lactococcus lactis* subsp. *lactis* Mer 1363. *Int. Dairy J.,* **5**: 699-713.

Stepaniak, L., Fox, P.F., Sorhaug, T. and Grabska, J.J. (1995). Effect of peptides from the sequence 58-72 of β-casein on the activity of endopeptidase, aminopeptidase form *Lactococcus. Agric. Food Chem.,* **43**: 849-853.

Stepaniak, L., Gobbetti, M., Sorhaug, T., Fox, P.F. and Hojrup, P. (1996). Peptides inhibitory to endopeptidase and aminopeptidase from *Lactococcus lactis* ssp. *Lactis* MG 1363, releases form bovine β-casein by chymosin, trypsin or chymotrypsin. *Z. Lebensm. Unters Forsch.,* **202**: 329-333.

Suetsuna, K., Chen, J.R. and Yamauchi, F. (1991). Immunostimulating peptides derived from sardine muscle and soybean protein, amino acid sequence, synthesis and biological properties. *Clin. Rep.,* **25** (15): 75-86.

Sutas, Y., Hume, M. and Isolauri, E. (1996a). Down regulation of anti CD3 antibody induced II-4 production by bovine caseins hydrolysed with *Lactobacillus* GG-derived enzymes. *Scand. J. Immunol.,* **43**: 687-689.

Sutas, Y., Soppi, E., Korjonen, H., Syvdoja, E.L., Saxelin, M. and Rokka, T. (1996b). Suppression of lymphocyte proliferation *in vitro* by bovine caseins hydrolysed with *Lactobacillus casei* GG-derived enzymes. *J. Allergy Clin. Immunol.,* **98**: 216-224.

Teschemacher, H. Koch, G. and Brantl, V. (1997). Milk protein derived opioid receptor ligands. *Biopolymer,* **43**: 99-117.

Teschemacher, H. and Brantl, V. (1994). Milk protein derived atypical opioid peptides and related compounds with opioid antagonist activity. In: *β-casomorphins and*

Related Peptides: Recent Developments (Eds. Brantl, V. and Teschemacher, H.), VCH, Weinheim, pp. 3-17.

Tirelli, A., DeNoni, I. And Resmini, P. (1997). Bioactive peptides in milk products. *Ital. J. Food Sci.,* **1**: 91-98.

Visser, S. (1993). Proteolytic enzymes and their relation to cheese ripening and l flavour: and overview. *J. Dairy Sci.,* **76**: 329-350.

West, D.W. (1986). Structure and function of the phosphorylated residues of casein. *J. Dairy Res.,* **53**: 333-353.

Yamamoto, N., Akino, A. and Takano, T. (1994). Antihypertensive effects of the peptides derived from casein by an extracellular proteinase from *Lactobacillus helveticus* CP790. *J. Dairy Sci.,* **77**: 917-922.

Yamamoto, N., Maeno, M. and Takano, T. (1999). Purification and characterization of an Antihypertensive peptide from a yoghurt-like product fermented by *Lactobacillus helveticus* CPN4. *J. Dairy Sci.,* **82**: 1388-1393.

Yamauchi, K. (1992).Biologically functional proteins of milk and peptides derived from milk proteins. *Bull. Intl. Dairy Fed.,* 272: 51-57.

Yoshikawa, M., Tani, E., Shiota, H, Usui, H., Kurahashi, K. and Chiba, H.D. (1994). Casoxin D, an opioid antagonist/ileumcontracting/vasorelaxing peptide derived from human α_{s1}-casein. In: β-*Casomorphins and Related Peptides: Recent Developments* (Eds. V. Brantl and H. Teschemacher), VCH. Weinheim, pp. 43-48.

2015, Dairy Product Technology: Recent Advances ***Pages 39–57***
Editors: **Subrota Hati, Surajit Mandal and Birendra Kumar Mishra**
Published by: **DAYA PUBLISHING HOUSE, NEW DELHI**

Chapter 3

Cheese: A Novel Milieu of Probiotics

S. Makhal, S. Hati, S. Bera and Pravin Dahale

Introduction

An important allure of the modern health conscious consumer is to consume functional foods to derive health-giving and curative functions, naturally imparted to foods. Likewise, this is the miracle of nature that there is a universe of life within the digestive tract. Microorganisms populate the intestines and colon in numbers 10 times greater than the total number of cells in the body itself - over 10 billion per gram of stool. One half the dry weight of stools is a microorganism. This population of organisms is being increasingly found to have profound effects on health. There is a delicate balance, however, between those organisms which contribute to health in assisting digestion, synthesizing nutrients, and inhibiting cancer-causing biochemicals, for example - and those which can cause disease. Beneficial (probiotic) organisms in the diet can help rebalance the digestive tract.

The potential health-promoting effects of dairy products, which incorporate these beneficial organisms, such as *Lactobacillus* and *Bifidobacterium* spp. have stimulated a foremost research effort in recent years. To date, the most fashionable food delivery systems for these cultures have been freshly fermented dairy products, such as yoghurts and fermented milk as well as unfermented milk with cultures added (Fernandes *et al.*, 1987 and Sanders *et al.*, 1996). The survey report conducted by Letherhead Food Research Association, Randalls showed that the global market for functional foods reached at $ 6.6 billion in 1994, with Japan accounting for just under half of that (LFRA, 1996). Some forecasts suggested that the market would reach $17 billion by the year 2000,with the fastest growth rates in the US (Young, 1996). Currently, the functional food market is flourishing at the rate of 15-20 per cent per annum and the functional food industry is claimed to be worth $ 33 billion (Hilliam, 2000). Growing public awareness of diet-related health issues has fuelled the claim for probiotic foods, which are currently restricted predominantly to fermented milk drinks and yoghurt harbouring beneficial probiotic cultures, such as bifidobacteria and lactobacilli associated with a plethora of health benefits. While cheese has long been associated with a high quality nutritional image, more recently research efforts devoted to the development of probiotic cheese with demonstrated health promoting properties, have been made to push cheese into the "functional foods" category. Generally, incorporation of probiotic bacteria into milk-based food systems, including cheese provides challenges in terms of maintaining viability and probiotic functionality during manufacture and shelf life. Therefore, there is a good potential to incorporate some probiotic attributes in cheeses to derive some health benefits beyond its basic nutrients.

In the recent endeavour to expand probiotic product range, a small number of researchers and companies have attempted to manufacture cheeses, which sustain a high viable count of probiotic organisms. Cheese will have a number of advantages over fresh fermented products as a delivery system for viable probiotics to the GIT (Stanton, *et al.,* 1998) having a higher pH and buffering capacity than the more traditional probiotic foods, a more solid consistency and a higher fat content, which may provide a more stable milieu to support the long-term survival of probiotic organisms. Given the increasingly competitive panorama of the European market in food products, the cheese industries in European countries are trying to derive benefits from a marketing advantage, such as added-value probiotic-containing cheese, which would afford a competitive edge over existing products (Stanton *et al.,* 1998). The development of probiotic cheeses would thus lead to a major economic advantage also.

Probiotics

The GI tract may be the least biologically appreciated organ in the human body. In addition to a role in nutrition, the high metabolic and endocrine bustles of the GIT result in a significant impact on health and well being of the host. The presence of bacteria in the intestinal tract has long been acknowledged. However, more recently the gut microbiota, in particular colonic bacteria, have been implicated as major determinants of health and disease in humans. As relationships between microbial community structure and the health of the host have been elucidated, interest in manipulation of gut bacterial populations by beneficial microorganisms for improved human health has increased. The word "Probiotics" is derived from the Greek meaning "for life" and has several different meanings over the years (Fuller, 1989). Lilly and Stillwell used it in 1965 to describe substances secreted by one microorganism, which stimulated the growth of another. In an endeavour to improve the definition, Fuller (1991) redefined probiotics as an "alive microbial feed supplements, which beneficially affect the host animal by improving its intestinal microbial balance". While the health benefits for the individual can only be inferred, the effect on prevention of spoilage would indubitably have beneficial effect on the health community. For human's use, a probiotics may be defined as a food or supplement containing concentrates of defined strains of living microorganisms that on ingestion in certain doses put forth health benefits beyond inherent basic nutrition. They are believed to contribute to the well being of the consumers by improving host's microbial balance in the GI tract (Chandan, 1999). The concept of probiotic foods is based on the fact that the micro flora in GI tract are having significant role in the health status of an individual, which is influenced by a diet consisting of the organisms (Guerin *et al.,* 1998). Oral probiotics are living microorganisms, which upon ingestion in certain numbers exert health benefits over intrinsic nutritional significance (Guarner and Schaafsma, 1998).

Probiotics are basically mono or mixed culture of live microorganisms which when applied to animal or man decreases the number of intestinal infections and/or improves the general health by contributing to a better GI environment (Fuller, 1992a, 1997; Nousianinen and Setala, 1998). Recently a European expert group proposed a definition that the probiotics are "live microbial food ingredients that have a beneficial effect on human health" (Salminen *et al.,* 1999). Probiotics have been reported to have several health benefits such as balancing of intestinal microflora, stimulation of the immune system, prevention of diarrhoea, and anticarcinogenic activity (Sanders, 1998 and Ziemer and Gibson, 1998).

Backdrop of Probiotics

Interest in the role of probiotics for human health goes back at least as far as 1908 when Metchnikoff suggested that man should consume milk fermented with lactobacilli to prolong lifespan (Hughes and Hoover, 1991; O'Sullivan *et al.,* 1992). The history recording the beneficial properties of live microbial food supplements, such as fermented milks dates back many centuries. At the beginning of the 20th Century, the Russian bacteriologist Eli Metchnikoff (Pasteur Institute, France) was the pioneer to give a scientific explanation for the beneficial effects of lactic acid bacteria present in fermented milk (Hughes and Hoover, 1991; O'Sullivan *et al.,* 1992). He attributed the good health and long life of the Bulgarians to their consumption of large amounts of fermented milk called Y*ahourth* (Metchnikoff, 1907). Almost at the same time, in 1899, Tissier (Pasteur Institute, France) isolated bifidobacteria from the stools of breast-fed infants and found that they were a predominant component of the intestinal flora in humans (Ishibashi and Shimamura, 1993). Tissier recommended the administration of bifidobacteria to infants suffering from diarrhoea believing that the bifidobacteria would displace putrefactive microbes responsible for gastric upsets, while-re-establishing themselves as the dominant intestinal microorganisms (O'Sullivan *et al.,* 1992).

The incorporation of intestinal species of lactobacilli in these products evolved from the work of Rettger in the 1930s (Rettger *et al.,* 1935). Their use in treatment of body ailments has been stated even in Biblical scriptures. Known scientists in early ages, such as Hippocrates and others considered fermented milk as not only a food product but a medicine as well. They used to prescribe sour milks for curing mayhems of the stomach and intestines (Oberman, 1985). The inclusion of bifidobacteria in yoghurt is a more recent incident, probably reflecting the realization that they are amongst the most predominant bacterial species that inhabit the human intestinal tract.

At present, it is generally recognized that an optimum 'balance' in microbial population in our gastrointestinal (GI) tract is associated with good nutrition and health (Rybka and Kailasapathy, 1995). In present day probiotic bacteria have more and more been integrated into foods as dietary adjuncts. The popular dairy products for the delivery of viable *Lactobacillus acidophilus* and *Bifidobacterium bifidum* cells yoghurt, fermented drinks and cheese.

Human Gastrointestinal Ecosystem

The human intestinal tract is virtually sterile, during the prenatal stage. However, during passage through the birth canal and succeeding exposure to the environment, a wide range of microorganisms marches into and takes up dwelling in the digestive tract. The human intestinal tract constitutes a complex ecosystem of microorganisms. The total mucosal surface area of the adult human GI tract is upto 300 $m^{2,}$ making it the largest body area interacting with the environment. It has been investigated that human intestinal tract contains between 1 and 2 kg microflora, and digestive track gives resort to about 10^{14} bacteria consisting of approximately 500 species (Ballongue, 1993). The digestive tract of human harbours a large and complex collection of microbes, mostly bacterial species, which forms part of the normal microflora of the healthy human. Harboured in the distal small bowel and in the large bowel, the intestinal microflora is composed of numerous bacterial species, some of which attain population levels of 10^{10} bacterial cells per g of large bowel contents (Tannock, 1995). There are thus, at a conservative estimate, about 10^{12} bacterial cells present in the large bowel of every human. At least four hundred bacterial species have been detected in faecal samples from humans, but thirty to forty species comprise 99 per cent of the intestinal microflora of one human (Drasar and Barrow, 1985). This complex and highly adapted community further benefits the host by providing resistance to pathogens, because the native gut microflora are so well adapted to their environment, it is difficult for other organisms including pathogens to colonize in the lumen.

The numerically predominant bacterial species are obligatory anaerobic. The huge surface area would suggest a great capacity for effective absorptive area, yet a dilemma for defensive exclusion of infectious, toxic and allergenic material from the internal milieu. Although sterile at birth, we rapidly acquire a commensally enteric microflora resulting in the creation of a complex ecosystem in the GI tract. This microflora adds an additional competitive component to defense capability through competitive exclusion. The bacterial populace in the large intestine is very high and achieves maximum counts of 10^{12} cfu g^{-1}. In the small intestine the bacterial content is considerably lower, 10^4-10^8 cfu g^{-1}., while in the stomach only 10^1-10^2 cfu g^{-1} are found due to the low pH (Hoier, 1992).

The intestine of a newborn infant is devoid of intestinal flora, but immediately after birth colonization by many bacteria commences. With weaning and ageing of the human being, gradual changes in the intestinal flora profile occur. Within one to two days, coliforms, enterococci, clostridia and lactobacilli are detected in the faeces within three to four days; bifidobacteria come out and become predominant around the fifth day. *Bifidobacterium* spp. is natural inhabitants of the gut of warm-blooded animals. The most common species in human infants are *Bifidobacterium infantis, Bifidobacterium breve* and *Bifidobacterium longum*. As both *Bifidobacterium infantis* and *Bifidobacterium. breve* are replaced with both *Bifidobacterium adolescentis* and *Bifidobacterium longum* (Scardovi, 1986). In adult humans, the Bacterioidaceae, at 86 per cent of the total flora, are the most prevalent genera of bacteria. The balance of the flora is eubacteria (6-19 per cent), *Bifidobacterium* (6-36 per cent) and *Peptococaceae* (2-14 per cent), *Megasphaerae* (0.3-0.8 per cent only found with bifidobacteria), *Enterobacteriaceae* (trace to 5.3 per cent), Streptococci and Lactobacilli (Modler *et al.*, 1990).

Bifidobacteria are the prime group in infants, but shift to the third largest group in adults. Normal hale and hearty adults also possess other anaerobes including clostridia, veillonellae, coliforms, streptococci and facultative anaerobic lactobacilli (Mitsuoka, 1982). He also observed somewhat low total counts in the small intestine, which increased from 10^4 to 10^6/g towards the distal end of the ileum. Lactobacilli are found primarily in the distal end of the small intestine while bifidobacteria reside principally in the large intestine. The coliforms and other bacteria are restricted and decrease in response to the boost of bifidobacteria. Bifidobacteria counts of 10^{10}-10^{11} cfu g^{-1} faeces are common in breast-fed infants (Modler *et al.*, 1990) representing 25 per cent of the intestinal bacteria. Lactococci, enterococci and coliforms correspond to less than 1 per cent of the intestinal inhabitants, and normally Bacteroides, clostridia and other organisms are absent (Rasic, 1983). Bottle-fed babies normally have 1-log count less of bifidobacteria (10^9-10^{10} cfu g^{-1}) present in their faecal samples than breast-fed babies to have higher levels of enterobacteriaceae, streptococci and other putrefactive bacteria (Yuhara *et al.*, 1983). This suggests that breast-fed infants are more resistant to infections than bottle-fed infants due to antibacterial substances produced by bifidobacteria.

The proportion of bifidobacteria declines to represent the third most common genus in the GI tract; *Bacteroides* predominates at 86 per cent of the total flora in the adult gut, followed by *Eubacterium* (Finegold *et al.*, 1977). In addition, infant type bifidobacteria, *Bifidobacterium. bifidum* are replaced with adult type bifidobacteria, *Bifidobacterium longum* and *Bifidobacterium adolescentis*. This change in profile may be facilitated by the intake of bifidogenic factors (Modler *et al.*, 1990). The adult type flora is rather stable but during the middle and again at an older age the intestinal flora changes again. Bifidobacteria decrease even further while certain kinds of harmful bacteria augment (Benno *et al.*, 1984). For paradigm, a dramatic decline in the number of bifidobacteria and an increase in *Clostridium perfringens* cause diarrhoea in elderly persons (Hoier, 1992).

The composite oeuvre of the intestinal flora is relatively stable in vigorous human beings. Any commotion in this balance results in transformation in the intestinal flora, which accordingly allows undesirable microorganisms to govern in the intestine and as a result leads to contagious diseases. Changes in the intestinal flora are not only affected by ageing but also by extrinsic factors, *e.g.* stress, diet, drugs, bacterial contamination and constipation (Hoier, 1992).

In 1987, Mitsuoka proposed a hypothetical scheme in which he illustrated the interrelationship between intestinal bacteria and human health (Ishibashi and Shimamura, 1993). The intestinal bacteria were classified into three categories, namely harmful, beneficial, or neutral with respect to human health. Newborn infants are devoid of intestinal flora but through breast-feeding the bifidobacterial population increases rapidly and accounts for more than 25 per cent (w/w) of the intestinal flora (Rasic, 1983). Among the beneficial bacteria are *Bifidobacterium* and *Lactobacillus.* Harmful bacteria are *Escherichia coli, Clostridium, Proteus* and types of *Bacteroides.* These bacteria generate a variety of harmful substances, such as amines, indole, hydrogen sulfide, or phenols from food components and cause certain intestinal troubles. These bacteria could also sporadically be potential pathogenic (Ishibashi and Shimamura, 1993).

Originally, it was thought that bifidobacteria emerged only in the faeces of breast-fed infants. However, work by Braun (1981) exposed that bottle-fed infants contained 10^9 to 10^{10} bifidobacteria per g or 1 log count less than breast-fed babies. Work by Yuhara *et al.* (1983) as well as Kim and Kang (1984) have revealed no significant difference in intestinal bifidobacteria counts between breast- and formula-fed infants. Bacterial counts for the two groups were $10^{10.7}$ and 10^{10}/g of faeces, correspondingly. There is however, a penchant for bottle-fed infants to have higher counts of enterobacteriaceae, streptococci and anaerobes other than bifidobacteria. Bifidobacteria have been observed to grow better in human milk than bovine milk. This maybe due to a lower protein and buffering capacity of human milk (Bullen *et al.,* 1977 and Faure *et al.,* 1984).

Probiotic Foods

Since now a day consumers are more diverted towards the health driven foods, hence recent advances in functional foods demonstrate much promise in new product development using probiotics to derive health benefits. Probiotic foods come under functional foods which, in addition to their basic nutritive value and natural being, will contain the proper balance of ingredients, which will help us to function better and more effectively in many aspects of our lives, including helping us directly in the prevention and treatment of infirmity and diseases. Functional foods serve to endorse health or help to prevent diseases, and in general the term is used to indicate a food that contains some health promoting components beyond traditional nutrients (Berner and O'Dannell, 1998). Probiotic foods are the most important discipline of functional foods, which are defined as "foods containing live microorganisms, which actively enhance the health of consumers by improving the balance of microflora in the gut when ingested live in sufficient numbers" (Fuller, 1992b). Studies relate the promising health benefits of consuming cultured and culture containing milks. There have been long-term interests in the use of cultured milks products with various strains of lactic acid and other probiotic bacteria to improve health of humans (Salminen *et al.,* 1998a,b).

Understanding the relationship between microbial populations in the colon and health continues to increase. Clearly, it is important to comprehend the effects of colonic bacteria on host health in order to fully exploit potential applications of prebiotics and probiotics. The colon can be both an organ of health and of disease, especially with regard to the microbiota. Thus the interest in probiotics and prebiotics arises, in part, from the craving to manipulate or enhance the 'beneficial' gut microbiota in an approach that decreases the peril of developing bacterial associated colonic diseases. This may occur by provision of substrates that subvert the toxigenic potential, *e.g.* switching a proteolytic species towards a more saccharolytic type of fermentation.

Recommendation for Effective Probiotic Foods

Currently, there are no legal recommendations for consumption of probiotics in foods. However, it is generally accepted that health benefits from consumption of particular strains should be demonstrated through controlled clinical trials, and that the manufacturer should provide advice on the minimum dose and duration of use of each individual strain or product (FAO/WHO, 2001). Adequate numbers of viable cells, namely the "therapeutic minimum" needs to be consumed regularly for transfer of the probiotic effect to consumers. The international Dairy Federation (1997) proposed that in probiotic foods, "the specific microorganisms shall be viable, active and abundant at the level of at least 10^7 cfu/g in the product to the date of minimum durability" (Ouwchand and Salminen, 1998). Dose-response studies conducted with *Lactobacillus* GG demonstrated that when administered in either freeze-dried powder or gelatin capsules, the minimum dose required to yield faecal recovery was 10 cfu/day (Saxelin *et al.,* 1991&1995) with lower doses (10^6–10^8 cfu/day) found to be ineffective.

An intake level of 10^9 cfu/day of *L. johnsonii Lal* elicited immune effects; a lower dose of 10^8 cfu did not, although faecal recovery was found in all subjects consuming the culture (Donnet-Hughes *et al.,* 1999). It is evident that there are difficulties involved in defining a general minimum effectual dose for all probiotic cultures, given that variations occur depending on the particular strain or delivery system used. It has been suggested that approximately 10^9 cfu/day of probiotic microorganisms is necessary to elicit health effects. Based on daily consumption of 100 g of a probiotic food, it has been suggested that a product should contain at least 10^7 cells/g, a level paralleling current Japanese recommendations (Ishibashi and Shimamura 1993), but considerably higher numbers have been proposed by others (Lee and Salminen, 1995).

In another report, it has been documented that consumption of more than 100 g per day of bio-yoghurt containing more than 10^6 cfu/ml (Rybka and Kailasapathy, 1995). Survival of these bacteria during shelf life and until consumption is, therefore, an important consideration. Kurmann and Rasic (1991) suggested to achieve optimal potential therapeutic effects, the number of probiotic organisms in a probiotic food should meet a suggested minimum of > 10^6 cfu/ml as satisfactory leveL. This criterion is referred to as the "therapeutic minimum" (Davis *et al.,* 1971 and Rybka and Kailasapathy, 1995). One should aim to consume 10^8 live probiotic cells per day. Regular consumption of 400-500 g/week of bio-yoghurt, containing 10^6 viable cells per ml would provide these numbers (Tamime *et al.,* 1995).

Fermented Milks and Lactic Acid Bacteria Beverage Association of Japan has developed a standard, which requires a minimum of 10^7 viable bifidobacteria cells/ml to be present in fresh dairy products (Ishibashi and Shimamura, 1993). The National

Yoghurt Association (NYA) of the United States specifies 10^8 cfu/g of lactic acid bacteria at the time of manufacture as a prerequisite to use the NYA "Live and Active culture" logo on the containers of products (Kailasapathy and Rybka, 1997). At the same time, attainment of pH 4.5 or below is also legally required to prevent the growth of any pathogenic contaminants (Micanel *et al.*, 1997). It has been claimed that only dairy probiotic foods with viable microorganisms have beneficial health effects.

Probiotics: A Plethora of Health Benefits

It is only recently, however, that the interrelationship between intestinal microorganisms and the health benefits deriving from it are beginning to be understood. The microorganisms primarily associated with this balance are lactobacilli and bifidobactera. Increasing evidence indicates that consumption of probiotic microorganisms can help maintain such a favourable microbial profile and results in several therapeutic benefits (Lourens-Hattingh and Viljoen, 2001). In recent years, probiotic bacteria have increasingly been incorporated into foods as dietary adjuncts with the aim to use diet related health strategies to target a number of chronic diseases.

The human intestinal tract constitutes a complex bionetwork of microorganisms. All these intestinal microflora exist in dynamic balance with one another. Some are useful, some detrimental and some neutral to the physiological functions of the body. Thus, intestinal microflora can influence health in a number of ways, both positive and negative (Sandine, 1979). These include impacts on nutrition and physiological functions. For instance, one of the short chain fatty acids produced by colonic bacteria (butyrate) is significant in determining the rate of colonic cell growth and differentiation, drug efficacy, carcinogenesis, immunological responses, resistance to infection and resistance to endotoxins and other stresses (O'Sullivan, 1996 and Buttriss, 1997). The beneficial bacteria tend to predominate during periods of good health. If the ecological balance of the gut is disturbed due to prolonged disease, dispossession from foods and water, travel (especially by air), antibiotics, radiation etc. (Hanevaar and Huis in't Veld, 1992). Certain microorganisms with negative roles in the human system may dominate. In such cases, there is an increase in the products of putrefaction, toxins and carcinogens. Pathogenic bacteria, which are normally present at low levels also boom if the resistance of the body is lowered for any reason, manifesting their pathogenicity and causing diseases. Living probiotic cultures help uphold the critical balance and can stabilize a disturbed intestinal flora.

The claimed beneficial effects from the consumption of fermented milks were one a very litigious issue. Research conducted since the turn of the past century has, however, enhanced the understanding of the resulting therapeutic effects and it is currently recognized as wholesome. The consumption of probiotic products is helpful in maintaining good health, restoring body vigour, and in skirmishing intestinal and other diseases (Mital and Garg, 1992). Current clinical applications of probiotic bacteria in the well document areas, such as treatment of acute rota virus diarrhoea, lactose maldigestion, constipation, colonic disorders and side-effects of pelvic radiotherapy and more recently food allergy including milk hypersensitivity and changes associated with colon cancer development (Salminen *et al.,* 1998). There are myriad evidences to support the view that oral administration of some *Lactobacillus* and *Bifidobacterium* species are competent to restore the normal balance of probiotic populations in the intestine. In addition to their established role in GI therapy, the probiotic foods are claimed to serve several nutritional and therapeutic benefits, such as antimicrobial properties (Shah, 2000), antimutagenic properties (Lankaputra and Shah, 1998), anticarcinogenic properties (Mitsuoka, 1989), improvement in lactose metabolism (Vesa *et al.,* 1996), reduction in serum cholesterol level (Fukushima and Nakano, 1996) and immune system simulation (Schiffrin *et al.,* 1994). Therefore, in the near future probiotic foods will be seen in many different markets beyond what is seen today

Fuller (1989) listed out the claimed beneficial effects and therapeutic application of probiotic bacteria in humans, which includes: (i) Beneficial effects, such as maintenance of normal intestinal realm, augmentation of immune system, reduction of lactose intolerance, reduction in serum cholesterol levels, anticarcinogenic activity and improved nutritional value of foods, and (ii) Therapeutic applications, such as prevention of urogenital infections, mitigation of constipation, protection against travellers diarrhoea, prevention of infantile diarrhoea, reduction of antibiotic of induced diarrhoea, prevention of hypercholesterolaemia, defence against colon/ bladder cancer and prevention of osteoporosis. Probiotic bacteria, thus, offer new dietary alternatives for the management of such conditions through stabilization of intestinal microflora, promotion of colonization resistance, regulation of the immune response and preservation of intestinal integrity (Salminen *et al.,* 1998a).

Cheese as Probiotic Carrier Food

For probiotic cheese to be beneficial for human health, the probiotic strains should maintain their viability in cheese until the time of consumption and be present in significant numbers, at levels of at least 10^7 cfu/ml (Ishibashi and Shimamura, 1993). When selecting strains for incorporation into cheese like other foods, factors requiring especial emphasis include their ability to survive passage through GIT in order to exert the beneficial effect, ability to survive the manufacturing process and capacity to grow and survive during the ripening and storage periods.

Taking benefits of the consumer interest in the improved therapeutic and nutritional attributes of probiotic cultures, some probiotic cheeses have been developed. Viability of a number of probiotic bacteria in Cheddar cheese during manufacture and ripening has been studied. These strains include lactobacilli (Gardiner *et al.,* 1998) that has been previously isolated from the healthy human GIT and commercial probiotic strains of enterococci (Gardiner *et al.,* 1999a) and bifidobacteria (McBrearty *et al.,* 2001). These studies have demonstrated that Cheddar cheese is suitable as a potential probiotic 'functional food' with a number of probiotic strains demonstrating the capability to survive cheese manufacture and ripening (Stanton *et al.,* 1998). A number of studies have addressed the development of probiotic cheese, using such cheese varieties as Cheddar (Dinakar and Mistry 1994; Gardiner *et al.,* 1998 and 1999a,Gomes *et al.,* 1995; Daigle *et al.,* 1999; McBrearty *et al.,* 2001), white brined (Ghoddusi and Robinson 1996), goats milk cheese (Gomes and Malcata 1998), Crescenza (Gobbetti *et al.*, 1998), Cottage (Blanchete *et al.*, 1996), Kariesh (Murad *et al.*, 2000), Tallaga (El Zayatt and Osman, 2001) and fresh cheeses (Roy *et al.*, 1997). Such studies highlight the importance of the food delivery system in determining the final numbers of beneficial bacteria that survive gastric transit, and demonstrate the suitability of cheese for this purpose. Though a number of studies already have been made to design probiotic cheese with promised *in vivo* health benefits, "Festivo"- the only commercial probiotic cheese, has evolved with great fanfare to the consumers in Finland. Examples of probiotic cheese are listed below.

Table 3.1: Examples of Probiotic Cheese Developments

Cheese Variety	*Probiotic Strain*	*Reference*
Bulgarian yellow	*L. paracasei* M3	Atanassova *et al.* (2001)
Canestrato Pugilese	*B. bifidum* Bb02	Corbo *et al.* (2001)
	B. longum Bb46	
Cheddar	*B. bifidum* ATCC 15696	Dinakar and Mistry (1994)
	B. infantis ATCC 27920G	Daigle *et al.* (1999)
	B. lactis Bb-12	McBrearty *et al.* (2001)
	L. paracasei NFBC 338	Gardiner *et al.* (1998)
	L. Helveticus |	Madkor *et al.* (2000)
	E. faecium PR 88	Gardiner *et al.* (1999)
Cottage	*B. infantis* ATCC 27920G	Blanchette *et al.* (1996)
	L. rhamnosus GG	Tratnik *et al.* (2000)
Fresco soft cheese	*B. bifidum* B3 and B4	Vinderola *et al.* (2000)
	L. acidophilus A1 and A2	Vinderola *et al.* (2000)
Gouda	*B. bifidum* (Bo)	Gomes *et al.* (1995)
	L. acidophilus Ki	Gomes *et al.* (1995)
Edam	*Bifidobacterium bifidum* ATCC 15696	Sabikhi (1999)
Ras	*B. lactis Bb-12*	Osman and Abbas (2001)
	L. acidophilus La-5	Osman and Abbas (2001)
Tallaga	*B. lactis Bb-12*	El-Zayat *et al.* (2001)
	L. acidophilus La-5	El-Zayat *et al.* (2001)

Presently, authors have undertaken a research work to develop probiotic Cottage cheese with the incorporation of *Lactobacillus casei* spp *casei.* The project aims to investigate hypocholesterolemic effect of the probiotic Cottage cheese *in vivo.*

Probiotic Biocultures Used in Cheese

Bifidobacteria

A number of studies have investigated to incorporate bifidobacteria, either singly or in combination with lactobacilli in a variety of hard and soft cheese types Including Cheddar (Dinakar and Mistry, 1994; McBrearty *et al.*, 2001), hard pressed (Cheddar-like) cheese (Daigle *et al.*, 1999), Gouda (Gomes *et al.*, 1995), white brined (Ghoddusi and Robinson, 1996), goats' milk cheese (Gomes and Malcata 1998; Tratnick 2000), Crescenza (Gobbetti *et al.*, 1998), Cottage (Blanchette *et al.*, 1996), Kariesh (Murad *et al.*, 1998), Canestrato pugliese (Corbo *et al.*, 2001), fresco (Vinderola *et al.*, 2000), Tallaga (El-Zayatt and Osman, 2001) and fresh cheeses (Roy *et al.*, 1997).

A European manufacturer has introduced two varieties of cheese with added bifidobacteria-medium nature and Cheddar cheese (Anon, 1995). Dinakar and Mistry (1992, 1994) developed probiotic Cheddar cheese supplemented with *Bifidobacterium bifidum* (ATCC 15696). Lyophilised powder of the immobilized cells of the organisms containing 1.9 per cent 10^8 cells/g was added along with the salt to low fat Cheddar cheese curd. The initial count of bifidobacteria in the cheese was 388 cfu/g which increased to 1 per cent 10^6 cfu/g after 4 months of ripening at 8°C. The organisms remained viable till the end of the study period of 24 weeks. Bifidobacteria have been found to survive at 1.4-2.6 x 10^7 cfu/g in Cheddar cheese (Dinkar and Mistry, 1994), 3 x 10^6 cfu/g in hard pressed (Cheddar-like) cheese (Daigle *et al.*, 1999), 6-18 x 10^8 cfu/g in Gouda cheese (Gomes *et al.*, 1995) and 7.5 x 10^6 cfu/g in goats' milk cheese (Gomes and Malcata, 1998). However, other studies have found that bifidobacteria declined significantly throughout the ripening period of such cheeses as white-brined cheese (Ghoddusi and Robinson, 1996), Cottage cheese (Blanchette *et al.*, 1996) and fresh cheese (Roy *et al.*, 1997).

A hard-pressed Chadder-like cheese was produced using microfiltered milk standardized with cream fermented by *Bifidobacterium infantis* (Daigle *et al.*, 1997 and Daigle *et al.*, 1999). In another study by Gobbetti *et al.* (1998), *Bifidobacterium bifidum, Bifidobacterium infantis and Bifidobacterium longum* were incorporated into Crescenza cheese individually or as multi-species mixture to a concentration of about Log-$_{10}$ 6 cfu/ml cheese milk. When added individually, the cell counts of *Bifidobacterium bifidum, Bifidobacterium infantis and Bifidobacterium longum* were $\log_{10}$ 8.05, $\log_{10}$ 7.12 and $\log_{10}$ 5.23 cfu/g, respectively. Presence of *Bifidobacterium* did not influence the aerobic microflora, the growth of *S. thermophilus* used as a starter culture or the composition of the cheese. In another experiment,goat milk's milk cheese has been manufactured containg *B.lactis* at 10^6 cfu/g cheese and exhibited good flavour and texture characteristics (Gomes and Malcata,1998)

Sabikhi (1999) conducted a study to develop probiotic Edam cheese by incorporation of *Bifidobacterium bifidum* ATCC 15696 into cheese milk. Ripening studies indicated that the presence of *Bifidobacterium* was synergistic towards attaining typical sensory and physico-chemical attributes of Edam cheese. The process ensured the presence of adequate numbers of the *Bifidobacterium* (~7.5 log cycles/g of Edam cheese) after three months of ripening needed to accomplish the requirement for a product to be labeled as probiotic. Animal bioassays exhibited that feeding of probiotic Edam cheese over a 5 day period to alino rats resulted in the intestinal colonization of *Bifidobacterium bifidum* (6.68 long cycles/g of faeces) and there was a concomitant reduction in the coliform counts indicating a distinctive antagonistic consequence on the microbial ecosystem of gut.

Malcata *et al.* (1995) manufactured a probiotic cheese from the milk of a native Portuguese goat breed inoculated with *Bifidobacterium* spp. strain Bo and *Lactobacillus acidophilus* strain Ki. A starter entirely composed of the same organisms, which were *Bifidobacterium* spp. Strain Bo and *Lactobacillus acidophilus* strain Ki was also used for the manufacture of a Gouda-type cheese. High rates of inoculums (3.5 per cent) were required to satisfy the technological requirements (Gomes *et al.*, 1995). After 9 weeks of storage, the numbers of viable bifidobacteria and lactobacilli were 6-18 per cent 10^8 cfu/g and 0.2-5 per cent 10^7 cfu/g, respectively. Two commercial probiotic bifidobacteria cultures, *B. lactis* Bb-12 and *B. longum* BB536 were evaluated for suitability as probiotic starter adjuncts at levels of 10^7 x cfu/ml of cheese milk during manufacture (McBrearty *et al.*, 2001). Interestingly, the *Bifidobacterium lactis* Bb-12 strain survived at high numbers (>=10^8 cfu/g cheese), while the viability of *Bifidobacterium longum* BB536 was reduced to 10^6 cfu/g cheese, following six months of ripening. The presence of this *Bifidobacterium* culture in the cheese did not adversely affect cheese composition and contributed to improve Cheddar flavour at an earlier stage of ripening compared with the control cheese. Gouda cheese was also developed using bifidobacteria in combination with *L. acidophilus* strain K_1 as the starter (Gomes *et al.*, 1995) and a significant effect on cheese flavour was obtained after nine weeks of ripening, possibly due to acetic acid production by the bifidobacteria. A recent study reported the production of low-fat cheese using cheese starter cultures in combination with *L. acidophilus* and bifidobacteria which was found to contain bioactive peptides with potential anti-hypertensive effect (Ryhanen *et al.*, 2001).

A number of soft cheeses have also been investigated as carriers for bifidobactefria, either alone or in combination with strains of lactobacilli. For example, when *B. infantis* and *B. bifidum* were incorporated into cheese (Blanchette *et al.*, 1996) and white-brined cheese (Ghoddusi and Robinson, 1996), they did not sustain high viability during storage. Blanchette *et al.* (1995) reported that supplementation of cream dressing with freeze-dried concentrates is a suitable method of incorporating *Bifidobacterium* in Cottage cheese. Incorporation of these organisms may also be a way of making β-galctosidase available for lactose-intolerant people. Viability of the *B. infantis* strain was sustained in cultured Cottage cheese dressing for 10 to 15 days at less than 10^7 cfu/g. In another, study, *B. breve* and *B. longum* survived up to 15 days in fresh cheese at greater than 10^6 cfu/g cheese (Roy *et al.*, 1997). Kariesh cheese, a popular soft cheese from Egypt was manufactured with *B. bifidum* (Murad *et al.*, 1998), which reached numbers greater than 3 x 10^{10} cfu/g of cheese after 5 days of storage at 7°C and retained high viability (>= 10^8 cfu/g) upto 10 days.

Lactobacilli

A number of studies have involved the introduction of lactobacilli strains. Such cultures include *L. helveticus* for the production of Swiss cheese (Valence *et al.*, 2000), *L. casei* (Madkor *et al.*, 2000) in Cheddar cheese and *L. acidophilus* in goats' milk (Gomes and Malcata, 1998), ras (Osman and Abbas, 2001) and tallaga (El Zayat and Osman, 2001) cheeses.

Gardiner *et al.* (1998) manufactured probiotic Cheddar cheese harbouring strains of human-derived *Lactobacillus paracasei* (From a commercial point of view, this system is particularly attractive because incorporation of probiotic lactobacilli into Cheddar cheese can be easily achieved using a low inoculum and without altering the cheese making technology. Furthermore, the inoculum for Cheddar cheese manufacture containing *L. paracasei* NFBC 338 could be prepared as a spray dried skim milk powder (Gardiner *et al.,* 2002). When this powder, containing 1x10^9 cfu/g of rifampicin-resistant *L. paracasei* NFBC 338 strain was introduced into the cheesemilk, probiotic numbers of ~ 10^8 cfu/g throughout cheese ripening were obtained, without adversely affecting product quality. In another study, creamed Cottage cheese has also been used as a carrier of probiotic lactobacilli (Tratnik, 2000), where viable counts doubled during 14 days storage at 8°C.

Some probiotic cultures (*e.g. L. rhamnoasus*) produce some antimicrobial agents that act specifically against undesirable bacteria, such as clostridia. The use of these organisms is a possible proxy for nitrate addition to asphyxiate the growth of gas-formers in cheese like Edam and Gouda (Anon, 1996). Thus, they persuade natural preservation and may safeguard human health from clostridial outbreaks.

Enterococci

In a study, probiotic enterococci for the manufacture of probiotic cheese have also been used (Gardiner *et al.,* 1999). More recently, an Expert Consultation convened by the FAO/WHO (2001) has recommended against the use of enterococci as probiotics for human use (FAO/WHO, 2001) despite the fact that some strains of *Enterococcus* display probiotic properties. Investigation to use of an *Enterococcus* strain for the development of probiotic cheese has been conducted. When added to cheesemilk at an inoculum of 2 x 10^7 cfu/ml, the enterococcal adjunct was found to maintain viability in Cheddar cheese at 3 x 10^8 cfu/g during nine months of ripening. At this level, the enterococcal adjunct strain did not affect cheese composition, but six month-ripened Cheddar cheese harbouring PR88 exhibited accelerated ripening and improved Cheddar flavour compared with the corresponding control. Feeding of 15-month-old cheese containing probiotic enterococci to pigs resulted in significantly higher probiotic excretion than feeding fresh 5-day-old yogurt containing the probiotic strain (Gardiner *et al.,* 1999).

Conclusion

The health benefits derived from the consumption of foods containing lactobacilli and bifidobacteria has enthused the food scientists to widen the range of probiotic foods and till today more than 90 commercial probiotic foods are available in the international markets. Besides yoghurt and fermented milk drinks, cheese provides an alternative vehicle to transport probiotic biocultures to the human GIT. We are still at the early state in the development of probiotic cheese for human application in terms of its viability and survivability. The development of spray dried skim milk powders harbouring probiotics has been proved to be useful as direct-vat inoculation, thereby providing more convenient way of incorporation of beneficial biocultures into cheese. One of the novel challenges in developing probiotic cheese is to get survivability during long ripening period without hampering the normal ripening cultures responsible for the development of proper body and texture, and flavour. Probiotic cheese may further be manufactured without altering the cheese manufacturing process, thus making this system attractive for commercial exploitation.

Microencapsulation is a process in which the cells are retained within the encapsulating membrane to reduce the cell injury or cell loss and may have promising application in developing probiotic cheese.

References

Anon. (1995). Cultured cheese aids rapid digestion. *Dairy Industries Int.,* 60(1) 6.

Anon. (1996). Waisby product information pamphlet. Waisby Gmbh and Co. KG, Gotteskoostrabe 40-42, D-25899, Niebull, Germany.

Ballongue,J.: Grill, J.P. and Baratte-Euloge, P. (1993). Effects of *Bifidobacterium* fermented milks on human intestinal flora. *Lait,* 73: 249-256.

Benno, Y., Sawada, K. and Mitsuoka, T. (1984). The intestinal microflora of infants: Composition of fecal flora in breast-fed and bottle-fed infants. *Microbiol. Immunol.,* 28(9): 975-986.

Berner, L.A. and O' Dannell, J.A. (1998). Functional foods and health claims legislation: Application to dairy foods. *Int. Dairy J.,* 8: 355-362.

Blanchette, L., Roy, D. and Gauthier, S.F. (1995). Production of cultured Cottage cheese dressing by bifidobacteria. *J. Dairy Sci.,* 78 (7): 1421-1429.

Braun, O.H. (1981). Effect of consumption of human and other formulas on intestinal bacterial flora in infants. In: *Gastroenterology and Nutrition Infancy* (Ed. E. Lebenthal), New York, Raven Press. pp. 247-251.

Bullen, C.L., Tearle, P.V. and Stewardt, M.G. (1977). The effect of "humanized" milks and supplemented breast-feeding on the faecal flora of infants. *J. Med. Microbiol.,* 10: 403.

Buttriss, J. (1997). Nutritional properties of fermented milk products. *Int. J. Dairy Technol.,* 50 (1): 21-27.

Corbo, M.R., Albenzio, M., De Angelis, M., Savi, A. and Gobbetti, M. (2001). Microbiolgoical and biochemical properties of Canestro Pugliese hard cheese supplemented with bifidobacteria. *J. Dairy Sci.,* 84 (3): 551-561.

Daeschel, A.M. (1989). Antimicrobial substances from lactic acid bacteria for use as food preservatives. *Food Technol.,* 43: 91-4.

Daigle, A., Roy, D., Belanger, G. and Vuillemard, J.C. (1999). Production of probiotic cheese (Cheddar like cheese) using enriched cream fermented by *Bifidobacterium infantis. J. Dairy Sci.,* 82: 1081-1091.

Daigle, A., Roy, D, Belanger, G. andVuillemard, J.C. (1997). Production of hard-pressed cheese (Cheddar cheese-like) using cream fermented by *Bifidobacterium infanits* ATCC 27920G. *J. Dairy Sci.,* 80: (Suppl. 1): 103.

Dinakar, P. and Mistry, V.V. (1994). Growth and viability of *Bifidobacterium bifidum* in Cheddar cheese. *J. Dairy Sci.,* 77: 2854-2864.

Dinakar, P.and Mistry, V.V. (1992).Bifidobacteria in Cheddar cheese. *J. Dairy Sci.,* 75: (Suppl.1): 131.

Drasar, B.S, and Barrow, O.A. (1985). *Intestinal Microbiology,* American Society for Microbiology, Washington DC, USA.

El-Zayatt, A.I. and Osman, M.M. (2001). The use of probiotics in Tallaga cheese. Egypt. *J. Dairy Sci.,* 9(1): 99-106.

FAO/WHO (2001). Evaluation of health and nutritional properties of powder milk with live lactic acid bacteria. *Report from* FA/WHO *Expert consultation* 1-4 Oct, 2001, Cordoba, Argentina.

Faure, J.C., Schellenberg, D.A., Bexter, A. and Wuerzner, H.P. (1984). Barrier effect of *Bifidobacterium longum* on a pathogenic *Eshcerichia coli* strain by gut colonization in the germ-free rat. *Z. Ernahrungswiss,* 23: 41.

Fernandes,C.F., Shahani,K.M. and Amer,M.A.(1987).Therapeutic role of dietary lactobacilli and lactobacillic dairy products. *FEMS Microbiol. Rev.,*46: 343-356.

Fukushima, M. and Nakamo, M. (1996). Effect of a mixture of organisms, *L. acidophilus* or *S. faecalis* on cholesterol metabolism in rats fed on a fat and cholesterol enriched diet. *Br. J. Nutr.,* 76: 857-867.

Fuller, R. (1989). Probiotics in man and animals. *J. Appl. Bacteriol.,* 66: 365-378.

Fuller, R. (1991). Probiotics in human medicine. *Gut,* 32: 439-442.

Fuller, R. (1992a). The effect of probiotics on the gut microecology of farm animals. In: *The Lactic Acid Bacteria* (Ed. B.J.B. Wood), Vol. 1, Chapman and Hall, London pp: 171-192.

Fuller, R. (1992b). *Probiotics-The Scientific Basis.* Chapman and Hall, London.

Fuller, R. (1997). *Probiotics* 2*: Application and practical Aspects*. Chapman and Hall, London.

Gardiner, G., Ross, R.P., Collins, J.K., Fitzgerald, G. and Stanton, C. (1998). Development of a probiotic Cheddar cheese containing human-derived *Lactobacillus paracasei* strains. *Appl. Environ. Microbiol.,* 64: 2192-2199.

Gardiner, G., Ross, R.P., Wallace, J.M., Scanlan, F.P., Jagers, P.P.J.M., Fitzgerald, G., Collins, J.K. and Stanton, C. (1999a). Influence of a probiotic adjunct culture of *Enterococcus faecium* on the quality of Cheddar cheese. *J. Agric. Food Chem.,* 47: 4907-4916.

Gardiner, G., Bouchier, P., O'Sullivan, E., Kelly, J., Fitzgerald, G., Ross, R.P. and Stanton, C. (2002). A spray dried culture for probiotic cheddar cheese manufacture. *Int. Dairy J.* (In press as cited in *Aus. J. Dairy Technol.,* Vol. 57, No. 2, July, 2002).

Ghoddussi, H.B. and Robinson, R.K. (1996). The test of time. *Dairy Industries Int.,* 61: 25-28.

Gobbetti, M., Corsetti, A., Smachhi, E., Zocchetti, A. and de Angelis, M. (1998). Production of Crescenza cheese by incoporation of Bifidobacteria. *J. Dairy Sci.,* 81 (1): 37-47.

Gomes, A.M.P. and Malcata, F.X. (1998). Development of probiotic cheese manufactured from goat milks: Response Surface Analysis via technological manipulation. *J. Dairy Sci.*, 81: 1492-1507.

Gomes, A.M.P., Malcata, F.X., Klaver, F.A.M. and Grande, H.J. (1995). Incorporation and survival of *Bifidobacterium* spp. strain Bo and *Lactobacillus acidophilus* strain Ki in a cheese product. *Neth. Milk Dairy J.*, 49: 71-95.

Guarner, F. and Schaafsma, G.J. (1998). Probiotics. *Int. J. Food Microbiol.*, 39 : 237-238.

Guerin, D.C., Chabanet, C., Pedone, C., Popot, F., Vaissade, P., Bunley, C., Szylit, O. and Andricux.(1998). Milk fermented with yoghurt cultures and *Lactobacillus casei* compared with yoghurt and gelled milk: Influence on intestinal microflora in healthy infants. *Am. J. Clin. Nutr.*, 67: 111-117.

Hanevaar, R. and Huis in't Veld, J.H.J. (1992). Probiotics: A general view. In: *The Lactic Acid Bacteria in Health and Disease* (Ed. B.J.B. Wood), Elsevier Applied Science, London, NY. pp. 151-170.

Hilliam, N. (2000). Functional food: How big is the market?. *World Food Ingredients*, 12: 50-53.

Hoier, E. (1992). Use of probiotic starter cultures in dairy products. *Food Aus.*, 44(9): 418-420.

Hughes, D. F., Rochat, P., Errant, S., Aeschlimann, J.M. and Schiffrin, E. (1999). Modulation of non-specific mechanisms of defense by lactic acid bacteria: effective dose. *J. Dairy Sci.*, 82: 863-869.

Hughes, D.B. and Hoover, D.G. (1991). Bifidobacteria: Their potential for use in American dairy products. *Food Technol.*, 45(4): 74-83.

IDF. (1997). Standards for fermented milks. International Dairy Federation. D-Doc 316.

Ishibashi, N. and Shimamura, S. (1993). Bifidobacteria: Research and development in Japan. *Food Technol.*, 47(6): 126-134.

Kailasapathy, K. and Rybka, S. (1997). *L. acidophilus* and *Bifidobacterium* spp.–Their therapeutic potential and survival in yoghurt. *Aus J. Dairy Technol.*, 52: 28-35.

Kim, H.S. and Gilliland, S. (1983). *Lactobacillus acidophilus* as a dietary adjunct for milk to add lactose digestion in humans. *J. Dairy Sci.*, 66: 959-966.

Kim, H.S. and Kang, K.H. (1984). Bifidobacteria in feces of Korean infants. *Korean J. Dairy Sci.*, 6: 126.

Lankaputra, W.E.V. and Shah, N.P. (1998). Antimutogenic properties of probiotic bacteria and organic acids. *Mutation Res.*, 397: 169-182

Lee, Y.K. and Salminen, S. (1995). The coming of age of probiotics. *Trends Food Sci. Technol.*, 6: 241-245.

LFRA.(1996).Letherhead Food Research Association,Randalls, Letherhead,Survey KT22 7RY,UK.

Lourens-Hattingh, A. and Vijoen, B.C. (2001). Yoghurt as probiotic carrier food. *Int. Dairy J.*, 11 : 1-17.

Madkor, S.A., Tong, P.S. and El Soda, M. (2000). Ripening of Cheddar cheese with added attenuated adjunct cultures of lactobacilli. *J. Dairy Sci.*, 83(8): 1684-1691.

Malkata, F.X., Gomes, A.M.P. and Coster, M.L., da. (1995). Probiotic goat cheese. Effect of ripening temperature and relative humidity on proteolysis and lipolysis. *J. Dairy Sci.*, 78 (Suppl.1): 99.

McBrearty, S., Ross, R.P., Fitzgerald, G., Collins, J.K., Aunty, M.A.E., Wallace, J.M. and Stanton, C. (2001). Influence of two commercially available bifidobacteria cultures on Cheddar cheese quality, *Int. dairy J.*, 11 (8): 599-610.

Metchnikoff, E. (1907). The prolongation of life: Optimistic Studies. William Heinemann, London, UK.

Micanel,N., Haynes,I.N. and Playne,M.J.(1997). Viability of probiotic cultures in commercial Australian yoghurts.*Aus.J.Dairy Technol.*, 52: 24 - 27.

Mital, B.K. and Garg, S.K. (1992). Acidophilus milk products: Manufacture and therapeutics. *Food Rev. Int.*, 8(3): 347-389.

Mitsuoka, T. (1982). Recent trends in research on intestinal flora. *Bifidobacteria Microflora*, 1: 13-24.

Mitsuoka, T. (1989). *Microbes in Intestine.* Yakult Honsha Co. Ltd., Tokyo, Japan.

Modler, H.W., McKellar, R.C. and Yaguchi, M. (1990). Bifidobacteria and bifidogenic factors. *Can. Inst. Food Sci. Technol. J.*, 23: 29-41.

Murad, H.A. *et al.* (1998). Production of bifidus Kariesh cheese, *Deutsche Lebensmittel Rundschan*, 94 (12), 409(Cited in Aus. *J. Dairy Technol.*, 2002,57(2): 71-78).

Nousiainen, J. and Setala, J. (1998). Lactic acid bacteria as animal probiotics. In: *Lactic Acid Bacteria* (Eds. S. Salminen and A. Von Wright), 2nd Ed., Marcel Dekker, Inc., NY, Hong Kong. pp. 437-473.

O'Sullivan, M.G. (1996). Metabolism of bifidogenic factors by gut flora -An overview. *Bull. Int. Dairy Fed.*, 313: 23-30.

O'Sullivan, M.G., Thornton, G., O'Sullivan, G.C. and Collins, J.K. (1992). Probiotic bacteria: myth or reality? *Trends Food Sci. Technol.*, 3: 309-314.

Oberman, H. (1985). Fermented milks. In: *Microbiology of Fermented Foods* (Ed. BJ.B. Wood), Vol. 1, Elsevier Applied Science Publishers, London, New York: pp. 167-186.

Osman, M.M. and Abbas, F.M. (2001). Fate of *Lactobacillus acidophilus* La-5 and *Bifidobacterium lactis* Bb-12 in Probiotic Ras cheese. Proceedings of 8th Egyptian Conference for Dairy Science and Technology. pp. 653-664.

Ouwchand, A.C. and Salminen, S. (1998). The Health effects of cultured milk products with viable and nonviable bacteria. *Int. Dairy J.*, 8: 749-758.

Rasic, J.L. (1983). The role of dairy foods containing bifido and acidophilus bacteria in nutrition and health. *North Eur. Dairy J.*, 4: 1-5.

Rettger, L.F., Levy, M.N., Weinstein, L. and Weiss, J.E. (1935). *Lactobacillus acidophilus and its Therapeutic Application.* Yale University Press, New Haven, USA.

Roy, D., Mainville, I. and Mondou, F. (1997). Selective enumeration and survival of bifidobacteria in fresh cheese, *Int. Dairy J.*, 7: 785-793.

Rybka, S. and Kailasapathy, K. (1995). The survival of culture bacteria in fresh and freeze dried. AB Yoghurt. *Aus. J. Dairy Technol.*, 50 (2): 51-57.

Ryhanen, E.L., Pihlanto, L.A. and Pahkala, E, (2001). A new type of ripened, low fat cheese with bioactive properties, *Int. Dairy J.* 11 (4-7): 441-447.

Sabikhi, L. (1999). *Biotechnological Studies on The Enhancement of Probiotic Attributes Through Bifidobacterium Bifidum in Edam Cheese. Ph.D. Thesis,* National Dairy Research Institute (Deemed University), Karnal, India.

Salminen, S., Deighton, M.A., Benno, Y. and Gorbach, S.L. (1998a). Lactic acid bacteria in health and disease. In: *Lactic Acid Bacteria* (Eds. S. Salminen and A Von Wright), 2nd Ed., Mared Dekker, Inc., NY, Hong Kong. pp. 24-253.

Salminen, S., Ouwchand, A., Benno, Y. and Lee, Y.K. (1999). Probiotics: How should they be defined ?. *Trends Food Sci., Technol.*, 10: 107-110.

Salminen, S., Ouwchand, A.C. and Isolauri, E. (1998b). Clinical applications of probiotic bacteria. *Int. Dairy J.*, 8: 563-572.

Sanders, M.E. (1998). Overview of functional foods: Emphasis on probiotic bacteria. *Int. Dairy J.* 8: 341-347.

Sanders,M.E., Walker,D.C, Walker,K.M., Aoyama,K.And Klaenhamer,T.R.(1996). Performance of commercial cultures in fluid milk application. *J.Dairy Sci.*,79: 943- 955.

Sandine, W.E. (1979). Role of *Lactobacillus* in the intestinal tract. *J. Food Prot.*, 42: 259-262.

Saxelin, M., Elo, S., Salminen, S. and Vapaatalo, H. (1991). Dose response colonization of faeces after administration of *Lactobacillus casei* strain GG. *Microb. Ecol. Health Dis.*, 4: 209-214.

Saxelin, M., Pessi, T. and Salminen, S. (1995). Fecal recovery following oral administration of *Lactobacillus strain* GG (ATCC 53103) in gelatin capsules to healthy volunteers, *Int. J. Food Microbiol.*, 25: 199-203.

Scardovi, V. (1986). *Bifidobacterium.* In: *Bergey's Manual of Systematic Bacteriology* (Eds.P.H. Sneath, N.S. Mair, M.E. Sharpe and J.G. Holt), 9th Ed., Volume 2, Williams and Wilkins Publishers, Baltimore, MD. p. 1418.

Schiffrin, E.J., Rochat, F., Link-Amster, J. and Aeschlimann, J.M. (1994). *REMS Immunol. Med. Microbiol.*, 10, 55-64.

Shah, N.P. (2000). Probiotic bacteria: Selective enumeration and survival in dairy foods. *J. Dairy Sci.*, 83: 894-907.

Stanton, C., Gardiner, G., Lynch, P.B., Collins, J.K., Fitzgerald, G. and Ross, R.P.(1998). Probiotic cheese. *Int. Dairy J.*, 8: 491-496.

Tamime, A.Y., Marshall, V.M.E. and Robinson, R.K. (1995). Microbiological and technological aspects of milks fermented by Bifidobacteria. *J. Dairy Res.*, 62: 151-187.

Tannock, G.W. (1995). *Normal Microflora: An Introduction to Microbes Inhabiting the Human Body*. Chapman and Hall, London.

Tratnik, L., Sukovic, J., Bozanic, R. and Kos, B. (2000). Creamed Cottage cheese enriched with *Lactobacillus GG. Mljekarstvo*, 50 (2): 113.

Valence, F., Deutsch, S.M., Richoux, R., Gagrair, V. and Lortal, S. (2000). Autolysis and related proteolysis in Swiss for two *Lactobacillus helveticus* strains. *J. Dairy Res.*, 67: 261-271.

Vesa, T.H., Marteau, P., Zidi, S., Potchart, P. and Rambaud, J.C. (1996). Digestion and tolerance of lactose from yoghurt and different semi-solid fermented dairy products containing *L. acidophilus* and *Bifidobacteria* in Lactose maldigesters: Is bacterial lactose important. *Eur. J. Clin. Nutr.*, 50: 730-733.

Vinderola, C.G., Prosello, W., Ghiberto, D. and Reinheimer, J.A. (2000). Viability of probiotic (*Bifidobacterium, Lactobacillus acidophilus* and *Lactobacillus casei*) and nonprobiotic microflora in Argentinean Fresco cheese, *J. Dairy Sci.*, 83 (9): 1905-1911.

Yahura, T., Isojima, S., Tsuchiya, F. and Mitsuoka, T. (1983). On the intestinal flora of bottle-fed infants. *Bifidobacteria Microflora*, 2: 33-39.

Young,J.N.(1996). Functional foods: Strategies for successful product development.FT Management Report.

Ziemer, C.J. and Gibson, G.R. (198). An overview of probiotics, probiotics and synbiotics in the functional food concept: Prospectives and future strategies. *Int. Duiry J.*, 8: 473-479.

2015, Dairy Product Technology: Recent Advances ***Pages 59–82***
Editors: **Subrota Hati, Surajit Mandal and Birendra Kumar Mishra**
Published by: **DAYA PUBLISHING HOUSE, NEW DELHI**

Chapter 4

Genetically Modified Cheese: A Novel Biotechnological Development

S. Makhal, S. Hati and S. Bera

Introduction

Cheese, the nature's wonder food and the classical product of biotechnology, is a highly nutritious food with good keeping quality, enriched level of pre-digested protein and fat, calcium, phosphorus, riboflavin and other vitamins, available in a concentrated form. Cheese is the most important category of fermented foods, which has been reported to have therapeutic, anticholesterolemic, anticarcinogenic and anticariogenic properties beyond their basic nutritive value (Renner, 1993). It is appreciated by the consumers for great interest and variety it adds to the eating experience. It has an excellent image, being perceived as healthy, natural and nutritious. It has, therefore, been truly classified as a value added product and is consumed in various other forms like dietetic foods, snacks fast foods and spreads. With the triumphal achievement in the arena of dairy science and biotechnology since the last two decades, lot of advancements have been made in cheese technology to provide ease in its processing and to gift the mankind with novel kind of cheese with improved flavour and textural characteristics achieved within short ripening period.

Now the genetic engineering approach based on the recent knowledge of the genome of many living cells and proteins or enzymes they express has made possible to modify the genes and to transfer and express these genes in the target organism. Gene technology, carefully applied to the production of food, will be beneficial for the consumers and the society as a whole in the form of healthier, better tasting and less expensive, high quality foods. Gene expression can be increased, decreased or totally eliminated (Johansen, 1999) to meet our requirements. It has now become feasible to genetically alter specific traits of a starter strain, or to transfer them to another, more preferable strain for a particular application. With the recent advancements in the field of biotechnology, cheese industry has taken a quantum jump to harvest the benefits of genetically modified starters for processing and nutritional benefits as well as for promoting health attributes of cheese. Lactic acid bacteria of the genus *Lactococcus* are used for the production of a variety of fermented milk products including various cheeses. The quality of cheese is highly dependent on the properties of the strains used, which are in turn dependent on the expression of relevant genes. Hence, gene technology can be used to modify/alter gene expression yielding strains to enhance, such as flavour, texture and health benefits to the cheese.

Genetically Modified Foods

The techniques of modern genetics have made possible the direct manipulation of the genetic makeup of organisms. In agriculture, genetic engineering allows simple genetic traits to be transferred to crop plants as well as animals from wild relatives, other distantly related plants and animals, or virtually any other organism. Recombinant DNA technology thus has brought a new precision to the process of crop and livestock development, which traditionally selects desired traits through crosses between crops and their wild relatives. Genetic modification can be used in many ways to control a variety of traits of organisms, and the consequences of one manipulation may be completely different from another based on the traits modified. The use of biotechnology in agriculture and related areas now has gained momentum in the last few decades owing to multifaceted orientations including production, quality improvement and nutritional enhancement. Foods obtained/processed by the application of direct or indirect DNA technology offer the opportunity not only of better control of existing food processing but also to improve the food quality with minimum production and post harvest losses, are referred to as GM foods. The initial efforts of biotechnology to develop GM foods focused on improving the agronomical traits mainly to guarantee food security for ever-growing world population now have been diverted to improve the pre as well post harvest characteristics, nutritional and processing characteristics. There is also notable success in the field of enzymes, food ingredients, and safety of processed foods through r DNA techniques. There are already a number of foods in the market, which are produced using GMOs or containing GM ingredients *e.g.* chymosin used in cheese making, use of GM tomatoes in paste, and GM soybean and corn products. There are over 50 approved varieties of GM foods developed and more than 100 millions acres of transgenic crops are grown world-over and more than 300 millions people consume GM foods in North America alone since 1994 (Banerjee, 2003).

Biotechnological Approaches in Dairying

Amid the explosion of fundamental knowledge generated from transgenic animal models, scientists have been producing transgenic livestock with goals of improving animal production efficiency and generating new products (Wall *et al.*, 1997). Besides the biotechnological approaches to augment the productivity of livestock, improvement of processing properties of milk through genetic engineering deduces a special implication for obtaining end products having superior functional, nutritional as well as aesthetic qualities. Recent researches have evidently demonstrated that the genetic variants of milk protein can influence on the processing quality of milk, such as heat stability, rennet coagulation time, rate of curd syneresis, firmness of curd and age related storage defects in processed milk products (Mathur *et al.*, 2003). This has led to a novel concept of "Designer Milk"-achieved through the genetic manipulation of milk constituents.

One of the major contributions of biotechnology in dairying is producing transgenic dairy cattle for improved production and post processing properties by altering the genetic make up of the inherent characteristics. This might include altered fatty acid profiles, for instance, or even the production of humanized milk for formula food applications (Yom and Bremel, 1993; Wall, 1999). Of special significance to the dairy industry is Mr. Jefferson. Mr. Jefferson is a cloned calf, opening up the way to GM dairy cattle (Rastall and Maitin, 2002). Cows, genetically modified to produce high-protein milk for the cheese industry, have been successfully created in New Zealand (Brophy *et al.*, 2003). It is the first time that cow's milk has been engineered to improve its quality, rather than to contain profitable pharmaceuticals. Transgenic cows now can produce milk having higher levels of milk proteins, mainly β- and κ-casein (Wong, 2003).

Producing drugs through transgenic animal has also been conceptualized through the use of biotechnology. Transgenic technique may also be employed for production of human pharmaceuticals in farm animals (Zuelke, 1998; Colman, 1999; Wilmut *et al.*, 2000). Even with a great deal of success, the recombinant immunoglobulins have been expressed in mammalian transgenic milk (Ishikawa *et al.*, 1992). The milk of transgenic cattle may also provide an attractive vehicle for large-scale production of biopharmaceuticals (Rosen *et al.*, 1996). It has been reported that the transgenic bull has been successfully produced that caries the gene for human lactoferrin, an iron-binding glycoprotein involved in innate host defense (Krimpenfort *et al.*, 1991). van Berkel *et al.* (2002) described the production of recombinant human lactoferrin (rhLF) at gram per liter concentrations in bovine milk. A GM bull named Herman has also been created in the Netherlands; his female offspring produce human lactoferrin in their milk (Ratall and Maitin, 2002). The expression of lysozyme in milk aids in reducing of rennet clotting time and achieving greater gel strength in the milk clot (Maga *et al.*, 1995). They concluded that the use of transgenic animals producing lysozyme in the milk is feasible and potentially useful to the dairy industry, particularly for the cheese industry. Another fact to be noticed is that cow milk allerginicity in children is often caused by the presence of β-Lg, which is absent in human milk. Transgenic technology may be employed to knock out the β-Lg gene form cow to yield milk devoid of β-Lg and thus bring it closer to human milk in composition (Sabikhi, 2003).

The interest in gene technology in dairy processing started initially as an interest in producing the milk clotting enzyme 'chymosin' from sources other than calf stomach. Today chymosin is produced by the members of the genus *Aspergillus*, which allows the cheese manufacturers a more constant source of chymosin. Genetically Modified Microorganisms (GMMs), the novel consequence of modern biotechnological intervention in manipulating dairy starters, are being used increasingly in the dairy industry. Genetically modified organisms are defined by law as entities capable of replication and/or transmission of hereditary material that had been altered by the insertion or removal of a DNA fragment (Drobnik, 2002). They may be starter cultures used in cheese, yoghurt, wine making, etc or fermentations to produce enzymes, colourants, organic acids, flavour, etc (Sabikhi, 2003).

Gene transfer technology has opened up a novel vista for the production of designer milk having defined make up of milk proteins and milk composition suitable for specific application for processing of specific product. By genetic modification, there is good potential for altering the milk composition for diet and human health cares achieving greater proportion of unsaturated fatty acids in milk fat, reducing lactose content in milk to serve the people with lactose intolerance, removing β-Lg from milk, increasing the casein content for processing purpose, etc. With the novel application of biotechnology, it has been conceptualized to remove α-La or to reduce lactose content by gene 'knock out' or by introducing the gene for producing lactase enzyme in to milk through mammary gland specific expression. Genetic engineering for secretion of low lactose milk - recently demonstrated - could make whole milk available to the majority of the world's adults currently excluded by lactose intolerance. Modification of milk composition though transgenesis is a promising way for improving existing products and extending the uses of milk components.

Most of the opportunities for dairy industry to be achieved through the genetic engineering lie in the alteration of primary structure of casein for improving technological properties of milk, production of milk with enhanced protein content to produce designer milk meant for cheese manufacturing *viz.* to accelerate curd clotting and improve the yield of cheese with more protein recovery and producing milk having nutraceuticals. Genetic selection of cows yielding milk with higher content of unsaturated fatty acids in milk fat may permit the potential manufacture of butter with improved spreadability at lower temperature. Production of low fat milk by the reduction of acetyl CoA carboxylase required for the synthesis of fat in the mammary gland and altering the composition of fat to suit the nutritional requirements of humans can be achieved by genetic modification of cows, which have been experimentally achieved on farm level (Wheeler *et al.*, 2002).

Genetically Modified Cheese: Philosophy and Development

As consumer preference for animal products is likely to continue, it would be important to modify animal products in such a way that dietary benefits are maximized while there would be simultaneous maximum benefits in respect processing of the products. Rapid development of genetic technology has placed the dairy processors open to improvement by modern biotechnology, while novel horizons beckon in nutrition, food technology and pharmacology. Biotechnology, in the form of cheese making, has been a part of the dairy industry for centuries. Now, genetic engineering is providing new technological advances including: more efficient milk production from genetically engineered bovine growth hormone supplements; cheese processing using a fermentation-derived coagulation aide-chymosin; enhanced cheese culture performance through the use of phage-resistant cheese cultures; and improved flavor development through the use of gene transfer technology. Cheese manufactured by the application of genetic engineering by adopting the following three approaches, such as modification in milk composition, addition of recombinant coagulating enzymes and application of modified starter culture can be called as GM cheese (Ghosh, 2003). These three approaches may be adopted individually and or in combination to obtain cheese with approved vegetarian sentiment, improved functional attributes, increased nutritional/therapeutic value, accelerated ripening activities and enhanced sensory quality.

The progress in molecular biology and genetic engineering have broadened the possibilities for using genetically modified LAB in cheese making and may also allow the improvement of existing cheeses and the development of novel products applications. Of various biotechnological approaches to manufacture GM cheese, considerable amount of work have been carried out on the application of GMMs to obtain improved starter culture and recombinant coagulating enzymes for cheese making. Work on alteration of milk composition by genetic manipulation in animals results in the improvement of milk production system which can even produce dairy products with functional characteristics. In this context, main emphasis for GM cheese is now on the processing aspects *i.e.* on starter culture and coagulating enzymes.

GM Cheese through Modification of Milk Composition

Today, milk represents an important food source consumed not only in its natural form, but also in a wide variety of processed products. Milk protein, 80 per cent of which consists of casein, is one of the most valuable components of milk because of its nutritional value and processing properties (Brophy *et al.,* 2003), specially for cheese making, which involves the clotting of casein. Therefore, casein is a prime target for the improvement of composition of cheese milk. Biotechnological tailoring of milk protein involving the selection of natural variants expressing different ratios of individual milk proteins, protein with different primary structure and milk with different composition has now been proposed to have great potential to suit the manufacture of cheese. Improved functional characteristics of milk protein can be achieved by varying the amino acid sequence by reorganizing the hydrophobic regions of milk proteins suitable for cheese manufacture (Mathur *et al.,* 2003). It is established that genetic variation in animal causes the different composition in milk. The κ-casein increases heat stability in the cheese-making process. The other, β-casein, improves the cheese making process by reducing the clotting time of rennet, which curdles the milk. It also increases the expulsion of whey; the watery part of milk, which remains after the cheese has formed (Mathur *et al.,* 2003). The protein content of livestock can also be increased dramatically by introducing foreign DNA into the germline. Exclusive expression of this DNA has been ensured by the presence of regulatory sequences from mammary gland-specific genes. In sheep > 50 per cent of the protein in milk can be encoded by a transgene and it appears that the foreign protein is additional to the normal complement of proteins (Colman, 1996).

These days the most thrilling interest of successful application of genetic engineering is to create transgenic animal that may produce designer milk (Enserink, 1998) to suit especially cheese manufacture. A transgenic animal has been defined as an animal that is altered by the introduction of recombinant DNA through human intervention (Murthy and Kanawjia, 2002). The process involves that a DNA construct is designed and built to express the desired protein in animal, the construct then introduced into a single cell embryo to allow incorporation of the transgene into animal genome (Pennisi and Vogel, 2000). There are several methods available for this purpose, including retroviral transmission, stem cell transfection, and microinjection into the pronucleus or cytoplasm. Transgenic cows with altered genetic make up to produce milk with 2 per cent fat with a greater proportion of unsaturated fatty acids in milk fat, higher levels of milk protein mainly β- and κ- casein and reduced lactose content in milk have been developed. Such milk is suited ideally for people suffering from lactose intolerance and also to produce improved varieties of specialty cheeses with effective cost price efficiency (Murthy and Kanawjia, 2002). Biotechnology has enabled dairy producers to use a genetically engineered, injectable protein bovine somatotropin [rBST] or bovine growth hormone [rBGH]) (Kane *et al.*, 1991), which improves cows' efficiency as milk producers. Genetically modified bovine somatotrophin (rBST) also play a role in regulation of milk yield, growth rate and protein to fat ratio of milk, which results in milk composition alteration. Improvement of protein to fat ratio of milk is directly associated with the ease and profit with cheese making.

Of special importance is k-casein, which is thought to coat the surface of the micelle. Increased k-casein content has been linked to a reduction of the micelle size and to improved heat stability and cheese making properties (Jimenez and Richardson, 1988). In the mammary gland, calcium-induced precipitation of α S_1-, α S_2-, and β-casein is prevented by their association with κ-casein as submicelles stably bound together through calcium bridges (Rollema, 1992). Higher κ-casein content was found to be associated with smaller micelle size. Furthermore, heat stability of milk was markedly improved by addition of κ-casein (Singh and Creamer, 1992). Because of the key roles of κ-casein, the relevant gene is one of the most obvious candidates for transgenesis. Transgenic mice carrying a caprine β-casein: κ-casein fusion gene produced milk containing up to 3 mg caprine κ-casein per ml (Persuy *et al.*, 1995). This hybrid gene is thus quite suitable for microinjection into eggs from farm animals. Mammary-tissue-specific expression of a caprine k-casein-encoding minigene driven by a β-casein promoter in transgenic mice has been made possible (Persuy *et al.*, 1995). In future, such mammary-tissue-specific expression of a foreign k-casein gene in transgenic cattle would be a new frontier to the cheese industry.

β-caseins that bind the otherwise insoluble calcium phosphate· As one of the predominant milk proteins, β-casein is thus implicated in determining milk calcium levels. Moreover, increased β-casein content has been correlated with improved processing properties, including reduced rennet clotting time and increased whey expulsion (Jimenez and Richardson, 1988) and thereby increases curd firmness by about 50 per cent. One modification that has been made to β-casein is the deletion of plasmin cleavage site causing prevention of bitter flavour in cheese due to plasmin cleavage. The second alteration is removal of cleavage site for chymosin done to mice milk. The last modification made is the addition of glycosylation sites to the molecule increasing its hydrophilicity. Also the native caprine β-casein gene successfully tested in mice, which produced upto 25 mg exogenous β-casein per ml of milk, is being used as such for generating transgenic goats producing β-casein. β-Lg causes aggregation and gelation of milk at higher temperature, and also is potent allergen. This gives the reason to remove this gene by ES cells.

GM Cheese from Cloned Cows

The production of the first transgenic livestock (Hammer *et al.*, 1985) stimulated discussions about the application of a transgenic approach to improve the milk composition of dairy animals (Jimenez and Richardson, 1988 and Wall *et al.*, 1997). Scientists in New Zealand have created the world's first cow clones that produce special milk that can increase the speed and ease of cheese making. The researchers in Hamilton led by Goetz Laible, New Zealand biotech company 'AgResearch' have reported that their herd of nine transgenic cows makes highly elevated levels of milk proteins *i.e.* casein-with improved processing properties and heat stability (Brophy *et al.*, 2003). Cows were previously been engineered to produce proteins for medical purposes, but this is the first time the chemical make up of milk itself has been genetically modified. The research group hopes this breakthrough will transform the cheese industry, and if widened, the techniques could also be used to "tailor" milk for human consumption.

To modify milk composition and enhance milk processing efficiency by increasing the casein concentration in milk, they have introduced additional copies of the genes encoding bovine β- and κ-casein (*CSN2* and *CSN3*, respectively) into female bovine fibroblasts (Brophy *et al.*, 2003). Nuclear transfer with four independent donor cell lines resulted in the production of 11 transgenic calves. The analysis of hormonally induced milk showed substantial expression and secretion of the transgene-derived caseins into milk. Nine cows, representing two high-expressing lines, produce milk with an 8 to 20 per cent increase in β-casein, a two fold increase in κ-casein levels, and a markedly altered κ-casein to total casein ratio (Brophy *et al.*, 2003). Two years old and living in New Zealand, the clones produce about 13 per cent more milk protein than normal cows. These results have showed that it is feasible to substantially alter a major component of milk in high producing dairy cows by a transgenic approach and thus to improve the functional and technological properties of milk to suit cheese manufacturing. Reporting their findings in the journal '*Nature Biotechnology*' the scientists said that controlling the levels of two proteins could offer big savings for cheese manufacturers (Brophy *et al.*, 2003). This should allow cheese-makers to produce more cheese from the same volume of milk. The manufacturing process should also be quicker, due to the faster clotting times associated with the higher protein levels. Protein-rich milk from cloned, genetically modified cows could cut cheese-making costs.

Recombinant Coagulating Enzymes in Cheese Making

One of the success stories for the application of genetic engineering is the manufacture of recombinant chymosin and its use as milk coagulating enzyme for commercial cheese production (Mohanty *et al.*, 1999). The product gained approval from the Vegetarian Society and from religious groups. Concern over the supply of chymosin from traditional source *viz.* suckling calves has led to efforts over the past three decades to develop a recombinant source. Cheese industry has been the major beneficiary of this biotechnological technological achievement. The first commercial application of agricultural biotechnology approved by the FDA in 1990 was the development of fermentation-derived chymosin, an enzyme used in cheese production to coagulate milk. Because natural chymosin must be extracted from the lining of the stomachs of slaughtered 10-day-old calves, supplies are limited. Advances in biotechnology have enabled scientists to create an unlimited, cheaper, and more consistent supply of chymosin by using a genetically engineered microorganism to produce the enzyme through fermentation. This technology, now used in over 90 per cent of all cheese manufacturing in North America, creates chymosin that is 40 to 50 per cent less expensive than the natural enzyme.

The DNA of calf chymosin has been successfully cloned into yeast (*Kluyveromyces marxianus* var. *lactis*) (Vandenberg *et al.*, 1990), bacteria (*Escherichia coli*) (Green *et al.*, 1985; Kawaguchi *et al.*, 1987) and molds (*Aspergillus niger*) (Berka *et al.*, 1991; Dunncoleman *et al.*, 1991) to give at least three commercial products, which are now being widely used in cheese manufacture. All three recombinant chymosins are currently available on the market produced by standard large-scale fermentation methods and extracted by down stream processing (Mathur *et al.*, 2003). The composition and activity of the genetically engineered chymosin is identical to that extracted from calves. All microbial host material is left behind once the chymosin enzyme is harvested and purified, thus no genetically modified material is found in the final cheese product

The enzymatic properties of the recombinant enzymes are identical from those of calf chymosin. The cheese making properties of recombinant chymosin produces very satisfactory results and its use in commercial plant have been approved by many countries. Three recombinant chymosins, such as Maxiren- from *K. marxianus* var. *lactis* produced by Gist Brocades, Netherlands; Chymogen-from *A. niger* produced by Hansen's, Denmark and Chymax-from *E. coli by* Pfizer, USA are now marketed commercially, which have taken more than 35 per cent share of the total market. All the three chymosins available are indistinguishable to calf chymosin, considered as vegetarian source and accepted by all groups. More knowledge on genetic engineering combined with better understanding of protein structure might one day give us chymosin with higher activities, lower cost and with flavour enhancing properties of cheese during ripening (Rastall and Maitin, 2002).

Genetically Modified Starter Culture: A Novel Cheese-Biotech Alliance

The production of cheese is one of the oldest manifestations of biotechnology. Many consumers consider cheese to be a delicacy with exquisite taste and aroma characteristics and would recoil at the thought of using GMMs in manufacturing process. Continual development in biotechnology and genetic engineering coupled with overabundance of information emerging form genome projects on LAB and gene functionality in lactic starter cultures has brought a renaissance in the cheese industry. By recombinant DNA technology, 'tailor-made' starter and protective cultures may be constructed so as to combine technically desirable features. A single strain which normally would fail to accomplish a given 'task' may now be improved so as to meet a set of requirements necessary for a specific production or preservation process (*e.g.* wholesomeness, no off-flavour production, overproduction of bacteriocins or particular enzymes) (Geisen and Holzapfel, 1996).

Phase Resistant Cheese Starters

Infections with bacteriophages are still a major problem during dairy fermentations, particularly in cheese making. The major cause of slow acid production in cheese plants today is bacteriophage (phage). Bacteriophages are viruses that prey upon bacteria causing them to burst, or lyse. This can significantly upset manufacturing schedules and, in extreme cases, result in complete failure of acid production or "dead vats" (Ghosh, 2003). A possible way to solve this problem is the introduction of phage insensitive starter cultures through the application of genetic engineering. It is possible, for instance, to transfer genetic control elements from the bacteriophage to the starter cultures, such that, upon infection, the bacteriophase does not reproduce itself. Different possibilities to generate phage insensitive mutants with different degrees by using genetic tools have been found out (Janzen *et al.*, 2003): (1) genetic engineering resulting in a GMO: inactivation of pip gene; (2) spontaneous mutant selection resulting in non-GMO variants: screening for spontaneous phage resistant mutants towards c2-type phages; (3) combination of genetic engineering tools with a classical method resulting in a new variant: introduction of a marker plasmid by electroporation, conjugative transfer of phage resistance plasmids and successive curing of the marker plasmid.

Microbiologists have enhanced cheese culture performance by genetically engineering bacteria increasing the viability of the culture during cheese making (Kim *et al.*, 1992; Coakley *et al.*, 1997; McGrath *et al.*, 2001). The new strain of bacteria resist phase contamination and are suitable for prolonged use in milk fermentation. A phage resistant starter culture of cheddar cheese, *L. lactis* DPC 5000 have been developed, which was shown to embody three effective phage resistant mechanisms (Ghosh, 2003). Cheddar cheese manufactured with DPC 5000 compared favorably in term of composition with cheeses manufactured using commercial starter. Two phage resistant thermophilic starter strains DPC 1842 and DPC 5099 have been developed in Cork, Ireland which performed well in commercial plants for Mozzarella cheese preparation. The new cultures provide more predictable performance and reduce the chance of vats failure (Ghosh, 2003). Researchers at Food Science Australia and the University of Melbourne have constructed plasmids that increase the resistance of cheese starter bacteria to bacteriophage infection-a major cause of slow fermentations and lower quality cheese (Anon,). Regions of the bacteriophage genome were randomly cloned into the starter bacteria and bacteriophage resistant colonies were obtained.

Accelerating Cheese Ripening

Cheese ripening is a complex process of concerted biochemical changes, during which a bland curd is converted into a mature cheese having the flavour, texture and aroma characteristics of the intended variety. The unprecedented accomplishments of biotechnology and genetic engineering since the past decade have added a new margin to the approaches of accelerated ripening. Controlled use of biotechnological products like genetically engineered proteolytic and lipolytic enzymes can accelerate cheese ripening, which are now in use. Modified/genetically tailored microorganisms and enzymes are now being used for enhancing flavour production in cheese. Enzyme addition is now one of the few preferred methods of accelerated ripening of cheese. The enzymatic reactions are specific and hence undesirable side effects caused by live microorganisms are avoided in the cheese.

The use of recombinant DNA technology to produce GMMs for accelerated ripening of cheese is one of the most important scientific advances of the 20th century. Through genetic engineering, specific genes can be implanted/removed to increase or decrease the activity of specific property of the existing strain of LAB. It can be stated from the utilization of genetically engineered starter as: i) Cloning of exogenous proteinases in starter cells leads to enhanced proteolysis, ii) Debittering action of aminopeptidase is now well recognized, iii) Starter peptidases and proteinases produce small peptides and amino acids in cheese but may not have a direct impact on flavour. It has great potential in industry, especially cheese industry, leading to processes and products that would be difficult to develop using conventional techniques. It is very significant that many commercially important traits of the starter bacteria are in reality plasmid encoded (Batish and Grover, 2003a). These embrace lactose and citrate utilization, phage resistance, and proteinase production and bacteriocin production/immunity. It has proved to be a blessing to the scientists that the plasmid-encoded genes are more accessible for manipulation and more easily analyzed than the chromosomal genes.

Recent accomplishments in gene transfer and cloning technologies have provided vast opportunities for the application of genetic approaches to novel starter strain development programme appropriate for their application in dairy industry, particularly in cheese industry. The application of gene cloning technology to LAB is the potential process in generation of improved starter cultures for manufacture of cheese. Increasing the expression of a gene is to increase the copy number by moving the gene from the chromosome to a multicopy plasmid (Henriksen *et al.*, 1999). Several food grade vectors have been developed. One of these, pFG1 (Dickely *et al.*, 1995) has been used to overexpress several genes and the resulting strains have been experimentally used in cheese making. These starter cultures are mainly made of species of *Lactococcus*, *Lactobacillus* and *Streptococcus*. Modifications of these microorganisms were achieved mainly on three directions for cheese making process. These are: development of phage resistant cheese culture, organisms with probiotic activity for cheeses and acceleration of cheese ripening.

In view of the fact that amino acids are widely thought to be major contributors, directly or indirectly, to flavour development in cheese, the use of a starter with increased aminopeptidase activity would appear to be attractive in achieving early maturation of cheese. Aminopeptidases have an effect on flavour development in cheese by their involvement in the degradation of casein to small peptides and amino acids, which are the precursors of flavour compounds. Two studies (McGarry *et al.*,

1995 and Christensen *et al.,* 1995) have been reported on the use of a starter genetically engineered to super produce aminopeptidase N; although the release of amino acids was accelerated, the rate of flavour development and its intensity were not, suggesting that the release of amino acids is not rate limiting. In an on-going study, engineered *Lactococcus* starters harbouring *Pep*G or *Pep*I genes from *Lb. delbrueckii* have shown increased proteolysis, especially at the level of amino acids; *Pep*G was the more effective. The cheeses have not yet been assessed organoleptically.

Several aminopeptidases have been overexpressed by insertion into pFG1. These include the general aminopeptidase *pepN,* dipeptidase *pepV,* and cysteine aminopeptidase *pep*C and endopeptidase *pep*O (Henriksen *et al.,* 1999). After four and six months, the cheese made with the genetically engineered starter culture overexpressing *pep*N and the culture overexpressing *pepC* had significantly reduced bitter flavour compared to the control cheese made with a strain containing only the cloning vector. The flavour preference was also significantly higher for the cheeses overexpressing *pepN* or *pepC.* Overexpression of certain aminopeptidases thus improved the organoleptic properties of cheese. Genetically engineered starter cultures with enhanced complements of proteinase and/or peptidases, which could be released early and evenly distributed in the curd would be an ideal method of accelerating cheese ripening and recent cloning of the genes for a dipeptidyl aminopeptidase and phase lysine would indicate that research is being directed to this end.

Food-grade controlled lysis of *Lactococcus lactis* for accelerated cheese ripening is an important approach. Controlled expression of the lytic genes *lytA* and *lytH,* which encode the lysin and the holin proteins of the lactococcal bacteriophage phi-US3, respectively, was accomplished by application of a food-grade nisin-inducible expression system. Simultaneous production of lysin and holin is essential to obtain efficient lysis and concomitant release of intracellular enzymes as exemplified by complete release of the debittering intracellular aminopeptidase N. Production of holin alone leads to partial lysis of the host cells, whereas production of lysin alone does not cause significant lysis. Model cheese experiments in which the inducible holin-lysin overproducing strain was used showed a four fold increase in release of L-Lactate dehydrogenase activity into the curd relative to the control strain and the holin-overproducing strain, demonstrating the suitability of the system for cheese applications. A cheese culture has also been prepared at Chr. Hansen A/S, Denmark, which gives a better flavour and possibly an earlier maturation of the cheese. They have expressed a bacteriophage gene, specifically a lysin gene in the food-grade cloning vector. It is, however, contentious whether or not the lysin gene belongs to the genus *Lactococcus.*

Genetically modified *L. lactis* subsp. *cremoris* strain AM2 is being extensively used in the cheese industry and has an accelerated autolysis caused by the presence of a prophage-encoded lysin (LePeuple *et al.,* 1998). Thus, phage lysin genes can help with flavour development of cheese by lysing some of the bacteria and allowing the release of flavour developing enzymes. Expressing the bacteriophage gene in the bacterium in a controlled manner can precisely regulate this lysis process. The lysis obtained by the use of the GM strain is a natural process and already occurs in cheese vats either due to bacteriophage infection or by prophage-induced autolysis. It is generally believed that cell lysis is necessary during cheese ripening to release aminopeptidases into the cheese matrix. To determine the effect of increasing cell lysis, the bacteriophage/vML3 lysin gene was inserted into pFG1 to produce a plasmid designated pFG7. The resulting strain has significantly increased autolysis. Cheeses made with a GMM that expresses the/vML3 lysin gene also has a significantly reduced bitterness and a significantly increased flavour preference (Henriksen *et al.*, 1999).

The characteristic butter flavour note of certain types of fresh cheese is due to the presence of low concentrations of diacetyl. Unfortunately, *Leuconostoc* also has a high affinity for diacetyl and therefore efficiently reduce the diacetyl into the flavourless components 'acetoin' and 'butanediol' causing the rapid disappearance of the butter flavour during storage of some fresh cheese. Two metabolic engineering strategies have been applied at Chr. Hansen A/S, Denmark for improving the flavour and flavour stability of buttermilk (Henriksen *et al.*, 1999). The 1st strategy is aimed at the isolation of *Leuconostoc* with strongly attenuated diacetyl reductase (DR) activity, while the other focuses on enhancing the diacetyl production of *lactis* subsp. *lactis* biovar. *diacetylactis*. The former was entirely performed through the use of classical mutagenesis techniques whereas the latter was strongly facilitated by the use of genetic engineering tools. The formation of diacetyl in milk is primarily observed during the cocatabolism of lactose and citrate by LAB strains harbouring the genes for citrate uptake and catabolism (Henriksen *et al.*, 1999)

The gene for the neutral proteinase (neutrase) of *B. subtilis* has been cloned in *Lc. lactis* UC317. Cheddar cheese manufactured with this engineered culture as the sole starter showed very extensive proteolysis, and the texture became very soft within 2 weeks at 8°C (Fox *et al.*, 2000). Cheddar cheese made with *Lc. lactis* subsp. *cremoris* SK11 (cloned with proteinase) revealed that starter proteinases are required for the accumulation of small peptides and free amino acids in cheddar cheese. The strain in which the proteinase remained attached to the cell wall appeared to contribute more to proteolysis than the strain that secreted the enzyme. Cheeses made with proteinase positive starter produced more pronounced flavour than those with proteinase negative strain during ripening. The inactivation of genes in a metabolic pathway can be used to alter end products that accumulate from a given pathway. Application of regulated promoters is the controlled expression of lytic genes resulting in autolysis of the starter culture. This would result in rapid release of enzymes (*i.e.* peptidase) into the cheese matrix and potentially accelerate cheese flavour development.

Dramatic advances in understanding the genetics of LAB, notably *Lactococcus lactis* and *Lactobacillus* species, have been driven by immediate application opportunities. In the modern cheese industry, tight production schedules and demands for consistently high quality rely on starter cultures with unpredictable behaviour. Spontaneous genetic change that spreads rapidly can soon shift growth rates, lactic acid production, flavour, and phage resistance. Historically, the dairy technologist has responded by screening for new strains with desirable traits, but genetic engineering offers a more efficient solution.

Enzyme controlled chemical pathways used by *L. lactis* to degrade the milk protein *i.e.*, casein has been mapped recently (Yvon and Rijnen, 2001; Tanous *et al.*, 2002;

Marilley and Casey, 2004). A cascade of reactions, which continue as the cheese matures, follows initial breakdown during fermentation. The reactions' products, a collection of smaller peptide fragments, are responsible for flavour, and some 200 have been identified together with unique genes that code for the controlling peptidase enzymes. By constructing strains with individual genes added or deleted, respectively to overexpress or minimize production of specific enzymes, their contribution to peptide profiles and flavour can be assessed with trial cheeses. Ultimately, the research should enable commercially useful strains to be constructed that efficiently develop desired flavour in cheese. Similar advances have been made in the field of accelerated cheese ripening, diacetyl synthesis for butter flavour, and phage resistance of cheese starter cultures, antimicrobials, and many related fields.

Apart from legal aspects, the principal technical problems when engineering starters with improved cheese ripening properties is the lack of knowledge on the key or limiting lactococcal enzymes involved in cheese ripening. A range of peptidase-deficient *Lactococcus* mutants (lacking 1 to 4 peptidases) has been developed (Mireau *et al.*, 1996) primarily with the objective of identifying the importance of the various peptidases, alone or in various combinations, to the growth of the organism in milk. Mutants deficient in 1 or 2 peptidases in various combinations were also used for the small-scale manufacture of Cheddar cheese (mutants deficient in 3 or 4 peptidases are unable to grow in milk at a rate sufficient for cheese manufacture). Perhaps surprisingly, cheeses made using the peptidases-deficient mutants, even those lacking both *Pep*N and *Pep*C, did not differ substantially from the control with respect to the level and type of proteolysis or flavour and texture. The results appear to suggest that there are alternative routes for amino acid production.

GM Cheese with Improved Flavour

One of the oldest applications of biotechnology is in the bacterial fermentation of dairy products in the production of cheese. The LAB responsible for this process have been the focus of intensive research worldwide aiming at improving the quality and flavour of several cheeses. Scientists have successfully developed gene transfer technologies to enhance the flavour potential in cheese cultures. An example of this research is the genetic manipulation of a strain of LAB to produce higher levels of diacetyl, a desirable flavour component of cheese. The modified bacteria produce approximately 3.5 times more diacetyl than the unmodified strain. The new cultures not only deliver better flavor, but also stop growing at a particular pH level. This self-limiting growth behaviour reduces the further proteolysis that prevents the development of bitterness and off-flavours.

GM Cheese with Reduced Bitterness

Cheddar cheese sometimes has a slightly bitter flavour. Most Western consumers are unaware of it, or actually prefer it. However, Asians generally have a lower tolerance of the bitter flavour, and Japanese consumers' preference for non-bitter cheddar limits exports of Australian cheddar in Japan. Bitterness has been a problem in cheese for decades, but modern consumer preference for mild-flavoured Cheddar has lent greater significance to the impact of bitterness on dairy economics. Bitterness is caused by the accumulation of hydrophobic peptides produced by some starter bacteria and chymosin (Broadbent *et al.,* 1998). Starter proteinase specificity is the primary determinant in whether or not a starter culture produces bitter peptides (Christensen *et al.,* 1999). Fortunately, bitter peptides produced by chymosin and starter bacteria can be by degraded by intracellular peptidases from starters and adjunct bacteria, but the relative contribution of individual peptidases to these reactions remains unknown (Steele *et al.,* 1998). Food Science Australia is studying the bacteria used in cheese making to identify genes involved in the synthesis of flavour compounds, with the aim of producing Australian cheddar tailored to different palates. They have identified a peptidase, a bacterial enzyme that breaks down the bitter flavour compounds. By selecting or developing genetically engineered microbes that make increased amounts of the enzyme, they aim to develop strains that would produce non-bitter cheddar cheese for the Japanese market, and for Australian consumers who also prefer more mellow flavours. It has been made possible to use gene technology to turbo-charge the peptidase gene to increase its output.

Some GM strains have been constructed, which reduce the bitterness of some cheeses. The resulting strains have extra copies of a specific gene, which is already present on the chromosome of the original strain (Henriksen *et al.,* 1999). Cheese manufactured with an amino peptidase N-negative clone strain of *Lactococcus* produced bitter off flavour (Baankreis, 1992). The possible role of aminopeptidase as a debittering agent confirmed by Prost and Chamba (1994) after making the Emmental cheese with *Lb. helveticus* strain L_1 (high amino peptidase activity), L_2 or strain L_3 (clones selected for lack of aminopeptidase activity). They also explained the bitterness in ripened cheese made with *Lb. delbrueckii* subsp. *lactis* to be due to their very low aminopeptidase activity.

Biotechnological Loom in Developing Probiotic Cheese

Cheese will have a number of advantages over fresh fermented products as a delivery system for viable probiotics to the GIT (Stanton, *et al.,* 1998) having a higher pH and buffering capacity than the more traditional probiotic foods, a more solid consistency and a higher fat content, which may provide a more stable milieu to support the long-term survival of probiotic organisms. Given the increasingly competitive panorama of the European market in food products, the cheese industries in European countries are trying to derive benefits from a marketing advantage, such as added-value probiotic-containing cheese, which would afford a competitive edge over existing products (Stanton *et al.,* 1998). The development of probiotic cheeses would thus lead to a major economic advantage also.

Plasmid biology of LAB have opened up new opportunity for exploring possibilities for using recombinant DNA technology and genetic engineering to improve the nutritional and therapeutic attributes of probiotic cheese. Expression of foreign genes in LAB and enhancing the expression of desired phenotypic properties have now come in the grip of scientists (van de Guchte *et al.,* 1992; Wegmann *et al.,* 2000; Luoma *et al.,* 2001). These might include the incorporation of attachment molecules to facilitate colonization, increased production of antimicrobial compounds directed against common foodborne pathogens, enhanced immune stimulation.

Molecular approaches are currently being used to improve the probiotic functions of lactobacilli and bifidobacteria particularly in respect of their colonization and antimicrobial potentials (Mital and Garg, 1995; Gomez *et al.*, 1997). As the LAB is associated both with dairy products and human intestinal tract, they are ideal candidates for overexpression of β-galactosidase to improve the digestion of lactose. Efforts are going on in several laboratories to create overexpression in probiotic lactobacilli.

Recently, Genome project on LAB could provide very useful information in determining the gene function, their regulatory mechanisms and interaction (Vaughan and Mollet, 1999). With the completion of these genome projects, the genomics and proteomics will undoubtedly enlighten the understanding of probiotic bacteria and the discovery of novel probiotic properties. Some of the future targets aimed in this direction include identification of the genetic determinants for probiotic function and regulatory signals to improve the colonization ability of these organisms in different hosts, inter and intraspecies conjugal transfer of this character and understanding the exact mode of action of probiosis (Batish and Grover, 2003b).

Bioactive Cheese: A Novel Biotechnological Approach

Genetic polymorphisms of milk proteins (from transgenic animals) influence the occurrence of bioactive peptides, which display physiologically significant extra nutritional attributes. Since during the microbial fermentations involved in manufacture of cheese, milk proteins undergo controlled proteolysis under the influence of microflora native to milk, starter bacteria and rennet, cheese is a natural source of physiologically active peptides. Cheese as a potential carrier of probiotical cultures and as a natural source of biological peptides, is a fascinating area for technological development. Scientists have fruitfully clasped gene transfer technologies to improve the flavour potential of cheese cultures (Henricksen *et al.*, 1999). The new genetically engineered cultures not only produce better flavour but also stop growing at a particular pH level. This self limiting growth behaviour may control the bitterness and further degradation of physiologically active peptides.

With the biotechnological approach to manipulate dairy starter cultures, there is a good potential to incorporate some health attributes in cheese by *in situ* synthesis of bioactive peptides at an elevated level to derive some health benefits beyond its inherent nutrients. This approach of enriching cheese with bioactive peptides seems to be more attractive in terms of its stability and maintaining functionality. Genetic engineering to produce genetically modified cheese starters can be judiciously exploited to modulate gene expression yielding strains with optimal properties for specific bioactive peptide production (Makhal *et al.*, 2004). With the modified stability, activity, and specificity via site-directed mutagenesis of well-known proteinases, proteolytic enzymes of LAB, due to their high regiospecificity and proven efficiency, may be assayed for *in situ* enzymatic synthesis of biopeptides in cheese. The liberation of bioactive peptides during cheese ripening, for example by use of specific bacterial enzymes or GMMs, is of interest for future research work.

Genetically Modified Cheese: The Future

Large ranges of crops, GMMs, ingredients and enzymes have now been produced using biotechnological tools, although most of these products have yet to reach the stage of commercial development. In future, the GM foods are expected to offer consumers many benefits including improved choice, quality, and flavour and keeping qualities at a lower price. Rapid development of genetic technology has placed the dairy industry at the dawn of revolution. Genes, with their orderly DNA structure, a code for enzymes and other proteins that control living processes, can be isolated, deciphered and used to modify the performance of microorganisms. Already, following one of the earliest applications of biotechnology in the cheese industry, chymosin (rennet) obtained from GMMs is used extensively for cheese production. However, the scope and benefits are much broader with many aspects of cheese manufacture open to improvement by modern biotechnology, while novel horizons beckon in manufacturing designer cheese with desired nutritional, functional and processing attributes. Turning to starter cultures, techniques are available to engineer LAB to produce health cheese with low cholesterol and antihypercholesterolemic property, for example.

Food R&D in the 21st century should support consumers oriented product development because it has the potential to become part of the consumer health care system. Researches are developing cheese with GM ingredients to achieve nutritional significance. The next generation of bacterial cultures to be used for cheese making will probably contain strains with properties modified by gene technology, for example with new properties created by changing the expressing level of one or a few genes. The use of GMMs in cheese manufacture is litigious due to a lack of acceptance by consumers; especially in Europe. Recent developments have opened many more possibilities for the use of gene technology. The global market for cheese products is very multifarious and competitive, with producers looking for new strains to differentiate their products. Hence, it would be interesting to develop GM cheese with the application of new starter strains giving novel flavour and textural characteristics to the cheese.

In future the cheese manufacturers would be able to manufacture 'designer cheese' having defined flavour, desired degree of ripening to suit the palate of the consumers as per their taste. Furthermore, genetic engineering, in the days ahead, would enable the manufacturers to produce cheese with defined health benefits. Genetic selection of milch animals to yield milk with desired polymorphic form of proteins in milk and alteration of genetic make up of DNA of the dairy cattle would, thus, enable to manufacture cheese with improved processing ease and enhanced yield at reduced cost. This is another matter of exhilaration that the food and Drug Administration (FDA) has released a draft risk assessment concerning the safety of food products derived from the animal clones, which indicates that they are as safe as non-clone counterparts. The draft builds on the findings of the National Academy of Science (NAS). As per their assessment, the cloned cows' milk is safe (Anon, 2004) and so the cheese made thereof will be safe to the consumers.

Genetically Modified Cheese: The Safety Aspects

Like the rest of the food chain, the use of crops, pharmaceuticals, and processing methods enhanced through biotechnology impacts the dairy industry. Currently, there are more questions than answers on whether global consumers will accept the use of biotechnology in the production of dairy products and how the dairy industry will be affected by consumer reaction. Application of gene technology for genetic modification of organisms involved in food production and processing is highly sensitive issue. Most of the applications of genetic engineering developed so far only provide economic benefits and business opportunities to the producers. Understanding the technology is likely to aid in gaining acceptance. However, it will depend on high quality risk assessment and convincing results from long term experiments whether the GMOs will find broad acceptance in the food industry. The potential of biotechnologically derived foods is mostly debated even after two decades of positive results. The technological breakthroughs in conventional food processing are still more acceptable to the consumers than genetic modifications for food uses. The strategy to be taken for popularizing gene technology is to help the consumers to make their transition to GM foods that have proven, desirable and popular health benefits that require minimum behavioural change, besides absolute assurance on safety of the product. Most scientists believe that milk from cloned cows is no different to normal milk. However they are less certain about the safety of milk from genetically modified cows and cheese made thereof.

Manufacture of GM cheese for human consumption has many complex problems, which must be carefully addressed. Allergenicity of GM food has become a public concern and international expert panels *e.g.* WHO/FAO have depicted decision trees for a rigorous assessment and testing for GM foods, especially where no history of safe use is available (Haslberge, 2003). The opponents have been expressing concern over GM cheese pointing out the long-term risk to human health. They are of the opinion that microorganisms, modified in laboratory may not behave as expected at the field level (Rastall and Maitin, 2002; Banerjee, 2003). Most scientists believe that milk from cloned cows and cheese made thereof are no different to normal milk and cheese made from it. However, they are less certain about the safety of milk from the cloned cows. One of the major anxieties in the public mind is whether cheese prepared with GMMs or recombinant coagulating enzymes or proteins are safe for consumption. Particularly, toxicity and food safety of GM cheese must be considered. Another aspect to be carefully considered is the risk of transfer of the new gene to the environment and allerginicity of cheese derived from genetically engineered starter cultures. The introduction on the market of food composed of or derived from (GMOs raises the question of their potential allergenicity (Adel-Patient and Wal, 2004). The GMMs to be used must be food grade in accordance with a general acceptability. Food grade GMMs must contain only DNA from the same genus and possibly small stretches of synthetic DNA. Artificial drug resistance markers in construction of genetically engineered organisms carrying drug resistant genes could pose serious health risks to the consumers.

Would Cheese Made with the Gene Technology be Safe? Scientists consider such concerns will be satisfied if modified LAB retains their original properties. Self-cloning, the improvement of strains using only genes and plasmid vectors from within the same species embodies this approach. As LAB are 'Generally Recognized As Safe' (GRAS) and no foreign DNA is used, the method is deemed intrinsically nonhazardous, as cheese related LAB have a history of safe use stretching back thousands of years, and evidence that they carry genes related to non-pathogenic mechanisms. An argument for the safety of modification using foreign genes can also be made where the donor species is another GRAS organism, and the inserted DNA is integrated into the host's chromosome. Many of the processes essential to cheese maturation are controlled by plasmid genes, lactic acid and flavour production, for instance. Safety concerns come up, because recombinant DNA lost in this way has the potential to modify other bacteria with unintended consequences. Chromosomal DNA is not vulnerable to such instability, and now, biotechnologists can incorporate plasmid-cloned genes into the chromosome. Food grade genetic modification also requires a different approach to selectable markers, which are used to isolate successfully engineered bacteria.

Studies on replacing drug resistant marker genes with some natural food-grade markers derived from LAB themselves are now underway. The application of genetically engineered starters with such food grade vectors in manufacturing GM cheese would not pose any dilemma in getting the approval of regulatory agencies. Lactic acid bacteria have a long history of safe use in fermented foods and drinks. While other transgenic microbes may be used to achieve desired and improved characteristics of cheese within short ripening period, such as flavour. Any transgenic starter culture developed for cheese manufacture will contain only DNA from other bacteria already used to produce fermented foods, preferably other *Lactococcus* and *Lactobacillus* bacteria. Food grade vector system where only naturally occurring genes, such as bacteriocin production/immunity or lactose/sucrose utilizing gene from LAB are incorporated into the food-grade vectors as a marker for the selection of the desired recombinants. DNA from other food microorganisms is also GRAS (Johansen, 1999). The use of starter cultures engineered with such food-grade vectors for different desirable characteristics in cheese will not raise any gale and cheese industry will have no vacillation to use such GMMs in manufacture of high quality cheese.

Most of the applications of genetic engineering developed so far only provide economic benefits and business opportunities to the producers. Acceptance is expected if the public is made attentive of the environmental benefits in terms of energy savings of GM foods. Although the existing situation might seem bleak, there are many opportunities for the cheese industry potentially arising from genetic engineering, some of which might take the industry towards a absolutely new direction. Realizing these opportunities will require more research, which should involve the dairy industry in a leading role, and it will require endurance and a sensitive consideration of a public concerns on the introduction of products. Consumer acceptance of GM cheese is likely to be increased when a direct consumer benefit is recognized. However, huge information gaps exist among the scientists and particularly between scientists and consumers. To match the promise offered by the biotechnological advances and optimize nutrition, the overcoming of barriers in psychological and as a consequence-political feasibility is required. Government agencies and the scientific community have the obligation to educate the public suitably so that the choice of accepting or rejecting GM foods is made on scientific basis rather than more conjectures. This needs not only research efforts but education in all stages of the food chain and in sector of communication because "if we fail to train we fail to convince, if we fail to convince we fail".

Conclusion

Today's research should be directed to demonstrate how the sorority of cheese technology with modern biotechnology and molecular biology can have a positive impact on the cheese making process through the generation of "state of the art" starter strains with novel properties, which in turn could potentially be exploited to produce cheeses with improved characteristics at a reduced cost. The emergence of molecular biology and gene technology has allowed a more through investigation and understanding of the role of LAB in transformation of milk to cheese. Molecular biology has fostered the development of genetically based selection tools to specifically screen hundreds or thousands of natural LAB strains or mutants to screen one meeting the desired characteristics.

With the triumphal accomplishment along with the successful application of genetic engineering, now it is possible to deliberately manipulate the proteolytic system of LAB and to determine the consequential effect on flavour as well as the body and texture development during cheese ripening. Controlled application of biotechnological products like genetically engineered proteolytic and lipolytic enzymes can speed up cheese ripening. GM starter cultures are being increasingly employed for manufacture of a variety of fermented products including cheese with intended intensity of flavour. Genetic manipulation of cheese cultures with respect to the important traits significant for cheese making has helped the modern cheese industry in efficient manufacture of high quality cheese with improved consumer acceptability. For the future, good potential is there for application of transgenesis to produce designer milk on farm level at reduced cost for manipulation of specific technological applications, especially for cheese making. The intelligent and responsible development as well as the application of biotechnology in dairying as a whole along with the use of genetically improved microorganisms will lead us into this new millennium bringing exciting new type of cheeses to the consumers.

References

Adel-Patient, K. and Wal, J. M. (2004). Animal models for assessment of GMO allergenicity: advantages and limitations. *Allerg. Immunol.* (Paris), **36** (3): 88-91.

Anon (2001). Cheese starter bacteria with increased resistance. CSIROnline, Commonwealth Scientific and Industrial Research Organization. http: // www.csiro.au/index.asp?

Anon (2004). News flashes from abroad. *Indian Dairyman*, **56** (1): 17.

Baankreis, R. (1992). The role of lactococcal peptidase in Cheese ripening. In: *Microbiology and Biochemistry of cheese and Fermented Milk,* ed. by B. A. Law. Published by Blackie Academic and Professional, London, U.K.

Banerjee, A. (2003). President's desk. *Indian Dairyman,* **55** (6): 3-4.

Batish, V.K. and Grover, S. (2003a). Genetically modified starters-Indian dairy industry geared up to reap the harvest. *Indian Dairyman,* **55**(6): 68-76.

Batish, V.K. and Grover, S. (2003b). Prospects of Biotechnological interventions in Dairy Processing and their impact on Dairy Industry. In: Lecture Compendium, 16th Short Course on *"Application of Biotechnology in Dairy and Food Processing"* Nov.4-24, 2003, Org. by Dairy Technology Division, NDRI, Karnal-132001.

Berka, R. M., Kodama, K. H., Rey, M. W., Wilson, L. J. and Ward, M. (1991). The development of *Aspergillus niger* var awamori as a host for the expression and secretion of heterologus-gene products. *Biochem. Soc. Transactions,* **19**: 681-685.

Broadbent, J. R., Strickland, M., Weimer, B., Johnson, M. E. and Steele, J. L. (1998). Peptide accumulation and bitterness in Cheddar cheese made using single-strain Lactococcus lactis starters with distinct proteinase specificities. *J. Dairy Sci.,* **81**: 327-337.

Brophy, B., Smolenski, G., Wheeler, T., Wells, D., L'Huillier, P. and Laible, G. (2003). Cloned transgenic cattle produce milk with higher levels of β-casein and k-casein. *Nature Biotech.,* **21** (2): 157 – 162.

Christensen, J. E., Dudley, E. G., Pederson, J. R. and Steele, J. L. (1999). Peptidases and amino acid catabolism in lactic acid bacteria. *Antonie van Leeuwenhoek,* **76**: 217-246.

Christensen, J. E., Johnson, M. E. and Steele, J.C. (1995). Production of Cheddar cheese using a *Lactococcus lactis* ssp *cremoris* SK11 derivative with enhanced aminopeptidase activity. *Int. Dairy J.,* **5**: 367-379.

Coakley, M., Fitzgerald, G. F. and Ross, R. P. (1997). Application and evaluation of phage-resistance and bacteriocin encoding plasmid pMRC01 for the improvement of dairy starter cultures. *Appl Environ. Microbiol.,* **63**: 1434-1440.

Colman, A. (1996). Production of proteins in the milk of transgenic livestock: problems, solutions, and successes. *Am. J. Clin. Nutr.,* **63**: 639S-645S.

Colman, A. (1999). Dolly, Polly and other 'Ollys': Likely impact of cloning technology on biomedical use of livestock. *Genetic Analysis-Biomol. Engg.,* **15**: 167-173.

Dickley, F., Nilsson, D., Hansen, E. B., and Johansen, E. (1995). Isolation of *Lactococcus lactis* nonsense suppressors and construction of a food-grade cloning vector. *Mol. Microbiol.,* **15**: 839-847.

Drobnik, J. (2002). Genetically modified organisms—problems and legislation. *Cas Lek Cesk.,* **141**(4): 107-11.

Dunncoleman, N.S., Bloebaum, P. and Berka, R. M. (1991). Commercial levels of chymosin production by *Aspergillus. Biotechnology,* **9**: 349-351.

Enserink, M. (1998). Dutch pull the plug on cow cloning. *Science,* **279**(5356): 1444.

Fox, P. F., Guinee, T. P., Cogan, T. M. and McSweeney, P. L. H. (2000). Fundamentals of Cheese Science, an Aspen Publication, Maryland.

Geisen, R. and Holzapfel, W. H. (1996). Genetically modified starter and protective cultures. *Int. J. Food Microbiol.,* **30** (3): 315-24.

Ghosh, B. C. (2003). Developments of genetically modified cheese: Generic to genetics. In: Lecture Compendium, 16th Short Course on "*Application of Biotechnology in Dairy and Food Processing*" Nov. 4-24, 2003, Org. by Dairy Technology Division, NDRI, Karnal-132001. pp: 101-106.

Gomez, S., Cosson, C. and Deschamps, A. M. (1997). Evidence for a bacteriocin like substance produced by a new strain of *Streptococcus* sp., inhibitory to Gram-positive foodborne pathogens. *Res. Microbiol.,* **148**: 757-766.

Green, M. L., Angal, S., Lowe, P. A. and Marston, F. A. O. (1985). Cheddar cheese making with recombinant calf chymosin synthesized in *Escherichia coli. J. Dairy Res.,* **52**: 281-286.

Hammer, R. E., Pursel, V. G, Rexroad, C. E. Jr., Wall, R. J., Bolt, D. J., Ebert, K. M., Palmiter, R. D., and Brinster, R. L. (1985). Production of transgenic rabbits, sheep and pigs by microinjection. *Nature,* **315**, 680-683

Haslberge, A. G. (2003). GM food: the risk-assessment of immune hypersensitivity reactions covers more than allergenicity. *J. Food Agric. Environ.,* **1** (1): 42-45.

Henriksen, C. M., Nilsson, D., Hansen, S. and Johansen, E. (1999). Industrial applications of genetically modified microorganisms: gene technology at Chr. Hansen A/S. *Intl. Dairy J.,* **9**: 17-23.

Ishikawa, H., Nishimori, K., Kohda, T., Saito, H. and Oishi, M. (1992). Production of an immunoglobulin gene product by the plasmid expression vector L-factor in mouse myeloma cells. *Plasmid,* **28** (2): 93-100.

Jimenez Flores, R. and Richardson, T. (1988). Genetic engineering of the caseins to modify the behavior of milk during processing: a review. *J. Dairy Sci.,* **71**: 2640-2654.

Johansen, E. (1999). Genetic engineering (b) Modification of bacteria. In: R. Robinson, C. Batt, P. Patel, *Encyclopedia of Food Microbiology*, Academic Press, London.

Kane, J. F., Balaban, S. M. and Bogosian, G. (1991). Commercial production of bovine somatotropin in *Escherichia coli. ACS Symposium Series,* **444**: 186-200.

Kawaguchi, Y., Kosugi, S., Sasaki, K., Uozumi, T. and Beppu, T. (1987). Production of chymosin in *Escherichia coli* cells and its enzymatic properties. *Agric. Biol. Chem.,* **51**: 1871-1877.

Kim, J.H., Kim, S. G., Chung, D. K., Bor, Y. C. and Batta, C. A. (1992). Use of antisense RNA to confer bacteriophage resistance in dairy starter cultures. *J. Indus. Microbiol.,* **10**: 71-78.

Krimpenfort, P., Rademakers, A., Eyesqtone, W., VanderSchans, A.V., VendenBroek, S., Kooiman, P., Kootwijk, E., Plantenburg, G., Piper, F., Strijker, R. and DeBoer, H. (1991). Generation of transgenic dairy cattle using *in vivo* embryo production. *Biotechnology,* **9**: 844-847.

LePeuple, A.S., van Gemert, E., and Chapot-Chartier, M.P. (1998). Analysis of the bacteriolytic enzymes of the autolytic *Lactococcus lactis* subsp. *cremoris* strain AM2 by renaturing polyacrylamide gel electrophoresis: identification of a prophage-encoded enzyme. *Appl. Environ.l Microbiol.*, **64**: 4142-4148.

Luoma, S., Peltoniemi, K., Joutsjoki, V., Rantanen, T., Tamminen, M., Heikkinen, I. and Palva, A. (2001). Expression of six peptidases from *Lactobacillus helveticus* in *Lactococcus lactis.Appl. Environ. Microbiol.*, **67**(3): 1232-8.

Maga, E. A., Anderson, G. B. and Murray, J. D. (1995). The effect of mammary gland expression of human lysozyme on the properties of milk from transgenic mice. *J. Dairy Sci.*, **78** (12): 2645-2652.

Makhal, S., Mandal, S. and Kanawjia, S.K. (2004). Development of bioactive cheese: a novel 'bio-boom' of cheese technology. *Indian Food Industry* (Submitted).

Marilley, L. and Casey, M. G. (2004). Flavours of cheese products: metabolic pathways, analytical tools and identification of producing strains. *Int J Food Microbiol.*, **90**(2): 139-59.

Mathur, B. N., Rao, K. H. and Sethi, S. (2003). Recent trends in processing genetically modified dairy foods. *Indian Dairyman*, **55**(6): 29-35.

McGarry, A., Law, J., Coffey, A., Daly, C., Fox, P. F. and Fitzgerald, G. F. (1995). Effect of genetically modifying the lactococcal proteolytic system on ripening and flavour development in Cheddar cheese. *Appl. Environ. Microbiol.*, **60**: 4226-4233.

McGrath, S., Fitzgerald, G. F. and van Sinderen, D. (2001). Improvement and optimization of two engineered phage resistance mechanisms in *Lactococcus lactis Appl. Environ. Microbiol.*, **67**: 608-616.

Mierau, I., Kunji, E.R., Leenhouts, K.J., Hellendoorn, M.A., Haandrikman, A.J., Poolman, B., Konings, W.M., Venema, G. and Kok, J. (1996). Multiple-peptidase mutants of *Lactococcus lactis* are severely impaired in their ability to grow in milk. *J. Bacteriol.*, **178**: 2794-2803.

Mital, B.K. and Garg, S.K. (1995). Anticarcinogenic, hopocholesterolemic and antagonistic activities of *Lactobacillus acidophilus. Crit. Rev. Microbiol.*, **21**: 175-214.

Mohanty, A.K., Mukhopadhyay, U.K., Grover, S. and Batish, V.K. (1999). Bovine Chymosin: Production by rDNA technology and its application in cheese manufacture. *Biotechnol. Adv.*, **17**: 205-217.

Murthy, G. L. N. and Kanawjia, S. K. (2002). Designer milk. *Indian Dairyman*, **54**: 49-58.

Pennisi, E. and Vogel, G. (2000). Animal cloning. Clones: a hard act to follow. *Science*, **288** (5472): 1722-7.

Persuy, M. A., Legrain, S., Printz, C., Stinnakre, M. G., Lepourry, L., Brignon, G. and Mercier, J. C. (1995). High-level, stage- and mammary-tissue-specific expression of a caprine k-casein-encoding minigene driven by a β-casein promoter in transgenic mice. *Gene*, **165**, 291-296.

Persuy, M. A., Stinnakre, M. G., Printz, C., Mahe, M. F. and Mercier, J. C. (1992). High expression of the caprine β-casein gene in transgenic mice. *Eur. J. Biochem.*, **205**: 887-893.

Prost, F. and Chamba, J. (1994). Effect of aminopeptidase activity of thermophilic lactobacilli on Emmental Cheese characteristics. *J. Dairy Sci.*, **77**: 24-33.

Rastall, R.A. and Maitin, V. (2002). Genetic engineering: threat or opportunity for the dairy industry?. *Int. J. Dairy Technol.*, **55**: 161-165.

Renner, E. (1993). Nutritional Aspects of Cheese. In: *Cheese: Chemistry Physics and Microbiology* (Ed. P. F. Fox), Chapman and Hall, 2nd Ed., 1993, Vol. 1. pp. 557-580.

Rollema, H.S. (1992). Casein association and micelle formation. In: P.F. Fox (Ed.). *Advanced Dairy Chemistry-1: Proteins*. Elsevier, London and New York, pp. 111-140.

Rosen, J. M., Li, S., Raught, B. and Hadsell, D. (1996). The mammary gland as a bioreactor: factors regulating the efficient expression of milk protein-based transgenes. *Am. J. Clin. Nutr.*, **63**: 627S-632S.

Sabikhi, L. (2003). Genetic modification in food science. *Indian Dairyman*, **55**(6): 45-53.

Singh, H. and Creamer, L. K. (1992). Heat stability of milk. In: P. F. Fox (Ed.). *Advanced Dairy Chemistry-1: Proteins.* Elsevier Applied Dairy Science, London, pp. 621-656.

Stanton, C., Gardiner, G., Lynch, P.B., Collins, J.K., Fitzgerald, G. and Ross, R.P.(1998). Probiotic cheese. *Int. Dairy J.*, **8**: 491-496.

Steele, J. L., Johnson, M. E., Broadbent, J. R. and Weimer, B.C. (1998). Starter culture attributes which affect cheese flavor development, pp. 157-170. In: Proc. LACTIC '97 conference, Which strains? For which products.

Tanous, C., Kieronczyk, A., Helinck, S., Chambellon, E. and Yvon, M. (2002). Glutamate dehydrogenase activity: a major criterion for the selection of flavour-producing lactic acid bacteria strains. Antonie Van Leeuwenhoek, **82**(1-4): 271-8.

van Berkel, P. H., Welling, M. M, Geerts, M., van Veen, H. A, Ravensbergen, B., Salaheddine, M., Pauwels, E. K, Pieper, F., Nuijens, J. H. and Nibbering, P. H. (2002). Large scale production of recombinant human lactoferrin in the milk of transgenic cows. *Nat. Biotechnol.*, **20**, 484-487.

van de Guchte, M., Kok, J. and Venema, G. (1992). Gene expression in *Lactococcus lactis. FEMS Microbiol Rev.*,**8** (2): 73-92.

Vandenberg, J.A., Vanderlaken, K. J. and Vanooyen, A. J. J. (1990). *Kluyveromyces* as a host for heterologus gene expression: Expression and secretion of prochymosin. *Biotechnology*, **8**: 135-139.

Vaughan, E. E. and Mollet, B. (1999). Probiotics in new millennium. *Nahrung*, **43**: 148-153.

Wall, R. J., Kerr, D. E. and Bondioli, K. R. (1997). Transgenic dairy cattle: genetic engineering on a large scale. *J. Dairy Sci.*, **80**, 2213-2224.

Wall, R. J., Kerr, D. E. and Bondioli, K. R. (1997). Transgenic dairy cattle: genetic engineering on a large scale. *J. Dairy Sci.,* **80**, 2213-2224.

Wegmann, U., Klein, J. R., Drumm, I., Kuipers, O. P. and Henrich, B. (2000).Introduction of peptidase genes from *Lactobacillus delbrueckii* subsp. lactis into *Lactococcus lactis* and controlled expression. *Appl Environ Microbiol.,* **66**(3): 1252.

Wheeler, T. T., Mao, J., Davis, S. R. and Seyfert, H. M. (2002). Examination of the relationship between milk fat content and levels of acetyl-CoA-carboxylase-alpha expression in the udder of the cow. Proceedings-of-the-7th-World-Congress-on-Genetics-Applied-to-Livestock-Production, Montpellier, France, Aug.,-2002, session-9. 2002, pp: 0-4.

Wilmut, I., Campbell, K. Tudge, C. (2000). The Second Creation: The Age of Biological Control by the Scientists who Cloned Dolly. London: Headline Book Publishing.

Wong, E. A. (2003). Transgenic cows overexpressing beta and kappa caseins in milk.*www.isb.vt.edu.*

Yom, H. C. and Bremel, R. D. (1993). Genetic engineering of milk composition: modification of milk components in lactating transgenic animals. *Am. J. Clin. Nutr.,* **58** (Suppl.): 299S-306S.

Yvon, M and Rijnen, L. (2001). Cheese flavour formation by amino acid catabolism. *Int. Dairy J.,* **11**: 185-201.

Zuelke, K. A. (1998). Transgenic modification of cows milk for value-added processing. *Reprod. Fertil. Dev.,* **10**(7-8): 671-676.

2015, Dairy Product Technology: Recent Advances ***Pages 83–120***
Editors: **Subrota Hati, Surajit Mandal and Birendra Kumar Mishra**
Published by: **DAYA PUBLISHING HOUSE, NEW DELHI**

Chapter 5

Value Addition to Traditional Dairy Products: Scope and Future Strategies

S. Makhal and S. Hati

Introduction

"*Let Food be Thy Medicine and Medicine be Thy Food*" espoused approximately 400 BC by Hipocrates who proclaimed that food should not only be a source of energy but also should have health benefits. A concept originally conceived in Japan, the development of 'functional foods' is relatively recent and is the leading food industry trend today. Nowadays, the food-pharma philosophy has received a greater attention in the name of functional foods or health foods or nutraceuticals. Consumer awareness of the crucial link between diet and health is growing rapidly, particularly the vital role that diet plays in combating heart disease, cancer, etc. People, who are committed to maintaining good health well into their golden years, are eager to learn about foods that can help them longer and happier lives. They are continually bombarded with both direct and indirect messages to focus on healthful foods and nutrition. As a result, diet and health have been inextricably linked. Consequently, interest in the field of value added foods and health benefits derived from them has grown enormously within the last few years. Hence, recent advances in functional foods demonstrate much promise in new product development or reformulation of existing foods using various nutraceuticals to achieve health benefits.

The consumer interest in natural health-promoting foods is almost assuredly increasing, perhaps dramatically because of the increased health consciousness of the modern people and renewed interest in exploring medicinal benefits from the natural origins. In India some R and D works have been carried out for development of functional foods or value added foods, which will be available on the near days to the consumers. Although functional foods were originated and are most popular in Japan, they are gaining acceptance internationally, especially in the wealthier nations of the world. Our domestic food market was valued to be Rs. 3,09,000 during 1993-94 and on the strength of value addition, the market was expected to reach 5, 99000 during 2000-01 (CII-Mckinsey Report). The market would be further driven by continuing consumer awareness towards healthier lifestyle, in conjunction the increasing need for better health maintenance. Dairy development in India has been acknowledged the world over as one of modern India's most successful developmental programme. Since time immemorial, a significant proportion of milk has been used in India for making a diverse sort of dairy delicacies comprising of an unending array of sweets and other specialties from different regions of the country. The incorporation of health attributes or value addition to our indigenous dairy products would obviously help in achieving marketing advantages as well as meeting the nutritional thinking of the modern consumers.

Food Industry and Dairying in India: A Mini Vision

Food processing industry is of enormous importance for development of our country because of the central linkages and synergies that it promotes between the two pillars of our economy, industry and agriculture. Fast growth in the food processing sector and progressive improvement in value addition sequence are also of great significance for achieving favourable terms of the trade for Indian agriculture both in the domestic and overseas markets. India is the world's second largest producer of foods next to China, and has the potential of being the largest with the food and agricultural sector, which contributes around 26 per cent of India's GDP. The total food production in India has been projected likely to double in the next ten years and there is a golden opportunity for large investments in food and food processing technologies, skills and equipment, especially in the areas of canning, dairy and food processing, specialty processing, packaging, frozen food/refrigeration and thermo-processing. Fruits and vegetables, fisheries, milk and milk products, meat and poultry, packaged/convenience foods, alcoholic beverages and soft drinks, etc are important sub-sectors of the food processing industry. Health food and health food supplements are rapidly growing segment of this industry, which is gaining vast popularity amongst the health conscious populace.

More than 2445 million people are economically involved in agriculture worldwide; probably two-third or even more are entirely or partly dependent on livestock farming. India, which has 66 per cent of economically active population, engaged in agriculture, derives 31 per cent of GDP from agriculture. The share of livestock product is estimated at 21 per cent of total agricultural sector. Dairying in India has made a radical progress during the last two decades and now day's dairy industry in India is playing a momentous role in amelioration of our national economy producing large gross agricultural output of Rs. 50,051 crores per annum exceeding that from paddy (Aneja and Puri, 1997). Undoubtedly no single economic development programme has given so much to so many and therefore, it is neither be a bafflement nor no longer a bizarre that just as the upshot of nearly three decades of toil since 1970 when Operation Flood Programme took its inception to herald a new genesis of the race to gain a foothold in the field of dairying, India has become the global leader in dairying. This is a matter of euphoria that India has reached at the zenith of world milk production surpassing the United States (Aneja and Puri, 1997). With the victorious voyage of "White Revolution" through the sustained efforts of development in the arena of dairying, India, as the global leader in milk production, is currently producing 84.5 MT milk (FAO, 2003), which means a hopping sum of Rs. 1050 billion (Aneja and Puri, 1997) being ploughed back each year into our national economy.

Status of Traditional Dairy Products

Indian dairy industry is presently having a growth rate of 5.6 per cent (Aneja and Puri, 1997). Consumption of traditional dairy products, especially the products like *paneer*, curd, *khoa*, etc grows at an annual growth rate of more than 20 per cent with an estimated market value of Rs. 650 billion (Thakur and Ray, 2000; Patil, 2004), which underlies the significance of our traditional dairy products in the national economy. The indigenous dairy products play an important role in economic, social, religious and nutritional needs of the Indian population. It is estimated that about 50 to 55 per cent of total milk produced in our country is converted into variety of traditional milk products. India's high-value, high-volume market for traditional dairy products and delicacies is all set to boom further under the technology of mass production as well as on the strength of value addition. This market is the largest in value after liquid milk with a $1 billion overseas market.

More and more dairy plants in the public, cooperative and private sectors in India are going in for the manufacture of traditional milk products. This trend will undoubtedly give a further stimulus to the milk consumption in the country and ensure a better price to primary milk producers. Simultaneously, it will also help to productively utilize India's growing milk surplus. The market for traditional dairy products in India exceeds Rs. 50,000 crores, and it is the largest and fastest growing segment of India's dairy industry. For example, the consumption of '*dahi*', a plain yoghurt-like traditional product, today exceeds 5 million tonnes. This is 50 times the amount of all types of yoghurt consumed in the United States. The per capita consumption of dairy products in India equals to ~66.1 kg per annum, which is among the highest in Asia. Consumption of milk and dairy products in India has continued to expand substantially. For example, the per capita annual consumption of milk and diary products in India increased from 39 kg in 1985 to 66 kg in 1990. However, in several other Asian countries the per capita consumption of milk and diary products has matured at relatively low levels. For example, the per capita consumption in some of the higher income Asian countries, such as Taiwan, Singapore and Hong Kong have matured at approximately 10 kg to 14 kg per annum.

As per a survey report, of the 30 per cent animal protein consumed in India, 70 per cent comes from milk primarily in form of milk products. Approximately 48 per cent of milk output in India is used to produce a variety of indigenous dairy products, such as ghee, curd, *makkhan* and *khoa, paneer* as well as a diverse category of sweets. The fact that most of these products have been received well will encourage more investment in this sector. Regarding the overseas demand for Indian milk products and sweets, surveys indicate that North America alone may be able to absorb these items worth $ 500 million. The mainstay of this market will, of course, be the Indian diaspora of around 20 million, over half of them living in the West. Besides, there is also potential to popularise these products among local consumers abroad.

Export Potential

The World Trade Organization (WTO) has provided for the Agreement on Agriculture (AoA). This agreement is also germane to the Global dairy trade and is bringing about noteworthy changes in it. Following the WTO development, the Indian Dairy industry gains challenging opportunity of trapping the more lucrative international food markets. To triumph in international trade, development of an effective production, value addition, quality assurance and marketing systems is crucial. The globalization of the product brand is also becoming increasing important for success in the international markets. The Indian dairy industry involved in manufacturing of traditional dairy products must adopts some steps to be evolved as a quality conscious productive enterprise, including (i) adaptation of Good Manufacturing Practice (GMP) for plant design, process control as well as for production of hygienic milk products. For victorious marketing of such products, it is vital that the milk used should be of fine microbiological and sensory quality, (ii) mechanization and use of improved packaging systems to conform to the global standards (iii) value addition in order to improve its nutritional as well as health attributes and (iv) labeling and quality assurance programmes in conformation with the international standards. Indian dairy industry must adhere stringent physical and chemical specifications to win consumer confidence in the overseas food market. The growth of traditional dairy products in domestic and export markets are attracting new entrepreneur-investors. It would also encourage the current producers to expand production and enlarge portfolio of dairy products.

Traditional dairy products are becoming increasingly important with the globalization of the ethnic food wave. Markets for traditional Indian milk products also exist overseas where the ethnic population of the Indian subcontinent has settled. However, to tap these markets, it is essential to conform to the international standards of food quality and safety. With the withdrawal of subvention to milk producers following the GATT agreement, a restructuring in the international food market has been observed. This offers an exclusive prospect for the Indian dairy industry to get the desired breakthrough and take a place in the global food market. Because in India's rapidly expanding dairy industry, large scale manufacture of traditional dairy products is gaining importance. It has now attracted the interest of dairy entrepreneurs following delicencsing of dairy industry in 1991. The next few years would see packaged *khoa* and *chhana* based dairy products on the shelves of the retail outlets even in several cities of neighbouring countries; as also new varieties, such as value added indigenous products with specified health benefits would soon be making their debut in the market. It is believed that competitive forces of free-market economy following the WTO agreement will govern the future global food market.

This mode will require market orientation in which production will be geared towards consumer preference and demand. Market orientation will gradually replace production orientation. This is particularly factual for our dairy products finding large demand in the global food market due the presence of a large number of Indian populations scattered in several Western countries. Again the demographic changes have created momentous implications for dairying, particularly in India with its phenomenal augmentation in milk production. Rapidly budding urban demand are determining the type of milk and milk products to be made.

Moreover reputed companies in food sector are now venturing into dairy products, keeping in view the huge profit margins as well as national and international market potential in these product ranges. Economic remuneration is waiting for those who will succeed in trapping the marketing potential for various dairy products. A significant demand for several Indian sweets in the international market, particularly in USA UK and Canada already exists. Production plant design as well as design of packaging systems conforming to the standards enunciated in GMP of United State Food and Drug Administration and extending shelf life as well as value addition would accelerate the growth of this industry in our country. Value addition, in particular to these products is likely to increase further opportunity to ensnare the overseas lucrative markets. The Indian dairy industry wishing to manufacture such products for export purpose will not only need to produce them safe to the consumers but also convince the consumers about the safety of their products through aggressive marketing and the use of external quality certification systems like HACCP or ISO 9000/14000.

Functional Foods: A Novel Concept of Value Addition to Existing Foods

Functional foods do not have an internationally agreed definition but they can be defined as foods that provide health benefits beyond their basic nutrition. The regular consumption of functional foods reduces the risk of specific diseases but is not intended to act as a vehicle for food fortification to address deficiency. Disease treatment is considered possible with natural health products. Whereas, according to the Health Canada definition, functional foods are used only to reduce the risk of some diseases. A functional food may contain an added ingredient that makes the traditional food functional, *e.g.* incorporation of probiotic bacteria into yoghurt. However, a food can also be naturally functional, *e.g.* oatmeal, which naturally contains β-glucan, which has been proven to reduce blood cholesterol levels. Functional foods contain the proper balance of ingredients, which help us to function better and more effectively, including helping us directly in the prevention and treatment of illness and diseases. They serve to promote health or help to prevent disease, and in general the term is used to indicate a food that contains some health-promoting components beyond traditional nutrients. Functional foods have been variously termed as *nutraceuticals, value added foods, designed foods, medicinal foods, therapeutic foods, superfoods, foodiceuticals, medifoods,* etc. Pharmafoods or medical foods or medifoods are another discipline of functional foods, which are especially designed for specific dietary management of disease or condition for which there are distinctive nutritional requirements and used under physician supervision during active or ongoing treatment for the disease. Marketing of such foods is stupendously increasing and worldwide health consciousnesses of the consumers are influencing the food industry to introduce such value added health foods, designed for specific purposes.

The goal of achieving an optimal or maximal state of nutrition and health is becoming an increasing challenge with the introduction of many nutraceuticals. The ascribed health benefits of nutraceuticals are legion. Various products are claimed not only to reduce the risk of cancer and heart disease but also to prevent or treat hypertension, high cholesterol, excessive weight, osteoporosis, diabetes, arthritis, macular degeneration (leading to irreversible blindness), cataracts, menopausal symptoms, insomnia, diminished memory and concentration, digestive upsets and constipation and not to mention headaches.

Functional Ingredients

In most cases, the term 'functional' refers to a food that has been modified or value-added. Modifications can be achieved by incorporation of active ingredients, such as, dietary fibre, antioxidants, natural isoflavone, plant sterols/stanols, other phytochemicals or phytonutrients, bioactive peptide, þ-3 PUFA and probiotics and/or prebiotics, etc. These ingredients play a pivotal physiological and nutritional role in human. A wide array of nutrient phytochemicals is available in the nature that provides a diverse sort of health beneficial effects. Ascorbic acid, tocopherol, β-carotene, lycopene, plant sterols and stanols, etc are the such bioactive ingredients, which are also recognized as natural antioxidants and protective phytonutrients and play an important role in reducing the risk of free radical related oxidative disorders like CVD, cancer, cataracts.

In the recent past, considerable attention has focused on the potential benefits of plant derived sterols and stanols. Typical diets may commonly include sitosterol, campesterol and stigmasterol along with smaller amounts of plant stanols (which are saturated plant sterols) like sitostanol. There has been a lot of interest in the health benefits of plant polyphenols-attributed with antioxidant, antiatherogenic, anti-inflammatory and anticarcinogenic activities. Cocoa and chocolate are particularly rich sources of polyphenols; however, conventional chocolate-making processes are thought to reduce the potential polyphenol content of the final product. Table 5.1 depicts some broad categories of functional foods.

Probiotic foods are the most important discipline of functional foods, which are defined as "foods containing live microorganisms, which actively enhance the health of consumers by improving the balance of microflora in the gut when ingested live in sufficient numbers". Studies have related the promising health benefits of consuming cultured and culture containing milks and milk products, some of which are in International supermarkets. There have been long-term interests of using cultured milk products with various strains of LAB and other probiotic bacteria to improve the health.

Table 5.1: Categories of different Functional Ingredients and their Functions

Food Modifications	*Functionality*
Incorporation of phytochemicals (as plant ingredients or extracts)	Antioxidant, lower risk of CHD, cancer, and lower blood pressure
Addition of probiotics	Improved gastrointestinal function, enhanced immune system, lower risk of colon cancer and of food allergy
Additional of prebiotics	Improved gastrointestinal function, lower risk of colon cancer, enhanced immune system
Addition of bioactive proteins or peptides	Enhanced immune function and bioavailability of minerals, hypertensive function
Addition of dietary fibers	Prevention of constipation, lower risk of colon cancer and lowering of blood cholesterol level
Addition of ω-3 PUFA	Lower risk of heart attack, lower risk of some cancers, enhanced immune system
Removal of allergens	Reduce or eliminate allergy to specific foods

Health Benefits of Some Potent Functional Ingredients for Value Addition to Dairy Products

Probiotics and Probiotic Foods

Functional dairy products incorporating probiotic bacteria with scientifically supported health claims have great potential for improving quality of life and are widely predicted to become one of the biggest dietary trends of the present decade. The concept of probiotic foods is based on the fact that the microflora in GI tract are having significant role in the health status of an individual, which is influenced by a diet consisting of the organisms. Probiotics can be defined as "a live microbial feed supplements, which beneficially affect the host animal by improving its intestinal microbial balance". Probiotics is "A preparation or a product containing viable, defined microorganisms in sufficient numbers, which alter the intestinal microflora by implantation or colonization in a compartment of the host and by that exert beneficial health effects in this host. Nevertheless, probiotics have been recognized to have several health benefits, such as balancing of intestinal microflora, stimulation of the immune system, prevention of diarrhoea, anticarcinogenic activity, etc. Interest in the field of probiotic foods and the health benefits derived from them has grown enormously within the last few years; hence, recent advances in functional foods demonstrate much promise in new product development, particularly development of probiotic dairy products using these live microorganisms.

Currently, there are over 70 bifidus and acidophilus-containing products produced worldwide, including sour cream, buttermilk, yoghurt, and powdered milk and frozen desserts. Out of 90 commercial probiotic products available worldwide, more than 53 different types of dairy products containing probiotic organisms are marketed in Japan (Hilliam, 2000). The probiotic food industry is flourishing with the European probiotic yoghurt market currently estimated to be worth around £520 million (Shortt, 1999). In many European countries, most notable are France and

Germany and the market is expanding with the result that probiotic yoghurts now account for over 10 per cent of all yoghurts sold in Europe (Stanton *et al.,* 2001). In 1997, these products was accounted for 65 per cent of the European functional foods market, and valued at US$889 million, followed by spreads, at US$320 million accounting for 23 per cent of the market. In a recent study undertaken by LFRA, it has been shown that the probiotic yoghurt market in nine countries (United Kingdom, France, Germany, Spain, Belgium, Netherlallds, Denmark Finland and Sweden) totaled > 250 million kg in 1997 (Hilliam, 1998), with France representing the largest market, having sales of ~90 million kg, valued at US$219 million. The German market for probiotic yoghurts is growing rapidly; for example, during 1996-1997, it increased by 150 per cent, whereas the UK market grew by a more modest 26 per cent during the same period.

In the year 2002, sales of probiotics in the US markets were estimated at $160 million, with a growth rate of 9 per cent from the year before. The probiotic food market of EU was valued at around $12.9 million in 2003 and the total market for probiotic ingredients in applications for human consumption is currently growing at an estimated rate of 14 per cent. Probiotic yoghurt is clearly the largest segment in the probiotic dairy foods market, accounting for 82 per cent of the total volumes in Australia. Netherlands is currently marketing fermented yoghurt '*Fysig*' made with one strain of *L. acidophilus,* which is promoted as being useful in helping maintain a healthy cholesterol level. 'Actimel Cholesterol Control' is another commercially available therapeutic milk product containing probiotic culture, which helps, if taken regularly as a part of a healthy and varied diet, reduces the cholesterol value (Coussement, 1997).

Newer Probiotic Milk Products

Food products containing probiotic bacteria are almost exclusively dairy products, capitalizing on the traditional association of LAB with fermented milk. Yoghurt and fermented milks have received the most attention as carriers of live probiotic cultures, whereas other dairy foods, such as some cheese varieties, frozen yoghurts, and ice cream have been investigated as potential carriers (Stanton *et al.,* 2001). Probiotic bacteria used in these products include *L. acidophilus* and *Bifidobacterium* species, among others. A variety of food products and supplements containing viable probiotic microorganisms are commercially available (Table 5.2) in the market. Many European countries are experiencing considerable growth in demand for existing probiotic products and there is a surge in the number of new products being launched. Dairy products, accounting for 65 per cent of the total European functional foods market, are at the forefront of probiotic developments (Hilliam, 1998). Within this sector, probiotic cultures have been incorporated in yoghurts and fermented milk products; of these, LCl (Nestle, Vevey, Switzerland), Vifit (Campina Melkunie, Zaltbommel, Netherlands), Actimel (Danone, Parisl), and Yakult have emerged as the market leaders (Hilliam, 1998). The introduction of probiotic products has fueled growth in the German yoghurt market, with sales of probiotic products increasing from DM130 million to DM300 million between 1996 and 1997. The leading brand in the sector is LC1 from Nestle with a 60 per cent market share, ahead of Actimel from Danone with a 25 per cent market share and followed by Mueller's Procult brand (Mueller, Germany).

Table 5.2: Some Commercial Probiotic Dairy Products

Product	*Culture*	*Country*
Acidophilus Bifidus yoghurt	A+B+C	Germany
Bifidus milk	*B. bifidum* or *B. longum*	German
Bifidus yoghurt	*B. longum* + C	Many countries
Bifighurt	*B. longum + S. thermophilus*	Germany
Bifilak (c) t	A + B	Russia
Biobest	*B. bifidum* or *B. longum + C*	Germany
Biomild	A + B	Germany
Mil-Mil	A + B + B.*breve*	Japan
Cultura	A + B	Norway, Denmark
Kyr	A + B + C	Italy
Ofilus	A + B + *S. thermophilus*	France
Biogarde	A + B + *S. thermophilus*	Germany
Actimell	*L. casei*	Germany
Vifit	*L. casei* GG	Germany, UK
Primo	Bacto Lab Cultures	Germany
Zabady	*B. bifidum* + C	Egypt
BA live	A + B + C	U.K.
Femilact	A + B + *Pediococcus acidilactici*	Czechoslovakia
Philus	A + B + *S. thermophilus*	Sweden
Fysig	*L. acidophilus*	Netherlands
Gaio	*E. faecium* + *S. thermophilus*	Demark
Stoneyfield Yoghurt (BioGaia)	*L. reuteri*	Sweden
Bio K+	*L. acidophilus*+ *L. casei*	Canada
Yakult	*L. casei* Shirota	Japan (23 countries)
LC1 (Nestle)	*L. johnsonii LJ*	European countries
French "bio" yoghurt	–	French
ProViva" (Probi AB)	*L .plantarum* 299v	Sweden
Culturelle", "Gefilus"(Valio)	*L. casei* GG	US
PrimaLiv" (Probi AB)	*L.rhamnosus* 271	–
Arla Acidophilus	*L. acidophilus* NCFB 1748	–
BIO Aloe Vera, Biotic Plus, Dairy FIT, Fyos, Jour après jour, PROAC, ProCult, Silhouette Plus	–	European countries

A: *L. acidophilus*, B: Bifidobacteria, C: Yoghurt culture (*S. thermophilus* and *L. bulgaricus*)

Source: Lourens-Hattingh and Viljoen (2001), Coussement (1997).

Nestlé's LC1, available either as set cultured milk or as a drinking product, contains the *L. acidophilus* Lal. This *Lactobacillus* strain, chosen for its probiotic characteristics, has been researched extensively by Nestle. On the basis of human studies, this culture is claimed to stimulate the immune system, leading to the statement, 'helps the body protect itself' (Young, 1996). The product was launched in France in 1994 and is currently available in most European countries. By 1996, LC1 had seized a 15 per cent share of the French bioyoghurt market and, even within the traditionally skeptical UK market, accounted for 20 per cent of the company's European trade in yoghurts and fermented milks (Young, 1996). In Switzerland a 'probiotic enhanced' yogurt called 'SymBalance' (ToniLait AG, Bern, Switzerland) was recently introduced that contains the prebiotic inulin as an enhancing agent and human probiotic strains, including *L. reuteri* (Casas *et al.,* 1998). Most yoghurt sold in the U.S. contains probiotic bacteria and represent the largest probiotic product category. All unpasteurized yogurts in the U.S. contain *S. thermophilus* and *L. bulgaricus* used as starter cultures to make the yoghurt. Danone has released nationwide a fermented dairy-based beverage containing 10^{10} live *L. casei* per serving. The product, called 'DanActive' has been sold in Europe for years under the Actimel brand (CDRF, 2004).

Prebiotics

Prebiotic are non-digestible oligosaccharides (NDO) used as food ingredients to modify the composition of endogenous gut microflora and more specifically defined as "a non-digestible food ingredient that beneficially affects the host by selectively stimulating the growth and/or activities of one or a limited number of bacteria in the colon, and thus improves host health". To be an effective prebiotic, it must (i) neither be hydrolyzed nor absorbed in the upper part of the GIT (ii) have a selective fermentation such that the composition of the large intestinal microbiota is altered towards a healthier one and (iii) induce luminal or systemic effects that are beneficial to the host. The most effective prebiotics identified are oligofructose and inulin. Other effective growth enhancers are GOS, lactulose, lactitol, lactosucrose, etc. However, the greatest scientific interest was focused on the nutritional and health benefits of oligofructose and inulin. Extensive *in vitro* and *in vivo* research on the effects of inulin and oligofructose on the human gut microbiota, reported selective fermentation by the beneficial flora, namely bifidobacteria and to lesser extent lactobacilli. In the small intestine, prebiotics have desirable influence on sugar digestion and absorption, glucose and lipid metabolism and protection against known risk factors of cardiovascular diseases. Due to the fermentation of prebiotics in the colon, short chain fatty acids (SCFAs) are produced and this is considered a major beneficial feature related to the primary prevention of colorectal cancer. These SCFAs help to protect intestinal wall health and stimulate repair and butyrate plays a key role in colon health and therefore, this is a preferred nutrient, indirectly prevents colonic mucosal reduction that develops within days of oral starvation. These stimulate cell division and apoptosis, the colon's mechanism to remove excess or abundant cells. Unlike the small intestine that derives energy from various bodily sources, the colon epithelial cells derive energy solely from SCFAs. Another important role prebiotics can play is in the prevention of intestinal permeability or leaky gut syndrome', which could lead to chronic bowel inflammation caused by mucosa's exposure to foreign bodies. Studies also revealed that, the changes in the intestinal microflora that occur with the consumption of prebiotic fibers may potentially mediate immune changed via the direct contact of lactic acid bacteria or bacterial products (cell wall or cytoplasmic components) with immune cells in the intestine, the production of SCFAs from fiber fermentation, or by changes in mucin production. Moreover, there is a high level of evidence of positive effects of some prebiotics to alleviate constipation and treat to hepatic encephalopathy. Already proven beneficial effects with reference to prebiotics include: non-digestible and low energy value, increase and modulation of stool volume by stimulation of beneficial bacteria (*Bifidobacterium, Lactobacillus* and *Eubacterium* spp.) and inhibition of "undesirable" bacteria (*Clostridium* and Bacteroides). A variety of foods containing inulin and/or oligofructose formulations, claiming to have beneficial effects on gut health and general well being are becoming prevalent in the market. Prebiotics, such as inulin can be used in almost all food categories and help in a systematic approach with other functional ingredients. Moreover, because of its chemical make up and functional properties it is quite unique in its effects on texture modification. Inulin has relatively low viscosity and is easily incorporated into foods. It helps to reduce fat and/or sugar content and improve the overall texture, mouth feel and flavour of foods while providing a selective prebiotic fiber source for the health promoting bacteria. Moreover, inulin is found to improve mineral absorption also. Prebiotics are incorporated in a number of products such as beverages, cereals, breads, yoghurts, frozen desserts, confectionaries, fermented and fluid "healthy milk" products, extruded snacks and baked foods.

Synbiotics and Synbiotic Foods

A synbiotic, a combination of probiotics and prebiotics is defined as a mixture of a probiotic and prebiotic that beneficial effects on host. This improves the survival and the implantation of live microbial dietary supplements in the GIT, and selectively stimulates the growth and/or by activating the metabolism of one or a limited number of health-promoting bacteria. Synbiotic is where probiotics and prebiotics are used in combination, to manage microflora due to the potential synergy between them and foods containing a combination of these ingredients are often referred to as synbiotic foods. A synbiotic may prove more efficacious than either the agent alone. However, the concept of synbiotic is in its infancy, and is yet to be applied to the full extent in developing new functional foods. Probiotic and a prebiotic in a single food product, the expected benefits are an improved survival during the passage of probiotic bacteria through the upper intestinal tract. Several reports have suggested the significant effect of synbiotics on the composition of faecal flora and survival of probiotic *B. longum, B. infantis* and *B. adolescentis* in various dairy products. Maximal counts and production and change in pH and biochemical metabolites were not influenced by the presence of fructo oligosaccharides in the infant formulae. Lyophilized synbiotic preparation containing *Bifidobacterium* species and inulin significantly improves the survivability of probiotics during storage and reduces fecal clostridial counts, coliforms and carcinogenic β-glucuronidase enzyme activity in mice. A synbiotic low-fat soft curd cheese has been developed with dietary fibres and added wheat fibre, the prebiotic oligofructose to the product and increased both the sensory and the nutritional quality of the product. Researches in the area of prebiotic oligosaccharides and synbiotic combinations with probiotics are leading towards a more targeted development of functional foods. Use of inulin and oligofructose in fruit yoghurts, milk-based drinks, milk, spreads, cheese and ice cream has also been reported. Number of synbiotic products containing Bifidobacteria and lactulose are already available in Japanese markets. Some of these products are Hounyu Milk Powder for adults [lactulose – 8.3 g/100g and Bifidobacteria >3 x 10^7], Sawayaka sour milk [lactulose 4g/100g and Bifidobacteria > 10^8], etc. Some synbiotic dairy products have been marketed in Europe also, *e.g.*, Symbalance, mixture of *Lactobacillus reuterii, L. acidophilus* and *L. casei* along with RAFTILINE, an inulin and John après Jour a UHT skimmed milk with ACTILIGHT, etc. Synbiotic yoghurt was prepared by incorporating *L. acidophilus* NCDC 13 and 3 per cent inulin with better acceptability on sensory evaluation and the higher viability of probiotics in the product during refrigerated storage.

Bioactive Peptides

The recent advances in research on bioactive peptides show much promise in new product development using these active biomolecules to derive multifarious health benefits. It is now evident that during fermentation of milk with certain dairy starters, peptides with various bioactivities are formed and are detected in an active form even in the final products, such as fermented milks and cheeses. Milk proteins have been identified as an important source of bioactive peptides, which can be released during hydrolysis induced by digestive or microbial enzymes. Researches, carried out during the last 15 years, have demonstrated that major milk protein groups, caseins and whey proteins, are the important source of these biologically active peptides. Functional or bioactive peptides are described as defined sequences of amino acids that are inactive within the native protein, but display specific properties once they have been released by enzymatic activity. These bioactive peptides usually contain 3 to 20 amino acid residues per molecule, except for glycomacropeptides (GMPs), which consists of 64 amino acid residues. The sequence of the amino acids in these small molecules is a crucial factor in their activity. The amino acid located in the C-terminal or N-terminal position is often significant. They serve to modulate metabolic processes like digestion, circulation, immunological responsiveness, cell growth and repair, nutrient intake, etc. These peptides display partial resistance to hydrolysis and can exert their effects either locally in the digestive tract or elsewhere in the body and could confer the a role of messenger molecules. Their contribution to the health of the newborn would be three-fold: an easily assimilated source of organic nitrogen, a good source of essential amino acids and a potential source of bioactive molecules. Several casein-derived peptides play a significant role in stimulation of the immune system. They have been found to exert a protective effect against microbial infections. The peptides derived from milk proteins have been shown to have a variety of different functions *in vitro* and *in vivo*, for example, antihypertensive, antimicrobial, antioxidative, antithrombotic, immunomodulatory, mineral-carrying and opioid, etc.

There is increasing commercial interest in the production of bioactive peptides with the purpose of using them as active ingredients for bioactive foods or in the development of fermented milk products with elevated level of these bioactive peptides by the manipulation of processing conditions. In an endeavour to expand the range of bioactive dairy products, a small number of researchers and companies have attempted to manufacture bioactive dairy products, such as cheese and fermented milks, which would contain a considerable amount of these wonder molecules *viz.* bioactive peptides.

Phytochemicals

Phytochemicals is simply a word that means plant chemicals. Hundreds of phytochemicals are currently being studied. Many are believed to have a major positive impact on human health. Some contribute to the bright and vivid colours found in fruits and vegetables. The term phytochemical is a fairly recent term that emphasizes the plant source of some of these health-protecting compounds. Currently, "phytochemical" and "phytonutrient" are being used interchangeably. Scientists are examining the antioxidant, immune-boosting and other health-promoting properties of these active compounds in plants. Chemically phytonutrients or phytochemicals that are being studied presently include terpenes, carotenoids, limonoids, and phytosterols.

Carotenoids are a group of more than 700 fat soluble nutrients that produce the colours in foods, such as carrots, pumpkins, sweet potatoes, tomatoes, and other deep green, yellow, orange, and red fruits and vegetables. Many are proving to be very important for health. The β-carotene is the most widely studied carotenoid, but others are proving to be of great interest. As with some, but not all, carotenoids, β-carotene is known as a provitamin A. Another phytochemical is lycopene, which is responsible for the red colour in fruits and vegetables, including tomatoes, red grapes, watermelon, and pink grapefruit. It is also found in papayas and apricots and does not convert to vitamin A but may have important cancer fighting properties and other health benefits. Xanthophylls are also a group of such plant derived chemicals, which contain oxygen and most are found in green vegetables, such as broccoli, cabbage, and kale. They are also present in yellow fruits and vegetables. Xanthophylls include lutein and zeaxanthin, which are both stored in the retina of the eye. They neither convert to vitamin A. They are also powerful antioxidants and may be very important for healthy eyes.

Polyphenols are also important phytochemicals, and flavonoids (or catechins) are members of the polyphenol family that may have significant health benefits. Laboratory studies have shown that specific flavonoids suppress tumor growth, interfere with sexual hormones, prevent blood clots, and have anti-inflammatory properties. In general, flavonoids are found in celery, cranberries, onions, kale, dark chocolate, broccoli, apples, cherries, berries, tea, red wine or purple grape juice, parsley, soybeans, tomatoes, eggplant, and thyme. Most common berries contain flavonoids and are particularly rich in potent antioxidants. Among the important flavonoids are resveratrol, quercetin, and catechin. Evidence suggests that resveratrol found in red wine, grapes, olive oil may be extremely potent. It increases cell survival and has been shown to increase the life span of worms and fruit flies. Catechins are the primary flavonoids in tea and may be responsible for its possible beneficial effects. Flavonoids in dark chocolate may also be health protective.

The phytochemicals 'saponins' are forms of carbohydrates that neutralize enzymes in the intestines that may cause cancer. They may also boost the immune system and promote wound healing. Saponins are found in ginseng, beans, including soy beans, and whole grains. Capsaicin, another important plant derived molecule and found in hot red peppers, seems to reduce levels of substance P, a compound that contributes to inflammation and the delivery of pain impulses from the central nervous system. Research suggests that it may inhibit cancer-generating substances.

Dietary Fibre

Dietary fiber consists of the structural and storage polysaccharides and lignin in plants that are not digested in the human stomach and small intestine. Dietary fibre exerts a wide range of physiological effects when consumed and its complex nature is responsible for a range of physical and chemical properties that are responsible for these physiological effects. They consist of complex soluble carbohydrates and soluble fibres, such as lignans, hemicellulose, amyiopectins, mucilage, gums and insoluble cellulose. As dietary fibre is better defined and understood, it is becoming more important in human health. It promotes satiety, provides roughage, slows digestion and reduces hunger, promotes desirable intestinal bacteria, reduces constipation and diverticular diseases, reduces haemorrhoids, bowel cancer and irritable bowel syndrome. It functions in conjunction with the monounsaturated oils, minerals, vitamins and phytonutrients, and plays a role in reducing the risk of cardiovascular diseases, cancers and diabetes. The majority of expert committees have recommended an increase in the fibre content because there is accumulating evidence that fibre is important in prevention of a large number of bowel disorders.

A wealth of information supports the American Dietetic Association position that the public should consume adequate amounts of dietary fiber from a variety of plant foods. Recommended intakes, 20-35 g/day for healthy adults and age plus 5 g/day for children, are not being met, because intakes of good sources of dietary fiber, fruits, vegetables, whole and high-fiber grain products, and legumes are generally low. Consumption of dietary fibers that are viscous lowers blood cholesterol levels and helps to normalize blood glucose and insulin levels, making these kinds of fibers part of the dietary plans to treat CVDs and type 2 diabetes. Recommendations for adult dietary fiber intake generally fall in the range of 20 to 35 g/day. Others have recommended dietary fiber intakes based on energy intake, 10 to 13 g of dietary fiber per 1000 kcal. Nutrition facts labels use 25 g dietary fiber per day for a 2,000 kcal/day diet or 30 g/day for a 2,500 kcal/day diet as goals for American intake.

Fibre is often divided into two broad classes: insoluble and soluble forms. Wheat bran, for instance, is an insoluble form that is a good stool-softener but a poor absorber of cholesterol, a function that the soluble form, such as oat bran, does better. Insoluble fibre makes stools heavier and speeds their passage through the gut. Like a sponge, it absorbs many times its weight in water, swelling up and helping to eliminate feces and relieve constipation. Wheat bran and whole grains, as well as the skins of many fruits and vegetables, and seeds, are rich sources of insoluble fibre. High-fibre diets have replaced bland, low-residue treatments for bowel problems, such as diverticular disease.

Soluble fibre includes pectin, gums (such as guar), β-glucans, some hemicellulose and other compounds and is found in oats, legumes (peas, kidney beans, lentils), some seeds, brown rice, barley, oats, fruits (such as apples), some green vegetables (such as broccoli) and potatoes. Soluble fibre breaks down as it passes though the digestive tract, forming a gel that traps some substances related to high cholesterol. There are several evidences that soluble fibre may lessen heart disease risks by reducing the absorption of cholesterol into the bloodstream. Studies find that people on high-fibre diets have lower total cholesterol levels and may be less likely to form harmful blood clots than those who consume less soluble fibre. A recent USA report found that, in sufficient amounts, fibre apparently reduced heart disease risks among men who ate more than 25 grams per day, compared to those consuming less than 15 grams daily.

Soluble fibre is also considered especially helpful for people with either form of diabetes. It helps control blood sugar by delaying gastric (stomach) emptying, retarding the entry of glucose into the bloodstream and lessening the postprandial (post-meal) rise in blood sugar. It also lessens insulin requirements in those with Type I diabetes as fibre slows the digestion of foods, it can help blunt the sudden spikes in blood glucose that may occur after a low-fibre meal. Such blood sugar peaks stimulate the pancreas to pump out more insulin. Some researchers believe that a lifetime of blood glucose spikes could contribute to Type II diabetes, which typically strikes after the age of 40, and more than doubles the risk of stroke and heart disease. The cholesterol-lowering effect of soluble fibres may also help those with diabetes by reducing heart disease risks.

Fibers that are incompletely or slowly fermented by microflora in the large intestine promote normal laxation and are integral components of diet plans to treat constipation and prevent the development of diverticulosis and diverticulitis. A diet adequate in fiber-containing foods is also usually rich in micronutrients and nonnutritive ingredients that have additional health benefits. In recent years, much attention has been focused on the possible protective role of soy fibre in the prevention of CVD, diabetes and cancer of the large bowel, etc. Due to a vast array of physiological significance and physicochemical properties of dietary soy fibre, a host of foods containing different levels of these bioactive ingredients has been developed and is being marketed for specific purposes. Dietary fibre - already known to reduce the risk of colon cancer - may also have the potential to replace antibiotics according to emerging new Australian research. Scientists at CSIRO Health Sciences and Nutrition in Adelaide are investigating new types of dietary fibre - including resistant starch and shorter chain oligosaccharides - that could help to fight disease and avoid chronic health conditions.

Several dietary fiber sources lower blood cholesterol levels, specifically that fraction transported by low density lipoproteins (LDL). Fibers that lower blood cholesterol levels include foods such as apples, barley, beans and other legumes, fruits and vegetables, oatmeal, oat bran and rice hulls; and purified sources, such as beet fiber, guar gum, karaya gum, konjac mannan, locust bean gum, pectin, psyllium seed husk, soy polysaccharide and xanthan gum. Two of these fibers, namely β-glucan in oats and psyllium husk, have been sufficiently studied for the FDA to authorize a health claim that foods meeting specific compositional requirements and containing 0.75 or 1.7 g of soluble fiber per serving, respectively, can reduce the risk of heart disease. Consequently, these two dietary fibers are specifically included in the most recent National Cholesterol Education Programme and American Heart Association guidelines. The mechanism by which these fiber sources lower blood cholesterol levels has been the focus of many investigations, and characteristics, such as solubility in water, viscosity, fermentability, and the kinds and amounts of protein and tocotrienols have been explored as possible bases for this physiological effect.

Phytoestrogens

Phytoestrogens, a special class of phytonutrients that include isoflavones and lignans, are found in plant-based foods, such as soybeans, flaxseeds and berries. Phytoestrogens are plant-derived nonsteroidal compounds that possess estrogen-like biological activity. In the past few years, phytoestrogens have received recognition as yet another unique health-promoting feature offered by whole foods. Most notably, a significant number of research studies have shown that Asian populations who consume high levels of soy foods (Tofu, soybeans, and soymilk) that contain isoflavones, have lower levels of breast, ovarian and endometrial cancers. Given that breast cancer is one of the most common cancers in women, and its development appears to be associated with diet, scientists have been very interested in how isoflavones might protect against breast cancers. Subsequent experimental research has clarified that isoflavones are converted in the body to hormone-like compounds that have the ability to modulate estrogen activity and dampen its potentially damaging effects in cells within the female reproductive organs.

Isoflavone

The role of isoflavones is widely appreciated and is currently the subject of intense research. It appears to protect against hormone-related disorders, such as breast cancer. Isoflavones do this by competing with body's own estrogen for the same receptor sites on cells. Some of the risks of excess estrogen can be lowered in this way. Isoflavones can also have estrogen activity. If during menopause the body's natural level of estrogen drops, isoflavones can compensate by binding to same receptor sites thereby easing menopause symptoms as a result.

Soy isoflavones are a group of compounds found in and isolated from the soybean. Besides functioning as antioxidants, many isoflavones have been shown to interact with animal and human estrogen receptors, causing effects in the body similar to those caused by the hormone estrogen. Soy isoflavones have both weak estrogenic and weak anti-estrogenic effects. They have been found to bind to estrogen receptors-alpha (ER-alpha) and beta (ER-beta). They appear to bind better to ER-beta than to ER-alpha. Soy isoflavones comprise three main isoflavones and their glycosylated forms. The three main isoflavones are the aglycones genistein, daidzein and glycitein. The malonyl glycosides of genistein are the major forms of the soy isoflavones that are found in soybeans. Soy isoflavones also produce non-hormonal effects. Isoflavones compounds, such as genistein and daidzein, are found in a number of plants, but soybeans and soy products like *Tofu* and textured vegetable protein are the primary food source. Isoflavones acts as antioxidants to counteract damaging effects of free radicals in tissues. Isoflavones can act like estrogen in stimulating development and maintenance of female characteristics or they can block cells from using other forms of estrogen. Isoflavones also have been found to have antiangiogenic effects (blocking formation of new blood vessels), and may block the uncontrolled cell growth associated with cancer, most likely by inhibiting the activity of substances in the body that regulate cell division and cell survival (growth factors). Fermented soy foods, such as *Tempeh* and *Miso,* are rich in the soy isoflavone aglycones. The most abundant of the soy isoflavones in soybeans are the genistein glycosides (~50per cent), followed by the daidzein glycosides (~40per cent). The least abundant of the soy isoflavones in soybeans are the glycitein glycosides (~5 to 10per cent). Soy protein derived from soybeans contains about 2 mg of genistin and daidzin per gram of protein. In soy germ, the order is different. Glycitein glycosides comprise about 40 per cent of soy germ, daidzein glycosides about 50 per cent and genistein glycosides about 10 per cent. Soy isoflavones, when marketed as nutritional supplements, are mainly present as the isoflavone glycosides genistin, daidzin and glycitin.

Epidemiological data suggest that higher intakes of foods containing soy isoflavones are significantly correlated with reduced incidence of heart disease and some forms of cancer. Soy proteins have been shown to lower plasma levels of cholesterol in animal models of hypercholesterolemia, and, subsequently, a meta-analysis of human studies has more recently established that soy consumption is significantly associated with reduction in plasma cholesterol levels in humans, as well. These effects are largely attributed to the isoflavone components of soy. Epidemiological data also indicate that consumption of soy is particularly associated with reduced risk of breast, lung and prostate cancers, as well as leukemia. Soy isoflavones have also been shown to prevent bone resorption and to help increase bone density in some *in vitro* and animal studies. The synthetic isoflavone ipriflavone, the major metabolite of which is the soy isoflavone daidzein, has demonstrated a significant ability to prevent osteoporosis in both animal models and in humans.

Mechanism of Action

Soy isoflavones have weak estrogenic activity. The order of activity in *in vivo* assays is glycitein greater than genistein greater than daidzein. They bind to estrogen receptors-alpha and beta. They appear to bind better to estrogen receptor-beta than to estrogen receptor-alpha. Genistein has been found to have a number of antioxidant activities. It is a scavenger of reactive oxygen species and inhibits lipid peroxidation. It also inhibits superoxide anion generation by the enzyme xanthine oxidase. In addition, genistein, in animal experiments, has been found to increase the activities of the antioxidant enzymes superoxide dismutase, glutathionine peroxidase, and catalase and glutathione reductase. Daidzein and glycitein also appear to have reactive oxygen scavenging activity.

Several mechanisms have been proposed for genistein's possible anticarcinogenic activity. These include upregulation of apoptosis, inhibition of angiogenesis, inhibition of DNA topoisomerase II and inhibition of protein tyrosine kinases.

Genistein's weak estrogenic activity may be involved in its putative activity against prostate cancer. Other possible anti-prostate cancer mechanisms include inhibition of NF (nuclear factor)-k B in prostate cancer cells, downregulation of TGF (transforming growth factor)-β and inhibition of EGF (epidermal growth factor)-stimulated growth. Genistein's anti-estrogenic action may be another possible mechanism to explain its putative activity against breast cancer. Additional possible anti-breast cancer mechanisms include inhibition of aromatase activity and stimulation of sex hormone binding globulin, both of which might lower endogenous estrogen levels. The possible anti-atherogenic activity of soy isoflavones may be accounted for, in part, by their possible antioxidant activity, particularly with regard to inhibition of lipid peroxidation and oxidation of LDL. Soy isoflavones may have some cholesterol-lowering activity, but the mechanism of this possible effect is unclear. Soy isoflavone's weak estrogenic effect may help protect against osteoporosis by preventing bone resorption and promoting bone density. However, the mechanism of this possible effect is entirely speculative at this time.

Health Benefits

Research in several areas of healthcare has shown that consumption of isoflavones may play a role in lowering post-menopausal disorders. Isoflavones can fight disease on several fronts. The following potential health benefits are attributed to isoflavones:

Ease of Menopause Symptoms

The benefits of soy go beyond reducing long-term cancer risk. Recent studies have found that soy isoflavones can reduce menopause symptoms such as hot flushes and increase bone density in women. Indeed, many menopausal and post-menopausal health problems may result from a lack of isoflavones in the typical Western diet. Although study results are not entirely consistent, isoflavones from soy or red clover may be helpful for symptoms of menopause.

Heart Disease Risk

Soy isoflavones also appear to reduce CVD risk via several distinct mechanisms. Isoflavones inhibit the growth of cells that form artery clogging plaque. These arteries usually form blood clots, which can lead to a heart attack. There is some evidence that isoflavones are the active ingredients in soy responsible for improving cholesterol profile.

Prostate Problems

Eating isoflavones rich products may protect against enlargement of the male prostate gland. Studies show isoflavones slowed prostate cancer growth and caused prostate cancer cells to die. Isoflavones act against cancer cells in a way similar to many common cancer-treating drugs.

Bone Health

Soy isoflavones help in preservation of the bone substance and fight osteoporosis. This is the reason why people in China and Japan very rarely have osteoporosis, despite their low consumption of dairy products, whereas in Europe and North America the contrary happens. Unlike estrogen, which helps prevent the destruction of bone, evidence suggests that isoflavones may also assist in creating new bone. Other studies are not entirely consistent, but evidence suggests that genistein and other soy isoflavones can help prevent osteoporosis.

Cancer Risk

Isoflavones act against cancer cells in a way similar to many common cancer-treating drugs. Population-based studies show a strong association between consumption of isoflavones and a reduced risk of breast and endometrial cancer. Women who ate the most soy products and other foods rich in isoflavones reduced their risk of endometrial cancer by 54 per cent. A recent study has demonstrated that isoflavones have potent antioxidant properties, comparable to that of vitamin E. The anti-oxidant powers of isoflavones can reduce the long-term risk of cancer by preventing free radical damage to DNA.

Vegetable Fats and Oils/ω-Fatty Acids

The quantity and quality of fat in the diet play a critical role in maintaining human health. Several studies have directly implicated the amount and type of fat intake to specific diseases, such as CVDs, hypercholesterolemia, cancer, high blood pressure and obesity. Many components naturally present in vegetable oils have been shown to have beneficial properties. Once isolated and concentrated, a number of these compounds have proven effective in treating a wide range of conditions ranging from irritable bowel syndrome to chronic liver disease. Similarly, many of the fatty acids and other compounds present in vegetable oils have long been known to benefit our health. There is clearly great potential for developing value added dairy products by supplementation of milk fat with vegetable oils.

The number of active ingredients so far identified in oil seeds is impressive. Vegetable oils are a major dietary source of vitamin E, a powerful antioxidant. Linoleic acid, found in several commonly used vegetable oils at high concentration, is a polyunsaturated fatty acid (PUFA) with cholesterol-lowering properties and linolenic acid is also linked to heart health. Linoleic acid is the active ingredient in castor oil and is a powerful stimulant laxative, whilst linolenic acid provides the main benefits of evening primrose oil, used among other things, to treat breast pain and atopic eczema.

Many phytosterols are found in vegetable oils, particularly germ oils. Margarines fortified with sterols have recently hit the headlines because their cholesterol lowering capacity is as effective as many drugs. It is now also suggested that natural levels of phytosterols found in many vegetable oils (maize oil: 968 mg/100g, wheat germ oil: 553 mg/100g and olive oil: 221mg/100g) may also make a significant contribution to cholesterol lowering. Plenty of other beneficial compounds are also extracted and concentrated from by-products of the refining process including β-carotene, Vitamin K, phosphatidylcholine, which is used in the treatment of liver conditions, phophatidylserine, used to prevent brain deterioration.

Functional Possibilities

Since many compounds in oil seeds already have proven nutritional benefits, there are great possibilities for using them to develop new functional vegetable oils. Vegetable oils containing enhanced levels of beneficial active ingredients could have a substantial impact on human health. In fact in Japan this is already happening and oils are now available with improved levels of vitamin E and phytosterols.

Polyunsaturated fatty acids form several double bonds between several carbon atoms. They remain liquid at room temperature, or when chilled. Polyunsaturated fat is an essential element in our diet because it includes the essential fatty acids called ω-3 and ω-6. They are essential because the human body cannot synthesize them, and must obtain them in the diet. Healthy food sources include: vegetable oils like safflower, canola (contains ω-3) corn, sunflower, flaxseed (contains ω-3) and soybean. Polyunsaturated fatty acids are better than saturated fatty acids because (like monosaturated fats) they lower LDL cholesterol. However, diet research indicates that (unlike monounsaturated fats), polyunsaturated fat also lowers HDL cholesterol. Polyunsaturated fat helps to reduce LDL cholesterol, thus benefiting heart health. When consumed in the correct ratio (3:1 ω-6 to ω-3), polyunsaturated essential fatty acids offer a wide range of specific benefits.

A monunsaturated fatty acid forms one double bond between two carbon atoms. It is liquid at room temperature. With the exception of essential fatty acids, monounsaturated fat is probably the healthiest type of dietary fat, although it should be consumed in small amounts. Monounsaturated fats have none of the adverse effects associated with saturated fats, hydrogenated fat, trans-fats or ω-6 polyunsaturated vegetable oils. Healthy food sources include: olive oil (73 per cent), canola/rapeseed oil (60 per cent), hazelnuts (50 per cent), almonds (35 per cent), Brazil nuts (26 per cent), cashews (28 per cent), avocado (12 per cent), sesame seeds (20 per cent), pumpkin seeds (16 per cent), etc. Monounsaturated fat is believed to lower cholesterol and may assist in reducing heart disease. For example, the high consumption of olive oil in Mediterranean countries is considered to be one of the reasons why these countries have lower levels of heart disease. It is also believed to offer protection against certain cancers, like breast cancer and colon cancer. It is also high in Vitamin E, the antioxidant vitamin, which is usually in short supply in many Western diets. Cold pressed extra virgin olive oil, if not over-heated, contains phytochemicals and phenols, which help to boost immunity and maintain good health.

Essential Fatty Acids

There are two basic types of essential fatty acids (EFAs): linoleic acid (ω-6) and α-linolenic acid (ω-3), both belonging to the PUFA family. When eaten in the right amounts (3 grams ω-6, to one gram of ω-3), these EFAs have powerful healing properties and may assist weight control. There are two important things to know about the health and nutritional effects of ω-6 and ω-3 essential fatty acids. Consumption of too much ω-6 can block these benefits. A healthy balanced diet plan must include the correct balance of ω-6 to ω-3.

Essential fatty acids have a huge range of nutritional and healths benefits, and are active throughout the body. First and foremost, they are oxygen 'magnets'. They help to transport oxygen from our red blood cells to our individual cells. In addition, they enhance growth in the body, increase metabolic rate and help to reduce elevated blood cholesterol. They regulate glandular processes, chromosome activity, and boost immunity to disease as well as resistance to infection. According to *Udo Erasmus*, a world authority on dietary fats, approximately 3-6 grams a day of EFAs is sufficient for most people. The richest natural source of ω-6 is safflower oil. Other good sources include sunflower, hemp and wheatgerm oil. The ω-3 is much less widely available, the best natural source being flax seed oil. This is also found in oily fish, such as mackerel, herring, salmon and tuna. Oily fish contain the ω-3 fatty acids- eicosapentaenoic acid (EPA) and docosahexaenoic acid (DHA).

ω-3 Fatty Acid

The main type of ω-3 fatty acid is α-linolenic acid and it is the essential fatty acid in shortest supply. According to dietary experts, our consumption of this fatty acid has shrunk to one sixth of 1850 levels. The best dietary sources of omega-3 essential fats include: flax seed oil (linseed oil) - the richest natural source, hemp oil, canola oil, pumpkin seeds, walnuts, etc. Oily fishes like salmon, herring, mackerel, sardines, eels and tuna, are another rich source of ω-3 essential fatty acids, namely EPA and DHA. Evidence indicates that a diet rich in ω—3 fatty acids helps with weight loss. First, these fats help regulate the body's blood sugar levels, which helps keep hunger at bay. Second, they raise our metabolic rate thus burning more calories. In the long term, it is believed that a diet rich in ω-3 might lower the risk of diabetes and obesity.

The ω-3 fatty acids appear to slow down tumour growth in humans and also help to prevent new tumours from starting/spreading. Health experts believe this is because ω-3s boost our immune system, which stops pre-cancerous cells from developing. Supplements of ω-3 have also been shown to improve the symptoms of rheumatoid arthritis. They may also reduce the risk of other inflammatory diseases, such as asthma and gingivitis.

ω-6 Fatty Acid

The main type of ω-6 fatty acid is linoleic acid. This is the essential fatty acid in greatest supply. According to diet experts, our consumption of linoleic acid has doubled from what it was in 1940. Excess intake of ω-6 can cause increased water retention, raised blood pressure and increased risk of blood clotting. The ω-6 and ω-3 EFAs are best consumed in a ratio of about 3:1. However, the average Western diet provides between 10-20 grams of ω-6 to each gram of ω-3, which is not good for health. The best dietary sources of omega-6 essential fats include: safflower oil- the richest natural source, sunflower oil- corn oil, sesame oil, hemp oil-best balance of ω-6 to ω-3), pumpkin oil, soybean oil, walnut oil, wheatgerm oil, evening primrose oil, etc.

The ω-6s and ω-3s help to regulate blood glucose levels by increasing insulin sensitivity. This helps to keep hunger at bay and may lead to a lower risk of both diabetes and obesity. They appear to relieve depression and improve mood. The ω-3 from fish oils seems to lower the risk of Alzheimer's disease and may also benefit conditions, such as attention deficit disorder, dyslexia and dyspraxia. The ω-6s and ω-3s are known to benefit healthy brain development in unborn children and infants. The ω-3s are also believed to benefit skin complaints, such as eczema and psoriasis.

Rice Bran Oil (RBO)

Studies have shown that RBO in the diet significantly reduces LDL cholesterol and triglycerides; it increases HDL cholesterol, inhibits platelet aggregation and prevents CVDs. Clinical studies have confirmed these results and named RBO as 'Health Oil'. In every 1 per cent reduction in cholesterol, there was a 2 per cent decrease in the risk of coronary heart disease (CHD). Thus RBO in the diet significantly reduces cholesterol without any side effects known to exist with pharmaceutical drugs and is the healthiest of all oils for human consumption.

Coconut Oil

Coconut oil has a unique role in the diet as an important physiologically functional food. The health and nutritional benefits that can be derived from consuming coconut oil have been recognized in many parts of the world for centuries. Additionally, coconut oil provides a source of antimicrobial lipid for individuals with compromised immune systems and is a nonpromoting fat with respect to chemical carcinogenesis. Perhaps more important than any effect of coconut oil on serum cholesterol is the additional effect of coconut oil on the disease fighting capability of the animal or person consuming the coconut oil.

Olive Oil

Olive oil is a natural juice, which preserves the taste, aroma, vitamins and properties of the olive fruit. It is the only vegetable oil that can be consumed as it is freshly pressed from the fruit. The beneficial health effects of olive oil are due to both its high content of monounsaturated fatty acids and its high content of antioxidative substances. Studies have shown that olive oil offers protection against heart disease by controlling LDL cholesterol levels while raising HDL levels. No other naturally produced oil has, as large, an amount of monounsaturated as olive oil -mainly oleic acid. Olive oil is also very well tolerated by the stomach. In fact, its protective function has a beneficial effect on ulcers and gastritis. It activates the secretion of bile and pancreatic hormones much more naturally than prescribed drugs. Consequently, it lowers the incidence of gallstone formation. Studies have shown that people who consumed 25 ml-about 2 tablespoons- of virgin olive oil daily for one week showed less oxidation of LDL cholesterol and higher levels of antioxidant compounds, particularly phenols, in the blood.

While all types of olive oil are sources of monounsaturated fat, extra virgin olive oil, from the first pressing of the olives, contains higher levels of antioxidants, particularly vitamin E and phenols, because it is less processed. Olive oil is clearly one of the good oils, one of the healing fats. Most people do quite well with it since it does not upset the critical ω-6 to ω-3 ratio and most of the fatty acids in olive oil are actually ω-9 oil, which is monounsaturated. Spanish researchers suggest that including olive oil in diet may also offer benefits in terms of colon cancer prevention.

Phytosterols

Sterols, a group of compounds, are alcoholic derivatives of cyclo-pentano-perhydro-phenanthrene and are an essential constituent of cell membranes in animals and plants. The sterol ring is common to all sterols; the differences are in the side chains. Phyto- sterols and stanol or their esters, belonging to the plant derived sterol group, are new bioactive and natural food ingredients having LDL cholesterol reducing property. Since, 1950s, plant sterols were recognized to lower serum cholesterol concentrations. These diminish the absorption of cholesterol from the gut by competing for the limited space for cholesterol in mixed micelles *i.e.* the "packages" in the intestinal lumen that deliver mixtures of lipids for absorption by the mucosal cells. These are occurred naturally in grain, nuts, maize, soya, vegetable oils, seeds, wood pulp, leaves etc. Structurally, these are very similar to the cholesterol which is exclusively an animal sterol. Over 40 plant sterols have been identified among these β-sitosterol, campesterol, and stigmasterol are the most abundant and are the hydrogenated forms are sitostanol and campestanol. Some plant sterols currently available are saturated, to form the stanol derivatives, sitostanol and campestanol, which after esterification form stanol esters. Stanols are saturated sterols and they have no double bonds in the sterol ring. Stanols are less abundant in nature than sterols. Plant stanols are produced by hydrogenating the sterols. Plant stanols pass into the intestine and prevent the absorption of cholesterol in the gastrointestinal tract and thus reduced serum cholesterol despite the compensatory increase in cholesterol synthesis, which occurs in the liver and other tissues. Like cholesterol plant sterols are potentially atherogenic but atherogenesis does not take place because so little of the plant sterols are absorbed *i.e.*, about 5 per cent of β-sitosterol, 15 per cent of campesterol, and less than 1 per cent of dietary stanols are absorbed.

The use of plant sterols as cholesterol lowering drugs has been limited; initially the market was small and later the greater efficacy of statins was evident. Soybean sterols work as part of the normal digestive process to help block absorption of dietary cholesterol. Clinical studies suggest that the consumption of about 2 to3 g/d of plant sterols/stanols decreases LDL cholesterol levels by between 9 and 20 per pent with little or no effect on HDL cholesterol or triglyceride levels and varies amongst individuals. The decline in LDL cholesterol levels has been found in both adults and children with hypercholesterolaemia, in those with normal blood cholesterol levels, in people with Type II diabetes and in postmenopausal women with CHD. FDA also reviewed phytostanol esters and determined that providing 3.4 g of phytostanol esters may reduce the risk of heart disease when used as part of a diet low in saturated fat and cholesterol (Anon, 2004b). The National Cholesterol Education Program guidelines released by the Heart, Lung, and Blood Institute, USA on May 15, 2001, encourage the use of phytostanol esters as part of the Therapeutic Lifestyle Changes (TLC) approach. The specific plant sterols and stanols, extracted from soybean oil or tall (pine tree) oil, may be incorporated into foods intended to lower blood cholesterol levels and in near future, additional sources will be available. The plant sterols or stanols, currently incorporated into foods, are esterified to unsaturated fatty acids which cause increased solubility of sterol or stanol esters in fats and lipids and allow maximal incorporation into a limited amount of fat. *Benecol* was the first fortified margarine.

Cereals

Humans have been enjoying grain foods for at least past 10,000 years. Grain foods, which include cereals, are dietary staples for many cultures around the world. Current research around the world is discovering the many and varied health benefits of cereal foods particularly in reducing the risk of diseases such as coronary heart disease, breast or colon cancers, etc. Common cereal foods include bread, breakfast cereals, cereal grains (such as oats, rice and barley), crackers, flours, pasta, etc. Wholegrain cereals contain various phytochemicals having health beneficial effects. These include, (i) Lignans - a phytoestrogen that can lower the risk of coronary heart disease, and regress or slow cancers in animals, (ii) Phytic acid - reduces the glycaemic index of food, which is important for people with diabetes, and helps protect against the development of cancer cells in the colon, (iii) Saponins, phytosterols, squalene, oryzanol and tocotrienols - have been found to lower blood cholesterol, (iv) Phenolic compounds - have antioxidant effects, etc.

Cereal fibre offers greater protection against the risk of heart attack than the fruits and vegetables fibre. Regular consumption of cereals, rich in soluble fibre, such as oats and psyllium, reduces the amount of cholesterol circulating in the bloodstream. For example, ingestion of just 3 gm of soluble fibre from oatbran lowers the blood cholesterol by as much as 2 per cent. Regular intake of whole-grain foods is associated with a reduction of about 26 per cent of coronary heart disease risk. In general, soluble fiber, such as that found in oats, is most often linked to reductions in cholesterol levels. However, diets with higher insoluble fiber, which is found in whole grains and vegetables and is mostly, unrelated to cholesterol levels, have been reported to correlate better with protection against heart disease in both men and women. A 20 gram of additional dietary fiber per day for several months' successfully lowers cholesterol. Whole grains (such as rye, brown rice, and whole wheat) contain high amounts of insoluble fiber—the type of fiber some scientists believe may help to protect against a variety of cancers. People who eat relatively high amounts of whole grains were reported to have low risks of lymphomas and cancers of the pancreas, stomach, colon, rectum, breast, uterus, mouth, throat, liver, and thyroid. Most research on the relationship between cancer and fiber has focused on breast and colon cancers. Consuming a diet, high in insoluble fiber is best achieved by switching from white rice to brown rice and from bakery goods made with white flour or mixed flours to 100 per cent-whole-wheat bread, whole-rye crackers, and whole-grain pancake mixes. Refined white flour is generally listed on food packaging labels as "flour," "enriched flour," "unbleached flour," "durum wheat," "semolina," or "white flour." Breads containing only whole wheat are often labeled "100 per cent whole wheat." A high-fiber diet, particularly soluble fiber (oats, psyllium seeds, fruits, vegetables, and legumes), is associated with decreased risk of both fatal and nonfatal heart attacks, probably because these fibers lower cholesterol levels. However, large trials separately studying men and women who were followed for years have linked the greatest protection to insoluble fiber (from whole grains, breads, and cereals), but the actual reason is remain to elusive. In spite of the details are better understood, doctors often recommend increasing intake of fruits, vegetables, beans, oats, and whole grains to maintain good health.

Eating of foods that are slowly digested and high in soluble fibre might reduce the risk of developing non-insulin dependent diabetes. Cereal fibre has been shown to have this type of activity, hence it is preferable for diabetics to consume wholegrain cereal products rather than refined cereals having higher glycaemic index. People who are obese tend to have energy-dense diets. High fibre foods, such as wholegrain breads and cereals, can be an effective part of any weight loss program as these require longer digestion period and thus create a feeling of fullness, which discourages overeating and also increase movement of food through the digestive tract. The result is increased stool bulk, softer, larger stools and more frequent bowel action and increased bowel action provides a good environment for beneficial bacteria and reducing the levels of harmful bacteria. There are many health benefits that have been linked to a diet high in grain foods as follows:

- ☆ A reduced risk of many different types of cancers, including those of the colon, stomach and breast.
- ☆ A strengthened immune system, because wholegrain cereals are high in vitamin E, zinc and certain phytochemicals.
- ☆ A reduction in the incidence of rectal polyps, particularly if oatbran is eaten regularly.
- ☆ Protection against the development of diverticular disease, which is characterised by herniated pockets in the intestines.

Fruits and Vegetables

For nearly a century, fruits and vegetables have been recognized as a good source of vitamins and minerals. They have been especially valuable for their ability to prevent vitamin deficiencies. Research of the past 20 years has shown that fruits and vegetables not only prevent malnutrition but also help in maintaining optimum health through a host of chemical components that are still being identified, tested, and measured. Fruits and vegetables are an important source of vitamins, minerals, flavonoids - plant chemicals that act like antioxidants, saponins - plant chemicals that have a bitter taste, phenols - organic compounds in foods, carotenoids - vitamin A-like compounds, isothiocyanates -sulfur-containing compounds, several types of dietary fiber ability to prevent vitamin C and vitamin A deficiencies. Eating fruits and vegetables rich in antioxidants translates into a lower incidence of cataracts-a clouding of the eye's lens that impairs vision. It is a consensus in the scientific field that a higher consumption of these healthy foods is associated with a reduced risk of myriad diseases and early aging. Works are now in progress on understanding the exact role of whole foods in health protection, and identification of compounds present in these foods that appear to be critical for prevention of chronic diseases. Among these, antioxidants, phytoestrogens, dietary fibers and resistant starches concentrated in foods such as fruits, vegetables studied well.

Research from the United States, United Kingdom, and The Netherlands suggests that the role of fruits and vegetables in preventing heart disease is a protective one. Risk reduction was estimated as high as 20-40 per cent among individuals who consumed substantial amounts of fruits and vegetables. People who were already diagnosed with coronary heart disease were able to reduce blockage modestly through exercise and an extremely low-fat, vegan-like diet rich in fruits and vegetables. A review by the World Cancer Research Fund and the American Institute for Cancer Research concluded in 1997 that "diets containing substantial and varied amounts of fruits and vegetables could prevent 20 per cent or more of all cases of cancer" The strongest evidence relates to stomach and lung cancer. Other areas that show convincing results are the mouth, pharynx, esophagus, colon, and rectum. Studies involving patients who were taking dietary supplements in place of fruits and vegetables were ended early due to a higher mortality rate among the supplement users. Researchers concluded that dietary supplements do not have the same positive effects as eating real fruits and vegetables.

Fruits and vegetables are high in cellulose—a type of insoluble fiber. Five studies have reported that high fruit and vegetable intake can reduce the risk of a stroke by up to 25 per cent. A 1997 study of 459 men and women found a high intake of fruits and vegetables could lower blood pressure in individuals with either high or normal blood pressure. The experimental diet included 8 - 10 daily servings of fruits and vegetables combined as well as low-fat dairy products. Diets that are high in fiber may be able to help in the management of diabetes.

Delayed development of cataracts is another beneficial effect of fruits and vegetables as indicated by some epidemiological reports. A fivefold reduction in cataract risk was found for individuals who consumed a minimum of 1 ½ servings of fruits and vegetables each day. Carotenoid- rich fruits and vegetables containing zeaxanthin and lutein proved the most beneficial because not all carotenoids offer equal protection. Examples of fruits and vegetables that contain lutein and zeaxanthin are spinach, collards, kale and sweet corn. Supplements of β-carotene did not reduce cataract risk. In one research study, asthmatic children in Great Britain who consumed fruit more than once a day had better lung function. The higher intake of fruits and vegetables seemed to increase the ventilation function of the lungs.

Other Phytonutrients including Natural Antioxidants

Natural antioxidants are the plant substances that protect the body by neutralizing free radicals, or unstable oxygen molecules, which can damage cells and lead to poor health. The most commonly known nutrients antioxidant are water and fat-soluble vitamins. Currently, the most important benefit claimed for vitamins A, C, E, and many of the carotenoids and phytochemicals is their role as *antioxidants*, which are scavengers of particles known as *oxygen-free radicals* (also sometimes called *oxidants*). These chemically active particles are by-products of many of the body's normal bio-chemical processes and their numbers are increased by environmental assaults, such as smoking, chemicals, toxins, stresses, etc. At higher levels, oxidants are very harmful. Free radicals and unstable oxygen species are known to promote the development of atherosclerosis, cancer, arthritis, diabetes and a host of other conditions. Therefore, we need more than just a high amount of a single antioxidants provided by a varied whole foods diet to keep the reactive oxygen species like free radicals "in check". Deficiencies in Vitamins A, C, E, and β-carotene have been linked to heart disease. These have antioxidant effects and other properties that should benefit the heart. However, a number of studies have found no reductions in heart disease in people who have taken antioxidant vitamins. A high intake of fruits and vegetables containing beta carotene, lycopene, and other carotenoids may reduce the risk of heart attack. For example, lycopene-poor diets (particularly lycopene in tomatoes) were associated with a significantly reduce the risk of heart disease and stroke and lutein having protective action against the early hardening of the arteries. The effects of antioxidant vitamins and carotenoids on stroke, dementia, or both are being studied. A very important study in 2001, reported no protection stroke with vitamins A, E or β-carotene. The vitamin B, folate (usually in the form of folic acid), may protect against stroke. People who have higher blood levels of folate have a lower than average risk for stroke. Its primary benefit in this case appears to be to reduce levels of homocysteine, an amino acid that has been strongly linked to an increased risk of coronary artery disease, stroke, and Alzheimer's disease. A major 2002 study suggested that lowering homocysteine levels with folic acid would reduce the risk for heart disease by 16per cent and stroke by 24 per cent.

Many fresh fruits and vegetables contain chemicals that may fight many cancers, including lung, breast, colon, and prostate cancers. Important cancer fighting foods include the cruciferous vegetables (*e.g.*, cabbage, Brussels sprouts, broccoli), tomatoes (which contain lycopene), carrots (which contain α-carotene), etc. There is some evidence that antioxidants may enhance the anticancer effects of chemotherapy. In a 2000 study, patients who maintained their antioxidant levels were better able to withstand the high stress caused by chemotherapy compared to those with low antioxidant levels. Antioxidant nutrients that may have properties that may help reduce the side effects of chemotherapy include vitamins E and C, β-carotene, genistein and daidzein (isoflavones found in soy), and quercetin (found in red wine a purple grape juice). Any protective effects of vitamins or specific phytochemical against cancer, however, appear to depend on the cooperative effort among them. Individual supplements of any vitamin or food chemical have not as yet shown any benefits. Lycopene, found in tomatoes, may have particular value in protection against prostate, colon, lung, and bladder cancer. Individual supplements, however, do not offer any advantage. In fact, evidence now strongly suggests that β-carotene supplements increase the risk for lung cancer in smokers.

Flavonoids in both black and green tea, dark chocolate, onions, red wine or red grape juice, and apples, appear to be strongly heart protective. In a recent study, people who consumed the most flavonoids in foods had a 20 per cent lower risk for heart disease. Flavonoids may protect against damage by cholesterol and help to prevent blood clots. A number of studies have now reported the heart protection from the flavonoid catechin, which is found in both black and green tea. Studies on tea-drinking however have been mixed. For example, the British consume a lot of tea but have high rates of heart disease. The flavonoid resveratrol, found in grape skin, appears to be responsible for the well-known heart protective effects in red wine and purple grape juice. A glass or two of red wine a day may be healthful. For people who cannot drink alcohol, juice from red grapes may be beneficial.

Global Scenario of Value Added/Functional Food Industry

Functional foods, as defined by the Institute of Medicine in Washington, are "those foods that encompass potentially healthful products including any modified food or ingredient that may provide a health benefit beyond the traditional nutrients it contains." Functional foods can include foods like cereals, breads and beverages which are fortified with vitamins, herbs or nutraceuticals. The genesis of the functional foods industry has occurred for a number of reasons. Today, consumers are aware of the possible positive role of diet that can play in disease risk management. In addition, the regulatory bodies have become increasingly cognizant and supportive of the public health benefits of the functional foods. In Western countries, governments looking at regulatory issues for functional foods are more aware of the economic potential of these products as part of public health saving strategies. Functional food market is a dynamic one and offers outstanding prospects for growth for well positioned food and drink manufacturers. During the period of 1998-2003, sale values of functional foods increased by a strong 60 per cent, and are set to rise by a further 40 per cent over the period of 2003-2008 (Anon, 2004). Japan is currently the global leader in production and development of functional foods. The value of the Japanese market for functional foods was estimated to be $ 3.0 billion corresponding to 5 per cent of the total processed foods marketed in Japan (PA Consulting Group, 1990). Another estimation prognosticated that the Japanese market for functional foods would grow at 8.5 per cent annually reaching a level of $ 4.5 billion by 1995 (Weitz, 1991). More than 100 companies are actively selling or developing functional foods in Japan. The market for functional foods in the United States was projected to reach a level of $ 7.5 to 9.0 billion by 1995, which was equivalent to an annual growth rate of 17-20 per cent (Weitz, 1991). The share of European market for functional food was estimated to approximately $ 8.52 billion in 1987 that was about 49 per cent of the US Market (PA Consulting Group, 1990).

The manufacture of functional foods has already captured consumer interest worldwide. The worth of the present global functional food industry was estimated to be $ 33 billion with a steady growth rate of 15-20 per cent per year (Hilliam, 2000). Total sales of functional foods increased by a record 29.4 per cent by value in 2000/2001 due to several major launches and relaunches (Anon, 2001). Japan has allowed more than 200 functional foods to be marketed under existing FOSHU (Foods for Specialized Health Use) legislation (Anon, 1998) and the United States where the Food and Drug Administration (FDA) permits health claims to be made for about 15 categories of food (Anon, 1999). Many products, such as *Yakult* and *Actimel Yoghurt Drinks, Benecol* and *Flora Proactiv Margarines, Tropicana Pure Premium, Calcium and Multivitamin* and *Juice Up Fruit Juices*, and *Onken, BioActivia, Müller Vitality Yoghurt,* etc. have become major functional food products. Probiotic yoghurts / fermented-milk drinks now represent 15-20 per cent of total yoghurt sales by value, and cholesterol-lowering margarines account for 10 per cent of total sales of margarines and spreads. Functional products are estimated to account for some 20 per cent of total breakfast cereal sales (Anon, 2001). One decade earlier, there were more than 180 medical foods available in the international market and produced by about 20 company worldwide (Hatten and Mackery, 1990). Estimation by the Federated Societies for Experimental Biology (Anon, 1992) showed that there were over 5 million patients on enteral medical foods in the United States. These foods comprise essentially sterile solutions or rehydratable powders with a value exceeding $ 5 billion.

Functional foods incorporating probiotic bacteria with scientifically supported health claims have great potential for improving quality of life and are widely predicted to become one of the biggest dietary trends of the present decade. Currently, there are over 70 bifidus and acidophilus-containing products produced worldwide, including sour cream, buttermilk, yoghurt, and powdered milk and frozen desserts. Out of 90 commercial probiotic products available worldwide, more than 53 different types of dairy products containing probiotic organisms are marketed in Japan alone (Hilliam, 2000). The probiotic food industry is flourishing with the European probiotic yoghurt market alone currently estimated to be worth around £520 million pound (Shortt, 1999). In many European countries, most notably France and Germany, the market is expanding with the result that probiotic yoghurts now account for over 10 per cent of all yoghurts sold in Europe (Stanton *et al.*, 2001).

Scope of Value Addition to Indigenous Dairy Products

Indian milk products have been in the news and are poised for explosive growth due to increased industrial production. The white revolution has already made India the world leader in milk production. Its second phase is balanced to transform the country into a global giant in the output of dairy products. This transformation will come about through the industrial production of desi milk-based products, including the vast range of *mithais* for which huge unmet demand exists in domestic and overseas markets. Today, the market for mass-produced and packaged indigenous milk-based products is expanding fast at both the domestic and global level. Even in the present day, the domestic market of ethnic milk products is estimated at Rs 5,000 crores. The diversity in Indian dairy products is relatively greater, offering the consumers as well as the manufacturing industry a wider choice. Many of the mass consumed products have already been commercialised by the organised sector. These include flavoured milks, *dahi, paneer, buttermilk, lassi, gulabjamun, shrikhand* and *kheer.* With rising income levels and the growing size of the middle and upper-middle class in urban and semi-urban centres, this market is bound to grow rapidly.

As per a recent CII-McKinsey report, the worth of the Indian food industry has gone up from Rs.3.09 trillion in 1993-94 to Rs.3.99 trillion in 2000-01. As a part of the agriculture sector, the value addition segment has recorded a 7.1 per cent growth in the last seven years, compared to 3.1 per cent for farm and livestock segment. The Food Processing Industry sector in India has been accorded high priority by the Government of India, with a number of fiscal relief and incentives, to encourage commercialization and value addition to agricultural produce. As per a study conducted by McKinsey and Confederation of Indian Industry (CII), the turnover of the total food market is approximately Rs.250, 000 crores (US $ 69.4 billion) out of which value-added food products comprise Rs.80, 000 crores (US $ 22.2 billion).

Since the liberalization in August, up-till February 2000 proposals for projects of over Rs.53, 800 crores (US.13.4 billion) have been proposed in various segments of the food and agro-processing industry. Besides this, the Government has also approved proposals for joint ventures, foreign collaboration, industrial licenses and 100 per cent export oriented units envisaging an investment of Rs.19,100 crores (US $ 4.80 billion) during the same period. Out of this, foreign investment is over Rs. 9100 crores (US $ 18.2 billion).

Value Addition

There is a phenomenal scope for innovations in indigenous product development, packaging and presentation. Given below are potential areas of value addition:

- ☆ Steps should be taken to introduce value-added products like *shrikhand, kulfi, paneer, khoa,* flavoured milk, dairy sweets, etc. This will lead to a greater presence and flexibility in the market place along with opportunities in the field of brand building.
- ☆ Addition of cultured products like *dahi, lassi,* etc. lend further strength - both in terms of utilization of resources and presence in the market place.
- ☆ A lateral view opens up opportunities in milk proteins through casein, caseinates and other dietary proteins, further opening up export opportunities.
- ☆ Another aspect can be the addition of infant foods, geriatric foods and nutritionals.

When increasing globalization is changing the ways in which consumers are looking at food and are constantly looking for new products and flavours, ethnic dairy delicacies would play an important role and over value addition, trading with these products would get a marketing debut.

In the recent years, much attention has been focused on the possible protective role of dietary fibres, natural isoflavone as well as other phytonutrients. One of the novel challenges that our food manufacturers can take is to develop value added dairy products, such as *dahi, shrikhand, lassi, paneer, kheer, rossmalai,* etc with such functional ingredients to achieve specific health benefits without hampering the normal body and texture as well as flavour of the products. Development of such value added foods with specific health benefits may shows great potential for food manufactures of our country and may continue to gain unprecedented momentum without facing any complex processing challenges. Therefore, different types of bioactive ingredients from plants, animals and microbial sources can be incorporated accordingly into the different types of indigenous milk products to improve their functional properties.

Value Addition to Indigenous Dairy Products Using Plant Sterols and Stanols: Scenario and Futuristic Vision

Cholesterol-lowering phytostanols and phytosterols or their esters have recently appeared in various foodstuffs, particularly margarines, yoghurts and salad dressings A new polyunsaturated margarine with added plant stanols, *'Benecol'* was introduced in several European countries in 1999 and a similar margarine with added plant sterols is waited to be introduced under the Flora label. Now, McNeil PPC in the US has filed a patent for the use of these ingredients in confectionery (Davis, 2001). Particularly confectionery represents an ideal vehicle for these health-promoting ingredients, as small amounts can be eaten at various times during the day. Of particular interest, these ingredients can be used to replace a proportion of the fat in chocolate, making a product that has not only a lower fat content, and also the added benefit of a cholesterol-lowering action. Incorporation of phytosterol and phytostanol esters into confectionery can also improve organoleptic properties of the product, resulting in a softer texture, faster melt-away and decreased adhesion to teeth. During processing, also acts as a lubricant, facilitating cutting, and imparting many of the properties traditionally achieved by including fats in the formulation. As well as in chocolate, phytosterols and phytostanols can be incorporated into a wide variety of confectionery, including toffee, nougat, fudge, caramels, hard candies and biscuit fillings.

Already, there is a precedent for such fortification, for instance in the United States; folic acid has been added to flour since 1997. In addition to the expected reduction in the incidence of neural tube defects, there has also been a significant reduction in the average serum concentration of homocysteine (Jacques *et al.,* 1999), which is likely to reduce mortality from heart disease. The *Benecol* yoghurt drink has been launched in European markets and can be integrated in the daily diet simply and easily. To achieve the optimum effect, it is sufficient to take one daily portion in the conveniently sized 65 ml bottle for maintaining a healthy heart. Several studies have used yoghurt with added steryl esters or free sterols. The tests demonstrated that consumption of sterol-supplemented yoghurts significantly lowers serum cholesterol.

Two new cholesterol-lowering margarines have been approved by the U.S. Food and Drug Administration as foods *'Benecol'* (McNeil Consumer Health Care) containing hydrogenated sterols, primarily sitostanol derived from pine tree wood pulp and *'Take Control®'* (Unilever, the parent company of Unilever Bestfoods and the world's leading manufacturer of margarines and spreads, with category expertise extending back more than seven decades) containing unsaturated sterols, primarily sitosterol from soybean oil. *Take Control®* is a delicious buttery-tasting spread. The plant sterols in *Take Control®* spread occur in low levels in fruits, vegetables, vegetable oils and margarines. The spread *Take Control®* actively reduces LDL cholesterol by 10 to 15 per cent in both patients who took cholesterol-lowering medication and in those who did not but HDL cholesterol levels remained unchanged (Neil *et al.,* 2001).

With the original technology developed in Finland, Europe has taken the lead on fortifying dairy products with stanol esters. More than 30 products have debuted in the world market since 1999, from cream cheese and process cheese, to yoghurt and yoghurt drinks, to butter and milk. One of the newer products on the market in Portugal, Spain and Switzerland is drinkable yoghurt with a full day's dose of stanols in one serving. There is also good potential to incorporate plant derived sterols or stanols in *paneer*, ice cream, cheese spread, Mozzarella cheese, Cottage cheese, *dahi*, *lassi*, dairy based candy, other fermented drinks, butter, cream cheese or in butter like spreads as well as in many of our traditional milk products to enrich them with the health attributes of these ingredients.

Future Research Goals

One of the novel challenges is to develop bioactive foods with functional ingredients to achieve specific health benefits without hampering the normal body and texture as well as flavour of the product. Development of functional dairy products with specific health benefits of phyto-sterols and stanols or other bioactive ingredients may shows great potential for food manufactures in our country and may continue to gain unprecedented momentum without facing any complex processing challenges.

Among the many health predictions for the new millennium, the most alarming is that of CVD- heart disease and stroke-topping the list for death and disability in our country. The great increase in rates of CVDs in developing countries, including India, will probably have grave implications for south Asia, which houses nearly a quarter of the world's population. Because of the association of some phytochemicals with hypocholesterolemic effect claimed, food and pharmaceutical companies in India may harvest interest, as a part of exercising modern dietary strategy to reduce the risk of CVD, in exploiting such bioactive ingredients as an opportunity for product development and enrichment of some of our exiting milk products with improved health attributes like cholesterol lowering ability. India is producing a diverse sort of milk products. So, when the unique role of functional foods enriched with phyto-sterols and stanols or other functional ingredients in maintaining serum lipid profile in a healthy levels as well maintaining a sound health has been well established by the scientific communities worldwide, our country may get rid of a vast economic toil, to an extent, disbursed every year in the treatment diseases, if some of our existing milk products consumed everyday can be enriched with target oriented and specific functional attribute, like hypocholesterolemic effect, anticarcinogenic property, etc, by using specific functional ingredients as food additives.

Cholesterol levels are affected by many factors, including what we eat, how quickly our body produces and gets rid of LDL cholesterol, physiological conditions, and eating habits; therefore anyone can be affected and should consider adjusting their diet. Foods are most easily modifiable facet of lifestyle without much effort by the subject. So some of our exiting dairy products consumed regularly can be reformulated to suit the requirements of the vulnerable segment of our society. While the dairy products have long been associated with a high quality nutritional image, more recently research efforts devoted worldwide to the development of value added foods or reformulation of existing foods with demonstrated health attributes some bioactive ingredients may also shove the existing dairy products into the "value added foods" category. Given the increasingly competitive landscape of the worldwide market for value added foods, the dairy industries in our country may also try to derive benefits from a marketing advantage, such as added value bioactive ingredient-containing dairy products, which would afford a competitive edge over existing products.

The subject of fortification in the dairy products can be a touchy one to push some of our existing products into functional food category. Dairy products are of superior nutritionally, but at the same time dairy products can be an attractive and affordable vehicle for delivery of nutraceutical ingredients to health conscious consumers. Today, many options exist to our dairy industry for creating a wide range of products to meet the many special nutritional needs of the sick and aging population, as well as more general public-health needs. Now is the time to capitalize on consumer interest and demand for special products with nutritional value-added ingredients that taste good, are affordable, and can play a significant role in a healthy lifestyle.

There is good potential to incorporate functional ingredients in *dahi, lassi, shrikhand* and other fermented drinks as well as in many of our traditional milk sweets to enrich them with the health attributes of these ingredients. India is producing a diverse sort of fermented milks and other cultured dairy products, *e.g., dahi* market in India alone is worth around 52,00,000 tonnes a year or about 15,000 crores annually, roughly the same as milk market growth (Anon, 2004). The inherent therapeutic attributes of *dahi* are due to the presence of live lactic acid bacteria (LAB) and according to the Eminent Russian scientist Elie Metchnikoff, pioneer worker in the field of probiotics, the regular consumption of live LAB through the fermented dairy products improve the intestinal microbial balance and reduced putrefactive fermentation by harmful bacterial population. So, addition of selective health beneficial microbial supplements *i.e.,* probiotics into *dahi* and other ferment indigenous milk products will obviously improve their therapeutic attributes beyond their traditional health beneficial effects. Besides the fermented indigenous dairy products, there is a huge scope of addition of probiotics in traditional heat desiccated products like *burfi, peda,* etc. by incorporating microencapsulated probiotics that will ensure the better viability of the live microorganisms in these products.

Recent trend, towards the incorporation of probiotics selectively and prebiotics in combination, as prebiotics improve the growth and activities of probiotics and the products specifically termed as 'Synbiotics'. So along with probiotics, different prebiotics, available commercially including inulin, lactulose, galactooligosaccharide, fructooligosaccharides, etc can be successfully incorporated in the indigenous fermented milk products to make them 'Synbiotics'. Consequently our traditional cultured milks and milk products can be enriched with target oriented and specific probiotic attribute by using specific probiotic strains as adjunct culture. In fermented indigenous dairy products, there is huge scope of addition of prebiotics and dietary fibres for 'Synbiotic' formulation.

However by the 1980s, it was established that as naturally occurring plant sterols and stanols could be added to foods. As fats are needed to solubilize sterols, margarines are an ideal vehicle for them; although *paneer,* fat rich indigenous products like *kulfi, ghee, makhann,* etc can also be used as vehicle of this phytochemical. A considerable part of milk fat in *paneer* can be replaced with vegetable oil like soy, ground nut, sunflower, etc. without hampering the quality of the products and which will reduce of cholesterol in such high fat indigenous dairy products. With changing dietary fashions, the current emphasis on low-fat items, which witness the vast array of 'light' and fat-reduced products lining supermarket shelves, has given dietary fibre a back seat. Although people may pay less attention to fibre, its health benefits have not vanished. Fibre remains an essential nutrient and a vital part of healthy eating for everyone, including those with diabetes. In fact, soluble forms of plant fibre may help to mute blood sugar swings, which can be easily incorporated in a number of our traditional dairy products.

Future Strategy

Diet is most easily modifiable factor of lifestyle without much effort by the subject. So existing foods need to be reformulated to suit the requirements of the vulnerable section of our society. The future prospects for value added traditional dairy foods to play a major role in improving the health and vitality of our population seem very bright indeed. The freedom to control our own destiny in terms of health and vitality as we age is now within our grasp. While the path already is filled with unmet research needs, economic barriers, and regulatory hurdles, the compelling need to achieve an improved health span to our increased life span will ensure success. It has been recognized that the cost of disease treatment is increasing at a too rapid rate and the population is getting older, now health promotion strategies focused upon functional medicine gifted by nature must be implemented. By the time we enter into the new eon of health therapy by dietetic strategy, functional dairy products will be commonplace in most market places in the developed world. Someday is coming when we will see one known product from the list our traditional dairy products that has been enriched with certain phytochemicals, to occupy a place in our medical store, targeted for AIDS therapy. The days are not far when people will consume value added sweets containing such phytochemicals as their staple diet.

According to the latest statistics as published (IDF, 2000), the average per capita annual consumption of fermented milk products is 22 kg in Europe. In total, this amounts to about 8.5 billion kg fermented milk. With an average microbial content in these fermented products of 10^8 bacteria g^{-1}, this amounts to a total of 8.5×10^{20} LAB. Assuming one bacterial cell weighs 4×10^{-12} g, this means that 3400 tonnes of pure LAB cells are consumed every year in Europe. In India, a number of fermented dairy products have been consumed since the time antiquity as mentioned in our old Testaments. Of them, *dahi*/curd, *lassi, shrikhand,* etc. finds special popularity throughout the country. There is a vast scope to incorporate some health benefits to these products using probiotic and prebiotics. The *in situ* production of bioactive peptides in fermented dairy products, such as *dahi*, yoghurt*, shrikhand, acidophilus milk, lassi* and other fermented milk beverages has also now been conceptualized as a novel approach to improve the health value of these products.

India produces some 900,000 tonnes of *ghee,* valued at Rs. 85,000 million. The value of the resultant *lassi* is Rs. 25,000 million. Butter-milk is a by-product in the preparation of *makkhan.* It is estimated that about 55 kg of butter-milk is produced for every kg of ghee. While most of it is consumed by villagers and their families, some quantity is either given away or fed to cattle. The reason for this is lack of any market for it in rural areas. Butter-milk is rich in milk protein, antioxidant phospholipids and calcium, and forms a nutritive and refreshing beverage. This buttermilk can be used to prepare refreshing beverages and addition of functional ingredients would obviously fetch extra marketing advantages. Buttermilk or buttermilk solids itself can be incorporated to a certain level to the base materials used for making variety of dairy delicatessens like *rasgulla, sandesh, dahi, milk cake, burfi, shrikhand,* etc to add some health benefits to the existing products.

The value of *khoa* and *chhana* produced in India is probably twice the value of all milk handled by the organized sector in the country. However, the traditional dairy products sector in India, like its agricultural counterpart, is grossly undermanaged. It, however, may provide economic opportunities that even the Western dairy world would be envious of. The value of *khoa* and *chhana*-based sweets could possibly exceed Rs 130,000 million. Approximately 1,200,000 tonnes of *chhana,* valued at Rs 6,000 million is produced in India, which is used in manufacturing a diverse sort of traditional sweets. Another major milk product in common use is *khoa,* obtained by rapidly evaporating milk in shallow pans to a total solids of about 70 per cent and capable of being preserved as such for several days. It is used as an ingredient in making different kinds of traditional mithais (sweets), such as *peda, burfi* and *gulabjamun.* Some 900,000 tonnes of *khoa* valued at Rs 45,000 million is produced in the country. Production of *chhana* and *khoa* in organised sectors as well as their use in preparation of these widely consumed sweets would strengthen the traditional dairy product sector. Upon value addition using product specific functional ingredients like microencapsulated probiotics, prebiotics, fibre, phytochemicals, etc., the nutritional image and the demand of the products would further increase and they would find place in the overseas market.

While our traditional dairy products like *Paneer, dahi, lassi, shrikhand, rasgulla,* etc have long been associated with a high quality nutritional image, more recently research efforts devoted worldwide to the development of value added foods or reformulation of existing foods with demonstrated health attributes may also shove these products into the "functional foods" category. Development of value added indigenous dairy products with specific health benefits may shows great potential for manufactures in our country and may continue to gain unprecedented momentum in the days ahead. The Table 5.3 outlines the possibility of using different functional ingredients in a number of our traditional dairy products.

Thus, future strategy of value addition will solely dependent on the extensive studies with the addition of these possible functional ingredients and evaluation of the products in terms of technological feasibility, products characteristics during processing and storage, textural properties as well as the evaluation of health benefits by both animal and human trials. Researches in this field has been started in both institutions as well as the field levels like R&D sections of companies and the resultant effort of both lab and field levels will ultimate determine the fate of the value addition to the indigenous dairy products in development of new functional products, which will be able to compete the markets both in domestic and overseas.

Table 5.3: Possible Functional Ingredients for Value Addition of Indigenous Dairy Products

Indigenous Dairy Products	*Possible Ingredients for Value Addition*
Dahi	Probiotics, prebiotics, buttermilk solids, plant sterols and stanols, dietary fibres, fruits, bioactive peptides, ω-3 and ω-6 PUFA, natural antioxidants, etc.
Shrikhand	Probiotics, prebiotics, buttermilk solids, plant sterols and stanols, dietary fibres, fruits, bioactive peptides, ω-3 and ω-6 PUFA, phytochemicals, natural antioxidants, soy isoflavone, etc.
Paneer	Buttermilk solids, CLA, plant sterols and stanols, prebiotics, soy fibres, ω-3 and ω-6 PUFA, phytochemicals, natural antioxidants, etc.
Ghee	CLA, plant sterols, ω-3 and ω-6 PUFA, phytochemicals, natural antioxidants, etc.
Makhann	Plant sterols and stanols, ω-3 and ω-6 PUFA, phytochemicals, natural antioxidants, etc.
Kulfi	Probiotics, prebiotics, buttermilk solids, plant sterols and stanols, dietary fibres, fruits, ω-3 and ω-6 PUFA, phytochemicals, natural antioxidants, etc.
Khoa and khoa based products: Burfi, peda, gulabjamun	Plant sterols and stanols, buttermilk solids, microencapsulated prebiotics, dietary fibres, ω-3 and ω-6 PUFA, natural antioxidants, etc.
Channa and channa based sweete like sandesh	Plant sterols and stanols, prebiotics, dietary fibres, ω-3 and ω-6 PUFA, soya isoflavone, natural antioxidants, etc.
Lassi	Probiotics, prebiotics, Plant sterols and stanols, soya fibres, fruits, ω-3 and ω-6 PUFA, phytochemicals, etc.
Cultured butter milk, acidophilus milk, etc.	Probiotics, prebiotics, plant sterols, fibres, fruits, bioactive peptides, ω-3 and ω-6 PUFA, natural antioxidants, phytochemicals, etc.

Conclusion

The demand for value added foods is being driven by a growing public understanding of the linkage between diet and health, and the interest in self-health maintenance, rising healthcare costs and advances in food technology and nutrition. Growing health consciousness and awareness for healthy nutrition have increased consumer demand for foods of superior health quality. Increasingly, medical and nutritional researchers have been linking food components to disease prevention and health enhancement. Due to the today's upward consumer awareness and interest to follow healthy nutrition and dietary strategy in achieving health benefits from foods beyond their basic nutrition, the market for value added foods has expanded manifolds. Today's consumers are incresingly seeking functional foods for their health and well being as a means of nutritional intervention in disease prevention. Dairy products enriched with the health attributes of functional ingredients would be safe and viewed as potiential novel foods for health promotion in the next few years. However, the the level of health claim with optimum sensory and textural properties of such foods has yet to be investigated.

For people who want to reduce their own risks of heart disease, choosing a dairy spreads or *dahi*, etc enriched with specific functional ingredients would make sense. The traditional dairy products fortified with probiotics, prebiotics, buttermilk solids, plant sterols and stanols, dietary fibres, fruits, bioactive peptides, þ-3 and ω-6 PUFA, natural antioxidants, etc would provide an additive effect on top of their medication when used regularly as part of a regular diet. The future viability and success of value added dairy products in the marketplace depend on several elements. The key issue is consumer acceptance of such products. For consumers to agree and pay the premium associated with value added foods, they must be convinced that their health claim messages are clear, truthful and unambiguous. Stimulation by government authorities to change legislation and approval procedures, encouraging involvements in the research, and gaining consumer credibility will foster more accepting commercial atmosphere for the development and introduction of such foods in the marketplace. It is evident that value added dairy foods will be seen in many different markets beyond what is known today. India is emerging as a mega dairy market of the 21st century. The recent liberalization has thrown a bagful of opportunities for dairy entrepreneurs to manufacture our traditional dairy products on industrial scale, which upon value addition and stringent quality assurance programme would possibly compete in the International food markets with the branded functional foods, now being seen in the shelves of supermarkets.

References

Aneja, R.P. and Puri, B.P.S. (1997). "India's dairy riddle unwanted". *Dairy India*, 5th Ed., International Book Distributing Co., Lucknow, pp: 4 -26.

Anon (1992). Federal Societies for Experimental Biology Symposium Development of Medicated Foods for Rare Diseases. Life Science Research Officer, Bethesda, MD.

Anon (1998). International Association of Consumer Food Organizations. Japan-The inventor of functional foods. Center for Science in the Public Interest report, 1998.
www.cspinet.org/reports/functional_foods/japan_recmnd.html

Anon (1999). Department of Health and Human Services, Food and Drug Administration. Food labeling: Use on dietary supplements of health claims based on authoritative statements. 1999 Jan 21. In: www.cfsan.fda.gov/

Anon (2001). Functional Foods Market Assessment. Research and Markets, Guinness Centre, Taylors Lane, Dublin 8, Ireland.
http: //www.researchandmarkets.com/

Anon (2004). The world market for functional foods and beverages Research and Markets, Guinness Centre, Taylors Lane, Dublin 8, Ireland.
http: //www.researchand markets. com/

Hilliam, M. (2000). Functional food. How big is the market? *World of Food Ingredients*, **12**: 50-53.

PA Consulting Group (1990). Functional foods: A new global added value market? London: PA Consulting Group 1990.

Roy, A. (2002). Changing profile of the Indian market. *Indian Dairyman*, **54** (2): 47-54.

Shortt, C. (1999). The probiotic century: Historical and current perspectives. *Trends Food Sci. Technol.,* **10**: 411-417.

Stanton, C., Gardiner, G., Meehan, H., Collins, J. K., Fitzgerald, G., Lynch, P. B., and Ross, R. P. (2001). Market potential for probiotics. *Am. J. Clin. Nutr.,* **73** (Suppl.): 476S-483S.

Thakur, A.K. and Ray, S. (2000). Marketing of dairy products in India. A proper presented at the seminar on "*Dairying in Eastern Region-Past Trends and Future Prospects*". Org. by IDA (EZ), Feb. 26th, 2000, pp: 23-28.

Weitz, P. (1991). *Nutraceutical Products and Functional Food Additives*. Falls Church, VA: Technology Catalyst International.

FAO (2003). *FAO Production Yearbook.* Food and Agriculture Organization, Vol., 2003.

Patil, G. R. (2002). Present status of traditional dairy products. *Indian Dairyman,* **54** (10): 35-46.

2015, Dairy Product Technology: Recent Advances ***Pages 121–136***
Editors: **Subrota Hati, Surajit Mandal and Birendra Kumar Mishra**
Published by: **DAYA PUBLISHING HOUSE, NEW DELHI**

Chapter 6

Health Benefits of Milk Derived Bio-active Peptides

P.V. Padghan, Bimlesh Mann, B.M. Narwader, Subrota Hati and N.S. Pawar

Milk is the liquid food secreted by the mammalians for the nourishment of their newborns. However, since ancient times, humans have exploited this natural food for their own benefits as a food. Milk is nearly complete natural food on earth. It contains near about all essential nutrients required for normal body growth like water, fat, protein, lactose and minerals as major constituents. Milk also contains several minor constituents. Among these, salts (Ca, PO_4, Cl, Na Mg, K, S, citrate) and trace elements are very important for physiological as well as nutritional point of view. It also contains several enzymes, non-protein nitrogenous substances, vitamins, pigment and so on. Milk is a very complex mixture of all these compounds that nature has given to us. Aside that, now days, there has been an increasing interest in understanding the relationship between food and health all over the world. The development of health promoting food is one of the set targets in food process engineering. Research during the last few decades has shown that milk proteins can be an excellent source for various biologically active peptides.

Bioactive Peptides

Bioactive peptides have been defined as specific protein fragments that have a positive impact on body functions or conditions and may ultimately influence health (Kitts and Weiler, 2003).

Biological active peptides or functional peptides are food derived peptides that in addition to their nutritional value exert a physiological effect in the body.

Defined sequences of amino acids which are inactive within the original protein, but which display specific properties once they are released by enzymatic hydrolysis. (Vermeirssen *et al.,* 2004)

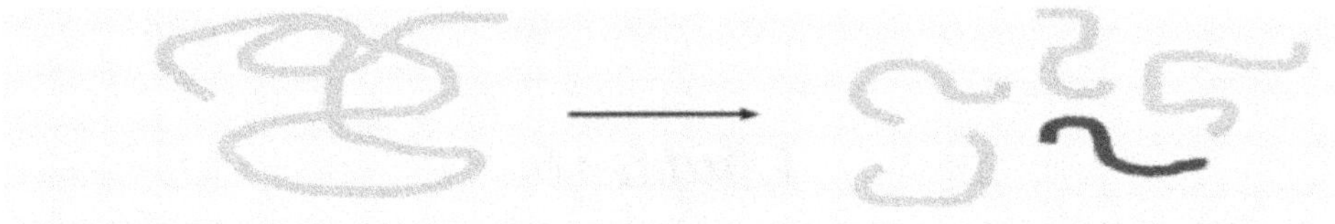

Bioactive peptides are specific protein fragments having a positive impact on body functions and conditions and may ultimately influence human health. Most of the biological activities are encrypted within the primary sequence of the native protein. Milk is a rich source of bioactive peptides which may contribute to regulate the nervous, gastrointestinal and cardiovascular systems as well as the immune system, confirming the added value of dairy products that, in certain cases, can be considered functional foods (Figure 6.1). The natural concentration of these biomolecules is quite low and, to date one of the main goals have been to realize products enriched with bioactive peptides that have beneficial effects on human health and proven safety.

Production of Bioactive Peptides

Bioactive peptides can be produced from milk proteins in the following ways: (a) enzymatic hydrolysis by digestive enzymes, (b) fermentation of milk with proteolytic starter cultures, (c) proteolysis by enzymes derived from microorganisms or plants. In many studies, a combination of above methods has proven effective in generation of short functional peptides. Examples of bioactive peptides produced by the above treatments are given below.

1. Enzymatic Hydrolysis

In the enzymatic hydrolysis, the enzyme likes pepsin, trypsin and chymotrypsin have been used to produce number of bio-active peptides shows antihypertensive, mineral binding/CPPs, (calcium and iron), antibacterial, immunomodulatory, angiotensin converting enzyme inhibitory and opioid activity both from different casein (α-, β- and δ-casein) and whey proteins, *e.g.,* α -lactalbumin (α -la), β -lactoglobulin (β -lg) and glycomacropeptide (GMP) from whole protein molecules. Apart from conventional production of peptides from natural protein sources by proteolytic enzymes, recombinant DNA techniques have been experimented successfully for the production of specific peptides or their precursors in Escherichia coli (Kim *et al.,* 1999).

2. Microbial Fermentation

Fermentation is easy and cost effective method to generate the bioactive peptides in fermented milk products. Many dairy starter cultures are highly proteolytic.

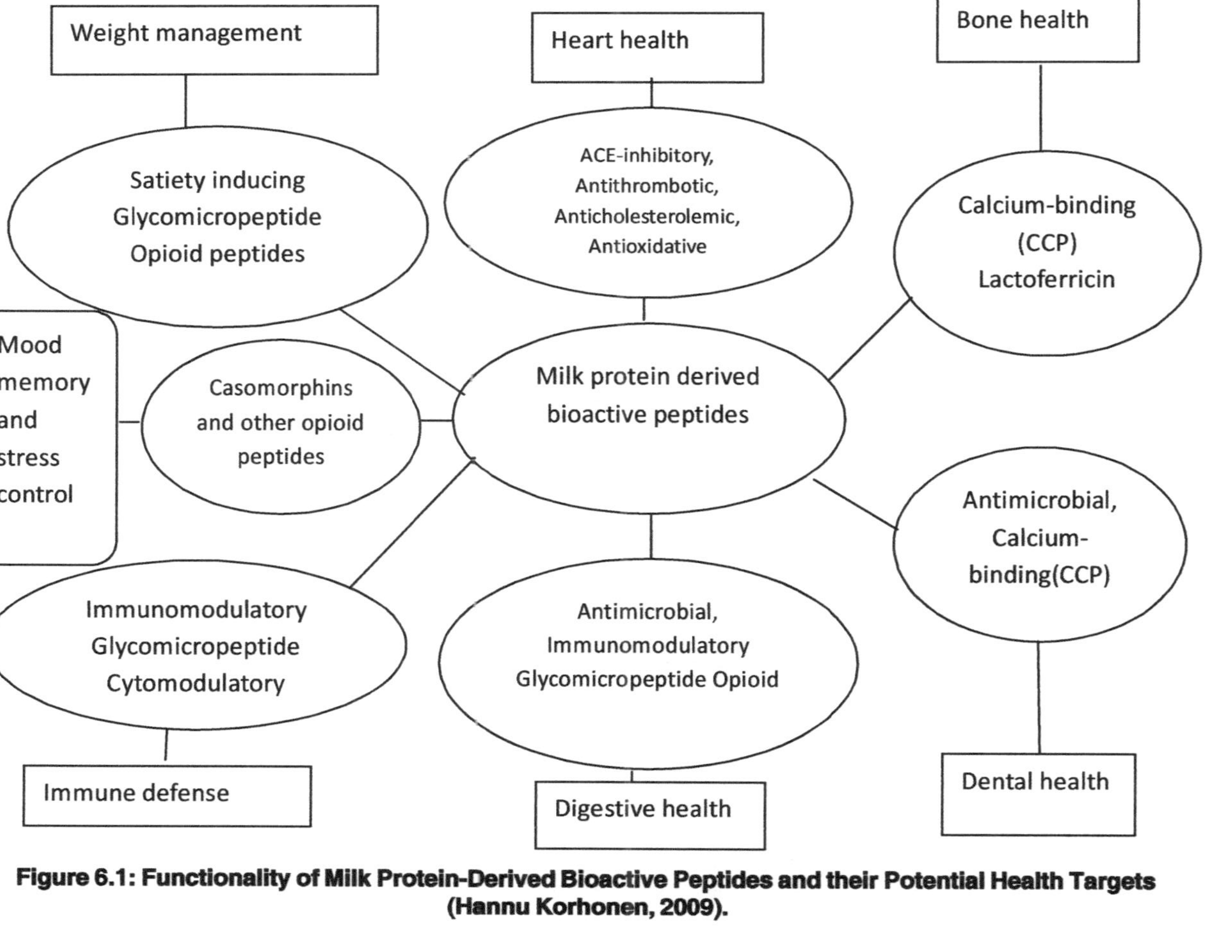

Figure 6.1: Functionality of Milk Protein-Derived Bioactive Peptides and their Potential Health Targets (Hannu Korhonen, 2009).

Formation of bioactive peptides can, thus, be expected during the manufacture of fermented dairy products. In fact, the release of different bioactive peptides from milk proteins through microbial proteolysis is now well documented. LAB commonly used to ferment milk into fermented milk products is thermophilic and mesophilic strains of *Streptococcus, Lactococcus* and *Lactobacillus* species.

3. Proteolysis by Enzymes Derived from Microorganisms or Plants

The protein are hydrolysed by the enzyme derived from microorganisms or plants. During fermentation of milk, the cell wall associated proteinases of lactic acid bacteria(LAB) hydrolyse casein into large peptides which then are transported into the cell and broken down by intracellular peptidases resulting in a range of bioactive peptides. In addition to live microorganisms, proteolytic enzymes isolated from LAB have been successfully employed to release bioactive peptides from milk proteins. Some commercially available proteolytic enzymes are used to produce ACE-inhibitory activity example enzymes (protease) isolated from *Aspergillus oryzae* showed the proteolysis activity.

4. Bulk Production of Bioactive Peptides

In the mass production of bioactive peptides used as ingredients for the manufacture of functional foods, it is necessary to have simple purification process. Starting materials or choice of enzymes can be an option for this purpose. Immobilised trypsin or glutamic acid-specific endopeptidase in fluidised bed bioreactor was used to produce caseinophosphopeptides from as-CN and beta-CN (Park and Allen 1998; Park *et al.,* 1998). This process does not require the removal of enzymes from the reaction mixture. Ellegard *et al.* (1999) combined diafiltration and anion-exchange chromatography for the large-scale production of caseinophosphopeptides. They achieved yield efficiency of 70 per cent when acid caseinate was trypsinated.

Sometimes, these processes may overlap since the protelytic action can start in food and continue in the organism. The activity is based on their inherent amino acid composition and sequence. The size of active sequences may vary from two to twenty amino acid residues, and many peptides are known to reveal multifunctional properties (Meisel and FitzGerald, 2003). In fact, some regions contain overlapping peptide sequences that exert different activities; these regions have been considered as "strategic zones" that are partially protected from further proteolytic breakdown. A strategic zone, for instance, is located in the sequence 60-70 of cow and human β-casein (Fiat *et al.,* 1993). The sequence is protected from proteolysis because of its high hydrophobicity and the presence of Proline residues. Proline, in fact, has an exceptional conformational rigidity compared to other amino acids; hence it loses less conformational entropy upon folding. Other examples of the multifunctionality of milk-derived peptides include the αs_1- casein fraction (f194-199) showing immunomodulatory and antihypertensive activity, the opioid peptides α- and β-lactorphin exhibiting also antihypertensive activity and the Caseinophosphopeptides, which possess mineral-carrier and immunomodulatory properties (Korhonen and Pihlanto, 2003).

Bioactive Peptides in Dairy Products

Milk naturally contains an array of bioactivity due to lysozyme, lactoferrin, growth factors, and hormones, which are secreted in their active form by the mammary gland. Colostrum is especially rich in nutrients and provides protection against pathogens due to its high concentration of antimicrobial proteins and, in particular, immunoglobulins (Pakkanen and Aalto, 1997). In addition, the evidence that milk proteins are the main source of many biopeptides, with different important physiological functions, proves that their role is not only to feed the neonate but also to regulate the complete growth of the body (Zabielski, 2007).

Bioactive Peptides in Fermented Milk and Milk Products

Fermented foods can be described as products whose physical, chemical and biological characteristics have been modified by microorganisms. They are known to contain specific microbial metabolites, such as alcohol, lactic acid, propionic acid, acetic acid, carbon dioxide and exopolysaccharides, as well as bio processed molecules derived from original food material like functional peptides from proteins. These derived products play a significant role in the biological activities of fermented foods (Table 6.1). Bioactive peptides are generated during food processing by chemical and physical but especially by enzymatic treatments as naturally happened during manufacture of fermented dairy products. The type of starter culture used is one of the main factors that influence the synthesis of peptides in fermented milks. The single most effective way to increase the number of bioactive peptides in fermented dairy products is to ferment or co-ferment with highly proteolytic strains of LAB. The challenges in this approach using LAB in dairy products lie in choosing the right

Table 6.1: Biological Activities Produced in Fermented Dairy Products (Maria *et al.*, 2007)

Fermented Product	*Observed Bioactivity*
Sour milk	Phosphopeptides, Antihypertensive properties
Yoghurt	ACE-inhibitory activity, Immunomodulatory, Antihypertensive properties, Antiamnesic, Microbiocidal, Antithrombotic
Quarg	ACE-inhibitory activity
Dahi	ACE-inhibitory activity
Parmesan, Reggiana cheeses	Opioid activity
Comte cheese	Phosphopeptides
Cheddar cheese	Phosphopeptides
Mozzarella, Italico cheeses	ACE-inhibitory activity
Crescenza, Gorgonzola cheeses	ACE-inhibitory activity
Edam, Emmental, 'Festivo' cheeses	ACE-inhibitory activity
Feta, Swiss, Cheddar, Edam, Camembert cheeses	ACE-inhibitory activity, Antiamnesic, Immunomodulatory, Opioid activity
Gouda cheese, Havarti cheese	ACE-inhibitory activity
Calpis® sour milk, Calpis Co. Japan	ACE-inhibitory activity, antihypertensive
Evolus® sour milk, Valio, Finland	ACE-inhibitory activity, antihypertensive

strains or combination of strains with optimal proteolytic activity and lysis tendency at the right time. The strain should not be too proteolytic to destroy the product and yet to give a high proteolysis, and with the right specificity to give higher concentrations of active peptides relative to other peptides, *i.e.*, bitter peptides. Thus milk fermentation has been described as a strategy to release bioactive peptides from caseins and whey proteins such as antihypertensive, antimicrobial, antioxidative, antithrombotic, immunomodulatory, mineral binding and opioid (Gobetti *et al.*, 2004).

At least two fermented sour-milk products with antihypertensive activity have been launched in Japan and Finland, respectively. The Japanese product "Calpis" is a soft drink, made from skim milk inoculated with starter cultures containing *Lactobacillus helveticus* CP790 and *Saccharomyces cerevisiae* (Takano, 1998), where the peptides Val-Val-Pro and Ile-Pro-Pro have been isolated and identified. In animal model studies, single oral administration of these tripeptides has been shown to have an antihypertensive effect in Spontaneously Hypertensive Rat (SHR). "Calpis" has also been demonstrated to prevent the development of hypertension in mildly hypertensive human subjects (Haque and Rattan, 2006). The Finnish milk product "Evolus" contains the tripeptide Ile-Pro-Pro and exerts a similar antihypertensive effect but it is produced by *Lactobacillus helveticus* LBK-16H strain as starter (Seppo *et al.*, 2002). PeptoPro® is a sport drink obtained by the cleavage of caseins, by means of a patented technology; it results rich in di- and tri-peptides that are stable, no longer bitter nor allergenic and supply energy and fast muscle refuelling by stimulating the production of insulin (Dutch State Mines, 2004). BioPure-Alphalactalbumin (Davisco©, 2007) is a Davisco product with a minimum of 90 per cent purified alpha-lactalbumin on a protein basis, containing the highest level of tryptophan naturally available from a protein source (4.4g tryptophan per 100g powder). Tryptophan is the precursor of serotonin in the brain and is associated with many health benefits, including improved sleep, memory, mood, etc. (Markus *et al.*, 2005). The presence of some of these fractions in food and beverages has given rise to the term "functional food" which is so called if it is satisfactorily demonstrated to beneficially affect one or more target functions in the body through active compounds, beyond adequate nutritional effects. According to this definition, functional food must remain food and cannot be made of pills or capsules. Another category of foodstuffs is labelled as "nutraceuticals" which, in proper cases, contain physiologically active components at a concentration significantly higher than the one naturally occurring in the original product (Childs, 1999). A nutraceutical is any substance that provides health or medical benefits, including the improved state of well-being and a reduction of risks related to certain diseases (DeFelice, 1995). Health-promoting food products are specifically aimed for weight management (prevention of obesity), natural defence (boosting of immunity), bone calcification (prevention of osteoporosis), digestion (prevention of intestinal disorders), cardiovascular health (prevention of heart diseases by lowering the cholesterol level or blood pressure).

Health Beneficial Effect

Milk derived bioactive peptides are considered as prominent ingredients for various health promoting functional foods targeted at heart, bone and digestive system health as well as improving immune defense, mood and stress control.

Effects on the Nervous System

Recent studies have shown that the consumption of dairy products causes interactions with the nervous system through the action of opioid peptides; basically they are receptor ligands with agonistic or antagonistic activities which are located in the nervous, endocrine and immune systems as well as in the gastrointestinal tract of mammals and can interact with their endogenous ligands (normally synthesized by the organism) or exogenous ligands (introduced by food). There are at least three types of opioid receptors: μ-type regulating the emotional behaviour and the intestinal mobility, δ-type involving the emotional behavior and the κ-type regulating calmness and appetite. They show different affinity, even though all of them present cross-interactions. The common structural feature of opioid peptides (except for α-casein opioids) is the presence of a Tyr residue at the N-terminal, coupled with the presence of another aromatic residue, such as Phe or Tyr, in the third or fourth position. This is an important factor that ensures fitting into the binding site of the receptors; furthermore, the negative potential, localized around the phenolic hydroxyl group of Tyrosine, seems to be essential for opioid activity (Silva and Malcata, 2005). The major and the first discovered opioid peptides, deriving from milk, are the so called β-casomorphins (Teschemacher, 2003) which are fragments of β-casein between the 60th and the 70th residues, mainly f60-63, f60-64, f60-65, f60-66 and f60-70, classifiable as μ-type ligands (Smacchi and Gobetti, 2000). The most potent seems to be the pentapeptide f60-64 (Fiat *et al.,* 1993) whose sequence appears similar in β-casein from sheep (Richardson and Mercier, 1979) and from water buffalo (Petrilli *et al.,* 1983) along with the fragment f60-63 of bovine β-casein called Morphiceptin (Chang *et al.,* 1981; Mierke *et al.,* 1990). The fragment f51-54 of human β-casein is also supposed to exert an agonistic opioid activity (Fiat *et al.,* 1993). Different bio- active peptides having opioid activity are presented in Table 6.2 along with their amino acid segments.

Table 6.2: Opioid Milk Peptides from Flavio *et al.,* 2009

Protein substrate	Bio-peptide	Amino acid segment	Reference
Bovine αs1-CN	Exorphin	f90-95, f90-96, f91-96	Loukas *et al.*, 1983
Human β-CN	β-Casomorphin (4,5)	f51-54, f51-55	Brantl, 1984
Bovine & Human α-LA	α-Lactorphin	f50-53	Chiba and Yoshikawa, 1986; Fiat and Jolles, 1989
Bovine β-Lg	β-Lactorphin	f102-105	Fiat et al.,1993; Yoshikawa *et al.*, 1986
Bovine β-CN	Mofphicetin	f60-63	Chang *et al.*, 1981; Mierke *et al.*, 1990
Bovine & Human k-CN	Casoxin A, B, C	f25-34, f35-41, f57-60	Yoshikawa *et al.*, 1986; Chiba *et al.*,1989
Lactotransferrin	Lactoferroxin A, B, C	f318-323, f536-540, f673-679	Tani *et al.*, 1990
Human αs1-CN	Casoxin D	f158-164	Yoshikawa *et al.*, 1994
Bovine serum albumin	Serorphin	f399-404	Tani *et al.*, 1994

Effects on the Immune System

The immune system is made up of specialized cells, antibodies and a lymphatic circulatory system since it protects the organism. Milk protein hydrolysates and peptides derived from caseins and the major whey proteins can enhance immune cell functions, measured as lymphocyte proliferation, antibody synthesis and cytokine regulation (Gill *et al.,* 2000). The physiological properties attributed to these diet related peptides have a common mechanism based on the inhibition of target enzymes which are somehow involved in essential processes like blood coagulation, phagocytosis and pathological infections. It is used to distinguish two main activities: the immunomodulatory and the antimicrobial peptides.

i) Immunomodulating Peptides

Breast feeding, especially at the beginning of lactation (colostrum), is the best way to provide the neonate with all the nutrients and, in particular, an adequate resistance against bacterial and viral infections. When gastrointestinal digestion occurs, many peptides with immunomodulating capacity are released from both the whey proteins and caseins. There are various hypotheses about the physiological action of such peptides: they might stimulate the proliferation and maturation of T-cells and natural killer cells for the defence of the newborn against different bacteria, especially enteric bacteria. The first isolated and sequenced peptide was the tryptic hydrolisate of human β-casein, Val-Glu-Pro-Ile-Pro-Tyr (corresponding to f54-59), that revealed immunostimulating activity (Jolles, 1981; Parker *et al.,* 1984). Later on several other peptides were identified, namely f63-68 and f191-193 from bovine β-casein and f194-199 from bovine αs1-casein (Migliore-Samour and Jolles, 1988) which stimulate phagocytosis in mice and humans in vitro and protect against Klebsiella pneumoniae infection in mice in vivo (Migliore-Samour *et al.,* 1989). Kayser and Meisel (1996) reported that di- and tri-peptides like Tyr-Gly and Tyr- Gly-Gly (partial sequences in the primary structure of bovine k-casein and α-lactalbumin respectively), significantly increased the proliferation of human peripheral blood lymphocytes in vivo. Recently these peptides were used for immunotherapy of human immunodeficiency virus infections, for example, to inhibit the development of infections in patients with pre-AIDS (Hadden, 1991). Furthermore, immunopeptides formed during milk fermentation have been shown to contribute to the antitumoural effects observed in many studies with fermented milks. Bioactive peptides present in yoghurt actually decreased tumour cell proliferation which may explain, at least partially, why consumption of yoghurt has been associated with a reduced incidence of colon cancer (Ganjam *et al.,* 1997).

ii) Antimicrobial Peptides

The antimicrobial properties of milk have been widely acknowledged for many years. The antimicrobial activity of milk is mainly attributed to immunoglobulins, and to nonimmune proteins, such as lactoferrin, lactoperoxidase, and lysozyme. It has been recognized for a long time that breast-feeding of infants provides protection from a range of enteric and respiratory infections. Antibacterial peptides are recognised as an important component of innate immunity, particularly at mucosal surfaces such as the lungs and small intestine that are constantly exposed to a range of potential pathogens. An amphiphilic and a positive charge are recognized as major structural motifs determining the interaction with bacterial membranes, which has been accepted as a common target in their mechanism of action. It has been demonstrated that some milk derived antibacterial peptides can reach intracellular targets. One of the most potent antimicrobial peptides described so far corresponds to a fragment of the whey protein lactoferrin, named lactoferricin(Bellamy *et al.*, 1992). The structure activity of lactofericcin fragment has been studied during last decade. It has been suggested that while the antimicrobial, antifungal, antitumor, and antiviral properties of lactofericcin can be related to tryptophan/arginine-rich proportion of the peptide, the anti-inflammatory and immunomodulating properties are more related to a positively charged region of the molecule.(Vogel *et al.*,2002) The peptides derived from LF hydrolysates can be useful for clinical applications because of their immunomodulatory effects or for chemoprevention of carcinogenesis. The use of LF-derivatives in oral care and as food preservative has also been proposed (Exposito and Recio, 2006). Lactoperoxidase catalyses the peroxidation of thiocyanate and some halides (I-,Br- but not Cl-) to generate products which are harmful for mammalian cells but kill or inhibit the growth of many species of microorganism (Boots and Floris, 2006). Lactoferrin, an iron-binding whey glycoprotein, shows indeed the most important antimicrobial activities (Chierici, 2001). However, this review deals chiefly with the antimicrobial peptides derived from milk proteins; several have been detected and some of them are listed in Table 6.3.

Effects on the Cardiovascular System

Bioactive peptides derived from milk or dairy products, mainly from caseins, have shown effects on the cardiovascular system, generally via antithrombotic and antihypertensive peptides.

1. Antithrombotic Peptides

The similarity between the clotting process of milk and the clotting of blood is well known since the undecapeptide (f106-116) from cow's kappa-casein involved in the coagulating mechanism presents a high structural homology with the human fibrinogen γ-chain (f400-411) (Jolles *et al.*, 1978). Casoplatelins, which are casein-derived peptides, behave like inhibitors of both the aggregation of ADP-activated platelets and the bound of human fibrinogen γ-chain to a specific receptor on the platelet surface (Fiat and Jolles, 1989). A blood anti-clotting effect is also displayed by the k-casein fragment f103-111 which can avoid the platelet aggregation, although is not able to affect fibrinogen binding to ADP-treated platelets (Fiat *et al.*, 1993). Furthermore k-caseinoglycopeptide, fragment f106-171 of sheep's k-casein, was shown to decrease thrombin- and collagen induced platelet aggregation in a dose-dependent manner (Qian *et al.*, 1995). Milk might also provide bioactive peptides with cholesterol-lowering effects. Nagaoka *et al.* (2001) isolated from milk β-lactoglobulin tryptic hydrolysate, a hypocholesterolemic peptide, which was identified to be the amino acid sequence Ile-Ile-Ala-Glu-Lys.

Table 6.3: Antimicrobial Milk Peptides.

Milk peptide fragment	Release protease	Gram (+) activity	Gram (-) activity	Yeast and fungi*
Casecidin κ-CN (f 17-21) αs1-CN	Chymosin and Tripsin	*Staphylococcus aureus Sarcina Bacillus subtilis Diplococcus pneumoniae Streptococcus pyogenes*		
Casocidin-I αs2-CN (f 165-203)	Synthetic peptide	*Staphylococcus carnosus*	*E. coli*	
Isracidin αs1-CN (f 1-23)	Chymosin and Tripsin	*Staphylococcus aureus*		*Candida albicans*
Caseicin αs1-CN A (f 21-29) B (f 30-38) C (f 195-208)	Synthetic peptide Synthetic peptide Synthetic peptide	*Listeria innocua*	*E. coli, E. sakazakii* *E. coli, E. sakazakii*	
Kappacin k-CN (f106-169)	Chymosin	*Streptococcus mutans*	*E. coli*	
Lactoferricin B Lactoferrin (f17-41)	Pepsin	*Bacillus* *Listeria* *Streptococci* *Staphylococci*	*E. coli 0111* *E. coli 0157H:7* *Klebsiella* *Proteus* *Pseudomonas* *Salmonella*	*Candida albicans* *Dermatophytes:* **Cryptococcus unigulattulus* **Penicillum pinophilum* **Trichophyton mentagrophytes*
Lactoferrampin Lactoferrin (f265-284)	Pepsin	*Streptococcus mutans*	*E. coli*	*Candida albicans*

2. ACE Inhibitory Peptides

Biologically active peptides derived from fermented foods proteins with an affinity to modulate blood pressure have been thoroughly studied. Angiotensin I converting enzyme (ACE; kinases II peptidyldipeptide hydrolase, EC 3.4.15.1) is important for blood pressure regulation. In the event where decreased blood volume or decreased blood flow to the kidneys is sensed, renin acts on angiotensinogen to form angiotensin I. ACE then catalyses the hydrolysis of the inactive prohormone angiotensin I (decapeptide) to angiotensin II (octapeptide). This result is an increase in blood pressure through vasoconstriction, via increased systemic resistance and stimulated secretion of aldosterone resulting in increased sodium and water absorption in the kidneys. ACE also inactivates the vasodilating peptide bradykinin (nonapeptide) and endogenous opioid peptide Met-enkephalin.

Two potent ACE inhibitory peptides from β casein, f(84–86), which corresponds to Val-Pro-Pro, and f(74–76), which corresponds to Ile-Pro-Pro and one from κ casein, f(108–110), which corresponds to Ile-Pro-Pro were purified from Japanese soft drink "Calpis" made from bovine skim milk fermented with *Lactobacillus helveticus* and *Saccharomyces cerevisiae.*(Nakamura *et al.,* 1995a) Single oral administration of sour milk containing these two tripeptides to spontaneously hypertensive rats (SHR) with dosage of 5ml/kg of body weight significantly decreased the systolic blood pressure from 6 to 8 h after administration.(Nakamura *et al.,* 1995b) Antihypertensive effect of these chemically-synthesized peptides was also observed from 2 to 8 h after administration and the effects were dose dependent. These two tripeptides have been also isolated from casein hydrolysate produced by extracellular proteinase enzyme of *L.helveticus* CP790 (Yamamoto *et al.,*1994). Using the same proteinase, Maeno *et al.,* identified a β casein-derived antihypertensive peptide from the casein hydrolysate (Maeno *et al.,* 1996). The antihypertensive effect of this peptide was dose dependent in SHR at a dosage level from 0.2– 2 mg of peptide per-kg body weight. This peptide did not show strong ACE inhibitory activity as such, but a corresponding synthetic hexa-peptide deleted by Gln (Lys-Val-Leu- Pro-Val-Pro-) exhibited strong ACE inhibitory activity as well as antihypertensive effect in SHR. This suggests possible activation of the peptides in the digestive tract. It has been demonstrated that a tetrapeptide isolated from β-lactoglobulin f(142–145), termed "β-lactosin B," had significant anti-hypertensive activity when administered orally to spontaneously hypertensive rats (Murakami *et al.,*2004).

Conclusion

Many food proteins can exert a physiological action, either directly or, after their degradation, in the form of fragments. Peptides represent a quite heterogeneous class of compounds and their characteristics deeply depend on the amino acidic composition and on the length of the chain. The acid-basic behaviour is determined by the free terminal residues and by the ionic lateral group of the residues in the chain; the reactivity of the terminal groups is also useful for their detection and quantification. Protein physico-chemical properties remarkably change after degradation and, consequently, some oligopeptides may play an important role in determining the rheological characteristics of a food. In fact they have been successfully used as additives as long as they are more soluble, less viscous and with greater emulsifying and foaming properties than the native proteins (Flavio *et al.,* 2009). The potential health benefits of milk protein-derived peptides have been a subject of growing commercial interest in the context of health-promoting functional foods. So far, some of peptides have been most studied for their physiological effects as antihypertensive, mineral-binding and anticariogenic peptides. The optimal exploitation of bioactive peptides for human nutrition and health possesses an exciting scientific and technological challenge, while at the same time offering potential for commercially successful applications. Bioactive peptides can be incorporated in the form of ingredients in functional and novel foods, dietary supplements and even pharmaceuticals with the purpose of delivering specific health benefits. Such tailored dietary formulations are currently being developed worldwide to optimize health through nutrition. This approach has been taken initially at target group level but will ultimately address individuals. Bioactive peptides offer an excellent basis for the novel concept of "personalized nutrition". Many scientific, technological and regulatory issues must, however, be resolved before these substances can be optimally harnessed to this end. Firstly, there is a need to develop novel technologies, such as chromatographic and membrane separation techniques, to enrich active peptide fractions from the hydrolysates of various food proteins. In addition to enzymatic hydrolysis, microbial fermentation provides a natural technology applicable for the production of bioactive peptides either from animal or plant proteins. The potential of this approach is already well demonstrated by the presence of bioactive peptides in fermented dairy products. Production of bioactive peptides from protein rich raw materials may be scaled up to industrial level using controlled fermentation in bioreactors with known LAB. Secondly, it is important to study the technological properties of active peptide fractions and to develop model foods that contain these peptides and retain their activity for a guaranteed period. It is recognized that, due to their lower molecular weight, peptides can be more reactive than proteins, and the peptides present in the food matrix may react with other food components. The interaction of peptides with carbohydrates and lipids, as well as the influence of the processing conditions (particularly heating) on peptide activity and bioavailability, should also be addressed. In this respect, the possible formation of toxic, allergenic or carcinogenic substances warrants intensive research. Modern analytical methods need to be developed to study the safety of foodstuffs containing biologically active peptides. Thirdly, molecular studies are needed to assess the mechanisms by which bioactive peptides exert their activities. For this approach, it is necessary to employ proteomics and associated technologies. By developing such novel facilities it will be possible to study the impact of proteins and peptides on the expression of genes and hence optimize the nutritional and health effects of these compounds. Bioactive peptides derived from milk proteins offer a promising approach for the promotion of health by means of a tailored diet and provide interesting opportunities to the dairy industry for expansion of its field of operation.

References

Bellamy, W., Takase, M., Wakabayashi, H., Kawase, K. and Tomita, M., 1992. Antibacterial spectrum of lactoferrin B, a potent bactericidal peptide derived from the N-terminal region of bovine lactoferrin. *J. Appl. Bacteriol.*, **73:** 472–479.

Boots, J.W. and Floris, R., 2006. Lactoperoxidase: From catalytic mechanism to practical applications. *Int. Dairy J.* **16:** 1272-1276.

Brantl, V., 1984. Novel opioid peptides derived from human beta-casein: human beta-casomorphins. *Eur. J. Pharmacol.* **106:** 213-1214.

Chang, K.J., Killian, A., Hazum, E., Cuatrecasas, P., 1981. Morphiceptin (NH_2-Tyr-Pro-Phe-$CONH_2$), a potent and specific agonist for morphine (μ) receptors. *Science* **212:** 75-77.

Chiba, H., Tani, F., Yoshikawa, M., 1989. Opioid antagonist peptides derived from β -casein. *J. Dairy Res.* **56:** 363-366.

Chiba, H., Yoshikawa, M., 1986. Biologically functional peptides from food proteins. In: R.E. Feeney (ed.). Protein tailoring for food and medical uses. Marcel Dekker Publ., New York, USA, 123-153.

Chierici, R., 2001. Antimicrobial actions of lactoferrin. *Adv. Nutr. Res.* **10:** 247-269.

Childs, N.M., 1999. Neutraceuticals industry trends. *J. Dietary* **2:** 73-85.

Davisco, 2007. Home page address: http: //www.daviscofoods. com/fractions/ alpha-beta.cfm.

DeFelice, S., 1995. The nutritional revolution: its impacts on food industry. *Trends Food Sci. Tech.* **6:** 59-61.

Dutch State Mines, 2004. DSM Home page address: http: //www.dsm.com/le/ en_US/peptopro/html/home_peptopro.htm.

Ellegard K H, Gammelgård-Larsen C, Sørensen E S and Fedosov S (1999) Process scale chromatographic isolation, characterization and identification of tryptic bioactive casein phosphopeptides. *International Dairy Journal* **9:** 639–652.

Exposito I. L. and Recio I.,2006. Antibacterial activity of peptides and folding variants from milk Proteins, *Int. Dairy J.*, **16:** 1294–1305.

Fiat, A.M. and Jolles, P., 1989. Caseins of various origins and biologically active casein peptides and oligosaccharides: Structural and physiological aspects. Mol. Cell. *Biochem.* **87:** 5-30.

Fiat, A.M., Migliore-Samour, D., Jolles, P., Drouet, L., Collier, C. and Caen, J., 1993. Biologically active peptides from milk proteins with emphasis on two examples concerning antithrombotic and immunomodulating activities. *J. Dairy Sci.* **76:** 301-310.

Flavio Tidona, Andrea Criscione, Anna Maria Guastella, Antonio Zuccaro, Salvatore Bordonaro and Donata Marletta., 2009. Bioactive peptides in dairy products. *Ital.J.Anim.Sci*, **8:** 315-340.

Gill, H.S., Doull, F., Rutherfurd, K.J., Cross, M.L., 2000. Immunoregulatory peptides in bovine milk. *Brit. J. Nutr.* **84:** 111-117.

Hadden, J.W., 1991. Immunoteraphy of human immunodefiency virus infection. *Trends Pharmacol. Sci.* **12:** 107-111.

Hannu Korhonen. 2009. Milk-derived bioactive peptides: From science to applications. *J. Functional Foods* **I**: 177–187.

Haque, E., Chand, R., 2006. Milk protein derived bioactive peptides. Home page address: http: //www. dairyscience.info/bio-peptides.htm.

Jolles, P., Parker, F., Floch, F., Migliore, D., Alliel, P., Zerial, A., Werner, G.H., 1981. Immunostimulating substances from human casein. *J. Immunopharmacology* **3:** 363-369.

Kayser, H. and Meisel, H., 1996. Stimulation of human peripheral blood lymphocytes by bioactive peptides derived from bovine milk proteins. *FEBS Lett.* **383:** 18-20.

Kim, Y. E., Yoon, S., Yu, D. Y., Lonnerdal, B., and Chung, B. H. (1999). Novel angiotensin-I-converting enzyme inhibitory peptides derived from recombinant human as1-casein expressed in Escherichia coli. *Journal of Dairy Research*, **66,** 431–439.

Kitts,D.D. and Weiler, K. (2003). Bioactive proteins and peptides from food sources. Applications of bioprocesses used in isolation and recovery. *Current Pharmaceutical Design,* **9**, 1309–1323.

Korhonen, H. and Pihlanto, A. 2006. Bioactive peptides: Production and functionality. *Int. Dairy J.*, **16**: 945–960.

Loukas, L., Varoucha, D., Zioudrou, C., Straty, R.A., Klee, W.A., 1983. Opioid activities and structures of alpha-casein-derived exorphins. *Biochemistry* **22:** 4567–4573.

Maeno, M., Yamamoto, N., Takano, T., 1996. Isolation of an antihypertensive peptide from casein hydrolysate produced by a proteinase from *Lactobacillus helveticus* CP790, *J. Dairy Sci.*, **79:** 1316–1321.

Markus, C.R., Jonkman, L.M., Lammers, J.H.C., Deutz, N.E.P., Messer, M.H. and Rigtering, N., 2005. Evening Intake of Alpha-Lactalbumin Increases Plasma Tryptophan Availability and Improves Morning Alertness and Brain Measures of Attention. Am. *J. Clin. Nutr.* **81:** 1026-1033.

Meisel, H., FitzGerald, R.J., 2003. Biofunctional peptides from milk proteins: mineral binding and cytomodulatory effects. *Curr. Pharm. Design* **9:** 1289-1295.

Mierke, D.F., Nobner, G., Schiller, P.W., Goodman, M., 1990. Morphiceptin analogs containing 2-aminocyclopentane carboxylic acid as a peptidomimetic for proline. *Int. J. Pept. Protein Res.* **35:** 34-35.

Migliore-Samour, D., Floch, F., Jollès, P., 1989. Biologically active casein peptides implicated in immunomodulation. *J. Dairy Res.* **56:** 357-362.

Migliore-Samour, D., Jolles, P., 1988. Casein Prohormone with an immunomodulating role for the newborn. *Experientia* **44:** 188-193.

Murakami, M., Tonouchi, H., Takahashi, R., Kitazawa, H., Kawai, Y., Negishi, H.et al., 2004. Structural analysis of a new anti-hypertensive peptide (β-lactosin B) isolated from a commercial whey product, *J. Dairy Sci.*, 87, 1967–1974.

Nakamura, Y., Yamamoto, M., Sakai, K., Okubo, A., Yamazaki, S., Takano, T., 1995a. Purification and characterization of angiotensin I-converting enzyme inhibitors from sour milk. *J. Dairy Sci.* **78:** 777–783.

Nakamura, Y., Yamamoto, N., Sakai, K., Takano, T., 1995b. Antihypertensive effect of sour milk and peptides isolated from it that are inhibitors to angiotensin I-converting enzyme, *J. Dairy Sci.*, **78:** 1253–1257.

Pakkanen, R., Aalto, J., 1997. Growth factors and antimicrobial factors of bovine colostrum. *Int. Dairy J.* **7:** 285-297.

Park O and Allen J C (1998) Preparation of phosphopeptides derived from as-casein and b-casein using immobilized glutamic acid-specific endopeptidase and characterization of their calcium binding. *Journal of Dairy Science,* **81:** 2858–2865.

Park O, Swaisgood H E and Allen J C (1998) Calcium binding of phosphopeptides derived from hydrolysis of a α_s-casein or β-casein using immobilized trypsin. *Journal of Dairy Science* **81:** 2850–2857.

Parker, F., Migliore-Samour, D., Floch, F., Zerial, A., Werner, G.H., Jollès, J., Casaretto, M., Zahn, H. and Jolles, P., 1984. Immunostimulating hexapeptide from human casein: Amino acid sequence, synthesis and biological properties. *Eur. J. Biochem.* **145:** 677-682.

Petrilli, P., Addeo, F., Chianese, L., 1983. Primary structure of water buffalo beta-casein tryptic and CNBr peptides. *Ital. J. Biochem.* **32:** 336-344.

Qian, Z.Y., Jolles, P., Migliore-Samour, D., Schoentgen, F. and Fiat, A.M., 1995. Sheep k-casein peptides inhibit platelet aggregation. *Biochim. Biophys. Acta* **1244:** 411-417.

Richardson, B.C., Mercier, J.C., 1979. The Primary Structure of the Ovine β-Caseins. *Eur. J. Biochem.* **99:** 285-285.

Seppo, L., Kerojoki, O., Suomalainen, T. and Korpela, R., 2002. The effect of a Lactobacillus helveticus LBK-16 H fermented milk on hypertension - a pilot study on humans. *Milchwissenschaft* **57:** 124-127.

Silva, S.V., Malcata, F.X., 2005. Caseins as source of bioactive peptides. *Int. Dairy J.* **15:** 1-5.

Smacchi, E. and Gobbetti, M., 2000. Bioactive peptides in dairy products: Syntesis and interactions with proteolyc enzymes. *Food Microbiol.* **17:** 129-141.

Takano, 1998. The impact of fermentation and *In vitro* digestion on the formation of Angiotensin-IConverting Enzyme inhibitory activity from pea and whey protein. *J. Dairy Sci.* **86:** 429-438.

Tani, F., Iio, K., Chiba, H., Yoshikawa, M., 1990. Isolation and characterization of opioid antagonist peptides derived from human lactoferrin. *Agric. Biol. Chem.* **54:** 1803–1810.

Tani, F., Shiota, A., Chiba, H., Yoshikawa, M., 1994. Serophin, an opioid peptide derived from serum albumin. In: V. Brantl and H. Teschemacher (eds.) β-casomorphins and related peptides: recent developments. VCH-Weinheim, Germany, pp. 49-53.

Teschemacher, H., 2003. Opioid receptor ligands derived from food proteins. *Curr. Pharm. Design* **9:** 1331-1344.

Vermeirssen, V., Van Camp, J., Verstraete, W. 2004. Bioavailability of angiotensin I converting enzyme inhibitory peptides. *Br. J. Nutr.* **92:** 357–366.

Vogel H. J., Schibli D.J., Weiguo J., Lohmeier-Vogel E.M., Epand R.F. and Epand R.M., 2002 Towards a structure-function analysis of bovine lactoferricin and related tryptophan and arginine containing peptides., *Biochem. Cell Biol.*, **80:** 49–63.

Yamamoto, N., Akino, A., Takano, T., 1994. Antihypertensive effect of the peptides derived from casein by an extracellular proteinase from *Lactobacillus helveticus* CP790. *J. Dairy Sci.*, **77:** 917–922.

Yoshikawa, M., Tani, F., Yoshimura, T., Chiba, H., 1986. Opioid peptides from milk proteins. *Agric. Biol. Chem.* **50:** 2419-2421.

2015, Dairy Product Technology: Recent Advances *Pages 137–148*
Editors: **Subrota Hati, Surajit Mandal and Birendra Kumar Mishra**
Published by: **DAYA PUBLISHING HOUSE, NEW DELHI**

Chapter 7

Clinical Studies for Novel Fermented Foods

Santosh Kumar Mishra, Amit Kumar, Gopika Talwar and K.K. Mishra

Introduction

Fermented foods have been associated with health benefits for many years. The health properties of these dairy products were a part of folklore until the concept of probiotics emerged, and the study of fermented milks and yoghurt containing probiotic bacteria has become more systematic. Milk alone is much more than the sum of its nutrients. It contains an array of bioactivities: modulating digestive and gastrointestinal functions, haemodynamics, controlling probiotic microbial growth, and immunoregulation. When fermented milk is enriched with probiotic bacteria and prebiotics it meets all the requirements of functional food.

Trial is from the Anglo–French *trier*, meaning *to try*. Broadly, it refers to the action or process of putting something to a test or proof. *Clinical* is from *clinic*, from the French *cliniqu'e* and from the Greek *klinike*, and refers to the practice of caring for the sick at the bedside. Hence, narrowly, a *clinical trial* is the action or process of putting something to a test or proof at the bedside of the sick. Clinical trials are sets of tests in medical research and drug development that generate safety and efficacy data for health interventions. Depending on the type of product and the stage of its development, investigators initially enroll volunteers and/or patients into small pilot studies, and subsequently conduct larger scale studies in patients that often compare the new product with others already approved for the affliction of interest. As positive safety and efficacy data are gathered, the number of patients is typically increased. Clinical trials can vary in size, and can involve a single research entity in one country or many such entities in multiple countries. Clinical trials hold enormous potential for benefiting patients, improving therapeutic regimens and ensuring advancement in medical practice that is evidence based. Efficacy and tolerance of any functional dairy product (as well as that of other fermented foods) and its establishment relies on clinical trials therefore different types of studies are usually performed during the product development or after commercialisation of food products. Research ideas are generally first tested in laboratory models, then in animals, and finally in humans. The main drawback of the clinical studies is the risk of biases, which can alter the results.

Clinical Trial: Protocols

Clinical trial design and objectives are written into a document called a clinical trial protocol. The protocol is the 'operating manual' for the clinical trial and ensures the researchers in different locations all perform the trial in the same way on patients with the same characteristics. The protocol describes the scientific rationale, objective(s), design, methodology, statistical considerations, and organization of the planned trial. Details of the trial are also provided in other documents referenced in the protocol, such as an investigator's brochure.

The major points that should be stated in a clinical trial protocol are shown in Table 7.1. and ideal regulatory framework for clinical trial is shown in figure 1.

Table 7.1: Topics that are Usually Stated in a Clinical Trial Protocol (Adapted from Marteau, 2003)

- ☆ Title – any amendment(s) and date(s).
- ☆ Sponsor – investigator(s), clinical laboratories and other institutions involved.
- ☆ Name and description of the product(s).
- ☆ Summary of findings from previous studies relevant to the trial (and references).
- ☆ Summary of the known and potential risks and benefits, if any, to human subjects.
- ☆ Description of the route of administration, dosage, dosage regimen, and treatment period(s).
- ☆ A statement that the trial will be conducted in compliance with the protocol, good clinical practice (GCP) and the applicable regulatory requirement(s).
- ☆ Inclusion criteria – exclusion criteria – withdrawal criteria.
- ☆ Description of the objectives and the purpose of the trial.
- ☆ A specific statement of the primary end-points and the secondary end-points.
- ☆ A description of the type/design of trial to be conducted (*e.g.* double-blind, placebo controlled, parallel design) and a schematic diagram of trial design, procedures and stages.
- ☆ Measures taken to minimise/avoid bias.
- ☆ Maintenance of trial treatment randomisation codes and procedures for breaking codes.
- ☆ Treatment(s) permitted and not permitted before and/or during the trial.

Contd...

Table 7.1–*Contd...*

- ☆ Procedures for monitoring compliance.
- ☆ Specification of the efficacy parameters.
- ☆ Methods and timing for assessing, recording and analysing safety parameters.
- ☆ Procedures for eliciting reports of and for recording and reporting an adverse event.
- ☆ Description of the statistical methods including the number of subjects planned to be enrolled and the reason for choice of sample size.
- ☆ Data handling and record keeping.
- ☆ Financing and insurance if not addressed in a separate agreement.
- ☆ Publication policy.

Increasing evidence

Meta-analysis RCTs

Randomised controlled trials

Open studies in humans

Animal studies

In vitro studies

Design

The most important design techniques for avoiding bias in clinical trials are randomisation and blinding, and these should be normal features of most controlled clinical trials intended to be included in a marketing application.

Randomized

Each study subject is randomly assigned to receive either the study treatment or a placebo. *Randomized trial* is a trial having a parallel treatment design in which treatment assignment for individuals (treatment units) enrolled is determined by a randomization process almost similar to coin flips or tossing of a die. Randomisation produces a deliberate element of chance into the assignment of treatments to subjects in a clinical trial. It provides a sound statistical basis for the quantitative evaluation of the evidence relating to treatment effects. It also tends to produce treatment groups in which the distributions of prognostic factors, known and unknown, are similar. The trialist's purpose in randomization is to avoid selection bias in the formation of the treatment groups. The bias is avoided because the treatment to which a person is assigned is determined by a process not subject to control or influence of the person being enrolled or those responsible for recruiting and enrolling the person. The randomisation schedule of a clinical trial documents the random allocation of treatments to subjects. In the simplest situation it is a sequential list of treatments (or treatment sequences in a crossover trial) or corresponding codes by subject number. The logistics of some trials, such as those with a screening phase, may make matters more complicated, but the unique pre-planned assignment of treatment, or treatment sequence, to subject should be clear.

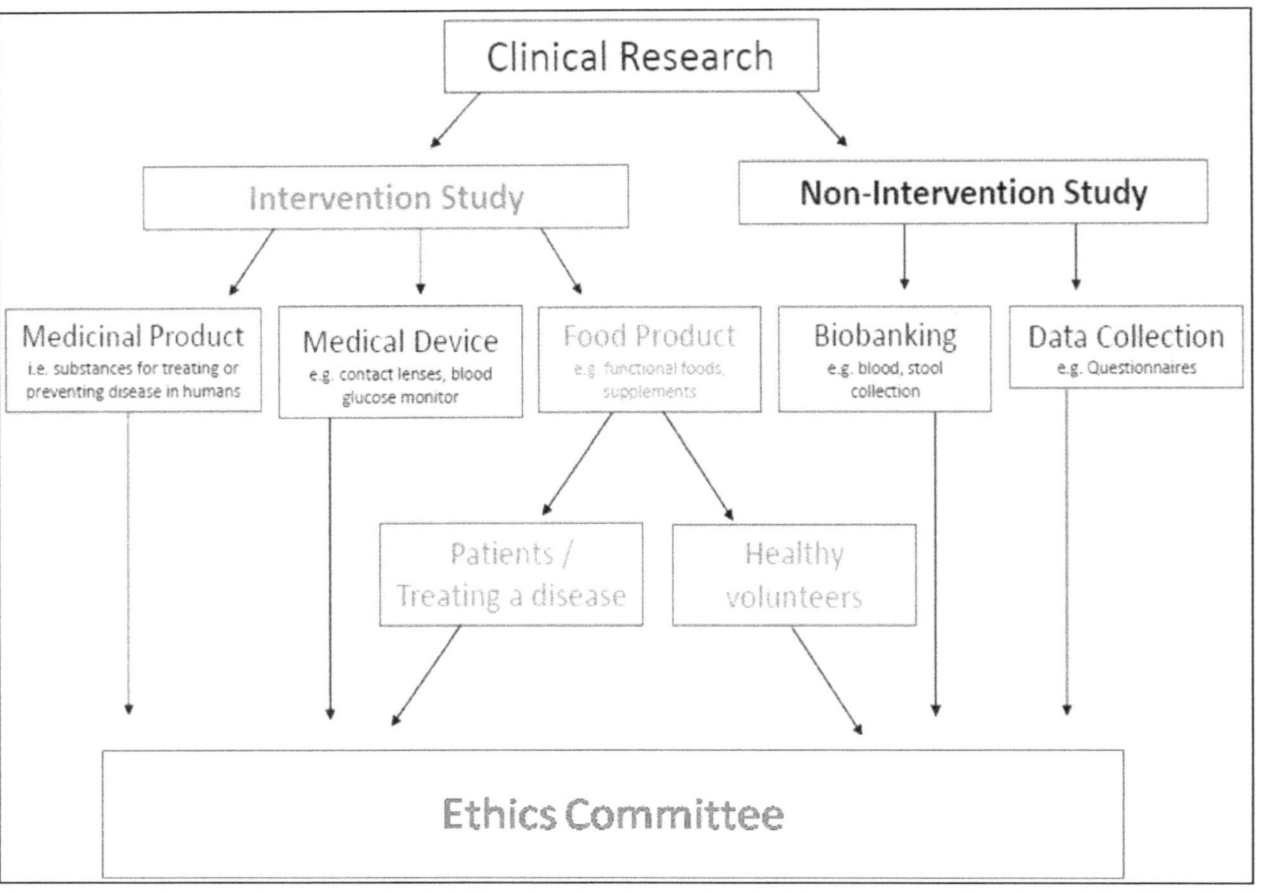

Figure 7.1: Ideal Regulatory Framework for Clinical Trial.

Blinding or Masking

Blinding or masking is intended to limit the occurrence of conscious and unconscious bias in the conduct and interpretation of a clinical trial arising from the influence which the knowledge of treatment may have on the recruitment and allocation of subjects, their subsequent care, the attitudes of subjects to the treatments, the assessment of end-points, the handling of withdrawals, the exclusion of data from analysis, and so on. The essential aim is to prevent identification of the treatments until all such opportunities for bias have passed.

The subjects involved in the study do not know which study treatment they receive. If the study is double-blind, the researchers also do not know which treatment is being given to any given subject. This 'blinding' is to prevent biases, since if a physician knew which patient was getting the study treatment and which patient was getting the placebo, he/she might be tempted to give the (presumably helpful) study substance to a patient who could more easily benefit from it. In addition, a physician might give extra care to only the patients who receive the placebos to compensate for their ineffectiveness. A double-blind trial is one in which neither the subject nor any of the investigator or sponsor staff who are involved in the treatment or clinical evaluation of the subjects are aware of the treatment received. This includes anyone determining subject eligibility, evaluating endpoints, or assessing compliance with the protocol. This level of blinding is maintained throughout the conduct of the trial, and only when the data are cleaned to an acceptable level of quality will appropriate personnel be unblinded. If any of the sponsor staff who are not involved in the treatment or clinical evaluation of the subjects are required to be unblinded to the treatment code (*e.g.* bioanalytical scientists, auditors, those involved in serious adverse event reporting), the sponsor should have adequate standard operating procedures to guard against inappropriate dissemination of treatment codes. In a single-blind trial the investigator and/or his staff are aware of the treatment but the subject is not, or vice versa. In an open-label trial the identity of treatment is known to all. The double-blind trial is the optimal approach. This requires that the treatments to be applied during the trial cannot be distinguished (appearance, taste, etc.) either before or during administration, and that the blind is maintained appropriately during the whole trial.

Difficulties in achieving the double-blind ideal can arise: the treatments may be of a completely different nature, for example, the daily pattern of administration of two treatments may differ. One way of achieving double-blind conditions under these circumstances is to use a 'double-dummy' technique. A form of double-blind study called a "double-dummy" design allows additional insurance against bias or placebo effect. In this kind of study, all patients are given both placebo and active doses in alternating periods of time during the study. This technique may sometimes force an administration scheme that is sufficiently unusual to influence adversely the motivation and compliance of the subjects.

If a double-blind trial is not feasible, then the single-blind option should be considered. In some cases only an open-label trial is practically or ethically possible. Single-blind and open label trials provide additional flexibility, but it is particularly important that the investigator's knowledge of the next treatment should not influence the decision to enter the subject; this decision should precede knowledge of the randomised treatment. For these trials, consideration should be given to the use of a centralised randomisation method, such as telephone randomisation, to administer the assignment of randomised treatment. In addition, clinical assessments should be made by medical staff who are not involved in treating the subjects and who remain blind to treatment. In single-blind or open-label trials every effort should be made to minimise the various known sources of bias and primary variables should be as objective as possible. The reasons for the degree of blinding adopted should be explained in the protocol, together with steps taken to minimise bias by other means. For example, the sponsor should have adequate standard operating procedures to ensure that access to the treatment code is appropriately restricted during the process of cleaning the database prior to its release for analysis. Breaking the blind (for a single subject) should be considered only when knowledge of the treatment assignment is deemed essential by the subject's physician for the subject's care. Any intentional or unintentional breaking of the blind should be reported and explained at the end of the trial, irrespective of the reason for its occurrence. The procedure and timing for revealing the treatment assignments should be documented.

Placebo-Controlled

The use of a placebo (fake treatment) allows the researchers to isolate the effect of the study treatment from the placebo effect. Although the term "clinical trials" is most commonly associated with the large, randomized studies, many clinical trials are small. They may be "sponsored" by single physicians or a small group of physicians, and are designed to test simple questions. In the field of rare diseases, sometimes the number of patients might be the limiting factor for a clinical trial. Other clinical trials require large numbers of participants (who may be followed over long periods of time), and the trial sponsor is a private company, a government health agency, or an academic research body such as a university.

Dose-Response Relationship Trials

Dose-response trials may serve a number of objectives, following are of particular importance: the confirmation of efficacy; the investigation of the shape and location of the dose-response curve; the estimation of an appropriate starting dose; the identification of optimal strategies for individual dose adjustments; the determination of a maximal dose beyond which additional benefit would be unlikely to occur. These objectives should be addressed using the data collected at a number of doses under investigation, including a placebo (zero dose) wherever appropriate. For this purpose the application of procedures to estimate the relationship between dose and response, including the construction of confidence intervals and the use of graphical methods, is as important as the use of statistical tests. The details of the planned statistical procedures should be given in the protocol.

Sample Size

Sample size plays an important role in any clinical trial. The number of subjects in a clinical trial should always be large enough to provide a reliable answer to the questions addressed. This number is usually determined by the primary objective of the trial. If the sample size is determined on some other basis, then this should be made clear and justified. For example, a trial sized on the basis of safety questions or requirements or important secondary objectives may need larger numbers of subjects than a trial sized on the basis of the primary efficacy question

Data Collection and Processing

The collection of data and transfer of data from the investigator to the sponsor can take place through a variety of media, including paper case record forms, remote site monitoring systems, medical computer systems and electronic transfer. Whatever data capture instrument is used, the form and content of the information collected should be in full accordance with the protocol and should be established in advance of the conduct of the clinical trial. It should focus on the data necessary to implement the planned analysis, including the context information (such as timing assessments relative to dosing) necessary to confirm protocol compliance or identify important protocol deviations. 'Missing values' should be distinguishable from the 'value zero' or 'characteristic absent'.

Ethical Conduct

Clinical trials are closely supervised by appropriate regulatory authorities. All studies involving a medical or therapeutic intervention on patients must be approved by a supervising ethics committee before permission is granted to run the trial. The local ethics committee has discretion on how it will supervise noninterventional studies (observational studies or those using already collected data).

International conference of harmonisation guidelines for good clinical practice is a set of standards used internationally for the conduct of clinical trials. The guidelines aim to ensure the "rights, safety and well being of trial subjects are protected". The notion of informed consent of participating human subjects exists in many countries all over the world, but its precise definition may still vary. Informed consent is clearly a 'necessary' condition for ethical conduct but does not 'ensure' ethical conduct. The final objective is to serve the community of patients or future patients in a best-possible and most responsible way. However, it may be hard to turn this objective into a well-defined, quantified, objective function. In some cases this can be done, however, for instance, for questions of when to stop sequential treatments, and then quantified methods may play an important role. Additional ethical concerns are present when conducting clinical trials on children (pediatrics).

In India the challenge faced is to apply universal ethical principles to biomedical research in the multicultural Indian society with a multiplicity of health-care systems of considerably varying standards. While on one hand, research involving human participants must not violate any universally applicable ethical standards, on the other hand, a researcher needs to consider local cultural values when it comes to the application of the ethical principles to individual autonomy and informed consent. In India, one will have to consider autonomy versus harmony of the environment of the research participant. In research on sensitive issues, this will have to be properly addressed in the research protocol to safeguard the human rights of the dependent or vulnerable persons and populations.

Analysis

The protection provided against treatment-related bias by the assignment process is futile if the analysis is biased. Treatment comparisons, to be valid, must be based on analyses that are consistent with the design used to generate them. In the case of the randomized trial, this means that the primary analyses of the outcomes of interest must be by assigned treatment. It means, for example, that observations relating to a morbid event are counted to a patient's assigned treatment regardless of whether or not the patient was still on the assigned treatment when the event occurred. Analyses involving arrangements of data related to treatment administered may be performed, but only as supplements to the primary analyses. They should not and cannot serve as replacements for those analyses.

Steps Involved in Evaluating Human Clinical Studies for Claim Substantiation

Regulatory bodies ask several questions in order to determine whether or not scientific conclusions can be made about the product and its relationship to the claim:

- ✰ Did the population of study subjects accurately portray the population relevant to the claim and marketing?
- ✰ Did the study include an appropriate control group?
- ✰ What type of biomarker of disease risk was measured?
- ✰ For how long was the study conducted?
- ✰ Does the intervention involve dietary advice? Was there proper follow up? How was this provided?
- ✰ Where were the studies conducted?

Evidence Required Supporting a Health Claim

Every claim has to follow specific criteria to ensure consumers are not misled or confused. Health and nutritional claims must be based on documented scientific information, validated, supported by evidence, complete, objective and verifiable. Claims must be clear and understandable by consumers (Codex Alimentarius, 1999). It is essential that claims for functional foods fulfil these criteria and special attention must be paid to objective scientific validation of functional and health claims. (Clydesdale, 1997).

Before functional and health claims are made for any fermented food, scientific evidence for the effects of fermented foods is required. Different types of evidence are needed, including data of biological/biochemical observations, epidemiological data and data from human intervention studies. The required data have to describe effects at the molecular, cellular, tissue and organ level, as well as effects on individuals, and also effects at the population level. The evidence must be based on studies in which accepted markers of improved functions or biological response, or markers of intermediate endpoints of disease, are used.

FDA task force on consumer health information for better nutrition in January 2003, prepared a report and recommendations for evaluating and ranking the scientific evidence for qualified health claims (FDA/CFSAN, 2003). The quality and strength of the scientific evidence for a proposed health claim are categorized through a grading system consisting of four levels, A, B, C and D, corresponding to high, moderate, low and extremely low levels of scientific support. The highest level 'A' means that the evidence is derived from well-designed studies and therefore there is significant scientific agreement about the health claim. This level is chosen as a point of reference and such a claim is referred to as an 'unqualified health claim'. The second level 'B' applies to claims that are supported by good scientific evidence, but the evidence is not conclusive. The third level 'C' refers to scientific evidence that is limited and inconclusive. The fourth level 'D' should be used when little scientific evidence supports the claim. Claims categorized as B, C or D are called 'qualified health claims' and they require a disclaimer or other qualifying language to ensure that they do not mislead consumers. All qualified health claims have to be reviewed by the FDA before they are used on the food label, a process that involves a careful review of all the available scientific data and may include detailed expert assessment. Such a review process and ranking system should protect consumers from unproved and misleading information.

Clinical Evidences for Novel Fermented Foods

Fermented milks are a rich source of whey proteins such as -lactalbumin, - lactoglobulin, lactoferrin, lactoperoxidase, immunoglobulins and variety of growth factors. These proteins have demonstrated a number of biological effects ranging from anti-carcinogenic activity to different effects on the digestive function (McIntosh *et al.*, 1998). The results of several clinical studies indicate that a regular administration of selected probiotics through fermented milk products may reduce the concentration of serum cholesterol, especially of LDL (Mikeš *et al.*, 1995; Agerholm-Larsen *et al.*, 2000). It has been shown in experiments which indicate that both *in vitro* and *in vivo* lactobacilli, bifidobacteria and other milk bacteria assimilate cholesterol, incorporate it into membranes, deconjugate and precipitate the bile acids (Tahri *et al.*, 1996; Pereira and Gibson 2002). Furthermore, fermented milks contain components with at least protective if not hypoholesterolemic effects; these include calcium, linoleic acid, conjugated linoleic acid (CLA), antioxidants, and lactic acid bacteria or probiotic bacteria (Rogelj, 2000). Numerous *in vitro* and animal studies have confirmed the anti-carcinogenic activity of CLA, as well as its role in preventing atherosclerosis and in modulating certain aspects of the immune system (Cook and Pariza, 1998, MacDonald, 2000). It has been found that lactic acid bacteria can effectively trap reactive forms of oxygen. The experiment with vitamin E-deficient rats has revealed that the intracellular extract from *Lactobacillus* sp. recovers this deficiency (Kaizu *et al.*, 1993). The classical yoghurt bacteria *Lactobacillus delbrueckii* ssp. *bulgaricus* and *Streptococcus thermophilus* inhibit peroxidation of lipids through scavenging the reactive oxygen radicals, such as hydroxyl radical, or hydrogen peroxide (Ling and Yen 1999). Controlled studies have indicated that consumption of fermented milk cultures containing lactic acid bacteria (LAB) can enhance production of Type I and Type II interferons at the systemic level. In animal models, LAB have been shown to promote interferon expression, and to reduce allergen-stimulated production of IL-4 and IL-5 in some cases (Cross *et al.*, 2001). A small pediatric nonrandomized pilot study suggested that *Lactobacillus* GG may improve gut barrier function and clinical status in children with mildly to moderately active, stable Crohn disease (Gupta, 2000).

Future Trends

Fermented foods have been expected to play an important role in modern nutrition, which is expected to promote health and reduce the risk of chronic diseases. Fermented foods must fulfil all standards of food safety assessment. However, for this type of food, the concept of benefits versus risk of long-term intake has to be elaborated, developed and validated. The safety of intake of low or high amounts of nutrients and high amounts of non-nutrients related to long- term consumption of fermented food, as well as interactions between food components and biological processes have to be monitored. Protocols for pre- marketing nutrition studies on fermented food and post-marketing monitoring are needed. The requirements of probiotic food safety indicated by FAO/WHO impose on the manufacturers the obligation to conduct placebo-controlled clinical studies and to evaluate their results in four phases, *i.e.* Phase 1 (safety), Phase 2 (efficiency), Phase 3 (effectiveness) and Phase 4 (surveillance). If the manufacturer obtains positive results from one research centre, they should also obtain the confirmation of such results from other researchers and the results have to be published in a peer-reviewed scientific or medical journal. It seems that similar requirements will have to be enforced in the future with respect to all fermented food. It should be borne in mind that during clinical studies we often discover new unforeseen so-called pleiotropic actions of functional food, as in the case of medicines.

Conclusions

Conducting a clinical trial in a regulatory environment that allows disease populations to be studied provides the opportunity to extrapolate the results into healthy populations thereby allowing one to navigate disease versus healthy population claims when a product may have indications for CVD, diabetes, etc. In addition, conducting a study in a country with a diverse population (*e.g.* India) opens doors for product application and marketing opportunities globally. There are many factors involved in conducting a clinical trial. A good study design and an experienced research partner are the key to providing sound results! As the health and nutrition market becomes more driven to succeed, clinical trials will become an absolute necessity and a routine part of product development and marketing efforts. It is advisable for the food and nutrition industry to carefully choose their research partner to achieve the maximum return on investment.

References

Agerholm-Larsen, L., Raben, A., Haulrik, N., Hansen, A.S., Manders, M., Astrup A. 2000. Effects of 8 Weeks Intake of Probiotic Milk Products on Risk Factors for Cardiovascular Diseases. *Eur. J. Clin. Nutr.*, **54:** 288–297.

Clydesdale F. M. 1997. A Proposal for the Establishment of Scientific Criteria for Health Claims for Functional Foods. *Nutr. Rev.*, **55:** 413±22.

Codex Alimentarius 1999. Proposed Draft Recommendations for the Use of Health Claims. WHO, Geneva.

Cook, M.E., and Pariza, M. 1998. The Role of Conjugated Linoleic Acid (CLA) in Health. *Int. Dairy J.*, **8:** 459-462.

Cross, M.L., Stevenson, L.M., Gill, H.S. 2001. Anti-Allergy Properties of Fermented Foods: An Important Immunoregulatory Mechanism of Lactic Acid Bacteria? *Inter. Immunopharmacol.*, **1(5):** 891–901.

FDA. 1997. Good Clinical Practice. *Federal Register*, **62:** 25691–25709.

FDA/CFSAN 2003. Consumer Health Information for Better Nutrition Initiative: Task Force Final Report (http: //www.cfsan.fda.gov/~dms/nuttftoc.html).

Gupta, P. 2000. Is *Lactobacillus* GG Helpful in Children with Crohn's Disease? Results of a Preliminary, Open-Label Study. *J. Pediatr. Gastroenterol. Nutr.*, **31:** 453Y7.

http: //www.ich.org/fileadmin/Public_Web_Site/ICH_Products/Guidelines/Efficacy/E9/Step4/E9_Guideline.pdf.

https: //www.pfizer.com/files/research/research_clinical_trials/ethics_committee_ guide.pdf

ICH Guideline for Good Clinical Practice: Consolidated Guidance (http: //www.ich. org/LOB/media/MEDIA482. pdf)

Kaizu, H., Sasaki, M., Nakajima, H. 1993. Effect Of Antioxidative Lactic Acid Bacteria On Rats Fed A Diet Deficient in Vitamin E. *J. Dairy Sci.*, **46:** 2493–2499.

Ling, M.Y. and Yen, C.L. 1999. Antioxidative Ability of Lactic Acid Bacteria. *J. Agric. Food Chem.* **47:** 1460–1466.

MacDonald, H.B. 2000. Conjugated Linoleic Acid and Disease Prevention: A Review of Current Knowledge. *J. Americ. Coll. Nutri.*, **19(2):** 111-118.

Marteau, P. 2003. Clinical trials, Chapter 14. *Functional Dairy Products,* Ed. By Mattila Sandholm, T. and Saarela, M. CRC Press New York Washington, DC. pp. 337-345.

McIntosh, G.H., Royle, P.J., LeLeu, R.K., Regester, G.O., Johnson, M.A., Grinsted, R.L., Kenward, R.S., Smithers, G.W. 1998. Whey Proteins as Functional Food Ingredients. *Int. Dairy J.*, **8:** 425-434.

Mikeš, Z., Fereněík, M., Jahnová, E., Ebringer, L., Èižnár, I. 1995. Hypocholesterolemic And Immunostimulatory Effects of Orally Applied *Enterococcus faecium* M-74 in Man. *Folia Microbiol.*, **40:** 639–646.

Pereira, D. and Gibson, G.R. 2002. Cholesterol Assimilation by Lactic Acid Bacteria and Bifidobacteria. *Appl.Environ.Microbiol.* **68:** 4689–4693.

Rogelj, I. 2000. Milk, Dairy Products, Nutrition and Health. *Food technol. Biotechnol.*, **38(2):** 143-147.

Sackett, D.L. 1979. Bias in Analytic Research. *J. Chronic. Dis.,* **32(1–2):** 51–63.

Tahri, K., Grill, J.P., Schneider, F. 1996. Bifidobacteria Strains' Behaviour Towards Cholesterol Co precipitation With Bile Salts Assimilation. *Curr. Microbiol.,* **3:** 187–193.

(www.fda.gov/Food/GuidanceComplianceRegulatoryInformation/GuidanceDocument)

2015, Dairy Product Technology: Recent Advances ***Pages 149–168***
Editors: **Subrota Hati, Surajit Mandal and Birendra Kumar Mishra**
Published by: **DAYA PUBLISHING HOUSE, NEW DELHI**

Chapter 8

Lactobacillus reuteri: A Multifaceted Lactic Acid Bacterium for Fermented Foods

Santosh Kumar Mishra, R.K. Malik, Amit Kumar and K.K. Mishra

Introduction

Lactic Acid Bacteria (LAB), comprising about 20 genera, are a heterogeneous group of microorganisms that metabolizes lactose to produce lactic acid. The prominent genera among them are *Lactobacillus, Lactococcus, Enterococcus, Streptococcus, Leuconostoc*. Since centuries, LAB have been used in industrial and artisanal food fermentations especially as starter cultures and thus, have been playing an important role in the food conservation by the production of a wide variety fermented foods. Furthermore, Elie Metchnikoff in the early 20thcentury explored their use as potential probiotics in extending the consumers life in 'theory of longevity' (Kimoto *et al.,* 2007; Ljungh and Wadstrom, 2007). Probiotics are defined as viable microorganisms that exhibit a beneficial effect on human health after ingestion in particular amount by improving its intestinal microbial balance (Fuller 1989; Lee and Salminen 1995). However, there is no widely accepted definition for Probiotics. Recently, FAO/WHO (2002) defined probiotics as, 'live microorganisms which when administered in adequate amounts confer a health benefit on the host'. During the last two decades probiotic researchers have shown a renewed interest in the health-promoting attributes of LAB and their successful transfer to human beings for their well being.

LAB are having Generally-Recognized-as-Safe (GRAS) status, Table 8.1 lists use of *L. reuteri* as a probiotic in different countries. Amongst all the probiotic LAB, genus *Lactobacillus*, has been most widely used. According to WHO/FAO, a probiotic organism must be isolated from the host source. Members of *Lactobacillus* genus are most often isolated from the human intestine. However, there exist technological problems while producing fermented products containing species/strains of genus *Lactobacillus*. For example, *Lactobacillus acidophilus* strains show reduced growth in milk due to poor mechanism of lactose uptake and making it difficult for production of fermented milk. However, to overcome this problem some additives such as compounds hydrolyzed by proteases or dairy lactic acid bacteria are added as accelerators during fermented milk manufacturing. Therefore, there has been search for new probiotic strains, which grow richly in milk and which are more technologically suitable (Kimoto *et al.*, 2007). *Lactobacillus reuteri* is considered as one of the few true autochthonous lactobacilli present regularly in host's large intestine (Reuter 2001). It has been isolated directly from GI tract or feces of human, monkeys, chicken, turkeys, doves, pigs, lambs, cattle, dogs and rodents (Casas and Dobrogosz 2000), and has also been found to be a major component of *Lactobacillus* species found in all hosts. In addition, it is the only *Lactobacillus* species which is believed to have established a symbiotic relationship with all hosts (Casas and Dobrogosz 2000). Therefore, all species of *L. reuteri* isolated from various hosts have exhibited probiotic efficacy to hosts, hence, convincingly supporting the probiotic concept. In animals and humans, experiments have shown that *L. reuteri* is transmitted from mothers to newborn animals or infants during birth and the nursing process via mainly the mammary duct (Casas and Dobrogosz 2000). Also, orally administrated *L. reuteri* has been shown to survive gastric acids and bile salts through the stomach and upper intestine, bind to the gut mucus and epithelial cells, and colonise the host intestine (Reuter 1965; Casas and Dobrogosz 2000). However, *L. reuteri* strains isolated from different hosts, exhibit host specific colonization characteristics, some of which are not crossable (Molin *et al.*, 1992; Casas and Dobrogosz, 1997).

Table 8.1: Use of *L. reuteri* as a Probiotic in different Countries (Adapted fromVollenweider, S. and Lacroix, C., 2003)

Country	*Company*	*Products*
Sweden	BioGaia AB	Tablets, capsules, powder and infant formula; milk, yogurt, juice, fresh cheese and ice-cream
Finland	Ingman Foods Oy Ab	Yoghurts, fermented milks, cottage cheese, ice-cream, and fruit drinks
Great Britain	Farm Produce Marketing Ltd.	Yogurt drink
Japan	Chichiyasu milk products company Ltd.	Milk products
Spain and Portugal	Kraft Jacobs Suchard Iberia	Fresh cheese
South Korea	Lotte Ham and Milk Co., Ltd.	Dairy products
Sweden	Milko	Dairy products
USA	Stonyfield Farm	Yoghurt products

Lactobacillus reuteri was recorded in scientific classification of lactic acid bacteria as early as the beginning of the 20th century (Orla-Jensen 1919), though at that time it was undistinguished from *L. fermentum*. In the 1960s, a German microbiologist Gerhard Reuter isolated *L. reuteri* from human fecal and intestinal samples, and subsequently separated it from *L. fermentum* and reclassified it as *L. fermentum* biotype II (Reuter 1965). Kandler *et al.* (1980) eventually identified *L. reuteri* as a distinct species based on the phenotypic and genptypic characteristics, and proposed it being a new species of heterofermentative lactobacilli. Later, modern technologies have further confirmed the identity and clearly separated the two species. Since 1980, *L. reuteri* has been classified as a distinct species in the *Lactobacillus* genus. *Lactobacillus reuteri* strains are Gram-positive, lactic acid-producing bacteria; their cells are slightly irregular, bent rods with rounded ends, generally 0.7-1.0 X 2.0-3.0μm in size (Kandler and Weiss 1986), occurring singly, in pairs and in small clusters. *Lactobacillus reuteri* belongs to the obligate heterofermentative group of lactobacilli, and uses the phosphoketolase based metabolic pathway to utilize available carbohydrates (Axelsson 1998; Casas and Dobrogosz, 2000). It can ferment glucose alone and produce lactate, ethanol and CO_2 as end products; but an essential characteristic of *L. reuteri* is its ability to utilize glycerol (Talarico *et al.*, 1988; El-Ziney *et al.*, 1998; Luthi-Peng *et al.*, 2002). When glycerol is added as a substrate, the end products change to more of acetate/less of ethanol, and the NADH formed during glycolysis is reoxidized by glycerol (Talarico *et al.*, 1990).

Lactobacillus reuteri as a Probiotic Organism

L. reuteri has been used as an important dairy starter since hundreds of years and has GRAS status too. In recent years, available literatures clearly add weight to prove potentials of Lactobacillus reuteri strains as potential probiotic and functional Lactic acid bacteria (LAB). According to these studies Lactobacillus reuteri strains can adhere, survive and pass through GI tract, have antimicrobial property, may remove cholesterol, modulate immune system and also showed anti-mutagenic and anti-diabetic activity. Table 8.2 lists different studies showing probiotic potentials of Lactobacillus reuteri

Survivability and Passage through GI Tract

It is an important property to establish an organism as probiotic and this requires the ability of the organism to survive in hostile conditions of GI tract *i.e.* low pH of stomach, bile in the intestine, proteolytic enzymes in oral cavity and intestine, etc. Administration of doses from 1.0×10^8 bacteria and higher have been reported to lead to efficient colonization in human trials, but the colonization is transient and gets lost after two months wash-out (Wolf *et al.*, 1995; Shornikova *et al.*, 1997; Valeur *et al.*, 2003; Wall *et al.*, 2007). In all these studies conventional cultivation methods were employed. However, in one of the studies *L. reuteri* was also identified in intestinal biopsy specimen by using fluorescence in situ hybridization with a molecular beacon probe (Valeur *et al.*, 2003). Recently, the survival of *L. reuteri* ATCC 55730 after a sudden shift in environmental acidity to a pH close to the conditions in the human stomach was examined. More than 80 per cent of the *L. reuteri* cells survived at pH 2.7 for one hour (Wall *et al.*, 2007). Some studies have also shown that they are bile resistant and survive passage through the human gastrointestinal tract (Gardiner *et al.*, 2002; Reid *et al.*, 2002). *L. reuteri* Probio-16 showed maximum bile resistance at 5.0 per cent concentration and formed cloudy zones of precipitate in the growth medium by producing bile-salt hydrolase (BSH) (Seo *et al.*, 2010). Thus, these reports proved the survivability of *L. reuteri* strain during the passage through a digestive system with typical conditions such as low pH in the stomach and the presence of bile in the intestine.

Table 8.2: Studies Showing Probiotic Potentials of *Lactobacillus reuteri*

Function	*Observation*	*Reference(s)*
Humans		
Prevention of diarrhea	Reduced duration and severity of diarrhea caused by rotavirus in children; reduced incidence of diarrhea in infants	(Shornikova and others 1997a; Shornikova and others 1997b; Weizman and others 2005)
Reduction of infant colic	Reduced colicky symptoms in 95 per cent of infants; improved gastric emptying and reduced crying time in premature infants	(Savino and others 2007;Indrio and others 2008)
Reduction of IgE-associated eczema and sensitization	Reduction of IgE associated eczema in 2 year old; reduced levels of TGF	(Böttcher and others 2008; Abrahamsson and others 2007)
Immune stimulation	Short-term survival of *L. reuteri* in the stomach and small intestine. Stimulation of CD4	(Valeur and others 2004)
Animals		
Immune stimulations	Transient increase in proinflammtory cytokines and chemokines in the intestinal tract	(Hoffmann and others 2008)
Immune regulation	Increased levels of regulatory T cells upon colonization of *Lactobacillus*-free free mice with *L. reuteri*	(Livingston and others 2010)
Prevention of experimental colitis	Reduced levels in animal models of colitis	(Møller and others 2005; Schreiber and others 2009; Madsen and others 1999; Peña and others 2005; Fabia and others 1993)
Immune cells		
Modulation of immune reactions in cultured macrophages, dendritic cells, and T cells	Reduction in TNF-α production in activated macrophages; reduced production of proinflammatory cytokines in dendritic cells, induction of regulatory T cells	(Lin and others 1984; Peña and others 2005; Christensen and others 2002; Smits and others 2005; Livingston and others 2010); Fink and others 2007), Thomas and others 2012

Adhesion to Intestinal Cells and Mucus

It is another important criterion for the selection of probiotic bacteria (Kimoto *et al.*, 2007). Colonization to the GI tract was improved when *L. reuteri* was exposed to mucin (Roos and Jonsson 2002). Mucin, the main component of the mucus secreted by the goblet cells helps *L. reuteri* to attach and replicate in the mucus layer over the epithelial cells in the GI tract and in the loose mucus layer in the lumen. Cell surface proteins involved in colonization and binding (adhesion) to mucus layers have been identified (Roos and Jonsson 2002). The gene corresponding to the mucus binding protein (*Mub*) has been identified in several *L. reuteri* strains and its presence correlates to a great extent with the adhesion of mucus *in vitro*. A gene from *Lactobacillus reuteri* 1063 encoding a cell-surface protein, designated *Mub*, that adheres to mucus components *in vitro* has been cloned and sequenced. The deduced amino acid sequence of Mub (358 kDa) shows the presence of 14 approximately 200 aa repeats and features typical for other cell surface proteins of Gram-positive bacteria (Roos and Jonsson 2002). In addition, the existence of a collagen binding protein (CnBP) involved in adherence to epithelia and/or mucus has also been reported in *L. reuteri* along with extracellular glucosyl transferase enzymes (commonly named glucansucrases, GTFs) and inulosucrase (Inu) that synthesize different homopolysaccharides which contribute to cell aggregation and *in vitro* biofilm formation under acidic conditions (Kralj *et al.*, 2005; Walter *et al.*, 2008).

Hypocholesterolemic Effects

It is well known that elevated serum cholesterol concentration is a risk factor associated with atherosclerosis and coronary heart disease (Taranto *et al.*, 1998, Katkamwar, 2012). Reduction of serum cholesterol is, therefore, very important to prevent cardiovascular disease. Numerous drugs that lower cholesterol have been used to treat hypocholesterolemic (HC) individuals (Suckling *et al.*, 1991). However, the undesirable side effects of these drugs have caused concerns about their therapeutic use (Erkelens *et al.*, 1988). It is, therefore, important to develop new ways of reducing serum cholesterol. The ingestion of probiotic lactic acid bacteria is possibly a more natural method to decrease serum cholesterol concentrations in humans (Taranto *et al.*, 1998; Katkamwar, 2012). *L. reuteri* administered to pigs and mice has shown to decrease the level of serum cholesterol (Smet *et al.*, 1998; Taranto *et al.*, 1998; Katkamwar, 2012). The administration of *Lactobacillus reuteri* CRL 1098 (10 cells/day) to hypercholesterolemic mice for 7 days decreased total cholesterol by 38 per cent, producing serum cholesterol concentrations similar to that of the control group. This dose of *Lactobacillus reuteri* caused a 40 per cent reduction in triglycerides and a 20 per cent increase in the ratio of high density lipoprotein to low density lipoprotein without bacterial translocation of the native microflora into the spleen and liver (Taranto *et al.*, 1998). Similarly, the administration of *Lactobacillus reuteri* to hypercholesterolemic mice for 60 days, a significant reduction in serum cholesterol level was noticed in group IV (probiotic fed group), with reduction in total cholesterol by 33.36 per cent (79.24mg/dl), increase in HDL cholesterol by 29.30 per cent (46.33mg/dl), reduction in triglycerides level by 25.25 per cent (59.24mg/dl), reduction in LDL cholesterol by 68 per cent (21.06mg/dl), reduction of VLDL by 25.23 per cent (11.85mg/dl) and reduction in Atherogenic index by 79.93 per cent (0.45) (Katkamwar, 2012).

Cholesterol lowering effects may be due in part to the deconjugation of bile salts by strains of bacteria that produce the enzyme bile salt hydrolase. As deconjugated bile salts are more readily excreted in the feces than conjugated bile salts, bacteria with BSH activity may effectively reduce serum cholesterol by enhancing the excretion of bile salts, with a consequent increase in the synthesis of bile salts from serum cholesterol; or by decreasing the solubility of cholesterol, and thus reducing its uptake from the gut (Nguyen *et al.*, 2007). Thus, these studies proved the probiotic effect of *L. reuteri* strains for the prevention of hypercholesterolemia and fatty liver disease.

Immunomodulatory Activity

Modulation of the immune system is the important beneficial health attribute of probiotic microorganisms. They have been found to enhance immune responses by proliferation of T-cell and B-cell, cytokine, immunoglobulin (IgA, IgE) production etc., which directly and indirectly help in the prevention of certain diseases like atopic eczema, allergy (a hypersensitive reaction), etc. (Kimoto *et al.*, 2007). Christensen *et al.* (2002), showed that *L. reuteri* had the ability to inhibit induction of proinflammatory cytokines interleukin (IL)-12, IL-6, and TNF-α in Murine Dendritic Cells (DCs). The priming of DCs by *L. reuteri*, which was initiated by the binding of C-type Lectin DC-specific intercellular adhesion molecule 3-grabbing nonintegrin (DCSIGN), resulted in an induction of regulatory T cells *in vitro* (Smits and others 2005). A similar down-regulation of proinflammatory cytokines (*e.g.*, TNF-α) by *L. reuteri* was also observed with macrophages, lipopolysaccharide-activated monocytes, and primary monocyte derived macrophages from children with Crohn's disease (Peña *et al.*, 2005; Lin *et al.*, 2008, Thomas *et al.*, 2012). The physiological relevance of the immune effects of *L. reuteri* has been quite recently demonstrated *in vivo* using *Lactobacillus*-free (LF) mice (Hoffmann *et al.*, 2008; Livingston *et al.*, 2010). In these animals, administration of *L. reuteri* resulted in a transient activation of proinflammatory cytokines and chemokines produced by intestinal epithelial cells in the jejunum and ileum (Hoffmann *et al.*, 2008). Also oral treatment with live *L. reuteri* can attenuate major characteristics of an asthmatic response in a mouse model of allergic airway inflammation (Forsythe *et al.*, 2006).

Anti-Mutagenic Activity

Cancer (malignant neoplasm) is a class of diseases in which a group of cells display uncontrolled growth (division beyond the normal limits), invasion (intrusion on and destruction of adjacent tissues) and sometimes metastasis (spread to other locations in the body via lymph or blood). A number of strains of *L. reuteri* have been tested under both *in vitro* and *in vivo* conditions to evaluate their anti-mutagenic effect. Epidemiological studies indicate an association between the risk of colon cancer development and consumption of high fat diets. It has been suggested that this effect is due to the raised level of bile acids being passed into the colon. To aid in the digestion of fats, bile salts, conjugated to taurine or glycine moieties are released into the small intestine and reabsorbed in this location. A proportion however, may pass into the colon, where bacterial action can metabolize them. The bile acids are initially deconjugated in the colon and they further undergo a dehydroxylation reaction of the 7α-hydroxyl group, forming the detrimental secondary bile acids, deoxycholic and lithocholic acids (Kitahara *et al.*, 2000). These acids are believed to exert a cytotoxic effect upon epithelial cells, thereby increasing cell proliferation and leading to a higher probability of colon cancer development (Ling *et al.,* 1995). Modulation of the intestinal microflora through probiotic consumption may affect the activity of 7α-dehydroxylase (Kitahara *et al.,* 2001). Whereas, earlier deBoever *et al.* (2000) suggested that the probiotics may physically bind the bile salts; using the hydrolase active *L. reuteri* in a simulated human intestinal microbial ecosystem and attempt to control cell toxicity in the presence of bile salts. Theoretically, *L. reuteri* should have increased the bioavailability of secondary bile acids; however, decreased cell toxicity was observed with the *L. reuteri* treatment, suggesting that, despite activating secondary bile acids, it also appears to protect against their toxic effects, potentially by binding the bile salts.

Antimicrobial Activity

Accumulated evidence has revealed probiotic efficacy of this universally indigenous species amongst various hosts, is especially due to its strong antimicrobial activity. It has been found that *L. reuteri* cells can strongly adhere to host GI epithelia via specific binding mechanisms and specialized surface proteins (Wadstrom *et al.,* 1987; Roos *et al.,* 1999; Roos and Johnsson 2002). This strong adhesion of *L. reuteri* has assured the exclusive competition capability against other microorganisms competing for the same niche, besides achieving a direct impact of secreted metabolites to the host's mucosal defenses. *L. reuteri* produces lactate, acetate, and perhaps a variety of other SCFAs along the heterofermentative metabolism pathway. Although it is certain that the acids partially contribute to the overall antimicrobial activity, this part of the review puts emphasis on those *L. reuteri*-specific substances and associated activities which are reported as unique to this species.

Reuterin

Reuterin is a pH neutral, water soluble, low molecular weight substance, which is non-bacteriocin and resistant to nuclease, protease and lipolytic enzymes. It is active over a wide range of pH values and capable of inhibiting growth of a wide spectrum of microorganisms, but it is labile to heat (100°C for 10 minutes) (Talarico *et al.,* 1988; Axelsson *et al.,* 1989; Dorogosz and Lindgren 1995, El-Ziney *et al.,* 1999, Mishra *et al.,* 2012). Reuterin was the first low molecular weight antimicrobial substance from *Lactobacillus* species ever to be chemically identified (Talarico and Dobrogosz 1989), and has been intensively studied and understood (Vollenweider *et al.,* 2003). Reuterin has been shown to exist in solution as an equilibrium mixture of three chemical compounds derived from glycerol dissimilation, namely the 3-HPA (3- hydroxy propionaldehyde) system, containing monomeric, hydrated monomeric and cyclic dimeric forms of 3-HPA (Talarico and Dobrogosz, 1989). Talarico and Dobrogosz (1989) were the first to chemically characterize reuterin. The unique and most attractive feature of reuterin is its strong antimicrobial activity. It has been found that concentrations of reuterin in the range of 15-30 μg/ml effectively inhibit growth of Gram-positive and Gram-negative bacteria, and lower eukaryotic organisms including yeast, fungi and protozoa; while much higher concentrations are required to kill lactic acid bacteria, including *L. reuteri* itself (Axelsson *et al.,* 1989; Chung *et al.,* 1989; Casas and Dobrogosz 2000, Mishra *et al.,* 2011, Mishra *et al.,* 2012).

Reutericin

Some strains of *L. reuteri* are also known to produce Reutericin 6, which is a cyclic class II bacteriocin that mainly targets Gram-positive bacteria and closely related *Lactobacillus* species (Kabuki *et al.*, 1997). Reutericin 6 was first reported by Toba *et al.* (1991) as produced by a *L. reuteri* strain LA6, isolated from human infant feces. Purified Reutericin 6 was reported to be hydrophobic and have a molecular weight of 5.6 kDa (Kawai *et al.*, 2001). The antimicrobial activity of Reutericin 6 has been tested against commercial strains including *L. acidophilus, L. delbrueckii* subsp. *bulgaricus* and, *L. delbrueckii* subsp. *lactis,* and found to be bacteriolytic. Similar to other bacteriocins, the antimicrobial mechanism of this extracellular protein is believed to be the formation of pores in the cell membranes of the target bacteria, causing membrane depolarization and efflux of small circular components resulting in cell death (Kabuki *et al.*, 1997; Kawai *et al.*, 2001; Kawai *et al.*, 2004). However, Reutericin 6 was also reported to be sensitive to proteolytic enzymes and low pH (pH 2-3) (Toba *et al.*, 1991, Kawai *et al.*, 2004).

Reutericyclin

Another low molecular weight compound isolated from *L. reuteri* LTH2584 is reutericyclin. It is a naturally occurring, amphiphilic, tetramic acid, and its optimal formation was observed between pH 4 and 5 (Holtzel *et al.*, 2000). It has been reported to have bactericidal and bacteriostatic activity against many Gram-positive species, but does not affect Gram-negative bacteria (Ganzle *et al.*, 2000). The mode of action of reutericyclin is to act as a proton ionophore, translocating proton across the cell membrane and dissipating the transmembrane pH potential (Ganzle and Vogel 2003). However, reutericyclin has not been identified in reuterin producing *L. reuteri* strains (Ganzle 2004).

Thus, it can be considered that the overall antimicrobial efficacy of *L. reuteri* results from contributions of all the above components. However amongst these compounds, reuterin has drawn the most attention and has been believed to have the most potential for industrial applications.

Functional Properties of *Lactobacillus reuteri*

Besides their several probiotic properties, it has been found that *Lactobacillus reuteri* strains also produce some functional biomolecules such as conjugated linoleic acid (CLA), glutamate, Vitamin B_{12} and exopolysaccharides, which have several therapeutic effects such as anti-microbial, anti-oxidative, anti-thrombotic, anti-hypertensive, anti-microbial and immunomodulatory.

Exopolysaccharides (EPS) Production

Exopolysaccharides (EPS) include a diverse range of molecules that play vital roles in a variety of biological processes. EPS-producing lactic acid bacteria have received growing attention in recent years because of their wide applications in food industry (as food stabilizers, gelling agents). Besides this, EPS also have several health benefits such as enhancement in adhesion of probiotics to intestinal mucosa (Nakajima *et al.*, 1992; Yadav *et al.*, 2007,) by acting as prebiotics (Yadav *et al.*, 2007), reduction of blood cholesterol level (Ruas-Madiedo *et al.,* 2006), anticarcinogenic ability (Kitazawa *et al.,* 1992) and immuno-stimulation (Kleerebezem *et al.,* 1999; deVos *et al.,* 1998; Kitazawa *et al.,* 1992; Cerning 1990). Bacterial EPS have a protective function in the natural environment against desiccation, phagocytosis and predation by protozoa, phage attack, antibiotics or toxic compounds and osmotic stress. Moreover they also play a role in cell recognition, adhesion to surfaces besides facilitating with the colonisation of various ecosystems (Looijestejn *et al.,* 2001). In the food processing industry, EPS from LAB can be used as viscosifying, stabilising, emulsifying, gelling, or water-binding agents. Many strains of lactobacilli produce homopolysaccharides (HoPS) (which in general are produced outside the cell) and oligosaccharides (OS) consisting of either glucose residues (glucans and gluco-oligosaccharides, GOS) or fructose residues (fructans and fructo-oligosaccharides, FOS) from sucrose by the extracellular enzymes glucosyltransferases and fructosyltransferases, respectively (Gänzle *et al.,* 2005). Strains of *L. reuteri* produce glucans and fructans of different linkage types and express the *gtf*A and *inu* genes encoding a Glucosyltransfrase (GTFA) and an inulosucrase, respectively (Schwab *et al.,* 2006). Besides GTFA, two other GTFs from different *L. reuteri* strains that have been characterized are GTF180 and GTFML1. These three enzymes are highly similar in terms of structures and amino acid sequences, but nevertheless synthesize different glucan products. The gene *gtfO* is present in *L. reuteri* ATCC 55730 and the homopolysacharide and the corresponding enzyme in *L. reuteri* ATCC 55730 have also been characterised as a reuteran and reuteransucrase (GTFO), respectively (Kralj *et al.,* 2005). The concentrations of sucrose needed to yield significant glucan polymer production may not normally be achieved in the gut. In the oral cavity on the other hand, GTFO may contribute to polymer formation and colonization on oral surfaces. Some *L. reuteri* strains possess glucan-binding proteins that contribute to coaggregation, even though they do not produce glucan. Inu is a glucan-binding protein and a receptor for the glucan produced by GTFA (Walter *et al.,* 2008). In a recent study by Wang *et al.* (2010) it was observed that exopolysaccharides produced by strains of *Lactobacillus reuteri* inhibited ETEC-induced hemagglutination of porcine erythrocytes

Production of Conjugated Linoleic Acid (CLA)

Conjugated linoleic acid (CLA) refers to a group of positional and geometric isomers of linoleic acid (LA; C_9C_{12}-$C_{18:2}$) (MacDonald, 2000). The major isomer, c9t11octadecadienoic acid has been suggested to be antioxidative, anticarcinogenic (Pariza and Hargraves 1985) and antiatherosclerotic (Lee *et al.,* 1994, Wilson *et al.,* 2000, Mitchell and Mcleod, 2008). Conjugated linoleic acid is produced by chemical synthesis (Christie and others 1997), by rumen bacteria (Bauman *et al.,* 1999) or by lactic acid bacteria (LAB) (Ogawa and others 2001, van Nieuwenhove *et al.,* 2007; Wall *et al.,* 2008). Pariza and Yang (1998) found a probiotic *Lactobacillus reuteri* strain, that very effectively enzymatically transforms LA into CLA. In an another study by Mireya (2005) he found ability of *L. reuteri* 55739 to produce CLA and it was primarily exogenous CLA (mainly c9t11-C18:2).

Glutamate Production

Monosodium glutamate (MSG) gives the taste "umami" which has been widely recognized as the fifth basic taste besides sweet, acid, salty and bitter (Kurihara *et al.*, 2009). It has been widely used as a flavor enhancer in the food industry. However, there is some disagreement about the safety of MSG as some people experience side effects such as wheezing, changes in heart rate and difficulty in breathing (Farombi and Onyema 2006). Interest in developing a natural flavor enhancer as the alternative to MSG has been increased. Among the yeast and fungi glutaminase is one of the most important enzymes used to enhance the flavor of food (Nandakumar *et al.*, 2003, Kurihara *et al.*, 2009). The deamidation of glutamine to glutamate by sourdough lactobacilli generates umami taste in sourdough bread, and improves growth of *L. reuteri* at low pH (Vermeulen *et al.*, 2007).

Also, glutaminase plays an important role in soy sauce fermentation hydrolyzing L-glutamine to produce L-glutamate which is a high flavor amino acid in foodstuffs (Masuo *et al.*, 2004). Especially, salt tolerant and heat stable glutaminase has been receiving much attention for applications in both the food and pharmaceutical industries (Yoshimune *et al.*, 2004).The glutaminase gene was cloned from *L. reuteri* KCTC3594 by PCR, and subsequently introduced into two Korean isolates of *Lactobacillus* species. All of the transformants harboring the glutaminase gene from *L. reuteri* KCTC3594 were able to elevate glutaminase activity when introduced into other lactobacilli (Jeon *et al.*, 2009). A gene coding for a putative glutamate decarboxylase, *gadB*, was identified in the genome of *L. reuteri* 100-23 (Su *et al.*, 2011).

Cobalamin (Vitamin B_{12}) Synthesis

Vitamin B_{12} (B_{12}), also termed cobalamin, is the generic name for a group of biologically active corrinoids that serves as a cofactor essential to cell growth across all branches of life. Only a few archaea and bacteria have the ability to synthesize this relatively large molecule. *De novo* synthesis starts with glutamate and requires over 30 gene products to produce an active form of B_{12} (Martens *et al.*, 2002). Although many prokaryotes require B_{12} as a cofactor, many do not have genes for de novo synthesis. A report suggests that of the sequenced bacterial genomes, only half that utilize B_{12} can synthesize it (Zhang *et al.*, 2009).

The ability to synthesize cobalamin *de novo* is not widely distributed among human intestinal bacteria or bacteria found in food fermentations (Roth *et al.*, 1996). *L. reuteri* is not only able to produce vitamin B_{12} (Taranto *et al.*, 2003), a unique property among lactobacilli that are generally characterized by multiple vitamin and amino acid auxotrophies, but also one of the few *Lactobacillus* species commonly associated with both habitats. In situ microbial B_{12} production is a convenient strategy to achieve natural enrichment of fermented foods, notably from vegetable sources. Strain CRL1098 and JCM1112 has been reported to produce different forms of B_{12} (Taranto *et al.*, 2003; Santos *et al.*, 2007; Santos *et al.*, 2008). *L. reuteri* CRL1098 encodes the complete machinery necessary for *de novo* synthesis of vitamin B_{12} in a single chromosomal gene cluster (Santos and others 2008; Santos and others 2011). This cluster was shown to be very similar to that present in various representatives of γ-Proteobacteria, standing out against canonical phylogeny. Complete genome sequence analysis of the type strain of *L. reuteri* revealed that the region immediately upstream of the vitamin B_{12} biosynthesis cluster maintains a gene order similar to that of *Salmonella* (Morita *et al.,* 2008). In one of the study, Vannini *et al.* (2011) cloned, expressed and characterized the gene in *Lb. reuteri* that codes for the S-adenosy l-methionine uroprophyrinogen III methyltransferase/synthase (CobA/HemD), a key bifunctional enzyme in the biosynthesis of cobalamin and other tetrapyrrols. Therefore, keeping all studies in mind *L. reuteri* should be considered as a potential source of cobalamin for human consumption.

Conclusions

During the last few years, the use of probiotics as health supplements/food has increased around the world. *Lactobacillus* and *Bifidobacteria* are the most commonly used commercial probiotics cultures. But in dairy fermentation industry, discovery of new probiotics strains is an exciting developments and isolation of probiotic strains of *Lactobacillus reuteri* could be suitable because they grow in milk easily as compared to other Lactobacilli and Bifidobacteria. Moreover, *L. reuteri* has been used as an important dairy starter since hundreds of years and has GRAS status too. In recent years, available literatures clearly add weight to prove potentials of *Lactobacillus reuteri* strains as potential probiotic and functional Lactic acid bacteria (LAB). According to these studies *Lactobacillus reuteri* strains can adhere, survive and pass through GI tract, may remove cholesterol, modulate immune system and also showed anti-mutagenic and anti-diabetic activity. Besides this, *Lactobacillus reuteri* strains produce some functional molecules such as conjugated linoleic acid (CLA), glutamate, vitamin B_{12} and exopolysaccharides. Therefore, we would like to propose *Lactobacillus reuteri* sp. as new potential probiotic bacteria having multifaceted characteristics for their use in the production of fermented dairy products.

References

Abrahamsson, TR, Jakobsson, T, Bottcher, MF, Fredrikson, M, Jenmalm, MC, Bjorkston, B, Oldaeus, G. 2007. Probiotics in Prevention of IgE-associated Eczema: A Double-Blind, Randomized, Placebo-Controlled Trial. *Journal of Allergy and Clinical Immunology* 119: 1174-1180.

Axelsson, LT, Chung, TC, Dobrogosz, WJ, Lindgren, SE. 1989. Production of a Broad Spectrum Antimicrobial Substance by *Lactobacillus reuteri. Microbial Ecology in Health and Disease* 2: 131-136.

Bauman, DE, Baumgard, LH, Corl, BA, Griinari, JM. 1999. Biosynthesis of Conjugated Linoleic Acid in Ruminants. *Journal of Animal Science* 77*:* 1-15.

Bottcher, MF, Abrahamsson, TR, Fredriksson, M, Jakobsson, T, Bjorksten, B. 2008. LowBreast Milk TGF-Beta2 is induced by *Lactobacillus reuteri* Supplementation and Associates with Reduced Risk of Sensitization during Infancy. *Pediatric Allergy and Immunology* 19: 497-504.

Casas, IA, Dobrogosz, WJ. 1997. *Lactobacillus reuteri*: An Overview of a New Probiotic for Humans and Animals. *Microecology and Therapy* 25: 221-231.

Casas, IA, Dobrogosz, WJ. 2000. Validation of the Probiotic Concept: *Lactobacillus reuteri* Confers Broad-Spectrum Protection against Disease in Humans and Animals. *Microbial Ecology in Health and Disease* 12: 247–285.

Cerning, J. 1990. Exocellular Polysaccharides Produced by Lactic Acid Bacteria. *FEMS Microbiology Review* 87: 113-130.

Christensen, HR, Frokiaer, H, Pestka, JJ. 2002. Lactobacilli Differentially Modulate Expression of Cytokines and Maturation Surface Markers in Murine Dendritic Cells. *Journal of Immunology* 168: 171-178.

Christie, WW, Dobson, G, Gunstone, FD. 1997. Isomers in Commercial Samples of Conjugated Linoleic Acid. *Lipids* 32: 1231-1232.

Chung, TC., Axelsson, L, Lindgren, SE, Dobrogosz, WJ. 1989. *In vitro* studies on Reuterin Synthesis by *Lactobacillus reuteri. Microbial Ecology in Health and Disease* 2: 137-144.

de Smet, I, de Boever, P, Verstraete, W. 1998. Cholesterol Lowering in Pigs through Enhanced Bacterial Bile Salt Hydrolase Activity. *British Journal of Nutrition* 79: 185-194.

de Vos, WM, Hols, P, Kranenburg, RV, Luesink, E, Kuipers, O, Oost, JVD, Kleerebezem, M, Hugenholtz, J. 1998. Making More of Milk Sugar by Engineering Lactic Acid Bacteria. *International Dairy Journal* 8: 227-233.

Dobrogosz, WJ, Lindgren, SE 1995. Stockholm, Sweden Patent No. 5413960 U.S. Patent.

El-Ziney, MG, Arneborg, N, Uyttendaele, M, Debevere, J, Jakobsen, M. 1998. Characterization of Growth and Metabolite Production of *Lactobacillus reuteri* during Glucose/glycerol Co-fermentation in Batch and Continuous Cultures. *Biotechnology Letters* 20: 913-916.

El-Ziney, MG, Van Den Tempel, T, Debevere, J, Jakobsen, M. 1999. Application of Reuterin Produced by *Lactobacillus reuteri* 12002 for Meat Decontamination and Preservation. *Journal of Food Protection* 62: 257-261.

Erkelens, DW, Baggen, MGA, Van Doormeal, JJ, Kettner M., Koningsberger, JC, Mol, MJTM. 1988. Clinical experience with simvastatin compared with cholestyramine. *Drugs* 39: 87–90.

Fabia, R, Ar'Rajab, A, Johansson, ML, Willen, R, Anderson, R, Molin, G, Bengmark, S. 1993. The Effect of Exogenous Administration of *Lactobacillus reuteri* R2LC and Oat Fiber on Acetic Acid-Induced Colitis in the Rat. *Scandinavian Journal of Gastroenterology* 28: 155-162.

FAO/WHO. 2000. *Health and Nutritional Properties of Probiotics in Food including Powder Milk with Live Lactic Acid Bacteria*. Report of a Joint FAO/WHO Expert Consultation on Evaluation of Health and Nutritional Properties of Probiotics in Food Including Powder Milk with Live Lactic Acid Bacteria.

Farombi, EO, Onyema, OO. 2006. Monosodium Glutamate-induced Oxidative Damage and Genotoxicity in the Rat: Modulatory Role of Vitamin C, Vitamin E and Quercetin: Human and Experimental. *Toxicology* 25: 251- 259.

Fink, LN, Zeuthen, LH, Christensen, HR, Morandi, B, Frokiaer, H, Ferlazzo, G. 2007. Distinct Gut-derived Lactic Acid Bacteria Elicit Divergent Dendritic Cell-Mediated NK Cell Responses. *International Immunology* 19: 1319-1327.

Forsythe, P, Wattie, J, Inman, M, Bienenstock, J. 2007. Oral Treatment with Live *Lactobacillus reuteri* inhibits the Allergic Airway Response in Mice. American *Journal of Respiratory and Critical Care Medicine* 175: 151-159.

Fuller, R. 1989. Probiotics in Man and Animals. *Journal of Applied Bacteriology* 66: 365-378.

Ganzle, MG, Holtzel, A, Walter, J, Jung, G, Hammes, WP. 2000. Characterization of Reutericyclin Produced by *Lactobacillus reuteri* LTH2584. *Applied and Environmental Microbiology* 66: 4325-4333.

Ganzle, MG, Vogel, RF. 2003. Studies on the Mode of Action of Reutericyclin. *Applied and Environmental Microbiology* 69: 1305-1307.

Ganzle, MG. 2004. Reutericyclin: Biological Activity, Mode of Action, and Potential Applications. *Applied and Environmental Microbiology* 64: 326-332.

Gardiner, G, Heinemann, C, Baroja, ML, Bruce, AW, Beuerman, D, Madrenas, J. 2002. Oral Administration of the Probiotic Combination *L. rhamnosus* GR-1 and *L. fermentum* RC-14 for Human Intestinal Applications. *International Dairy Journal* 12: 191-196.

Hoffmann, M, Rath, E, Holzlwimmer, G, Quintanilla-Martinez, L, Loach, D, Tannock, G, Haller, G. 2008. *Lactobacillus reuteri* 100-23 Transiently Activates Intestinal Epithelial Cells of Mice that have a Complex Microbiota during Early Stages of Colonization. *The Journal of Nutrition* 138: 1684-1691.

Holtzel, A, Ganzle, MG, Nicholson, GJ, Hammes, WP, Jung, G. 2000. *The First Low Molecular Weight Antibiotic from Lactic Acid Bacteria: Reutericyclin, a New Tetramic Acid*. Angewandte Chemie International Edition 39: 2766-2768.

Indrio, F, Riezzo, G, Raimondi, F, Bisceqlia, M, Cavallo, M, Francavilla, R. 2008. The Effects of Probiotics on Feeding Tolerance, Bowel Habits, and Gastrointestinal Motility in Preterm Newborns. *The Journal of Pediatrics* 152: 801-806.

Jeon, J, Lee, H, So, J. 2009. Glutaminase Activity of *Lactobacillus reuteri* KCTC3594 and Expression of the Activity in other *Lactobacillus* spp. by Introduction of the Glutaminase Gene. *African Journal of Microbiology Research* 3: 605-609.

Kabuki, T, Saito, T, Kawai, Y, Uemura, J, Itoh, T. 1997. Production, Purification and Characterization of Reutericin 6, a Bacteriocin with Lytic Activity Produced by *Lactobacillus reuteri* LA6. *International Journal of Food Microbiology* 34: 145-156.

Kandler, O, Stetter, KO, Kohl, R. 1980. *Lactobacillus reuteri* sp. nov., a new species of heterofermentative lactobacilli. Zentralblattfür Bakteriologie, Mikrobiologie, und Hygiene, 1 *Abt Origin*(C): 264-269.

Kandler, O, Weiss, N. 1986. Regular, nonsporing Gram-positive rods. In Sneath, PHA, Mair, NS, Sharpe, ME, Holt, JG (Ed.), *Bergey's Manual of Systematic Bacteriology* (Vol. 2). New York: Williams and Wilkins.

Katkamwar, S. 2012. Evaluation of probiotic properties of *Lactobacillus reuteri* strain using animal model. *Master's Thesis*, National Dairy Research Institute, Karnal, Haryana, India.

Kawai, Y, Ishii, Y, Arakawa, K, Uemura, K, Saitoh, B, Nishimura, J, Kitazawa, H, Yamazaki, Y, Tateno, Y, Itoh, T, Saito, T. 2004. Structural and Functional Differences in Two Cyclic Bacteriocins with the Same Sequences Produced by *Lactobacilli. Applied and Environmental Microbiology* 70: 2906-2911.

Kawai, Y, Ishii, Y, Uemura, K, Kitazawa, H, Saito, T, Itoh, T. 2001. *Lactobacillus reuteri* LA6 and *Lactobacillus gasseri* LA39 Isolated from Faeces of the Same Human Infant Produce Identical Cyclic Bacteriocin. *Food Microbiology* 18**:** 407-415.

Kimoto, H, Mizumachi, K, Masaru, N, Miho, K, Yasuhito, F, Okamoto, T, Ichirou, S, Noriko, MT, Kurisaki, J, Sadahiro,O. 2007. *Lactococcus* sp. as Potential Probiotic Lactic Acid Bacteria. *Japan Agricultural Research Quarterly* 41: 181-189.

Kimoto, H. 2000. *In vitro* studies on probiotic properties of *Lactococci. Milchwissenshaft* 55: 245-249.

Kitahara, M, Takamine, F, Imamura, T, Benno, Y. 2000. Assignment of *Eubacterium* sp. VPI 12708 and Related Strains with High Bile Acid 7 α-dehydroxylating Activity to *Clostridium scindens* and proposal of *Clostridium hylemonae* spp. Isolated from Human Faeces. *International Journal of Systematic and Evolutionary Microbiology* 50: 971-978.

Kitahara, M, Takamine, F, Imamura, T, Benno, Y. 2001. *Clostridium hiranonis* sp. nov., a Human Intestinal Bacterium With Bile Acid 7α-dehydroxylating Activity. *International Journal of Systematic and Evolutionary Microbiology* 51: 39-44.

Kitazawa, H, Yamaguchi, T, Itoh, T. 1992. B-cell Mitogenic Activity of Slime Products Produced from Slime-Forming, Encapsulated *Lactococcus lactis* ssp. *cremoris. Journal of Dairy Science* 75: 2946-2950.

Kleerebezem, M, Kranenburg, RV, Tuinier, R, Boels, IC, Zoon, P, Looijesteijn, E, Hugenholtz, J, and DeVos, WM. 1999. Exopolysaccharides Produced by *Lactococcus lactis*: from Genetic Engineering to Improved Rheological Properties. *Antonie Van Leeuwenhoek* 76: 357-365.

Kralj, S, Stripling, E, Sanders, P, vanSchutten, G, Dijkhuizen, L. 2005. Highly Hydrolytic Reuteransucrase from Probiotic *Lactobacillus reuteri* strain ATCC 55730. *Applied Environmental Microbiology* 71: 3942-3950.

Kurihara K. 2009. Glutamate: from discovery as a food flavor to role as a basic taste (umami). *American Journal of Clinical Nutrition* 90: 719S–22S.

Lee, KN, Kritchevsky, D, Pariza, MW. 1994. Conjugated Linoleic Acid and Atherosclerosisin Rabbits. *Atherosclerosis* 108: 19-25.

Lee, YK, Salminen, S. 1995. The Coming of Age of Probiotics. *Trends in Food Science Technology* 6: 241-245.

Lin, JHC, Savage, DC. 1984. Host Specificity of the Colonization of Murine Gastric Epithelium by *lactobacilli*. *FEMS Microbiology Letters* 24: 67-71.

Lin, YP, Thibodeaux, CH, Pena, JA, Ferry, GD, Versalovic, J. 2008. Probiotic *Lactobacillus reuteri* Suppress Proinflammatory Cytokines via C-Jun. *Inflammatory Bowel Disease 14*: 1068-1083.

Ling, WH. 1995. Diet and Colonic Microflora Interaction in Colorectal Cancer. *Nutrition Research* 15: 439-454.

Livingston, M, Loach, D, Tannock, GW and Baird, M. 2010. Gut Commensal *Lactobacillus reuteri* 100-23 stimulates an Immunoregulatory response. *Immunology and Cell Biology* 88: 99-102.

Ljungh, A, Wadstrom, T. 2007. Lactic Acid Bacteria. *Current Issues Intestinal Microbiology* 7: 73-90.

Looijestejn, PJ, Trapet, L, deVries, E, Abee, T, Hugenholtz, J. 2001. Physiological Function of Exopolysaccharides Produced by *Lactococcus lactis*. International *Journal of Food Microbiology* 64: 71-80.

Lüthi-Peng, Q, Dileme, FB, Puhan, Z. 2002. Effect of Glucose on Glycerol Bioconversion by *Lactobacillus reuteri*. *Applied Microbiology and Biotechnology* 59: 289-296.

MacDonald, HB. 2000. Conjugated Linoleic Acid and Disease Prevention: A Review of Current Knowledge. *Journal of the American College of Nutrition* 19: 111S-8S.

Madsen, KL, Doyle, JS, Jewell, LD, Tavernini, MM, Fedorak, RN. 1999. *Lactobacillus* species Prevents Colitis in Interleukin 10 Gene-Deficient Mice. *Gastroenterology* 116: 1107-1114.

Martens JH, Barg H, Warren MJ, Jahn D. 2002. Microbial production of vitamin B_{12}. *Appllied Microbiology and Biotechnology* 58: 275-285.

Masuo, N, Yoshimune, K, Ito, K, Matsushima, K, Koyama, Y, Moriguchi, M. 2005. *Micrococcusluteus* K-3-type Glutaminase from *Aspergillus oryzae* RIB40 is Salt-Tolerant. *Journal of Bioscience and Bioengineering* 100: 576-78.

Mireya, R.N., 2005, Production of Conjugated Linoleic Acid by *Lactobacillus reuteri*. *Master's Thesis*, Oklahoma State University, USA.

Mishra, S.K., Malik, R.K., Kaur, G., Manju, G., Pandey, N. and Singroha, G. 2011. Potential Bioprotective Effect of Reuterin Produced by *L. reuteri* BPL-36 Alone and in Combination with Nisin against Food Borne Pathogens. *Indian Journal of Dairy Science* 64(5): 406-411.

Mishra, S.K., Malik, R.K., Manju, G., Pandey, N., Singroha, G., Bahare, P. and Kaushik J.K. 2012. Characterization of a Reuterin-producing *Lactobacillus reuteri* BPL-36 Strain Isolated from Human Infant Fecal Sample. *Probiotics and Antimicrobial Proteins*. **4**(3): 154-161.

Mitchell, P.L and Mcleod, R.S. 2008. Conjugated linoleic acid and atherosclerosis: studies in animal models. *Biochemical Cell Biology* 86(4): 293–301.

Molin, G, Johansson, M.L, Stahl, M, Ahrne, S, Andersson, R, Jeppsson, B, Bengmark, S. 1992. Systematics of the *Lactobacillus* Population on Rat Intestinal Mucosa with Special Reference to *Lactobacillus reuteri*. *Antonie Van Leeuwenhoek* 61: 175-183.

Møller, PL, Paerregaard, A, Gad, M, Kristensen, NN, Claesson, MH. 2005. Coliticscid Mice Fed *Lactobacillus* spp. show an Ameliorated Gut Histopathology and an Altered Cytokine Profile by Local T cells. *Inflammatory Bowel Diseases* 11: 814-819.

Morita H, Toh H, Fukuda S, Horikawa H, Oshima K, Suzuki T, Murakami M, Hisamatsu S, Kato Y, Takizawa T. 2008. Comparative genome analysis of *Lactobacillus reuteri* and *Lactobacillus fermentum* reveal a genomic island for reuterin and cobalamin production. *DNA res*, 15(3): 151-161.

Nakajima, H, Suzuki, Y, Kaizu, H, Hirota, T. 1992. Cholesterol Lowering Activity of Ropy Fermented Milk. *Journal of Food Science* 57: 1327-1329.

Nandakumar, R, Kazuaki, Y, Mamoru, W, Mitsuaki, M. 2003. Microbial Glutaminase: Biochemistry, Molecular Approaches and Applications in the Food Industry. *Journal of Molecular Catalysis B: Enzymatic* 23: 87-100.

Nguyen, TDT, Kang, JH, Lee, MS. 2007. Characterization of *Lactobacillus plantarum* PH04, a Potential Probiotic Bacterium with Cholesterol-Lowering Effects. *International Journal of Food Microbiology* 113: 358-361.

Ogawa, J, Matsumura, K, Kishino, S, Omura, Y, Shimizu, S. 2001. Conjugated Linoleic Acid Accumulation via 10-Hydroxy-12-Octadecaenoic Acid during MicroaerobicTransformation of Linoleic Acid by *Lactobacillus acidophilus*. *Applied and Environmental Microbiology* 67: 1246-52.

Orla-Jensen, S. 1919. *The Lactic Acid Bacteria*, (*Copenhagen*: A.F. Host and Son).

Pariza MW, Yang X-Y, inventors; Wisconsin Alumni Research Foundation, assignee. 1998 Oct 13. Method of Producing Conjugated Fatty Acids. U.S. Patent 6060304.

Pariza, MW, Hargraves,WA. 1985. A Beef-derived Mutagenesis Modulator inhibits Initiation of Mouse Epidermal Tumors by 7, 12-dimethylbenz[a]anthracene. *Carcinogenesis* 6: 591-593.

Pena, JA, Rogers, AB, Ge, Z, Ng, V, Li, SY, Fox, JG, Versalovic, J. 2005. Probiotic *Lactobacillus* spp. Diminish *Helicobacter hepaticus* Induced Inflammatory Bowel Disease in Interleukin-10-Deficient Mice. *Infection and Immunity* 73: 912-920.

Reid, G, Charbonneau, D, Gonzalez, S, Gardiner, G, Erb, J, Bruce, AW. 2002. Ability of *Lactobacillus* GR-1 and RC-14 to Stimulate Host Defenses and Reduce Gut Translocation and Infectivity of *Salmonella typhimurium*. *Nutraceutical Foods* 7: 168-173.

Reuter, G. 1965. Das vorkommen von laktobazillen in lebensmitteln und ihrverhaltenimmenschlichenintestinaltrakt. *Zbl Bak Parasit Infec Hyg I Orig* 197(Supl): 468-487.

Reuter, G. 2001. The *Lactobacillus* and *Bifidobacterium* Microflora of the Human Intestine: Composition and Succession. *Current Issues in Intestinal Microbiology* 2: 43-53.

Roos, S, Jonsson, H. 2002. A High-Molecular-Mass Cell-Surface Protein from *Lactobacillus reuteri* 1063 Adheres to Mucus Components. *Microbiology* 148: 433-442.

Roos, S, Lindgren, S, Jonsson, H. 1999. Autoaggregation of *Lactobacillus reuteri* is mediated by a Putative DEAD-Box Helicase. *Molecular Microbiology* 32: 427-436.

Roth, JR, Lawrence, JG, Bobik. TA. 1996. Cobalamin (Coenzyme B_{12}): Synthesis and Biological Significance. *Annual Review of Microbiology* 50: 137-181.

Ruas-Madiedo, P, Gueimonde, M, Reyes-Gavila, CG, Salminen, S. 2006. Effect of Exopolysaccharide Isolated from "Villi"on the Adhesion of Probiotics and Pathogens to Intestinal Mucus. *Journal of Dairy Science* 89: 2355-2358.

Salminen, S, VonWright, A, Ouwehand, A. 1998. *Lactic Acid Bacteria: Microbiological and Functional Aspects,* (New York: Marcel Dekker Inc).

Santos F, Spinler JK, Saulnier DMA, Molenaar D, Teusink B, de Vos WM,Versalovic J, Hugenholtz J. 2011. Functional identification in *Lactobacillus reuteri* of a PocR-like transcription factor regulating glycerol utilization and vitamin B_{12} synthesis. *Microbial Cell Factories* 10: 55.

Santos, F, Vera, JL, Lamosa, P, de Valdez, GF, de Vos, WM, Santos, H, Sesma, F, Hugenholtz, J. 2007. Pseudovitamin B_{12} is the corrinoid produced by *Lactobacillus reuteri* CRL1098 under anaerobic conditions. *FEBS Letter* 581**:** 4865–4870.

Santos, F, Wegkamp, A, de Vos, WM, Smid, EJ, Hugenholtz, J. 2008. High-Level Folate Production in Fermented Foods by the B_{12} Producer *Lactobacillus reuteri* JCM1112. *Applied and Environmental Microbiology* 74: 3291-3294.

Savino, F, Pelle, E, Palumeri, E, Oggero, R, Miniero, R. 2007. *Lactobacillus reuteri*(ATCC 55730) versus Simethicone in the Treatment of Infantile Colic: A Prospective Randomized Study. *Pediatrics* 119: E124-E130.

Schreiber, O, Petersson, J, Phillipson, M, Perry, M, Roos, S, Holm, M. 2009. *Lactobacillus reuteri* Prevents Colitis by Reducing P-selectin Associated Leukocyte- and Platelet-Endothelial Cell Interactions. *American Journal of Physiology Gastrointestinal and Liver Physiology* 296: G534-G542.

Schwab, C, Gänzle, MG. 2006. Effect of Membrane Lateral Pressure on the Expression of Fructosyltransferases in *Lactobacillus reuteri. Systematic and Applied Microbiology* 29: 89-99.

Seo, BJ, Mun, MR, J, RK, Lee, I, Park, YH, Kim, CJ, Chang, YH. 2010. Bile Tolerant *Lactobacillus reuteri* Isolated from Pig Feces Inhibits Enteric Bacterial Pathogens and Porcine Rotavirus. *Veterinary Research Communication* 34: 323-333.

Shornikova, AV, Casas, IA, Isolauri, E, Mykkänen, H, Vesikari, T. 1997a. *Lactobacillus reuteri as* a Therapeutic Agent in Acute Diarrhea in Young Children. *Journal of Pediatrics Gastroenterology and Nutrition* 24: 399-404.

Shornikova, AV, Casas, IA, Mykkänen, H, Salo, E, Vesikari, T. 1997b. Bacteriotherapy with *Lactobacillus reuteri* in rotavirus gastroenteritis. *Pediatric Infectious Disease Journal* 16: 1103-1107.

Smits, HH, Enqering, A, van der Kleij, D, de Jong, EC, Schipper, K, van Capal, TM, Zaat, BA, Yazdanbakhsh, M, Wierenga, EA, van Kooyk, Y, Kapsenberg, ML. 2005. Selective Probiotic Bacteria Induce IL-10-Producing RegulatoryT cells *in vitro* by Modulating Dendritic Cell Function through Dendritic Cell-Specific Intercellular Adhesion Molecule 3-Grabbing Nonintegrin. *The Journal of Allergy and Clinical Immunology* 115: 1260-1267.

Suckling, KE, Benson, GM, Bond, B, Gee, A, Glen, A, Haynes, C, Jackson, B. 1991. Cholesterol Lowering and Bile Acid Excretion in the Hamster with Cholestyramine Treatment. *Atherosclerosis* 89: 183-190.

Talarico, T, Axelsson, L, Novotny, J, Fiuzat, M, Dobrogosz, W. 1990. Utilization of Glycerol as a Hydrogen Acceptor by *Lactobacillus reuteri*: Purification of 1, 3-Propanediol: NAD+ Oxidoreductase. *Applied and Environmental Microbiology* 56: 943-948.

Talarico, TL, Casas, IA, Chung, TC, Dobrogosz WJ. 1988. Production and Isolation of Reuterin, a Growth Inhibitor Produced by *Lactobacillus reuteri. Antimicrobial Agents and Chemotherapy* 32: 1854-1858.

Talarico, TL, Dobrogosz, WJ. 1989. Chemical Characterization of an Antimicrobial Substance Produced by *Lactobacillus reuteri. Antimicrobial Agents and Chemotherapy* 33: 674-679.

Tannock, GW. 2005. *Probiotics and Prebiotics: Scientific Aspects*, (Wymondham: Horizon Scientific Press).

Taranto, MP, Fernandez, MML, Lorca, G, deValdez, GF. 2003. Bile Salts and Cholesterol induce Changes in the Lipid Cell Membrane of *Lactobacillus reuteri. Journal of Applied Microbiology* 95: 86-91.

Taranto, MP, Medici, M, Perdigon, G, Ruiz holgado, AP, deValdez, GF. 1998. Evidence for Hypocholesterolemic Effect of *Lactobacillus reuteri* in Hypercholesterolemic Mice. *Journal of Dairy Science* 81: 2336-2340.

Taranto, MP, Vera, JL, Hugenholtz, J, deValdez, GF and Sesma, F. 2003. *Lactobacillus reuteri* CRL1098 Produces Cobalamin. *Journal of Bacteriology* 185: 5643-5647.

Thomas CM, Hong T, van Pijkeren JP, Hemarajata P, Trinh DV, Hu, W, Britton, RA, Kalkum, M, Versalovic, J. 2012. Histamine Derived from Probiotic *Lactobacillus reuteri* Suppresses TNF via Modulation of PKA and ERK Signaling. PLoS ONE 7(2): e31951. doi: 10.1371/journal.pone.0031951

Toba, T, Samant, SK, Yoshioka, E, Itoh, T. 1991. Reutericin 6, a New Bacteriocin Produced by *Lactobacillus reuteri* LA 6. *Letters in Applied Microbiology* 13: 281-286.

Valeur, N, Engel, P, Carbajal, N, Connolly, E, Ladefoged, K. 2004. Colonization and Immunomodulation by *Lactobacillus reuteri* ATCC 55730 in the Human Gastrointestinal Tract. *Applied and Environmental Microbiology* 70: 1176-1181.

van Nieuwenhove CP, Oliszewski R, Gonzalez SN, Perez Chaia AB. 2007. Conjugated linoleic acid conversion by dairy bacteria cultured in MRS broth and buffalo milk. *Letter in Appllied Microbiology* 44(5): 467–474.

Vannini, V, Rodríguez, A, Vera, J, Valdéz, G, Taranto, M, Sesma, F. 2011. Cloning and heterologous expression of *Lactobacillus reuteri* uroporphyrinogen III synthase/ methyltransferase gene (cobA/hemD): preliminary characterization. *Biotechnology Letters* 33(8): 1625-1632

Vermeulen N, Ganzle MG, Vogel RF. 2007. Glutamine deamidation by cereal-associated lactic acid bacteria. *Journal of Applied Microbiology* 103: 1197-1205.

Vollenweider, S, Grassi, G, König, I, Puhan, Z. 2003. Purification and Structural Characterization of 3-Hydroxypropionaldehyde and its Derivatives. *Journal of Agricultural and Food Chemistry* 57: 3287-3293.

Vollenweider, S, Lacroix, C. 2004. 3-Hydroxypropionaldehyde: applications and perspectives of biotechnological production. *Applied Microbiology and Biotechnology* **64**: 16–27.

Wadström, T, Andersson, K, Sydom, M, Axelsson, L, Lindgren, S, Gullmar, B. 1987. Surface Properties of Lactobacilli Isolated from the Small Intestine of Pigs. *Journal of Applied Bacteriology* 62: 513-520.

Wall, R., Ross, R.P., Fitzgerald, G.F., Stanton, C. 2008. Microbial congjugated linoleic acid production—a novel probiotic trait? *FST Bulletin*, 4(8): 87–97.

Wall, T, Bath, K, Britton, R.A, Jonsson, H, Versalovic, J, Roos, S. 2007. The Early Response to Acid Shock in *Lactobacillus reuteri* involves the ClpL Chaperone and a Putative Cell wall-Altering Esterase. *Applied and Environmental Microbiology* 73: 3924-35.

Walter, J, Schwab, C, Loach, DM, Gänzle, MG, Tannock, GW. 2008. Glucosyltransferase A (GtfA) and inulosucrase (Inu) of *Lactobacillus reuteri* TMW1.106 Contribute to Cell Aggregation, *in vitro* Biofilm Formation and Colonization of the Mouse Gastrointestinal Tract. *Microbiology* 154: 72-80.

Wang, Y, Ganzle, MG, Schwab, C. 2010. Exopolysaccharide Synthesized by *Lactobacillus reuteri* decreases the Ability of Enterotoxigenic *Escherichia coli* to bind to Porcine Erythrocytes. *Applied and Environmental Microbiology* 76: 4863-4866.

Weizman, Z, Asli, G, Alsheikh, A. 2005. Effect of a Probiotic Infant Formula onInfections in Child Care Centers: Comparison of Two Probiotic Agents. *Pediatrics* 115: 5-9.

Wilson, TA, Nicolosi, RJ, Chrysam, M, Kritchevsky, D. 2000. Conjugated Linoleic Acid Reduces Early Aortic Atherosclerosis Greater than Linoleic Acid in Hypercholesterolemic Hamsters. *Nutrition Research* 20: 1795-805.

Wolf, BW, Garleb, KA, Ataya, DG, Casas, IA. 1995. Safety and Tolerance of *Lactobacillus reuteri* in Healthy Adult Male Subjects. *Microbial Ecology in Health and Disease* 8: 41-50.

Yadav, H, Jain, S, Sinha, PR. 2006. Effect of Dahi Containing *Lactococcus lactis* on the Progression of Diabetes induced by a High-fructose Died in Rats. *Bioscience Biotechnology and Biochemistry* 70: 1255-1258.

Yoshimune, K, Yamashita, R, Masuo, N, Wakayama, M and Moriguchi, M. 2004. Digestion by Serine Proteases Enhances Salt Tolerance of Glutaminase in the Marine Bacterium *Micrococcus luteus* K-3. *Extremophiles* 8: 441-46.

Zhang Y, Rodionov DA, Gelfand MS, Gladyshev VN. 2009. Comparative genomic analyses of nickel, cobalt and vitamin B_{12} utilization. *BMC Genomics* 10: 78.

2015, Dairy Product Technology: Recent Advances *Pages 169–181*
Editors: **Subrota Hati, Surajit Mandal and Birendra Kumar Mishra**
Published by: **DAYA PUBLISHING HOUSE, NEW DELHI**

Chapter 9

Aspects of Infant Formulations: An Overview

Anamika Das, Dipanjan Palit and Viral Ray Sinh Chavda

Introduction

Feeding of mothers milk is best for the baby but in the event of lactation failure, insufficient milk secretion and where mothers are suffering from transmittable diseases, human milk substitutes serve as nutrient provider in early stages of infancy. Bovine milk based dried formulations have become a prominent feature of infantile dietetics because of easy availability of bovine milk. But with advancement in technologies and scientific research, infant formulations have been tailor made to meet the nutritional requirement of different types of infants *viz.*, normal babies, preterm infants, and infants born with different physiological disorders.

Greater emphasis were laid on the manufacture of formulations having compositional and biochemical characteristics similar to human milk. The U.S.Federal Food, Drug, and Cosmetic Act (FFDCA) defines infant formula as "a food which purports to be or is represented for special dietary use solely as a food for infants by reason of its simulation of human milk or its suitability as a complete or partial substitute for human milk".

Modifications to infant formulas are continually being made as the nutrient needs of diverse groups of infants are being identified. Remarkable improvements have been made in infant formulations over the past five decades. Previously formula-fed infants were usually given bovine milk–based infant formulas or homemade formulations of evaporated milk, sugar, and water. Today, options include soy formulas, lactose-free formulas, formulas with added fiber or rice starch, protein hydrolysates, preterm infant formulas, and formulas for infants with special conditions such as allergies to certain food components. As newer components of human milk are being characterized and their physiologic functions are determined, existing formulas undergo modifications to give rise to new formulations. Although human milk contains hormones, immune factors, growth factors, enzymes, etc. and most of which cannot be made avilable to infant formulas yet efforts are being done to modify the formulations so as to simulate the compositions and functions of human milk. Furthermore, before adding any component, the complex interactions of the bioactive substances should be completely understood.

Types of Infant Formula

The three major classes of infant formulas are:

1. Milk-based formulas These formulas are prepared from cow milk with added vegetable oils, vitamins, minerals and iron. These formulas are suitable for most healthy full-term infants and should be the feeding of choice when breastfeeding is not used, or is stopped before one year of age.
2. Soy-based formulas These formulas are made from soy protein with added vegetable oils (for fat calories) and corn syrup solids and/or sucrose (for carbohydrate). These formulas are suitable for infants who are lactose intolerant. These formulas are not recommended for low-birth-weight or preterm infants or for the prevention of colic or allergies.
3. Special formulas for low-birth-weight infants These formulas include low-sodium formulas for infants who are restricted from salt intake, and "predigested" protein formulas for infants who cannot tolerate or are allergic to the whole proteins (casein and whey) in cow milk and milk-based formulas
    a. Hypoallergenic formulas: Hypoallergenic formulas are intended for use by infants with existing allergic symptoms. Extensively hydrolyzed proteins derived from cow's milk, in which most of the nitrogen is in the form of free amino acids and peptides,1500 kDa, have been used in formulas for.50 years for infants with severeinflammatory bowel diseases or cow's milk allergy.
    b. Anti-regurgitation formulas: Gastro-oesophageal reflux (GOR) is the spontaneous movement of gastric content into the oesophagus due to an inappropriate relaxation of the lower oesophageal sphincter. It can occur with or without regurgitation and vomiting, and is a normal physiological process that occurs in many healthy infants
    c. Lactose free formulas: With the exception of extremely rare cases, all infants are born with the ability to digest lactose, because there is lactose in breast milk. However, lactose intolerance can occur in infancy after acute gastroenteritis, or due to severe intestinal diseases. Lactose-free infant formulas are designed for infants from 0 to 12+ months of age

with lactose intolerance. They are cow's milk protein-based, but do not contain any lactose. Lactose is the natural sugar found in milk. Ask your Healthcare professional before using a lactose-free infant formula.

Table 9.1: Different Types of Infant Formula Commercially available in the Market

Infant Formula	*Examples of Commercially available Formula*	*Age of Infants*
Starter formulas Whey dominant	Infacare® 1 Infacare® Gold 1 Infacare® Nurture 1 NAN® 1 Protect Start S26® S26® Gold	From birth- can be continued to one year of age
Casein dominant	SMA® 1 Lactogen® 1 Novalac® SD	May be used from birth, but generally recommended from three to six months
Follow-on formulas	Infacare®2 Infacare® Gold 2 Infacare® Nurture 2 S26 Promil® 2 S26 Promil® Gold 2 Lactogen® 2	From six months
Soy-based formulas	Infacare® Soya 1 and 2 Infacare® Gold Soya 1, 2 and 3 Similac® Isomil Advance Isomil® 1 and 2 Infasoy® 1 and 2	1: From birth 2: From six months
Hypoallergenic formulas Amino acid mixture	Neocate®	From birth
Extensively hydrolysed	Alfaré® (whey) Similac® Alimentum (casein)	
Partially hydrolysed	Infacare® Nurture HA Comfort NAN® HA 1 and 2 (whey) Novalac® HA 1 and 2 (whey) Similac® Advance HA (whey)	1: From birth 2: From six months
Lactose-free formulas	Infacare® Nurture LF NAN® Lactose Free S26® LF	From birth
Anti-regurgitation formulas	Infacare® AR (casein) Infacare® Nurture AR (casein) Novalac® AR 1 and 2 (casein) NAN® AR (whey)	0-12 months
Acidified cow's milk formulas	Melegi® Acidified NAN® Pelargon®	From birth

d. Extensively hydrolyzed infant formulas: Extensively hydrolyzed infant formulas are intended for feeding infants who have a diagnosed allergy to cow's milk or soy proteins. These types of formulas contain cow's milk protein that has been broken down into very small particles that can be tolerated by most infants suffering from cow's milk protein allergy.

Considerations with Respect to Composition for Infant Formulations

In order to achieve appropriate growth and maintain good health, infant formulas must include proper amounts of water, protein, fat, carbohydrate,vitamins, and minerals.

Protein and Amino Acid Profile

The protein content of infant formulations must provide the total amino acids required for growth. The quantity of nitrogen in human milk keeps on changing during the lactation period; from about 400 mg/100 mL in colostrum to about 180 mg/100 mL in mature milk several days after parturition. Also the concentration of different proteins changes with the duration of lactation: the whey-casein ratio is 90:10 in colostrum and becomes 55:45 in mature milk and 50:50 in late lactation(Thompkinson and Kharb, 2007). The decrease in total protein content is mainly due to a decrease in secretory immunoglobulin A (sIgA) and in lactoferrin. Together with lysozyme, sIgA and lactoferrin comprises of about 30 per cent of the total protein in mature human milk.

Human milk has a lower casein content (40 per cent) and smaller amounts of casein in breast milk help produce softer and more flocculent curd, which leads to faster gastrointestinal digestion.Human milk provides all the 11 essential amino acids as per infants' requirements (Dewey and others, 1996). (Histidine, cystine and tyrosine are also considered to be essential amino acids for infants as these are not synthesized in infants). Taurine is the predominant free amino acid in human milk (4 to 5 mg/100 mL or 0.3 to 0.4 mmol/L) (Agostoni and others, 2000) which cannot be synthesized in infants and therefore human milk is the only source for supply of taurine requirement.

Lipid Profile

The dietary fat is the predominant source of energy for breast-fed and formula-fed infants. Besides energy, dietary lipids provide fatty acids and fat-soluble vitamins. Furthermore, fats are carriers of flavors in the diet and contribute to its satiety value (Carey and Hernell 1992). The estimation of fat requirement in infancy is, however, based on the types of fat, triglycerides, and fatty acid make-up. Human milk is better absorbed in comparison to bovine milk fat, which may be attributed to the presence of palmitic acid in the beta-position in the triglyceride configurations. Human milk fat also contains adequate amounts of linoleic acid or linoleates and other essential fatty acids, which are required for maintaining the integrity of skin and normal growth.

The short-chain or unsaturated fatty acids of human milk are absorbed more readily than the more predominant long-chain or saturated fatty acids in cow milk fat. In contrast to cow milk, fat absorption from infant formulae with vegetable oil is similar to that of human milk. Monoglycerides with palmitic acids in the beta-position are more efficiently solubilized and absorbed by human infants than any other glycerides with palmitic acid or stearic acid or free acids. Polyunsaturated fatty acids (PUFAs) have a number of important biological functions, such as their role as components of membrane lipids and as precursors of prostaglandins. The PUFAs, including linoleic acid, must have sufficient amounts of vitamin E for utilization to maintain the integrity of skin and normal growth.

In human milk, PUFA levels vary widely depending on the type of fat consumed by the mother. However, blends of vegetable oil provide an adequate amount of PUFAs, including linoleic acid and others. It is necessary to provide 1.4 per cent of total calories in the form of linoleic acid to avoid essential fatty acid deficiency in infants.

The *trans* fatty acid content of formulae should be as low as feasible.

The concentration of DHA is higher in plasma, erythrocyte membranes and even in the brain in infants that are fed human milk or infant formulae with added DHA as compared to infants fed formulae containing only the precursors LA and ALA but no LCPUFAs. The supply of LA and ALA in the diet modulates the synthesis of AA and DHA. DHA has a potential benefit on visual acuity, but there is as yet no consensus that DHA or AA or both are indispensable nutrients for term infants, nor that an exogenous supply is truly beneficial, at least not after the 1st few months of life (Lucas and others 1999; Lauritzen and others 2001; Jensen and Heird 2002).

Carbohydrates Profile

The higher carbohydrate level in human milk plays a significant role in infant nutrition. Though human milk contains both digestible and indigestible carbohydrates but infant formulae and follow-on formulae mostly conatins only digestible carbohydrates.

Lactose also helps in the synthesis of certain vitamins and increases the absorption of calcium and iron. There exist differences in the composition of mono- and oligosaccharides (galactose, glucose, fucose, N-acetyl glucosamine, N-acetyl lactosamine, and others) of the carbohydrates in human and bovine milks. Galactose plays an important role in the synthesis of galactosides and cerebrosides for myelin formation and for the synthesis of collagen. Among the oligosaccharides, N-acetyl glucosamine and N-acetyl lactosamine possess bifidus-stimulating activity. These substances are present in substantial amounts in human milk, about 40 times more than the amount found in bovine milk. The beneficial effects of bifidus bacteria in the intestinal tract of infants include increased absorption of certain nutritionally important minerals like calcium, phosphorus, and iron, inhibition of enteropathogenic organisms, synthesis of "B" group vitamins, and detoxification influence in chronic liver diseases as reported by several workers (Pahwa and Mathur, 1982).

For infant formulae only lactose, maltose, sucrose, maltodextrins, corn-syrup solids, and precooked and gelatinized starch are permitted. Starches must be free of gluten by nature. Digestible carbohydrates serve as essential sources of energy in the diet and moreover, provide structural elements for the synthesis of glycolipids and glycoproteins. Disaccharides and polysaccharides from the diet are hydrolyzed to monosaccharides, which after absorption in the upper small intestine are converted to glucose in the liver. Human milk contains predominantly lactose (galactose-β(1-4) glucose), 55 to 70 g/L or 8.2 to 10.4 g/100 kcal, and in addition oligosaccharides, about 20 g/L in colostrum and 10 to 13 g/L in mature milk (Coppa and others, 1993). The more than 130 different oligosaccharides in human milk identified so far consist of glucose, galactose, N-acetyl glucosamine fucose and sialic acid (N-acetyl neuraminic acid), and in most cases lactose at their reducing end (Kunz and others 2000). These oligosaccharides are for a small part absorbed intact by the breast-fed child and excreted in the urine, minimally hydrolyzed in the small intestine, and the majority reaches the colon and are substrates for bacterial hydrolysis and fermentation (Kunz and others, 2000).

Human milk contains a complex mixture of more than 130 different oligosaccharides comprising a total concentration of 15 to 23 g/L in colostrum and 8 to 12 g/L in transitional and mature milk (Kunz and others 1999, 2000). Human milk oligosaccharides were shown to be resistant to enzymatic digestion in the upper gastrointestinal tract (Engfer and others 2000). Among other functions human milk oligosaccharides may serve as substrates for colonic fermentation. It has been shown that human milk oligosaccharides induce an increase in the number of bifidobacteria in the colonic flora in breast-fed infants, accompanied with a significant reduction in the number of potential pathogenic bacteria (Kunz and others, 2000). Complex oligosaccharides have the ability to inhibit the binding of pathogens to cell surfaces because they act as competitive receptors on the host cell surface, thereby preventing adhesion of a number of bacterial and viral pathogens.

Oligo-fructosyl-saccharose (oligo-fructose; fructo-oligosaccharides, FOS) and oligo-galactosyl-lactose (oligo-galactose; galacto-oligosaccharides, GOS) have been used in dietetic products for infants. Oligogalactose is found only in trace amounts in human milk. Many fermented milk products contain oligo-galactosyl-lactose obtained from lactose by galactosidases of lactic bacteria. Lactose-free formulae may also contain trace amounts of oligo-galactosyl-lactose. In addition to oligo-galactose, the preparation used in dietetic products for infants contains some 40 per cent (wt/wt) of mono- and disaccharides (Wiedmann and Jager, 1997).

Vitamins Profile

The vitamins A, E, D, and K are lipid soluble; hence their absorption from infant formulae is related to the efficacy of fat absorption. Human milk contains retinol and retinyl esters, which are effectively hydrolyzed in the infant gut by the action of human milk bile salt-stimulated lipase (BSSL) and pancreatic enzymes (Hernell and Blackberg 1994). The total concentrations of preformed vitamin A in mature milk have been reported as about 150 to 1100 μg/L (Gebre-Medhin and others 1976; Jensen 1995).

Vitamin D sufficiency is of particular importance during phases of rapid growth and bone mineralization, as is the case in infancy (Csaszar and Abel 2001). Vitamin D in human milk are reported to be in the order of 4 to 110 IU/L (0.015 to 0.4 μg/100 kcal), with up to tenfold higher values in the summer than in the winter (Fomon 1993; Jensen, 1995).

Vitamin E acts as a chain-breaking antioxidant in tissues and is considered essential for the protection of unsaturated lipids in biological membranes against oxidative damage (Traber and Sies, 1996). Prior to absorption, vitamin E esters need to be hydrolyzed by the action of human milk bile salt-stimulated lipase or pancreatic lipolytic activity (Hernell and Blackberg 1994). Vitamin E absorption appears to be a passive, nonsaturable process without a specific carrier.

Mature human milk contents of vitamin E are approximately 2 to 5 mg/L (about 0.5 to 1.6 mg α-TE/g PUFAs) (Fomon, 1993; Jensen, 1995).

Human milk contains low concentrations of vitamin K provided mostly by phylloquinone. Vitamin K concentrations between about 0.6 and 10 μg/L have been reported (von Kries and others 1987; Fomon 2001). There is general agreement that the supply with human milk does not suffice to meet the requirements of all young infants. Therefore, vitamin K supplementation in addition to the supply with breast milk is generally recommended in early infancy. In newborn and young infants, vitamin K deficiency may induce a severe coagulopathy with neonatal hemorrhagic disease or a late-onset bleeding, which can lead to intracranial bleeding,

Thiamine(Vitamin B1) is a water-soluble compound made up of pyrimidine and thiazole nucleus linked by a methylene bridge. The thiamine content during early lactation has been reported as 20 to 133 μg/L (3 to 20 μg/100 kcal) (Nail and others 1980; Dostalova and others 1988), and for mature milk as 200 μg/L (30 μg/100 kcal) (Picciano, 1995).

Riboflavin is a water-soluble vitamin and the reported riboflavin content of mature human milk varies widely, from 274 to 580 μg/L (41 to 87 μg/100 kcal) (Nail and others 1980; Thomas and others 1980; Dostalova and others 1988; Roughead and McCormick, 1990).

The concentration of niacin in human milk is much higher falling in the range of 1100 to 2300 μg/L of niacin (164 to 343μg/100 kcal).

Pantothenic acid performs multiple roles in cellular metabolism and in the synthesis of many essential metabolites. The mean concentration of pantothenic acid in human milk has been reported as 6.7 mg/L (1 mg/100 kcal) (Johnston and others 1981).

Pyridoxine embraces a group of chemically related compounds, including pyridoxamine and pyridoxal, which are found in animal products, and pyridoxine, which is found in plants.

The pyridoxine content of milk increases as the maternal intake of the vitamin increases and correlates with measures of pyridoxine status, which in turn is affected by stage of lactation, length of gestation, and oral contraceptive use.

Folate is a general term for compounds that have a common vitamin activity and includes the synthetic form of the vitamin folic acid (pteroyl glutamic acid, PGA).

Cobalamin is a water-soluble vitamin and a member of a family of related molecules known as corrinoids, which contain a corrin nucleus made from a tetrapyrrolic ring structure. Cobalamin is stored in the liver and at birth the liver of a full-term infant contains 25 to 30 μg cobalamin with the total body content being 30 to 40 μg. Formula-fed infants had higher cobalamin concentrations and lower urinary methyl malonic acid excretion than breast-fed infants (Specker and others 1990a, 1990b). L-ascorbic acid is transported in plasma as the free anion and is primarily lost to the body through urinary excretion. The minimum biotin content of infant formulae should be 1.5 μg/100 kcal (0.4 μg/100kJ), while maximum level be 7.5 μg/100 kcal.

Minerals and Trace Elements Profile

Iron is essential for virtually every living organism. The dominating function of iron in the human body is as the oxygen-binding core of hemoglobin, the red pigment of blood transporting oxygen from the lungs to all tissues. For this reason, most infant formulae are fortified with iron, but the optimal level of added iron is still an open question.

The concentration of iron in human milk is 0.3 mg/L, which is approximately the same as in cow milk, but the difference in bioavailability is reported to be at least fivefold in favor of human milk. The iron content of commercially available infant formulae varies widely, ranging from about 1 mg/L in unfortified formula to as much as 15 mg/L in some iron-fortified formulae. In mature human milk, the relation between whey protein and casein is about 60:40, whereas in cow milk, it is 20:80. Lack of iron in infant food often leads to iron-deficiency anemia both in normal and lactase-deficient infants (Lozoff and others 1998).

At birth about 99 per cent of the calcium in the body is part of the structural matrix of bone, with the remainder being physiologically active as a free calcium pool within cells and the extracellular fluid (ECF). Within cells, calcium acts as a 2nd messenger, modulating the transmission of hormonal signals and regulating enzyme function. Although net calcium absorption and calcium retention may be related to calcium intake, there are a number of other factors, either in the diet (lactose and phytate) or related to metabolic regulation, which influence the absorption, around 30 per cent to 50 per cent of dietary calcium, and retention of calcium (Arnaud and Sanchez 1996).

The minimum calcium content in infant formulae should be maintained at 50 mg/100 kcal, and the maximum level should be 140 mg/100 kcal. The ratio of calcium to available phosphorus (based on measured bioavailability, or calculated as 80 per cent of total phosphorus in cow milk protein based formulae and as 70 per cent of total phosphorus in soy protein based formulae) should be not be less than 1.0 nor greater than 2.0.

Phosphorus is an integral part of the inorganic matrix of bone, has an important homeostatic function as a metabolic buffer, and is an integral part of a range of compounds which play central roles as structural cellular components and in intermediary metabolism (phospholipids and phosphoproteins, nucleic acids and adenosine triphosphate). The phosphorus content of human milk has been reported to range from 107 to 164 mg/L or 16 to 24 mg/100 kcal (Fomon and Nelson, 1993), peaking in early lactation and decreasing as lactation progresses (Fomon and Nelson, 1993; Atkinson and others, 1995).

The minimum bioavailability for phosphorus should be 20 mg/100 kcal, and the maximum 70 mg/100 kcal.

With the objective to enhance utilization of dietary Ca and, consequently, fat absorption in infants, the ratio of Ca:P of soy infant formulae was adjusted to simulate human milk. In human milk, the Ca:P ratio is about 2:1 and 1.61:1 in buffalo milk. With unmodified infant formulae, about 0.79 g of Ca and 0.49 g per 100 g of spray-dried product were observed. Therefore, 2.11 kg of pharmaceutical grade calcium gluconate that would contribute 18 g of Ca is to be added per 100 kg of dried LIF to simulate human milk.

The metabolism of magnesium is tightly regulated. It is an integral constituent of bone and as the second most abundant intracellular cation plays a fundamental role in many aspects of intermediary metabolism, as a cofactor for many different enzymes, and as a modulator of physiological processes. The concentration in human milk appears to be regulated within narrow limits, 31.4 to 35.7 mg/L (Atkinson and others, 1995) or 17 to 28 mg/d (Lonnerdal, 1997). The minimum level for magnesium in infant formulae and follow-on formulae should be 5 mg/100kcal and the maximum 15 mg/100 kcal (SCF, 2003b).

Sodium is the principal cation in ECF and is present in the body mainly in the ionized form, which represents around 96 per cent. The amount and concentration of sodium determine the volume of

ECF. Potassium is the major intracellular cation, contributes to the intracellular osmotic activity, and in part determines the intracellular fluid volume. Sodium and potassium are lost from the body by extrarenal routes, but the body content of both cations is regulated predominantly by renal excretion. Chloride is the principal anion in the ECF and together with sodium contributes more than 80 per cent of the osmotic activity.

The Infant Formula Directive has set for infant formulae a minimum for sodium 20 mg/100 kcal and a maximum of 60 mg/100 kcal, and a minimum for potassium of 60 mg/100 kcal and a maximum of 145 mg/100 kcal, and a minimum for chloride of 50 mg/100 kcal and a maximum of 125 mg/100 kcal.

With a view to maintain extracellular and intracellular concentrations of water at normal levels, the Na:K ratio was modified in infant formulae to simulate human milk. The Na:K ratio of human milk is about 1:3.67 which is much lower than that of buffalo skim milk (1:2.18). With unmodified Na:K ratio 0.24 g Na and 0.53 g K per 100 g of infant formulae were observed.

Copper is required as an essential dietary trace element. It is absorbed from the diet in the upper jejunum by active and passive processes, stored in the liver and kidney, and excreted in the bile to be lost in feces. The minimum copper content of infant formulae and follow-on formulae should be 35 μg/100 kcal and the maximum 100 g/100 kcal. (SCF 2003b).

Zinc is a constituent of more than 200 metalloenzymes and plays a key role in the synthesis of genetic material and the regulation of gene expression as well as in cell division, epithelial integrity, cellular immunity, and sexual maturation. The young infant has a high zinc requirement to support the very rapid growth of early infancy. A large variability of zinc concentration in human milk (from 0.5 to 4.7 mg/L) has been shown during the course of lactation and among individuals. Zinc supplementation does not influence zinc concentrations of human milk in well-nourished lactating women (Krebs and others 1995). The concentrations of zinc in human milk decrease during early lactation from 4 mg/L at 2 wk to 3 mg/L at 1 month to 1.2 mg/L at 6 mo (Krebs and others 1995).

The minimum zinc content should be 0.5 mg/100 kcal for infant and follow-on formulae containing cow milk protein and protein hydrolysates only. The maximum zinc content should be 1.5 mg/100 kcal in both infant formulae and follow-on formulae containing cow milk protein and protein hydrolysates only. The minimum zinc content should be 0.75 mg/100 kcal for infant and follow-on formulae containing soy protein. The maximum zinc content should be 2.4 mg/100 kcal in both infant formulae and follow-on formulae containing soy protein (SCF 2003a).

Manganese is involved in the formation of bone and in amino acid, cholesterol, and carbohydrate metabolism. Average manganese content of human milk has been estimated as 3.5 μg/L, with a slight decrease over the course of lactation (Casey and others 1985). Thus, manganese intake of a breastfed infant is estimated to be 2.5 to 3 μg/d. The minimum manganese content should be 1 μg/100 kcal and the maximum should be 100 μg/100 kcal for both infant formulae and follow-on formulae (SCF 2003a).

Molybdenum acts as a cofactor of 3 enzymes in humans: xanthine oxidase, aldehyde oxidase, and sulfite oxidase. Molybdenum has an average concentration of 2 g/L in human milk.

The primary role of fluoride is improving caries resistance. It has also been suggested that fluoride may play a role in bone mineralization and maintenance of peak bone mass, as well as for normal growth in humans (Bergmann and Bergmann 1991). Mean fluoride content of human milk ranges from 0.007 to 0.011 mg/L (FNB 1997). Thus, mean fluoride intake of a breast-fed infant is estimated to be 0.005 to 0.01 mg/d. The maximum fluoride content in both infant formula and follow on formula should be 100 μg/100 kcal (SCF 2003a).

Iodine is an essential component of the thyroid hormones thyroxine (T4), containing 65 per cent by weight of iodine, and its active form triiodothyronine (T3), containing 59 per cent by weight of iodine.

The iodine content of human milk varies markedly as a function of the iodine intake of the population. The minimum iodine content should be 10 μg/100 kcal and the maximum iodine content should be 50 μg/100 kcal for both infant formulae and follow-on formulae (SCF2003a).

Selenium is necessary for the activity of glutathion peroxidase, which protects against oxidative damage in intracellular structures. The minimum selenium content should be 3 μg/100 kcal and the maximum selenium content should be 9 μg/100 kcal for both infant formulae and follow-on formulae (SCF 2003a).

Conclusion

Human infants should ideally be nursed on mother's milk, which constitutes nature's best food. Through ages, changes are continuously being made in infant formulas and these changes generally result in products with compositions and functions closer to those of human milk. New formulas are being developed to meet the needs of infants with special nutrient needs. Formulas for preterm infants after hospital discharge are designed to meet the needs of a population at risk of developing growth and nutrient deficiencies and several studies reported that growth was higher in infants fed these new formulas than in those fed standard term-infant formulas. Studies will continue to characterize the complex components of human milk and the specific nutrient needs of diverse groups of infants. Long-term studies with large groups of infants are needed to determine the beneficial effects associated with modifications in infant formulas and their functional outcomes.

References

Agostoni C, Carratu B, Boniglia C, Riva E, Sanzini E. 2000. Free amino acid content in standard infant formulas: comparison with human milk. *J Am Coll Nutr* (19): 434–438.

Arnaud CD, Sanchez SD. 1996. Calcium and phosphorus. In: Ziegler EE, Filer LJ Jr, editors. *Present Knowledge in Nutrition.* 7th ed. Washington, D.C.: ILSI Press, 245–255.

Atkinson SA, Alston-Mills B, Lonnerdal B, Neville MC, 1995. Major minerals and ionic constituents of human and bovine milks. In: Jensen RG, editor. *Handbook of milk composition*. New York: Academic Press Inc.593–622.

Bergmann RL, Bergmann KE. 1991. Fluoride nutrition in infancy—is there a biological role of fluoride for growth? In: Chandra RK, editor. *Trace elements in nutrition of children—II.* Nestl´e Nutrition Workshop Series, Vol 23. New York: Raven Press. 105–116.

Casey CE, Hambidge KM, Neville MC. 1985. Studies in human lactation: zinc, copper, manganese and chromium in human milk. *Am J Clin Nutr* (49): 773–785.

Csaszar A, Abel T. 2001. Receptor polymorphisms and diseases. *Eur J Pharmacol* 414: 9–22.

Fomon SJ. 2001. Feeding normal infants: rationale for recommendations. *J Am Diet.Assoc* (101): 1002–1005.

Carey MC, Hernell O. 1992. Digestion and absorption of fat. *Semin Gatrointest Dis* (3): 189–208.

Coppa GV, Gabrielli O, Pierani P, Catassi C, Carlucci A, Giorgi PL. 1993. Changes in carbohydrate composition in human milk over 4 months of lactation. *Pediatrics* (91): 637–641.

Dewey KG, Beaton G, Fjeld C, L¨onnerdal B, Reeds P. 1996. Protein requirements of infants and children. *Eur J Clin Nutr* 50(S1): S119–150.

Dostalova L, Salmenpera L, Vaclavikova V, Heinz-Erian P, Schuep W. 1988. Vitamin concentrations in term milk of European mothers. In: Berger H, editor. *Vitamins and minerals in pregnancy and lactation*. Nestle Nutrition Workshop Series. Vol 16, New York: Raven Press. 275–298.

Engfer MB, Stahl B, Finke B, Engfer MB, Stahl B, Finke B, Sawatzki G, Daniel H. 2000. Human milk oligosaccharides are resistant to enzymatic hydrolysis in the upper gastrointestinal tract. *Am J Clin Nutr* (71): 1589–1596.

Fomon SJ. 1993. In: Craven L. editor. *Nutrition of normal infants*. St. Louis, Mo.: Mosby-Book Inc. 366–394.

[FNB] Food and Nutrition Board. 1997. *Dietary reference intakes for calcium, phosphorus, magnesium, vitamin D, and fluoride*. Washington, D.C.: Institute of Medicine. National Academy Press.

Fomon SJ, Nelson SE. 1993. Calcium, phosphorus, magnesium and sulfur. In: Craven L, editor. *Nutrition of normal infants*. St. Louis, Mo.: Mosby-Year Book Inc. 192–211.

Gebre-Medhin M, Vahlquist A, Hofvander Y, Uppsall L, Vahlquist B. 1976. Breast milk composition in Ethiopian and Swedish mothers. I: Vitamin A and beta-carotene. *Am J Clin Nutr* (29): 441–451.

Hernell O, Bl¨ackberg L. 1994. Molecular aspects of fat digestion in the newborn. *Acta Paediatr* 405(S1): 65–69.

Johnston L, Vaughan L, Fox HM. 1981. Pantothenic acid content of human milk. *Am J Clin Nutr* (34): 2205–2209.

Jensen RG. 1995. *Handbook of milk composition*. New York: Academic Press.67–79.

Jensen CL, Heird WC. 2002. Lipids with an emphasis on long-chain polyunsaturated fatty acids. *Clin Perinatol* (29): 261–81.

Kunz C, Rodriguez M, Koletzko B, Jensen R. 1999. Nutritional and biochemical properties of human milk, part I. General aspects, proteins and carbohydrates. *Clinics in Perinatology* (26): 307–333.

Kunz C, Rudloff S, Baier W, Klein N, Strobel S. 2000. Oligosaccharides in human milk: lactation. *Am J Clin Nutr* (40): 1103–1119.

Krebs NF, Reidinger CJ, Hartley S, Robertson AD, Hambidge KM. 1995. Zinc supplementation during lactation: effects on maternal status and milk zinc concentrations. *Am J Clin Nutr* (61): 1030–1036.

Lauritzen L, Hansen HS, Jorgensen MH, Michaelsen KF. 2001. The essentiality of long-chain n-3 fatty acids in relation to development and function of the brain and retina. *Prog Lipid Res* (40): 81–94.

Lucas A, Stafford M, Morley R, Abbott R, Stephenson T, MacFadyen U, Elias-Jones A, Clements H. 1999. Efficacy and safety of long-chain polyunsaturated fatty acid supplementation of infant-formula milk: a randomized trial. *Lancet* (354): 1948–1954.

Lonnerdal B. 1997. Effects of milk and milk components on calcium, magnesium and trace element absorption during infancy. *Physiol Rev* (77): 643–669.

Lozoff B, Klein NK, Nelson EC, McClish DK, Manuel M, Chacon ME. 1998. Behavior of infants with iron-deficiency anemia. *Child Dev* (69): 24–36.

Nail PA, Thomas MR, Eakin R. 1980. The effect of thiamin and riboflavin supplementation on the level of those vitamins in human breast milk and urine. *Am J Clin Nutr* (33): 198–204.

Picciano MF. 1995. Water-soluble vitamins in human milk. In: Jensen RG, editor. *Handbook of milk composition*. New York: Academic Press. 189–194.

Pahwa A, Mathur BN. 1982. Influence of oral ingestion of bifidobacteria on the intestinal microflora and mineral absorption among rats. *Ind J Nutr Dietet* (19): 267–272.

Roughead ZK, McCormick DB. 1990. Flavin composition of human milk. *Am J Clin Nutr* (52): 854–857.

Specker BL, Black A, Allen L, Morrow F. 1990a. Vitamin B12: low milk concentrations are related to low serum concentrations in vegetarian women and to methylmalonic aciduria in their infants. *Am J Clin Nutr* (52): 1073–1076.

Specker BL, Brazerol W, Ho ML, Norman EJ. 1990b. Urinary methylmalonic acid excretion in infants fed formula or human milk. *Am J Clin Nutr* (51): 209–211

[SCF] Scientific Committee on Food. 2003a. Report of the Scientific Committee on Food on the Revision of Essential Requirements of Infant Formula and Follow-on Formula. Adopted 4 April 2003.

[SCF] Scientific Committee on Food. 2003b. Opinion on the tolerable upper intake level of copper. *Reports of the Scientific Committee for Food*. European Commission, Luxembourg.

Thompkinson DK, Suman K.2007.Aspects of Infant Food Formulation. *Comp. Rev. Food Sci. and Food Safety* (6): 79-102.

Thomas MR, Sneed SM,Wei C, Nail PA,Wilson M, Sprinkle EE. 1980. The effects of vitamin C, vitamin B6, vitamin B12, folic acid, riboflavin and thiamin on the breast milk and maternal status of well-nourished women at 6 months post-partum. *Am J Clin Nutr* (33): 2151–2156.

Von Kries R, Shearer M, McCarthy PT, Haug M, Harzer G, Gobel U. 1987. Vitamin K1content of maternal milk: influence of the stage of lactation, lipid composition, and vitamin K1 supplements given to the mother. *Pediatr Res* (22): 513–517.

Wiedmann M, Jager M. 1997. Synergistic sweeteners. *Food Integr Anal* (11): 51–56.

2015, Dairy Product Technology: Recent Advances ***Pages 183–202***
Editors: **Subrota Hati, Surajit Mandal and Birendra Kumar Mishra**
Published by: **DAYA PUBLISHING HOUSE, NEW DELHI**

Chapter 10

Nutritional Significance of Milk

Ahesanvarish Shaikh, Amit Kumar Jain and Satish Parmar

Introduction

Consumption of milk and milk products is practiced by humans since time immemorial. Milk is known as nature's most complete food, and dairy products are considered the most nutritious foods. It is a source of essential nutrients not only for the neonate of any mammalian species, but also for growth of children and nourishment of adult humans. This itself shows importance given by primitive humans to milk and milk products. Even according to FAO (Food and Agriculture Organization) milk is a major source of dietary energy, protein and fat, contributing on average 134 kcal of energy/capita per day, 8 g of protein/capita per day and 7.3 g of fat/capita per day.

Milk and milk products are nutrient rich foods supplying energy and significant amounts of protein and micronutrients. The inclusion of dairy products adds diversity to plant-based diets (Weaver *et al.,* 2013). Further, proteins, lipids and certain minerals in milk and fermented milk products exhibit bioactive properties with the potential to improve long-term human health (Shingfield *et al.,* 2008).

Nutritional Aspects of Major Milk Constituents

Milk imparts several beneficial effects as it possess high content of bioavailable proteins and minerals. Additionally absence of any anti-nutritional factors as in case of plant-based foods makes it an ideal component of human diet. It is also a good source of calcium, zinc and vitamins like A, B_2 (riboflavin) and B_{12}. It makes a valuable contribution to intakes of iodine, niacin and vitamin B_6. Moreover, milk fat is unique in that it contains several short chain fatty acids in significant concentration. They are important not only from health point of view but also for sensory properties of food. Milk fat also acts as a carrier of several fat soluble constituents like vitamin A, vitamin E (tocopherol), Vitamin D, Vitamin K, carotenoids, phospholipids, cholesterol, etc.

Milk, if taken at the recommended level, provides approximately 51 per cent of calcium, 31 per cent of phosphorus, 27 per cent of zinc, 41 per cent of iodine, 15 per cent of selenium, 39 per cent of vitamin B_2, 42 per cent of retinol and 20 per cent of vitamin B_{12}.

Though, nearly 1 lakh compounds have been identified in milk, Table 10.1 lists few of them.

Table 10.1: Composition of Human, Buffalo and Cow Milk

Sl.No.	Constituent	Unit	Human	Buffalo	Cow
				Amount per 100g milk	
1.	Energy	kcal	64.6	95.7	77.8
2.	Fat	g	3.4	6.5	5.0
3.	Cholesterol	mg	20	20	13
4.	Lactose	g	7.4	5.0	4.8
5.	Protein	g	1.1	4.5	3.4
6.	Minerals	g	0.1	0.8	0.7
7.	Calcium	mg	28	180	120
8.	Magnesium	mg	3	18	11
9.	Zinc	mg	0.3	0.5	0.4
10.	Iron	μg	80	80	20
11.	Copper	μg	40	30	2
12.	Phosphorus	mg	11	82	85
13.	Potassium	mg	–	100	140
14.	Vitamin A	RAE	80	960	500
15.	Vitamin D	μg	0.1	0.2	0.2
16.	Vitamin E	mg	0.4	33	31
17.	Thiamin, Vitamin B_1	μg	20	40	50
18.	Riboflavin, Vitamin B_2	μg	40	100	190
19.	Vitamin B_6	μg	6	330	65
20.	Folic acid	μg	1.3	5.6	8.5
21.	Vitamin B_{12}	μg	0.01	0.25-0.9	0.3-0.4
22.	Ascorbic acid	mg	3	2.3-3	2.0

Milk Proteins

Milk is recognized as an excellent source of high-quality protein. The primary nutritional role of proteins is to provide essential amino acids. Protein content of buffalo milk varies from 4-5 per cent ; whereas that of cow milk varies from 3.2-3.4 per cent. The major proteins found in milk are casein and whey proteins, with casein (αs_1-, αs_2-, β-, and κ-casein) accounting for approximately 78 percent of the total protein while whey proteins accounting for about 17 percent of the total protein. The main whey proteins are β-lactoglobulin, α-lactalbumin, serum albumin, immunoglobulins and Glycomacropeptide; minor proteins include lactoferrin, insulin growth factor (IGF), etc.

Milk protein is considered to be complete protein because it contains all essential amino acids in right proportions to meet body requirements. The nutritional value of the mixture of casein and serum protein is higher than that of each one separately because they are complementary in their concentration of some essential amino acids. While casein is relatively high in tyrosine and phenylalanine, whey proteins are higher in sulfur containing cysteine and methionine. Addition of little quantity of milk protein to vegetable diet enhances the nutritional quality of the food considerably. The digestibility of milk protein is close to 100 per cent, whereas digestibility of most plant proteins' is considerably lower. This is particularly significant for vegetarian population as milk is sole source of good quality proteins for them.

Caseins

Casein is generally defined as the protein precipitated at pH 4.6. Casein's biological function is to carry calcium and phosphate and to form a curd in the stomach for efficient digestion. About 1 mM casein in milk binds 20 mM calcium and 10 mM phosphate, while solubility of calcium phosphate in aqueous solution is only 10^{-6}M (Gebhardt *et al.*, 2011). Due to their unique structure caseins are readily digested and absorbed by the human gastrointestinal tract. Casein has a Protein Digestibility Corrected Amino Acid Score (PDCAAS) rating of 1.23 (generally reported as a truncated value of 1.0) indicating completeness of protein (Deutz *et al.*, 1998).

Whey Proteins

Though whey proteins are present in lesser amount in comparison to caseins, its nutritional value can never be overstated. Whey proteins contains high level of Branched Chain Amino Acids (BCAA) (about 26 per cent) *i.e.*, leucine, isoleucine, and valine.

Whey proteins are a good source of sulfur-containing amino acids such as cysteine and methionine. Whey proteins contains at least 4 fold more cysteine than other high quality proteins, and provide an excellent source of a key precursor for glutathione production. Glutathione is a metabolite found throughout our body essential for good immune system.

It contains high levels of arginine and lysine, which may stimulate release of growth hormone, and thus help increase muscle mass and a decline in body fat. Whey protein contain good amount of glutamine which helps in muscle glycogen replenishment and prevents decline in immune function from over exercising.

Table 10.2: Primary Components of Whey Protein and their Benefits

Whey Component	*% of Whey Protein*	*Benefits*
β-Lactoglobulin	50-55 per cent	Excellent source of essential amino acids especially branched-chain amino acids
α-Lactalbumin	20-25 per cent	Excellent source of essential amino acids; High in Tryptophan which helps regulate sleep, mood and stress
Immunoglobulins	10-15 per cent	IgA, IgD, IgE, IgG, IgM – primarily IgG with immune enhancing benefits
Lactoferrin	1-2 per cent	Antioxidant, anti-viral, anti-bacterial, anti-fungal; promotes beneficial bacteria; regulates iron absorption
Lactoperoxidase	0.5 per cent	Inhibits bacterial growth
Bovine Serum Albumin	5-10 per cent	Good profile of essential amino acids
Glycomacropeptide	10-15 per cent	Inhibits formation of dental plaque and cavities

Source: (Anon, 2008).

α-lactalbumin is calcium binding protein and thereby enhances calcium absorption. It is an excellent source of essential amino acids tryptophan and cysteine. Purified α-lactalbumin is used commercially in infant formula because it is structurally and compositionally similar to the protein in human milk.

Buffalo milk contains about 10mg of immunoglobulins per 100 ml of milk. In colostrum, their concentration is very high, being 43 and 32 per cent of the total proteins in the first and second milking, respectively. Immunoglobulins in milk are important for imparting immune defense to the host. Approximately 75 per cent of the total immunoglobulin is IgG_1 which provides active immunity.

Several milk protein components like α-lactalbumin, lactoferrin and glycomacropeptide possesses bifidogeninc activity (enhancing growth of friendly bacteria in the human gut).

Furthermore, during cheese making a protein fraction known as Glycomacropeptides (GMPs) is formed. It performs several functions including growth factors for bifidus bacteria in the intestine, antiviral activity, modulating digestion, improved calcium absorption, antibacterial properties and improvement of the immune system.

Bioactive Peptides

Milk proteins are precursors of several bio-active peptides; which plays role in neonatal and subsequent development. Fermented milk products like dahi, yoghurt, cheese, etc. are good source of bioactive peptides.

Caseins and whey proteins provide several bioactive peptides like opoid, immunomodulating, antihypertensive, antithrombotic, antimicrobial, caseinophospo-peptides, etc. with vital biological significance. Table 10.3 gives a brief of some bioactive peptides.

Table 10.3: Bioactive Peptides Derived from Milk Proteins

Bioactive Peptide	*Protein Precursor*	*Bioactivity*
Casomorphins	α-, β-casein	Opoid agonist
α-Lactorphins	α-lactalbumin	Opoid agonist
β-Lactorphins	α-lactalbumin	Opoid agonist
Lactoferroxins	Lactoferrin	Opoid antagonist
Casoxins	κ-casein	Opoid antagonist
Casoxinins	α-, β-casein	Antihypertensive
Casoplatelins	κ-casein, Transferrin	Antithrombotic
Immunopeptides	α-, β-casein	Immunostimulant
Phosphopeptides	α-, β-casein	Mineral carrier

Milk proteins thus have protective and physiological role besides primary function of providing essential amino acids.

Moreover it is worth emphasizing that the Digestible Indispensable Amino Acid Score (DIAAS) of milk protein (A protein quality evaluation measure adopted by FAO; this measure indicates presence and utilization of essential amino acids by the human body) is 30 per cent higher than vegetable source of protein. The score of milk protein being 1.31 against 0.98 of Soya Protein Isolate. Table 10.4 provides a comparison of DIAAS and PDCAAS value of milk proteins with other protein.

Table 10.4: Comparison of DIAAS[8] and PDCAAS[9] Value of Milk Proteins with other Protein

Essential Amino Acid	*WPI[1]*	*RPC[2]*	*MPC[3]*	*WPC[4]*	*SPI[5]*	*PPC[6]*	*Pearl Barley[7]*
Threonine	1.66	0.93	1.44	2.32	1.04	1.18	0.77
Methionine/Cystine	2.5	1.35	**1.31**	1.88	1.02	**0.93**	1.34
Valine	1.5	1.39	1.92	1.60	1.27	1.55	1.25
Isoleucine	2.55	1.34	2.09	2.7	1.59	1.88	1.01
Leucine	2.57	1.11	1.78	1.93	1.13	1.38	0.84
PHE/TYR	1.42	1.52	1.98	1.19	1.37	1.65	0.94
Histidine	**1.25**	1.18	1.81	**1.10**	1.34	1.52	1.08
Tryptophan	2.66	1.14	1.59	2.19	1.37	0.94	0.84
Lysine	2.47	**0.38**	1.75	2.01	**0.98**	1.49	0.58
DIAAS[8]	**1.25**	**0.38**	**1.31**	**1.10**	**0.98**	**0.93**	**0.58**
PDCAAS[9]	**1.00**	**0.38**	**1.00**	**1.00**	**0.97**	**0.93**	**0.52**

1: Whey Protein Isolate; 2: Rice Protein Isolate; 3: Milk Protein Concentrate; 4: Whey Protein Concentrate; 5: Soya Protein Isolate; 6: Pea Protein Concentrate; 7: Pearl Barley; 8: DIAAS: Digestible Indispensable Amino Acid Score; 9: PDCAAS: Protein Digestibility Corrected Amino Acid Score.

The high level of essential amino acids with good digestibility along with high bioavailability of milk proteins make them a vital food ingredient for managing protein energy malnutrition and supporting growth in young children. It is evident from the above comparison that milk is significantly superior to most vegetable proteins from the nutrition point of view. This is an excellent way to make the best use of current resources for enhancing nutritional quality of human diet.

Lactoferrin

Lactoferrin (formerly known as lactotransferrin) is a glycoprotein and a member of a transferrin family. It belongs to those proteins capable of binding and transferring Fe^{3+} ions (Metz-Boutique *et al.,* 1984). Lactoferrin works as an antioxidant by binding iron and preventing it from participating in free radical formation. Lactoferrin also plays a role in the cellular defense system, most likely by regulating the macrophage activity and stimulating the proliferation of lymphocytes. Lactoferrin has been reported to stimulate the growth of Bifidobacteria; indicating gut health promoting ability of this particular protein. Lactoferrin has a role non-specific defense of the host against invading pathogens. It is active against several Gram-positive and Gram-negative bacteria, yeasts, fungi and viruses.

Both buffalo milk and cow milk contains significant quantities of lactoferrin. Lactoferrin content of buffalo milk is 0.320mg/ml and that of cow milk 0.1 mg/ml. Thus, buffalo milk is richer in this particular protective protein than cow milk.

Colostrum contains high level of lactoferrin (mg/ml) as shown in Table 10.5.

Table 10.5: Colostrum Contains High Level of Lactoferrin (mg/ml)

Cow		*Buffalo*		*Human*	
Colostrum	*Milk*	*Colostrum*	*Milk*	*Colostrum*	*Milk*
0.70	0.10	0.75	0.32	10.0	1.0

Source: Bhatia (1997).

Lysozyme

Lysozyme is low molecular weight enzyme present in milk and colostrum. Human milk is the richest source of lysozyme containing as high as 30-40 mg/100ml. Lysozyme is a widely distributed enzyme that lyses cell wall of certain bacteria by hydrolyzing the β (1-4) glycosidic linkage from N-acetyl muramic acid and N-acetylglucosamine. It is found in many body fluids and the lysozyme in milk is usually isolated from whey. Lysozyme has bactericidal and bacteriostatic properties, which may help to protect children against intestinal infections (Chiavari *et al.,* 2005).

Lysozyme content of buffalo milk is 15.2 μg/100 ml and that of cow milk about 18.0 μg/100 ml. More than 75 per cent of the lysozyme activity in bovine milk survives heating at 75°C during 15 min or 80°C during 15 s and therefore it is not affected by HTST pasteurization.

Milk Carbohydrate

Lactose

Lactose, the major carbohydrate in human and animal milk, is a disaccharide sugar composed of the glucose and galactose. Although lactose is not a key nutritional component for adults, it is the main source of energy during the first year of a human's life. Lactose undergoes slow hydrolysis during digestion and generates a prolonged energy supply (Vesa *et al.*, 2000). Lactose acts as ready source of energy for neonate; providing 30 per cent of the energy in bovine milk; nearly 40 per cent in human milk and 53–66 per cent in equine milks (Fox, 2008).

Lactose content of buffalo milk varies from 4 to 5.5 per cent ; whereas Indian cow contains about 4.4 per cent and that of Western breed of cow 4.9 per cent.

In general, carbohydrates increase intestinal calcium absorption, and lactose contained mainly in dairy products seems to be the most effective carbohydrate supporting calcium absorption. Lactic acid, the metabolic product of lactose, lowers the pH in the intestine, which increases the solubility of calcium making it more available for absorption (Pérez *et al.*, 2008). Even, European Food Safety Authority (EFSA) has accepted the claim regarding lactose's possible effect on calcium absorption. Lactose is also reported to increase lactobacilli population in the intestinal tract supporting gut health.

Some individual are not able to digest lactose due to deficiency of enzyme lactase. Such persons are termed as lactose intolerant. However; for such people lactose reduced or lactose hydrolyzed products can be helpful in imparting the other health benefits of milk. In fact, several traditional fermented products like dahi, yogurts, cheese, etc. are better tolerated by lactose intolerant people.

Minor quantities of glucose, galactose and oligosaccharides are also present in milk. It is speculated that galactose may have a unique role in the rapidly developing infant brain.

Oligosaccharides

Oligosaccharides are strictly defined as carbohydrates which contain 3-10 monosaccharide residues covalently linked by glycosidic bonds (Gopal and Gill, 2000). Human milk contains good amount of oligosaccharides which are found to be potent bioactive molecules. The oligosaccharides found in bovine milk are structurally similar to those in human milk, but their concentration in milk is quite low, particularly in mature milk compared with colostrum. Bovine milk contains only 1-2 gm of oligosaccharides per liter in comparison to human milk which contains about 10-25 gm/liter.

Oligosaccharide plays vital role for human infants as it enhances growth of bifidobacteria in human gut. Some oligosaccharides prevent the adhesion of pathogenic bacteria in human gut thereby protects human infants from infectious diseases. Some oligosaccharides are also important for cognitive development of human infants.

Milk Fat

Milk fat contributes unique characteristics to the sensory properties and satiability of dairy foods, and is a source of energy, essential fatty acids, fat-soluble vitamins, and several other potential health promoting components.

Milk fat, the most complex of dietary fats, exists in microscopic globules in an oil-in-water emulsion in milk. It is a complex mixture of lipids, but the bulk (97-98 per cent of total lipids) is contained in the triglyceride form. Milk fat also contains diacylglycerols and monoacylglycerols, free fatty acids, phospholipids, sterols, carotenoids, fat-soluble vitamins and several flavor compounds. More than 400 fatty acids have been identified in milk fat however only about 20 of these make up approximately 95 per cent of the total. About 64 per cent of the fatty acids found in milk fat are saturated; of these about 12 per cent are short-chain fatty acids (C4:0-C10:0). Stearic acid (C18:0) contributes about 12 per cent of total saturated fatty acids (SFA). The remaining SFA represent a mixture of lauric acid (C12:0), myristic acid (C14:0) and palmitic acid (C16:0). Approximately 30 per cent of the fatty acids in milk are mono-unsaturated fatty acids (MUFA) and 4 per cent are poly-unsaturated fatty acids (PUFA).

As indicated earlier, milk fat provides essential fatty acids like linoleic acid and linoleic acids. These essential fatty acids have many roles; including being structural components of cell walls and helping regulate blood pressure and nerve transmissions. Buffalo milk fat contains about 1.6 per cent linoleic and 0.50 per cent linoleic acid whereas cow milk contains about 1.50 per cent linoleic and 0.60 per cent linolenic acid.

Milk fat is also good source of phospholipids (PLs). Milk fat, on an average, contains about 1 per cent phospholipids. PLs like gangliosides have been demonstrated to play important role in neonatal brain development, receptor functions, allergies, for bacterial toxins etc. Sphingomyelin shows a strong anti-tumour activity, influences the metabolism of cholesterol and exhibits an anti-infective activity. Glycerophospholipids protect against mucosal damage.

Milk fat is viewed with certain degree of suspicion specifically owing to presence of saturated fat and trans fat with possible role in heart ailments. Though milk fat contributes significant amount of lauric, myristic, palmitic acid and trans fatty acid to human diet; it is not prudent to advocate a decrease in milk, cheese, and butter consumption (Shingfield *et al.*, 2008).

In fact, the content and isomeric profile of Trans fatty acids found in dairy products differs significantly from those in partially hydrogenated vegetable oils. Growing evidence suggests that trans fatty acids from natural sources may not have the same adverse effect on human health as those from industrial sources. Thus, even though milk fat contains about 5 per cent Trans fat it is different than industrially hydrogenated vegetable fat with respect to health outcomes (FAO and WHO, 2010).

While it is true that certain saturated fatty acids in milk fat can raise LDL levels (Mensink *et al.*, 2003), it is most likely that this elevation is offset by an equal ability to raise HDL levels (Hu and Willett, 2000).

Several studies have indicated that milk even if taken in relatively high amounts; it does not have any significant adverse effects on health. On the contrary, it may be protective against several diseases like cardiovascular disease, metabolic syndrome, and colorectal cancer. While other dairy products (mainly fat rich) certainly contribute to consumption of saturated fatty acids, evidence for negative, or positive, effects on health is limited (Salter, 2013).

A certain amount of fat is required in the diet and inclusion of milk fat as part of a balanced diet should be beneficial rather than detrimental (Parodi, 2004).

Conjugated Linoleic Acid

Conjugated Linoleic Acid (CLA) refers to a mixture of positional and geometric isomers of linoleic acid with two conjugated double bonds at various carbon positions in the fatty acid chain. The CLA isomers are found in many foods but are predominant in products derived from ruminant sources because of the process of bacterial biohydrogenation of polyunsaturated fatty acids (PUFAs) in the rumen (Corl *et al.,* 2001).

The most biologically active is the diene of configuration cis-9, trans-11 (octadecadienoic). It is claimed to inhibit the occurrence and development of cancer of the skin, breast, colon, and stomach (Parodi, 1999), while its isomer trans-10, cis-12 is thought to prevent obesity (Bawa, 2003; Wang and Jones, 2004).

Additionally, CLA reduces the levels of triglycerides, total cholesterol, including LDL, and thus improves the ratio of LDL/HDL in plasma, which is a crucial factor in the prevention of coronary heart disease and (Tricon *et al.,* 2004). CLA is also said to inhibit the development of osteoporosis (Watkins and Seifert, 2000), to improve the metabolism of lipids, to reduce the blood glucose level, and to stimulate the immune system (O'Shea *et al.,* 2004).

Milk fat is the richest natural dietary source of CLA. Buffalo milk contains relatively higher levels of conjugated linoleic acid (CLA) than cow's milk. CLA content of cows' milk ranges from 3.38 to 6.39 mg CLA/g fat; whereas average content of CLA in buffalo milk ranges from 4.4 to 7.6mg/g fat (Lin *et al.,* 1995; Ahmad, 2013).The CLA contents of most dairy products range from 2.5 to 7.0 mg/g of lipid (Lin *et al.,* 1995).

Butyric Acid

Milk fat supplies considerable amount of butyric acid. It is unique to milk fat it and ranges between 75 and 130 mmol/mol of fatty acids. Butyrate is known to exhibit anticarcinogenic effects, inhibit cell growth, promote differentiation, and induce apoptosis in various human cancer cell lines (Parodi, 1999).

Cholesterol

Cholesterol is found in all body cells and plays a key role in the formation of brain, nerve tissue and is a precursor for some hormones and vitamin D. It is synthesized in the body and hence it is not an essential dietary component.

For most population groups, the amount of cholesterol ingested in the diet has a small effect on the concentrations of circulating cholesterol. Observational studies in humans have not demonstrated the cholesterol content of the diet as a major factor in increased risk of coronary heart disease (Kratz, 2005).

Buffalo milk fat has a lower concentration of total (275 mg per 100g) and free cholesterol (212 mg per 100g) than bovine milk fat (330 and 280 mg100g, respectively).

Fat Soluble Vitamins

The major fat soluble vitamin contained in milk is vitamin A, milk also contain β-carotene which acts as a provitamin A. However milk does not contain vitamin D in significant amount and hence milk is most often fortified with vitamin D. Appropriate amount of vitamin D in milk is necessary for absorption of calcium and milk is fortified at the level of about 100 IU per serving.

Vitamin A

It is one of the oldest known dietary factor identified for preventing blindness (Egyptians knew consumption of beef liver can prevent blindness). Retinoids is a collective term for the biologically active forms of vitamin A. Vitamin A (retinoids) exists in three forms showing metabolic activity: the alcohol (retinol), the aldehyde (retinal or retinaldehyde) and the acid (retinoic). Vitamin A has a key role in vision, proper growth, reproduction, and immunity, cell differentiation, in maintaining healthy bones as well as skin and mucosal membranes.

Buffalo milk contains about 3200 IU (960 RAE) of vitamin A whereas cow milk contains 1500 IU (500 RAE) per 100 ml milk. Not only milk is good source for vitamin A; it is one of the most important sources as there is considerable degree of prevalence of vitamin A deficiency. Vitamin A deficiency affects 190 million preschool-aged children and 19.1 million pregnant women (WHO, 2012).

Vitamin E

Vitamin E is a family of 8 naturally occurring compounds including 4 tocopherols (alpha, beta, gamma, delta) and 4 tocotrienols (alpha, beta, gamma, delta) with different biological activity. α-tocopherol is the most active form of this vitamin. Principal form of vitamin E in milk is α-tocopherol though it also contains γ- tocopherol and γ-tocotrienol. Vitamin E is the most important fat soluble antioxidant from human health consideration. It protects polyunsaturated fatty acids from oxidation in cell membranes and in plasma lipoproteins (Graulet *et al.*, 2013). Buffalo milk contains about 33 mg/100 ml tocopherol in comparison to 31-33 mg/100 ml in cow milk.

Vitamin K

Vitamin K was first discovered in the 1930s through its anti-haemorrhagic properties (Garulet *et al.*, 2013). The family of compounds known as vitamin K, or the quinones, include phylloquinones (vitamin K_1) from plants and menaquinones (vitamin K_2) found in animal products including milk (Carol Byrd-Bredbenner, 2009). Phylloquinone, is considered biologically most active form.

Cow milk contains low amounts of phylloquinone (0.6 µg/100 g) and very low concentrations of other menaquinones (Graulet *et al.*, 2013). However; fermented dairy products like cheese contain good amount of menaquinones (eg. 5–10 µg/100g of menaquinone-8 and 10–20 µg/100g of menaquinone-9) (Shearer *et al.*, 1996).

Vitamin D

Vitamin D has important role in maintaining blood calcium and phosphorus balance and assists calcium metabolism. Vitamin D is considered to have hormone like activity. Milk fat contains a number of vitamin D-related compounds that exhibit variable vitamin D activity. They include cholecalciferol (vitamin D_3), 25-hydroxyvitamin D_3 (25[OH] D3), 1,25- dihydroxyvitamin D_3 (1,25[OH]2D_3), 24,25-dihydroxyvitamin D_3, 25, 26-dihydroxyvitamin D_3 and ergocalciferol (vitamin D_2) (Parodi, 2004).

Cow milk contains about 0.1-0.3 µg per 100g milk. The vitamin D content of buffalo milk has been reported to be quite low, being of the same order as that of cow's milk and human milk.

Water Soluble Vitamins

Milk is excellent source of certain vitamins like riboflavin, thiamin and good source of many water soluble vitamins like pantothenic acid, folic acid, pyridoxine, biotin and vitamin B_{12} (Institute of Medicine, 1998).

Riboflavin

Riboflavin is also known as B_2, its name comes from its color as flavin means "yellow" in Latin. Riboflavin plays key role in energy metabolism of the body; as it is a component of two coenzymes *i.e.* FMN (flavin mononucleotide) and FAD (flavin adenine dinucleotide). About two cups of cow or buffalo milk can fulfill daily requirement of riboflavin. Milk is a rich source of riboflavin. Cow milk contains about 170-200 µg/100ml milk; whereas buffalo milk contains average 160 µg/100ml.

Vitamin B_6

Vitamin B_6 is a family of 3 compounds namely pyridoxal, pyridoxine, and pyridoxamine. It functions as coenzymes and has specific role in amino acid metabolism. It is thus vital in growth and maintenance of body. Buffalo milk is an excellent source of this vitamin as 100ml milk contains about 0.4mg against the RDA of 2mg per day.

Biotin

Biotin participates in fatty acid, amino acid and energy metabolism. Deficiency of biotin can lead to dermatitis, conjunctivitis, hair loss, nervous system abnormalities. 100g of buffalo milk can easily fulfill approximate human requirement of 8 µg/day. Buffalo milk contains about 10-13µg/100 ml whereas cow milk contains about 2-3.5 µg/100 biotin.

Thiamin

This vitamin is also known as vitamin B_1 or aneurin. Thiamin is crucial to the energy generating reactions involving carbohydrates, fatty acids and amino acids.

Buffalo milk contains about 35 µg/100 ml and cow milk contains about 30 µg/100 ml against the RDA of 1.4 mg per day for moderately working man and 1.1 mg per day for moderately working women.

Niacin

Niacin is also known as vitamin B_3. This vitamin has also been called vitamin PP for 'pellagra preventive', because pellagra occurs as a consequence of its deficiency. It exists in 2 forms–nicotinic acid (niacin) and nicotinamide (niacinamide). Both forms are used to synthesize the niacin coenzymes.

The amino acid tryptophan is also present in milk which can be used for synthesis of niacin. In the synthesis of niacin from tryptophan, 60 mg of dietary tryptophan is needed to make about 1 mg of niacin. Considering both niacin and tryptophan, milk can act as good source of niacin. Cow milk and buffalo milk contains about 13 and 0.17 mg/100g niacin respectively.

Pantothenic Acid

The name pantothenic acid was taken from the Greek word 'pantothen', meaning "from every side," because it is present in all body cells and present in wide variety of foods. As a consequence of the absence of dietary deficiency in humans, there is no RDA specified. Pantothenic acid content in cow milk varies from 0.34–0.58 mg/100 g; whereas that of buffalo milk contains about 0.37 mg/100g.

Folate (Vitamin B_9)

Folate is the generic name, referring to the various forms of the vitamin found naturally in foods; synthetic form of this vitamin is known as folic acid. Folic acid is essential for the synthesis of haemoglobin and its deficiency leads to macrocytic anaemia. Especially, pregnant women need more folic acid than other group of population. Deficiency of folic acid increases homocysteine levels in blood increasing the risk for heart disease.

Though milk is not good source of this particular vitamin, its contribution can be significant as milk and milk products are consumed in good amounts. Cow milk contains about 5–8 µg/100g; whereas buffalo milk contains about 6 µg per 100g of folate in milk. RDA for moderately working man and women is about 200 mg/day; however for pregnant women it is 500 mg/day.

Vitamin B_{12}

This particular vitamin is unique in that neither animal nor plant can synthesize it. Indeed, only bacteria are able to produce cobalamins. Even if B_{12} is synthesized by bacteria in human gut it not absorbed. Humans have to depend on dietary sources like milk and dairy products, meat, poultry, eggs and fish. Considering the vegetarian society as ours; milk and milk products are indispensable sources. Many epidemiological studies have reported a positive relationship between consumption of dairy products and vitamin B_{12} status indicating milk is indeed important dietary source (Graulet *et al.,* 2013). Buffalo milk and cow milk contain about 0.3-0.4 µg and 0.4 µg/100 ml respectively.

Deficiency of this vitamin can cause pernicious anemia, macrolytic anemia, etc. B_{12} deficiency produces nerve degeneration, which can be fatal.

Vitamin C

Vitamin C can occur in either reduced form *i.e.* ascorbic acid or oxidized form dehydroascorbic acid. Both forms possess vitamin activity. The ability of this vitamin to interconvert in either form makes it one of the good antioxidant. Milk is generally considered deficient in vitamin C. Humans are not able to synthesize this particular vitamin and have to rely on other dietary sources. The ascorbic acid content of buffalo milk is about 23–30 mg/kg and that of cow milk about 20mg/kg (Singh and Gupta, 1986; Guo, 2010).

Vitamin Losses during Processing

Various factors affect the stability of vitamins; important being heat, moisture, oxygen, pH and light. Milk and milk products are subjected to various processing. Generally heat is involved in most dairy operations. As a general rule the severe the heat treatment the more are the losses.

Vitamin A, riboflavin and pyridoxine are few most susceptible vitamins to light. Exposure to direct sunlight (in glass or certain plastic containers) can causes 20-40 per cent loss in riboflavin within an hour and 40-70 per cent loss in 2 hours. Since milk is one the important source of riboflavin, it is important that suitable care be taken during the processing and subsequent handling of milk and milk products.

Milk suffers a loss of about 10 per cent of vitamin A when exposed to sunlight for about 6 hours. However vitamin A is quite stable in butter *e.g.* only 5 per cent loss occurs in 9 months if stored at -18°C. However, it is generally considered to be heat stable thus pasteurization, ultra high temperature (UHT) heating or sterilization does not lead to significant losses.

Vitamin B_1 losses in milk with an average content of 0.04 mg per 100 g are normally less than 10 per cent for pasteurized milk; 5-15 per cent for UHT milk and 30 - 40 per cent for sterilized milk. During processing of evaporated milk thiamin loss is about 30 -50 per cent.

Both forms of niacin *i.e.* nicotinic acid and nicotinamide are normally very stable in milk and milk products because they are not affected by atmospheric oxygen, heat and light.

Biotin is lost to the tune of about 10 per cent when milk is subjected to pasteurization and sterilization processes.

Losses of pantothenic acid are about 10 per cent during typical milk processing operations.

Folic acid is decomposed by oxidizing and reducing agents. Sunlight, and particularly ultraviolet radiation, has a serious effect on the stability of folic acid. Pasteurization of milk causes about 5 per cent loss in folic acid, however in sterilized milk, losses to the tune of 40-50 per cent can occur (Berry Ottaway, 2012).

About 8 hours exposure of milk to sunlight can result in the loss of nearly 21 per cent vitamin B_6 (Ryley and Kajda, 1994). Pyridoxine is stable in milk during pasteurization but about 20 per cent can be lost during sterilization (Berry Ottaway, 2012). During the UHT processing about 27 per cent pyridoxine is lost.

Presence of oxygen causes conversion of ascorbic acid to dehydroascorbic acid. This form of vitamin C though remains bioavailable, its susceptibility to heat losses increases. About $3/4^{th}$ of the original amount of ascorbic acid is lost due to this reason.

Normally less than 5 per cent of folic acid is lost during the pasteurization of milk. Losses in the region of 20 per cent can occur during UHT treatment and about 30 per cent loss is found after sterilization. During the storage of UHT milk, more than 50 per cent of losses can occur at the end of three months (Berry Ottaway, 2002).

Vitamin B_{12} is normally stable during pasteurization of milk but up to 20 per cent can be lost during sterilization, and losses of 20–35 per cent can occur during spray drying of milk (Berry Ottaway, 2002).

Vitamin K is relatively stable to heat processing, but it can be destroyed by exposure to light (Carol Byrd-Bredbenner, 2009).

Minerals

Milk supplies virtually all of the minerals, major and minor, required by humans. Undoubtedly calcium is the most significant mineral supplied by milk. It also contains several essential minerals like phosphorus, potassium, magnesium and sodium. Most trace elements like zinc, iron, copper, iodine, fluoride, selenium and to lesser extent molybdenum, manganese, chromium are present in minute quantities. However, zinc is present in nutritionally significant quantities.

Calcium

Calcium is an essential nutrient which has a number of important roles in the body. It is needed to help build and maintain healthy bones, normal blood clotting, nerve and muscle function, maintenance of teeth, energy metabolism and for enzymes involved in digestion. Thus getting enough calcium in the childhood period is important for healthy skeleton apart from several other functions. Bones not only grow in length at this stage but it gets strengthened as well; around 90 per cent of the adult skeleton is formed by the age of 18.

Studies have indicated that postmenopausal women who practiced habitual dairy consumption during childhood and adolescence had higher bone density later in life (Sandler *et al.*, 1985; Nieves *et al.*, 1995). Both calcium and dairy products may protect against the development of colon cancer. Dietary calcium may also protect against some forms of kidney stones. Consumption of milk and milk products as a way to meet calcium requirements also help increase intake of potassium and magnesium, which have been linked with reduced risk of hypertension.

Recently, it has also been shown that dairy calcium may partly offset the cholesterol raising effect of the SFA in milk, perhaps through the formation of calcium soaps of fatty acids and increased fecal excretion (Lorenzen and Astrup, 2011).

On an average cow milk contains 112mg/100g calcium whereas buffalo milk contains about191 mg calcium per 100g milk. Thus, calcium is one of the most important essential nutrients present in milk of both cow and buffalo. Bioavailability of calcium from milk is high compared with that in other foods in the diet (Weaver,

Proulx and Heaney, 1999). Additionally absorption of calcium from milk and milk product is better than that of from other sources like Ragi and Rajgara due to better bioavailability. A person can utilize calcium present in skimmed milk powder to the level of 85 per cent in comparison to 22-27 per cent from vegetable sources.

Magnesium

Magnesium has many functions in the body, participating in more than 300 reactions. Magnesium deficiency has been linked to atherosclerosis as studies have shown that deficiency may give oxidative stress. Magnesium is also one of the important mineral from the point of view of bone health and skeletal growth. Buffalo milk contains more magnesium 18mg/100g than cow milk 11 mg/100g. Of the commonly eaten foods, in the portions usually consumed, dairy products are the best sources of magnesium.

Phosphorus

Phosphorus is an essential mineral required by every cell in the body. Phosphorus in grains and legumes is mainly in the form of phytates and is poorly absorbed because we lack enzymes that release the phosphorus. In addition to being a major component of bones and teeth, phosphorus is critical to the function of every cell in the body. This mineral also is a part of DNA and RNA, phospholipids in cell membranes and numerous enzyme and cellular message systems. Phosphorus is particularly important for promoting the growth of undernourished children.

Average phosphorus content of buffalo milk and cow milk is about 82 and 85mg/100g respectively.

Potassium

Potassium is the major cation inside the cell and performs many of the same functions as sodium. It has a role in carbohydrate metabolism, enhancement of protein synthesis and muscle contraction and nerve impulse conduction. Potassium has role in reducing blood pressure, although recently it has also come to light that it can alleviate osteoporosis.

Milk acts as a one of the good source of potassium. Cow milk contains about 148 mg/100ml whereas buffalo milk contains 106.6 mg/100 ml.

Sodium and Chloride

Sodium and chloride are other essential elements required by human body. Sodium has three main functions in body *viz.* it is involved in absorption of glucose and some amino acids. It is required for normal muscle and nerve function and in maintaining water balance.

Chloride is involved in nerve signal transmission, maintaining electrolyte balance and is component of acid produced in stomach.

Sodium content of buffalo milk is about 45 mg/100ml and that of cow milk about 50 mg/100ml. Chloride content of cow is nearly double than buffalo milk being about 120 mg/100ml in comparison to 64 mg/100ml. Intake of both of these elements through consumption of common salt is considerably higher than our daily requirements.

Zinc

Zinc is functioning in association with more than 300 different enzymes of various classes. Zinc contributes to DNA and RNA synthesis, alcohol metabolism, heme synthesis, bone formation, acid-base balance, immune function, reproduction, growth and development and the antioxidant defense network (as a component of superoxide dismutase enzyme). Zinc also plays role in stabilization of cell membrane structures.

Contribution of zinc from the milk to total requirement of the human body is less; however bioavailability of zinc from milk is higher in comparison to other vegetarian sources. Additionally, bioavailability of zinc from the vegetable sources increases if milk is included in the diet. Zinc content of buffalo milk varies from 3.2-7.3 mg/litre; whereas in case of cow milk it varies from 3-4 mg/litre.

Selenium

Selenium is recognized as an essential mineral hence it is required to be taken externally by humans. The selenium concentration in body fluids and tissues are directly related to selenium intake. Selenium is important in human health as it has a role in the immune and antioxidant system and in DNA synthesis and DNA repair. Selenium is also said to protect against several types of cancer like prostate, lung and other cancers. Selenium is also important in iodine metabolism because of its presence in one enzyme (iodothyroninedeiodinase enzymes) responsible for forming active T_3 (Thyroid) hormone. Reportedly selenium content of Indian buffalo milk varies from 1.536 to 1.684 µg/100g milk (Janabai *et al.,* 1988). Selenium content of cow milk varies from 1 to 3.7 µg/100g milk.

Other Trace Minerals

The human body contains about 4 and 2.5 g of Fe in adult men and women, respectively, and it plays essential roles in many biological functions (Gaucheron, 2013). The iron content of buffalo milk has been reported to be twice that of cow milk being 0.08 mg/100ml in comparison to 0.04 mg/100ml. The average Cu concentration in cow milk is about 0.1 mg/L; whereas copper content of buffalo milk varies between 0.07 to 2.6 mg/L.

Conclusion

Milk and milk products no doubt plays significant role in the diet of the people universally. Milk is found to have profound influence on wellbeing. Milk can be considered as complete package full of nutrient available conveniently and at reasonable cost. On the other hand; one would require consuming several foods in order to derive so many vital nutrients which milk provides. Further there is no valid evidence of any harm to health from milk consumption in terms of the risks of disease, rather, there are consistent evidences with reductions in vascular disease, diabetes, certain cancers and possibly all-cause death, in the subjects who drink the most milk.

World Health Organization (Villar *et al.,* 2006) has concluded that pregnant women globally are failing to consume the recommended daily supply of 1200 mg of calcium. Dairy products are the easiest and most reliable source of the mineral.

References

Ahmad S (2013). Buffalo Milk, In: Milk and Dairy Products in Human Nutrition, pp. 519-546, Ed. Young W. Park and George F.W. Haenlein. In: *Milk and Dairy Products in Human Nutrition*, pp. 200-215, Park, Young W./Haenlein, George F. W. (eds.), John Wiley and Sons, Inc.

Anon (2008). Whey proteins. *Alternative Medicine Review* 13: (4), 341-347.

Bawa S (2003). An update on the beneficial roles of conjugated linoleic acid (CLA). in modulating human health: mechanism of action, a review. *Pol J Food Nutr Sci* 3: 3–13.

Berry Ottaway P (2002). The stability of vitamins during food processing: In: *'The Nutrition Handbook for Food Processors'*, Ed. Henry, C. J. K and Chapman, C., Woodhead Publishing Limited, Abington, England.

Bhatia KL (1997). Protective proteins of milk, *Indian Dairyman*, 49: (6), 11-18.

Carol Byrd-Bredbenner, Donna Beshgetoor, Gaile Moe, Jacqueline Berning (2009). *Wardlaw's Perspectives in Nutrition*, 8th Ed., McGraw-Hill Publication, New York.

Chiavari C, Coloretti F, Nanni M, Sorrentino E, Grazia L (2005). Use of donkey's milk for a fermented beverage with Lactobacilli. *Lait* 85(6): 481–90.

Connie Weaver1, RamaniWijesinha-Bettoni, Deirdre McMahon and Lisa Spence (2013). Milk and dairy products as part of the diet, In: *Milk and Dairy Products in Human Nutrition*. FAO Publication, Rome.

Corl BA, Baumgard L, Dwyer H, Briinari DA, Phillps JM, Bauman E (2001). The role of delta (9). desaturase in the production of cis 9, trans 11. *J Nutri Biochem* 12: 622-630.

Deutz NEP, Bruins MJ and Soeters PB (1998). Infusion of soy and casein protein meals affects interorgan amino acid metabolism and urea kinetics differently in pigs. *Journal of Nutrition* 128: 2435-2445.

FAO and WHO (2010). Interim summary of conclusions and dietary recommendations on total fat and fatty acids. From the Joint FAO/WHO Expert Consultation on Fats and Fatty Acids. Available at: http: //www.who.int/nutrition/topics/ FFA_summary_ rec_conclusion.pdf. Accessed 10 October 2012.

Fox PF (2008). Milk: an overview. In: A. Thompson, M. Boland and H. Singh, eds. *Milk Proteins: From Expression to Food*, pp. 1–54. San Diego, CA, USA, Academic Press.

Gaucheron F (2013). Milk Minerals, Trace Elements, and Macroelements, In: *Milk and Dairy Products in Human Nutrition*, pp. 172-199, Park Young W./Haenlein, George F. W. (eds.), John Wiley and Sons, Inc.

Gebhardt R, Takeda N, Kulozik U, Doster W (2011). Structure and stabilizing interactions of casein micelles probed by high-pressure light scattering and FTIR, *J Phys Chem* B. 115(10): 2349-59. doi: 10.1021/jp107622d.

Gopal PK and Gill HS (2000). Oligosaccharides and glycoconjugates in bovine milk and colostrums. *British J. Nutrition*, 84, Suppl 1, S69-S74.

GrauletBenoît, Bruno Martin, Claire Agabriel and Christiane L. Girard (2013). Vitamins in milks, In: *Milk and Dairy Products in Human Nutrition*. pp. 200-215, Ed. Park, Young W./Haenlein, George F. W. (eds.), John Wiley and Sons, Inc.

Guo M (2010). Improving buffalo milk, In: *Improving the Safety and Quality of Milk, Vol. 2: Improving Quality in Milk Products*, Ed. Mansel W. Griffiths, Woodhead Publishing Limited, Cambridge, UK.

Hu FB and Willet W (2000). letter to the editor, 'Reply to O H Holmqvist', *Am J Clin Nutr*, 71(3): 848-849.

Institute of Medicine (1998). Dietary reference intakes for thiamine, riboflavin, niacin, vitamin B6, folate, vitamin B12, pantothenic acid, biotin and choline, Washington, DC, National Academy Press.

Janabai G, Jeyanthi GP and Selvi S (1988). *Indian J. Nutr. Dietet*. 25: 140-143.

Kratz M (2005). Dietary cholesterol, atherosclerosis and coronary heart disease. In: Atherosclerosis: Diet and Drugs (ed. A. von Eckardstein), *Handbook of Experimental Pharmacology* Vol. 170, pp. 195–213. Springer, Berlin.

Lin H, Boylston D, Chang MJ, Luedecke LO, Shultz TD (1995). Survey of the conjugated linoleic acid contents of dairy products. *J Dairy Sci*; 78: 2358-2365.

Lorenzen JK and Astrup A (2011). Dairy calcium intake modifies responsiveness of fat metabolism and blood lipids to a high fat diet. *Brit. J. Nutr*. 105: 1823– 1831.

Mensink RP, Zock PL, Kester AD and Katan MB (2003). Effects of dietary fatty acids and carbohydrates on the ratio of serum total to HDL cholesterol and on serum lipids and apolipoproteins: a meta-analysis of 60 controlled trials'. *Am J Clin Nutr*, 77(5): 1146-1155.

Metz-Boutique MH, Jolles J, Mazurier, J, Schoentgen, F, Legrand D, Spik, G, Montreuil, J, Jolles, P (1984). Human lactotransferrin: amino acid sequence and structural comparisons with other transferrins. *European Journal of Biochemistry*, 145: 659–676.

Moughan P (2012). Dietary Protein Quality – New Perspectives, In : *IDF World Dairy Summit*, Accessed: http: //www.asuder.org.tr/asudpdfler/mevzuat/idfsunumlari/moughan_paul.pdf (25th Mar, 2014)

Nieves JW, Golden AL, Siris E, Kelsey JL and Lindsay R (1995). Teenage and current calcium intake are related to bone mineral density of the hip and forearm in women aged 30-39 years. *Am J Epid* 141: 342-351.

O'Shea M, Bassaganya-Riera J, Mohede ICM (2004). Immunomodulatory properties of conjugated linoleic acid. *Am J Clin Nutr* 79(Suppl): 1199–206.

Parodi PW (1999). Conjugated linoleic acid and other anticarcinogenic agents of bovine milk fat. *J Dairy Sci* 82: 1339–49.

Parodi PW (2004). Milk fat in human nutrition. *Australian Journal of Dairy Technology*, 59: 3–59.

Parodi PW (2006). Nutritional significance of milk lipids. *Advanced Dairy Chemistry Vol. 2: Lipids*, (eds. P. F. Fox and P. L. H. McSweeney), 3rd edn., pp. 601–639, Springer, New York.

Pérez AV, Picotto G, Carpentieri AR, Rivoira MA., Peralta López ME and Tolosa de Talamoni NG (2008). Minireview on regulation of intestinal calcium absorption. *Digestion* 77, 22–34.

Peter Elwood, Katherine Livingstone (2012). Dairy is more that saturated fat and calcium: Addressing misconceptions about dairy and cardiovascular disease. *World Dairy Summit*, IDF.

Rodrigues Lígia R (2013). Milk minor constituents, enzymes, hormones, growth factors, and organic acids, In: *Milk and Dairy Products in Human Nutrition*, pp. 220-256, Park Young W./Haenlein, George F. W. (eds.), John Wiley and Sons, Inc.

Ryley J and Kajda P (1994). Vitamins in thermal processing, *Food Chem.*, 49, 119–29.

Salter AM (2013). Impact of consumption of animal products on cardiovascular disease, diabetes, and cancer in developed countries. *Animal Frontiers* 3 (1): 20-27.

Sandler RB, Slemenda CW, LaPorte RE, Cauley JA, Schramm MM, Barresi ML and Kriska AM (1985). Postmenopausal bone density and milk consumption in childhood and adolescence. *American Journal of Clinical Nutrition* 42: 270-274.

Shearer MJ, Bach A and Kohlmeier M (1996). Chemistry, nutritional sources, tissue distribution and metabolism of vitamin K with special reference to bone health. *Journal of Nutrition* 126, 1181S–1186S.

Shingfield, KJ, Chilliard Y, Toivonen V, Kairenius P and Givensm DI (2008). Trans fatty acids and bioactive lipids in ruminant milk, In: *Bioactive Components of Milk*. Ed. Zsuzsanna Bo sze, Springer Publication.

Sindhu JS and Arora S (2011). Buffalo Milk, In: *Encyclopedia of Dairy Science*, Vol. 3: 503-511. Elsevier.

Singh SP and Gupta MP (1986). Influence of certain treatments on the ascorbic acid content of buffaloes' and cows' milk. *Indian Dairyman*. 38: 379–381.

Tricon S, Burdge GC, Kew S, Banerjee T, Russell JJ, Jones EL, Grimble RF, Williams CM, Yaqoob P, Calder PC (2004). Opposing effects of cis-9,trans-11 and trans-10,cis-12 conjugated linoleic acid on blood lipids in healthy humans. *Am J Clin Nutr* 80: 614–20.

Vesa TH, Marteau P and Korpela R (2000). Lactose intolerance. *Journal of the American College of Nutrition* **19** (Suppl. 2), 165S–175S.

Villar J, Abdel-Aleem H, Merialdi M, Mathai M, Ali M, Zavaleta N, Purwar J, Hofmeyr N, ThiNhu Ngoc N and Campoâ Donico L (2006). World Health Organization randomized trial of calcium supplementation among low calcium intake pregnant women', *Am J Obs Gyn*, 194(3): 639-649.

Wang Y, Jones PJH (2004). Dietary conjugated linoleic acid and body composition. *Am J Clin Nutr* 79: 1153–8.

Watkins BA, Seifert MF (2000). Conjugated linoleic acid and bone biology. *J Am Coll Nutr* 19(4): 478–86.

Weaver CM, Proulx WR and Heaney, RP (1999). Choices for achieving dietary calcium within a vegetarian diet. *Am. J. Clin. Nutr.*, 70: 543S–548S.

Weaver Connie, Wijesinha-Bettoni Ramani, McMahon Deirdre, Spence Lisa (2013). Milk and dairy products as part of the diet, In: *Milk and Dairy Products in Human Nutrition*, Food and Agriculture Organization of the United Nations, Rome.

WHO (2012). The vitamin and mineral nutrition information system (VMNIS). Available at: http: //www.who.int/vmnis/database/anaemia/countries/en/ index.html. accessed 16 October 2012.

2015, Dairy Product Technology: Recent Advances ***Pages 203–216***
Editors: **Subrota Hati, Surajit Mandal and Birendra Kumar Mishra**
Published by: **DAYA PUBLISHING HOUSE, NEW DELHI**

Chapter 11

Fermented Dairy Products

Jagbir Rehal and Gagan Jyot Kaur

Fermentation is one of the oldest methods of food processing. Bread, beer, wine and cheese originated long before Christ. A variety of fermented foods are very popular with consumers because of their attractive flavor and their nutritional value. Milk originating from cows, buffaloes, sheep and occasionally other animals has high moisture content; contains proteins, minerals and vitamins and has a neutral pH. Milk is also rather voluminous; nomadic tribes developed methods to curdle milk and separate the coagulated casein from the residual liquid (whey). The coagulate represents only approximately 10 per cent of the original milk volume. Nearly every civilization has developed fermented milk products of some type. The terms *dahi, butter milk, yogurt, kumiss,* and *acidophilus milk* are familiar to many people, but those who first produced these foods did not know that they were fermented by bacteria.

In 1957, Louis Pasteur was the first to show that bacteria are involved in milk fermentation, a first step in elucidating the chemistry of fermentation. The understanding of the role of enzymes in fermentation reactions followed the experiments carried out in 1896 by the German scientists Hans and Eduard Buchner. In 1907, *Lactobacillus* was isolated from fermented milk by the Russian microbiologist Ellie Metchnikoff and in 1930; Hansvon Eular obtained the Nobel Prize for his work on the fermentation of sugars and fermentation enzymes.

Production of fermented milks is mentioned in early Sanskrit and Christian works, while recipe of both sweet and savory fermented milks date back to roman times around AD 200 (Oberman, 1985). The Vedas and Upnishads mentioned the origin of *dahi,* one of the oldest fermented milk products of the Hindus and fermented milk products during 6000-4000 BC (Yegna Narayan Aiyar, 1953). Buttermilk and ghee (clarified butter) were widely consumed milk products during Lord Krishna's time around 3000 BC (Prajapati and Nair, 2003).

Sir Joseph Lister first isolated the milk bacterium now known as *Lactococcus lactis* in 1878. After his discovery, attempts were made to use selected cultures of lactic acid bacteria for making cheese, butter, and fermented milks. The first instance of the use of a selected culture to make fermented milk is reported in 1890 by the Danish scientist Storch (1898) who used selected strains for souring cream for butter making.

Milk is a global drink that is a polyphasic emulsion having physical, chemical, and biological properties (Huria, 2002). Milk can be fermented into a wide range of different products with different flavors, consistencies and structure. The use of starter cultures is based on the development of lactic acid and other acids to decrease the pH resulting in microbiologically stable and safe fermented products. A rapid acidification by lactic acid bacterial fermentation to pH values of less than 4.5 strongly inhibits the survival and growth of spoilage causing or disease associated bacteria. Milk contains ample amounts of the fermentable carbohydrate lactose, which is an essential ingredient to enable this fermentation. Campbell-Platt (1987) defined fermentation as those foods that have been subjected to the action of microorganisms or enzymes so that desirable biochemical changes cause significant modification in the food. Lactic Acid Bacteria (LAB) species are among the best suited organisms to grow rapidly in milk with concomitant production of lactic acid. Not surprisingly, the fermentation process is mainly brought about by a series of different LAB species (FAO/WHO 2003). LAB are non-spore forming, gram positive, catalase negative without cytochromes, non-aerobic or aero-tolerant, fastidious, acid-tolerant, and strictly fermentative bacteria with lactic acid as the major end-product during sugar fermentation (Axelsson, 1998). The term LAB embraces a diverse set of bacteria, producing lactic acid as the major end product from carbohydrate utilization which reduces the pH of the substrate to a level where the growth of pathogenic, putrefactive and toxinogenic bacteria are inhibited (Holzapfel *et al.,* 1995). Typical LAB members belong to genera *Lactococcus, Lactobacillus, Leuconostoc,* and *Pediococcus* (nowadays *Propionibacterium* and *Bifidobacterium* also included). Development of LAB types to high cell densities during fermentation modifies milk constituents (protein and fats) through their proteolytic and lipolytic complex systems (Smit *et al.,* 2005). This contributes to the final rheological and sensorial characteristics of fermented products. Moreover lactic acid and other bacterial metabolites produced during the growth (H_2O_2, diacetyl, bacteriocins) further improve stability and safety to fermented products by inhibiting spoiling and pathogenic microorganisms.

As per IDF (1969) specifications, fermented milks are defined as the products prepared from milks- whole, partially or fully skimmed, concentrated milk or milk substituted from partially or fully skimmed dried milk, homogenized or not, pasteurized or sterilized and fermented by means of specific organisms. The application of lactic fermentation to modify the rheological properties of proteins is most common in the manufacturing of fermented milk products, since the main process occurring during their production process is agglomeration of milk proteins into three dimensional network structure (Haque, Richardson and Morris, 2001). The gel formation by milk proteins is critical step in the manufacturing of fermented milk products.

Milk proteins are a heterogeneous group of compounds that differ in composition and properties. Milk proteins may be divided into a casein fraction and whey proteins, based on their behavior under the influence of changes in H^+ ion concentration. At pH 4.6, casein precipitates, whereas whey proteins remain in the solution. In the case cheese, a major protein undergoing coagulation is casein, whereas in yogurt and other products made from milks subjected to heating; both caseins and whey proteins are subject to gelatinization (van Vliet, Lakemond and Visschers, 2004).

Starters in Fermented Milk Products

The primary function of almost all starter cultures is to develop acid in the product. The secondary effects of acid production include coagulation, expulsion of moisture, texture formation and initiation of flavor production. In addition to these, the starters also help in imparting pleasant acid taste, conferring protection against potential pathogens and providing a longer shelf life to the product. Depending on the principal function, *i.e.* promote acidification, added microorganisms are referred to as starters or primary cultures; for promoting flavor, aroma and maturing activities they are referred to as secondary cultures (Topisirovic *et al.*, 2006).

There are two groups of lactic starter cultures:

1. Simple or defined: single strain, or more than one in which the number is known
2. Mixed or compound: more than one strain each providing its own specific characteristics

Starter cultures may be categorized as mesophilic, for example:

- *Lactococcus lactis* subsp. *cremoris*
- *L. delbrueckii* subsp. *lactis*
- *L. lactis* subsp. *lactis* biovar *diacetylactis*
- *Leuconostoc mesenteroides* subsp. *cremoris*

or thermophilic:

- *Streptococcus salivarius* subsp. *thermophilus* (*S. thermophilus*)
- *Lactobacillus delbrueckii* subsp. *bulgaricus*
- *L. delbrueckii* subsp. *lactis*
- *L. casei*
- *L. helveticus*
- *L. plantarum*

Mixtures of mesophilic and thermophilic microorganisms can also be used as in the production of some cheeses (www.uoguelph.ca)

Types of Fermented Milks

A variety of traditional as well as industrialized fermented milk products are available nowadays. These can be classified as under:

I. On the basis of dominant microorganisms

Lactic fermentation: here LAB species lead to the production of fermented milk products. These can be

i. mesophillic- cultured buttermilk, natural acidified milk, cultured cream

ii. thermophillic-yoghurt, *dahi*

iii. therapeutic- acidophilus, bifidus milk

Fungal lactic fermentation: here LAB species combine with yeast species to give the final product.

i. alcoholic milk-*kefir, kumiss*, acidophilus yeast milk

ii. moldy milk- *viili*

II. On the basis of acid content of the product:

i. Acid alcohol- *kefir, kumiss*

ii. High acid- Bulgarian sour milk

iii. Medium acid- Acidophilus milk, yoghurt

iv. Low acid- culture butter milk, cultured cream

Kefir

It is a viscous, acidic, mildly alcoholic and distinctly effervescent product produced by fermentation of milk with a kefir grain as starter culture. The word 'kefir' is said to have originated from the Turkish word 'keif' which means 'good feeling'. Its country of origin is caucasusian China. It has been sold in Europe under a variety of names including *kephir, kiaphur, kefyr, képhir, kéfer, knapon, kepi*, and *kippe* (Kemp 1984). Traditionally kefir was made in specially designed leather bags made from skin of goat. It was customary to hang the bag near the door so that the persons passing through could kick or push the bag to keep the contents mixed. The kefir grain is a white, yellow irregular, cauliflower floret like grain having a elastic consistency but a firm texture which is constituted of inert polysaccharide matrix and are a product of a strong and specific symbiotic association of different lactic acid bacteria, acetic acid bacteria and yeast species. A crude analysis of the grains shows that they are a mass of bacteria, yeasts, polysaccharides, and proteins with a chemical composition of 890 to 900 g/kg water, 2 g/kg lipid, 30 g/kg protein, 60 g/kg sugars, and 7 g/kg ash (Zourari and Anifantakis 1988). When added in milk the grains swell and can be easily divided in portions for further propagation. The most active part of kefir starter microflora constitutes of *Lactococcus lactis* subsp. *lactis* and *L. lactis* subsp. *cremoris* and are mostly responsible for acidification during the initial acidification. *Lactobacillus kefiri, Lb. delbrueckii* grow much slower resulting in slow production of aroma compounds, alcohol and carbon dioxide. Yeasts such as *Kluyveromyces marxianus* var. *lactis, Torulaspora delbrueckii, Saccharomyces cerevisiae* are responsible for carbon dioxide formation, production of alcohol for imparting characteristic taste and aroma to kefir. Acetic acid bacteria like *Acetobacter aceti* and *Acetobacter resens* play a role in maintaining the symbiosis among the kefir grain microflora as well as increasing the consistency of kefir by increasing its viscosity.

The traditional method of kefir production is by incubating milk with kefir grains to pasteurized, cooled milk at 20°C-25°C for 18-24 hours. At the end of fermentation, grains are rinsed properly with water and subsequently used for a new batch. Here the final product cannot be used to inoculate new milk to produce kefir; kefir grains are essential to the process. The chemical composition varies but generally the pH of traditional kefir is between 4.2-4.6. It has 0.7-1 per cent lactic acid and 0.5-1.5 per cent ethanol along with sufficient carbon dioxide to cause effervescence as well as traces of acetaldehyde, diacetyl and acetoin (Duitschaever, Kemp, and Emmons, 1988).

Kefir has a long history of health benefits. Kefir can reduce symptoms of lactose intolerance by providing an extra source of β-galactosidase, an enzyme responsible for hydrolysis of lactose into glucose and galactose and which is lacking in people with lactose intolerance. Lactic and acetic acid bacteria produce a wide range of anti microbial compounds including organic acids (lactic and acetic), CO_2, H_2O_2, ethanol, diacetyl, and peptides (bacteriocins) exerting inhibitory action against food-borne pathogens and spoilage microorganisms (Leroy and de Vuyst, 2004). Kefir is known to have antitumor activity due to antioxidative properties.

Kumiss

Kumiss is an effervescent, acidic, alcoholic fermented, milky white/grayish liquid made primarily from mare's milk. It is a popular national beverage of Kazakhstan and its origin was in central Asia-Russia and Mongolia. It is also produced under other names such as *Kumis, Kumys, Koumiss,* etc. The mare's milk lends it its characteristic nature as it contains less casein and fatty matter and more sugars than cow's milk.

Kumiss is similar to Kefir but is produced from a liquid starter culture, in contrast to the solid kefir grains. It has higher though still mild, alcohol content compared to kefir. Since nowadays mare's milk is a limited commodity so cow's milk fortified with sucrose is used to get kumiss. Kumiss has 0.7-1.8 per cent lactic acid, 1.3 per cent ethanol and 0.5-0.88 per cent of carbon dioxide. Lab and yeast species are both responsible for acidification and final sensorial properties of the produce. These are identified as *Lactobacillus delbrueckii* ssp. *bulgaricus, Lactobacillus acidophilus, Lb. casei* and *S. unisporus, K. marxianus* and *S. cerevisiae.*

Acidophilus Milk

It is defined as highly acidic product made by fermentation of milk by *Lactobacillus acidophilus. Lb. acidophilus* strains are considered to fulfill most of the basic criteria of probiotics: survival in the gastro intestinal transit, bile and acid tolerance, and production of antimicrobials. The milk is heated to high temperature *i.e.,* 95°C for 1 hour to eliminate the microbial load and favor the slow growing culture. Milk is inoculated at a level of 2.5 per cent and incubated at 37°C until coagulated. The acidity is high as 1 per cent lactic acid. Another variation nowadays is the introduction of sweet acidophilus milk. Here the *Lactobaccillus acidophilus* cells are separated by centrifugation from the broth culture and added to pasteurized whole milk but there is no incubation. It is thought that the culture will reach the GI tract where its therapeutic effects will be realized, but the milk has no fermented qualities, thus delivering the benefits without the high acidity and flavor considered undesirable by some people. The product remains sweet and most of the cells remain viable for 1-2 weeks under refrigerated conditions.

Yoghurt

Yoghurt is produced using active cultures of bacteria to ferment milk. It is defined as a coagulated milk product obtained by lactic acid fermentation of milk by *Streptoccus thermophilus* and *Lactobacillus delbrueckii* subsp. *bulgaricus.* Yoghurt products may also have added ingredients such as sugar, sweeteners, fruits or vegetables, flavoring compounds, sodium chloride, coloring, stabilizers and preservatives. Hence it is available in various forms ranging from plain-traditional type with sharp acidic taste, fruit-with addition of fruits and sweetening agents, flavored- addition of synthetic flavoring and coloring compounds, stirred-the coagulum of the bulk yoghurt is broken before it's cooling and packaging, dried-its total solids are between 90-94 per cent, frozen-has higher level of stabilizers for stability during freezing, and smoked.

The fermentation process involves the inoculation of pasteurized milk that has been enriched in milk protein with concentrated cultures of bacteria, which is then incubated at 40-44°C for 4-5 hours. During fermentation, lactic acid is produced from lactose by the yogurt bacteria, the population of which increases 100- to 10,000-fold to a final concentration of approximately 109/mL. The reduction in pH, due to the production of lactic acid, causes a destabilization of the micellar casein at a pH of 5.1 to 5.2, with complete coagulation occurring around pH 4.6. At the desired final pH, the coagulated milk is cooled quickly to 4–10°C to slow down the fermentation process (Water and Naiyanetr 2008). Industrial yoghurt manufacture involves a preliminary milk fortification with non-fat powdered milk, milk protein concentrates, or condensed skim milk to increase total solids (up to 12 per cent to 13 per cent) and get a thicker consistency. Mixing is followed by homogenization (at 15-20MPa) and pasteurization (85-88°C for 30 min).

There has been a significant increase in the production of "bio-yogurts," which contain *Lb. acidophilus* and *Bifidobacterium* spp. ;known as AB-cultures (Dave and Shah 1998) *Lb. acidophilus, Bifidobacterium* spp., and *Lb. casei*;known as ABC-cultures (Maiocchi 2001), in addition to the traditional yogurt starter cultures, *S. thermophilus* and *Lb. bulgaricus.* For the manufacture of these probiotic yogurts, it is particularly attractive that conventional yogurt processing procedures can be applied with the probiotic bacteria added prior to fermentation along with the yogurt starter cultures, or after fermentation to the cooled product before packaging (Maiocchi, 2001). The methods used to manufacture stirred yogurt and drink yogurt, in particular, are well suited to the addition of probiotics after fermentation (Heller, 2001).

Cultured Butter Milk

It is low acid fermented milk popular in Scandinavian and European countries. It is the fluid remaining after ripened cream or sour cream is churned into butter. It is obtained by the souring the skimmed milk with mesophillic lactic acid bacteria as starter culture. The original buttermilk is really the liquid left after removing the butter from the butter churn which still contained butter flakes carried over from the butter making process. Nowadays cultured butter milk is made from skim or low fat milk which is pasteurized at 82-88°C for 30 min. The milk is then cooled to 22°C and inoculated with *Lactococcus lactis* subsp. *lactis* and subsp. *cremoris* for lactic acid production. For aroma and flavor production *Lactococcus lactis* biovar *diacetylactis* and *Leuconostoc mesenteroides* subsp *cremoris* are used (Marshall, 1986).The balance between aroma and acid producers is very important, and not more than 20 per cent of the total bacterial population should consist of aroma producers. Diacetyl, at a concentration of 2 to 5 mg/kg is responsible for the characteristic aromatic flavor of cultured buttermilk (Walstra *et al.,* 2010). The ripening process takes about 12-14 hours. After fermentation the resulting curd is broken and stirred slowly, cooled and slightly salted (Kosikowski and Mistry, 1997). Butter flakes can also be added to it. After packing it is stored under refrigeration. The acidity of the finished product is 0.8-0.85 per cent lactic acid. The product has a shelf life of approximately 10 days at 5°C.

Bulgaricus Butter Milk

It is a sour milk produced using *Lactobacillus delbrueckii subsp. bulgaricus* alone (Marshall 1984).The steps in the manufacture are the same as that of cultured butter milk. Heat treated milk at 85°C for 30 min is inoculated at 42°C for 10-12 hours. This has a 'clean' flavor and a sharp acidic taste reminiscent of yoghurt suggesting that the lactobacilli metabolize some of the milk components to acetaldehyde. It is high acid milk in which the total acidity (as lactic) may reach from 2.0 to 4.0 per cent (Kosikowski, 1977).

Dahi

It is a popular Indian fermented milk product that is quite analogous to plain yogurt in appearance and consistency. It is popular with consumers due to its distinctive flavor and because it is believed to have good nutritional and therapeutic value. It is utilized in various forms in many Indian culinary preparations. It is defined as a product obtained by lactic fermentation of cow or buffalo milk or mixed milk through the action of single or mixed strains of lactic acid bacteria or by lactic fermentation accompanied by alcoholic fermentation by yeast. Traditionally it is made from the starter known as *jamun* or *khatta* which is the left over *dahi* from the previous lot. The organisms commonly found in the inoculums are *L. lactis* subsp. *cremoris, L. lactis* subsp. *lactis, S. thermophilus, L. acidophilus, L. delbrueckii* subsp. *bulgaricus.* A good quality *dahi* is of firm and uniform consistency with a sweet aroma and clean acid taste. The surface is smooth and glossy and a cut surface is trim and free from cracks and air bubbles (Batish *et al.,* 1999). *Chakka* is a concentrated product obtained after draining the whey from *dahi.* When it is blended with sugar and other condiments, it becomes *shrikhand,* referred to as *shikhrini* in old Sanskrit literature. This has been a very popular dessert in Western India for several hundred years.

Cheese

It is a quintessential convenience food, which is ready to eat (though it may also be heated/cooked), nutritious and satiating. About 1870, Hansen in Denmark put a commercial rennet preparation on the market, and in the beginning of 1900, he put commercial cultures for cheese making on the market. This gave a boost to the manufacture of cheese on a wider scale (Davis, 1964). There are about 2000 names assigned to cheeses based on the area of the cheese's origin, country, source of milk, method of production, moisture content, cultures used, inventor, method of ripening, etc. Of these, about 800 varieties have been well established. They can be classified into 18 distinct types. Description of more than 400 varieties has been given by the U.S. Department of Agriculture (USDA).

Prior to the use of cultures, cheese was made by:

1. Natural souring with adjustment of temperature
2. Addition of milk of sour whey or buttermilk
3. Adding homemade starter.

Cheese is defined as 'a product made from the curd obtained from milk by coagulating the casein with the help of rennet or similar enzymes in the presence of lactic acid produced by added or adventitious microorganisms, from which part of the moisture has been removed by cutting, cooking and/pressing, which has been shaped in and then ripened by holding it for sometime at suitable temperatures and humidities' (Davis 1965).

For the production of cheese from milk, two key steps are essential:

1. Concentration of the milk casein and fat through coagulating the casein by proteolytic enzymes or lactic acid
2. Drainage of the whey after mechanical disruption of the coagulated casein

Variety is brought about by altering different aspects of cheese manufacture: type of starter culture, additional cultures, fermentation conditions, renneting, cutting the curd, scalding, and drainage of whey, forming of green cheese, salting, addition of spices, and ripening. According to moisture levels, at least three types of cheeses are generally defined: hard cheeses (moisture 20 per cent to 42 per cent), semihard/semisoft cheeses (moisture 45 per cent to 55 per cent), and soft cheeses (moisture > 55 per cent). All three types are consumed after a certain ripening period in contrast to fresh cheeses (moisture > 70 per cent), which are consumed after draining. Cheese is a concentrated protein gel, which occludes fat and moisture. Its manufacture essentially involves gelation of cheese milk, dehydration of the gel to form a curd treatment of the curd (*e.g.* dry stirring, cheddaring, texturisation, salting, moulding, and pressing). The moulded curd may be consumed fresh (shortly after manufacture, for example within 1week) or matured for periods of <"2 weeks to 2 years to form a ripened cheese.

The gelation of milk may be induced by (Guinee and O'Callaghan, 2010):

- Selective hydrolysis of the -casein at the phenylalanine105–methionine106 peptide bond by the addition of acid proteinases, referred to generically as rennets (chymosin, pepsin);
- Acidication (using starter cultures or food-grade acids and/or acidogens), at a temperature of 20–40°C, to a pH value close to the isoelectric pH of casein, *i.e.* ~4.6; and/or
- A combination of acid and heat, for example heating milk at pH ~5.6 to ~90°C.

Bacteria used as starter cultures generally belong to the genera *Lactococcus, Streptococcus, Leuconostoc,* and *Lactobacillus*. The typical mesophilic starter culture consists of *Lactococcus lactis* sp. *lactis* (*L. lactis*) and *L. lactis* ssp. *cremoris* (*L. cremoris*). These starters are used for cheeses with moderate scald temperatures (< 40°C, Gouda, Edam, etc.). For cheeses requiring higher scald temperatures, thermophilic starter cultures are used (*Streptococcus thermophilus, Lactobacillus helveticus, Lactobacillus delbrueckii* ssp. *bulgaricus,* etc.). When "eyes" are required, the gas-forming species *L. lactis* ssp. *lactis* biovar. *diacetylactis, Leuconostoc mesenteroides,* or *Propionibacterium shermanii* can be used (Heller, 2008).

Health Benefits

Convincing evidence from several studies indicate that lactose intolerant individuals suffer fewer symptoms if milk in the diet is replaced with fermented dairy products and functional prebiotic containing foods. Due to partial hydrolysis of lactose during fermentation, the reduced levels of lactose in fermented products relative to milk may contribute to the greater tolerance of yoghurt. (de Vrese *et al.,* 2001). Most of the lactose, the principal carbohydrate in milk, is lost in whey during cheese manufacture and hence most cheeses contain only trace amounts of carbohydrate. The residual lactose in cheese curd is usually fermented to lactic acid by the starter bacteria. Thus, cheeses can be consumed with-out ill-effects by lactose-intolerant individuals who are deficient in the intestinal enzyme, β-galactosidase (O'Brien and O'Connor, 2004).

Clinical studies have shown that *Lactobacillus* and *Bifidobacterium* species reduce lactose intolerance, alleviate some diarrhea, lower blood cholesterol, increase immune response and prevent cancer (Marteau and Ramband 1996, Gilliland 1996, Salminen *et al.,* 1998). LAB can be effective in preventing gastrointestinal disorders and in the recovery of diarrhea of miscellaneous causes (Marteau *et al.,* 2001), cause relief of constipation (Oyetango and Oyetango 2005). They also help in decreasing *Heliobacter pyroli* infection, reduction in cholesterol and certain allergies etc. (Sanders, 2012). Certain strains of bifidobacteria and lactobacillus produce conjugated linoleic acid(CLA) isomers and enhance the health properties by reducing the risk of antherosclerosis (Yadav, Jain and Sinha, 2007). The inhibitory effect of LAB to various pathogenic bacteria and spoilage organisms is due to the rapid growth and acid production by LAB, the consequent decrease in pH and the formation of other microbial factors associated with LAB, such as bacteriocins, hydrogen peroxide, ethanol and diacetyl (Caplice and Fitzerald, 1999).

The protective properties of LAB due to antimicrobial activities are useful in food fermentation, making foods safe to eat. The consumption of LAB in fermented foods without any adverse health effect confers a GRAS status, and, therefore, their bacteriocins might have potential as biopreservatives (Adams 1999). LAB compete with other microbes by screening antagonistic compounds and modify the micro-environment by their metabolism (Lindgren and Dobrogosz 1990). *L. acidophilus* is able to survive and grow in the intestinal tract. Here the increased permeability of the bacterial cells allows for the permeation of lactose and subsequent hydrolysis by β-galactosidase (Gilliland, 2001).It also helps in lowering serum cholesterol levels (Ouwehand *et al.,* 2002). The organisms in Kefir grains are reported to assimilate cholesterol (Kitazawa *et al.,* 1991).

Many benefits are attributed to fermentation. It can preserve food (*i.e.* increase its shelf life), improve digestibility, enrich food, and enhance taste and flavor. It is also an affordable technology and thus is accessible to all population. Furthermore, fermentation has the potential to enhance food safety by controlling a great number of pathogens in foods. Thus, it makes an important contribution to human nutrition, particularly in developing countries where economic problems are a major barrier in ensuring food safety.

Safety Concerns of Fermentation

Although it may seem simple, fermentation is delicate and complex technology. The safety and quality of the fermented product, besides being dependent on the quality and safety of starter cultures also depends on the processing conditions, the level of hygiene, the processing conditions, the final acidity of the product and the quality of the raw materials. In case of cottage industry or household application of fermentation it may be difficult to control these factors adequately. Over the past few decades, many countries have seen a significant rise in incidence of disease caused by *Salmonella* spp., *Campylobacter* spp., and *Escherichia coli* (Adams and Nout, 2008). Apart from the desired fermentation cultures, there may also be pathogens that can survive in an optimal environment for fermentation cultures. Cheeses of different categories have also been implicated in outbreaks of various types of food borne disease (James *et al.,* 1985). When basic GHP (good hygienic practices) requirements are not met with, say during the milk pasteurization step then a potential source of pathogenic cultures such as *Listeria monocytogenes, E coli, Clostridium* spp., *Salmonella* spp., and *Staphylococcus* spp. can survive. Although LAB can inhibit the growth of certain food –borne pathogens but it is important not to rely only on fermentation step to eliminate the contaminating organisms.

Research Needs

Research needs in the area of fermentation are extensive and relate to several areas

- ✰ Identification of new organisms and technological developments leading to a new products with enhanced nutritional or organoleptic quality

- ☆ Assessment of safety of starter cultures, be they traditional organisms selected for their specific features or genetically modified organisms
- ☆ Potential of fermentation technology to control pathogens
- ☆ Identification and selection of organisms with potential health benefits
- ☆ Implementation and evaluation of health education intervention carried out to ensure safe application of fermentation.

Conclusion

Fermentation is and will remain an important technology. The fermented dairy market is growing by leaps and bounds throughout the world. Fermented dairy beverages render tremendous potential as carrier of functional ingredients required for health and wellness of the human beings. Consumption of these products is also likely to be increased as the potential benecial effects of fermented foods on human health become better established, scientically and clinically. The challenge will be assessment of safety of starter cultures, be they traditional organisms selected for their specific features or genetically modified organisms. Potential of fermentation technology to control pathogens, in particular acid resistant pathogens needs to be researched. In developing world where refrigeration facilities are lacking, the technology can make great contribution to public health by preventing growth of pathogens. The technology, if applied in an appropriate manner will lead to safe and nutritious products.

References

Adams M R 1999. Safety of industrial lactic acid bacteria. *J Biotech* 63: 17-78.

Adams M R, Nout M J R 2008. *Fermentation and Food Safety*. Springer India Pvt Ltd, New Delhi, 290 p.

Axelsson L 1998. Lactic acid Bacteria: Classification and physiology. In: *Lactic Acid Bacteria Microbiology and Functional Aspects*. 2nd Edn., Salminen S, Wright AV (eds). Marcel Dekker, New York, pp. 1-72.

Batish VK, Grover S, Pattnaik P, Ahmed N 1999. Fermented Milk Products. In: Biotechnology: *Food Fermentation (Microbiology, Biochemistry and Technology) Vol II: Applied* Joshi VK, Pandey A (eds). Educational Publishers and Distributers, Kerala, pp. 781-864.

Campbell-Platt G 1987. *Fermented foods of the world: A Dictionary and Guide*. Butterworths, London, U K.

Caplice E, Fitzerald GF 1999. Food Fermentation: role of microorganisms in food production and preservation. *Intl J Food Microbiol* 50: 131-149.

Dave RI, Shah NP 1998. Ingredient supplementation effects on viability of probiotic bacteria in yogurt. *J Dairy Sci* 81: 2804–2816.

Davis JG 1965. *Cheese Volume 1: Basic Technology*. J and A Churchill Ltd., London p. 1-16.

deVrese M, Stegelmann A, Richter B, Fenselau S, Laue C, Schrezenmier J 2001. Prebiotics- compensation for lactase inefficiency. *Am J Clin Nutr.* 73: 5421-5429.

Duitschaever CL, Kemp N, Emmons D 1988. Comparative evaluation of five procedures for making kefir. *Milchwissenschaft* 43: 343–345.

FAO/WHO 2003. CODEX standard for fermented milks. Codex Stan 243-2003. Reviewed 2008. http: //www.codexalimentarius.net/download/standards/ 400/CXS_243e.pdf

Gilliland S E 1996. Special additional cultures. In: *Dairy Starter Culture.* Cogan TM Accolas JP (eds). VCH Publishers, New York, pp. 25-46.

Gilliland SE. 2001. Probiotics and Prebiotics. In: *Applied Dairy Microbiology.* 2nd ed. Marth EH and Steele JL (eds). Marcel Dekker, New York, pp. 327-344.

Guinee TP, O'Callaghan DJ 2010. Control and Prediction of Quality Characteristics in the Manufacture and Ripening of Cheese. In: *Technology of Cheesemaking.* Law BA, Tamime AY (eds). A John Wiley and Sons, Ltd., Publication: Wiley-Blackwell.

Haque A, Richardson RK, Morris ER 2001. Effect of fermentation temperature on the rheology of set and stirred yoghurt. *Food Hydrocolloids* 15: 593-602.

Heller KJ 2001. Probiotic bacteria in fermented foods: product characteristics and starter organisms. *Am J Clin Nutr* 73(2 Suppl.): 374S–379S.

Heller KJ, Bockelmann W, Schrezenmeir J, deVrese M 2008. Cheese and its Potential as a Probiotic Food. In: *Handbook of Fermented Functional Foods.* Farnworth ER (ed). CRC Press, Taylor and Francis Group, Boca Raton, pp. 243-266.

Holzapfel WH, Giesen R, Schillinger U 1995. Biological preservation of foods with reference to protective cultures, bacteriocins and food grade enzymes. *Intl J Food Microbiol* 24: 343-362.

https: //www.uoguelph.ca/foodscience/dairy-science-and-technology/dairy-microbiology/starter-cultures

Huria VK 2002. Milk: The classic food of modern civilization. *Ind Food Industry* 21(4): 24-32.

International Dairy Federation 1969. International standards, 47.

James SM, Fannin SL, Agee BA, Hall B, Parker E, Vogt J, Run G, Williams J, Lieb L, Salninen C, Prendergast T, Werner SB, Chin J 1985. Listeriosis outbreak associated with Mexican style cheese in California. *Morbid Mortal Weekly* Rep 34: 357–359.

Kemp N 1984. Kefir, the champagne of cultured dairy products. *Cultured Dairy Prod* J: 29–30,

Kitazawa H, Toba T, Itoh T, Kumano N, Adachi S, Yamaguchi T 1991. Antitumoral activity of slime-forming, encapsulated *Lactococcus lactis* subsp. *cremoris* isolated from Scandinavian ropy sour milk,"viili". *Japan J Anim Sci Tech* 62: 277-283.

Kosikowski F 1977. *Cheese and Fermented Milk Foods.* 2nd Ed. Edward Brothers Inc. Printers and distributers, Ann. Arbor, Michigan, USA.

Kosikowski FV and Mistry VV 1997. *Cheese and Fermented Milk Foods*. 2nd Edn. Kosikowski FV(ed). Westport,CT: LLC.

Leroy BA, de Vuyst L 2004. Lactic acid bacteria as functional starter cultures for the food fermentation industry. *Trends Food Sci Technol* 15: 67-78.

Lindgren SE, Dobrogosz WJ 1990. Antagonistic activities of Lactic acid bacteria in food and feed fermentations. *FEMS Microbiology Review* 87: 149-164.

Maiocchi G 2001. YOMO ABC: Functional food for consumers well-being and satisfaction. *Industria del Latte* 37(1-2): 94–98.

Marshall RJ 1986. Increasing cheese yields by high heat treatment of milk. J Dairy Res 53: 313. Marteau PR, de Verese M, Cellier CJ, Schrezenmier. 2001. Protection from gastrointestinal diseases the use of probiotics. *Am J clinical Nutr* 73: 4305-4365.

Marteau P, Rambaud JC 1996. Therapeutic applications of probiotics in humans. In: *Gut Flora and Health: Past, Present and Future*. Leeds AR, Rowland IR (eds). The Royal Society of Medicine Press Ltd, London, pp. 47-56.

Oberman H 1985. Fermented milks. In: *Microbiology offermented Foods*. Vol I Wood BB (ed). Elsevier Applied Science, London, UK, pp 155-191.

O'Brien NM, O' Connor TP 2004. Nutritional aspects of cheese. In: *Cheese: General* aspects Volume 1. Fox PF (ed). Academic Press, pp . 573-582.

Oyetayo VO, Oyetayo FL 2005. Potential of probiotics as biotherapeutic agents targeting the innate immune system. *African J Biotechnol* 4: 123-127.

Ouwehand AC, Salminen S, Isolauri E 2002. Probiotics: an overview of beneficial effects. *Antonie Van Leeuwenhoek* 82: 279–289.

Prajapati JB, Nair BM 2003. The history of fermented foods. In: *Handbook of Fermented Functional Foods*. Farnworth R (ed). CRC Press, New York, pp1-25.

Sanders ME 2012. Probiotics for healthy consumer. *Functional Food Reviews* 4(4): 144-151.

Salminen S, Deighton M, Gorbach S 1993. Lactic acid bacteria in health and disease. In: *Lactic Acid Vacteria*. Salimen S, Wright A (eds). Marcel Dekker, New York, pp. 237-294.

Smit G, Smit BA, Engels EJ 2005. Flavor formation by lactic acid bacteria and biochemical flavor profiling of cheese products. *FEMS Microbiology Reviews* 29: 591-610.

Storch V 1890. *Nogle Undersogelser over Flodens, Syrning,* 18 Beretn. Fra Forsoglab, Cited in Galloway and Crawford, 1985. (reference 21).

Topisirovic L, Kojic M, Fira D, Golic N, Strahinic I, Lozo J 2006. Potential of Lactic acid bacteria isolated from specific natural niches in food production and preservation. *Intl J Food Microbiol* 112: 230-235.

USDA 1978. *Cheese varieties and descriptions*, USDA, Washington, DC, pp. 1–140.

Van Vliet T, Lakemond CMM, Visschers RW 2004. Rheology and structure of milk protein gels. *Current opinion in Colloid and Interface Science* 9: 298-304.

Walstra P, Wouters JTM, Geurts TJ 2010. Fermented Milks. In: *Dairy Science and Technology*, Second Edition, CRC Press.

Water J Van de, Naiyanetr P 2008. Yogurt and Immunity: The Health Benefits of Fermented Milk Products that Contain Lactic Acid Bacteria. In: *Handbook of Fermented Functional Foods*. Farnworth ER (ed). CRC Press, pp 129-164.

Yadav H, Jain S, Sinha PR 2007. Production of free fatty acids and conjugated linoleic acid in probiotic *dahi* containing *Lactobacillus acidophilus* and *Lactobacillus casei* during fermentation and storage. *Intl Dairy* J 17: 1006-1010.

Yegna Narayan Aiyar AK 1953. Dairying in ancient India. *Indian Dairyman* 5: 77-83.

Zourari A, Anifantakis EM 1988. Le kefir. Caracteres physico-chimiques, microbiologiques et nutritionnels. Technologie de production. Une revue, *Le Lait*, 68: 373–392.

2015, Dairy Product Technology: Recent Advances *Pages 217–249*
Editors: **Subrota Hati, Surajit Mandal and Birendra Kumar Mishra**
Published by: **DAYA PUBLISHING HOUSE, NEW DELHI**

Chapter 12

Recent Status of Dairy Waste in India

Kanchan Mogha, J.B. Prajapati and Subrota Hati

Introduction

In India, the dairy sector plays an important role in the country's socio-economic development, and constitutes an important segment of the rural economy. Dairy industry provides livelihood to millions of homes in villages, ensuring supply of quality milk and milk products to people in both urban and rural areas. India is the world's largest milk producers, accounting for around 17 per cent of the global milk production and is known as 'oyster' of the global dairy industry. Besides it is one of the largest producers as well as consumers of dairy products. Due to their nutritional qualities, the consumption of dairy products has been growing exponentially in the country. Considering such facts and figures it is anticipated that the milk production in India will grow at around 4 per cent during 2011-2015.

In the year 2012, the total milk production in the country was over 121.8 million tones with a per capita availability of 281 gm/day (GOI, 2012). Milk production in India is around 35 per cent, of which the organized dairy industry account for 13 per cent of the milk produced, while the rest of the milk is either consumed at farm level, or sold as fresh, non-pasteurized milk through unorganized channels. Dairy Cooperatives account for the major share of processed liquid milk marketed in India. Milk is processed and marketed by 170 Milk Producers' Cooperative Unions, which federate into 15 State Cooperative Milk Marketing Federations. Over the years, several brands have been created by cooperatives like Amul (GCMMF), Vijaya (AP), Verka (Punjab), Saras (Rajasthan). Nandini (Karnataka), Milma (Kerala) and Gokul (Kolhapur). Exports of dairy products have been growing at the rate of 25 per cent per annum in the terms of quantity and 28 per cent in terms of value since 2001. Significant investment opportunities exist for the manufacturing of value-added milk products like milk powder, packaged milk, butter, ghee, cheese and ready-to-drink milk products.

Pollution caused by industrial and dairy effluents is a serious concern throughout the world. Of all industrial activities, the food sector has one of the highest consumptions of water and is one of the biggest producers of effluent per unit of production besides it also generates a large volume of sludge in biological treatment (Ramjeawon, 2000). Consequent to the increased milk production and processing, wastewater generation has also increased. The dairy industry is one of the most polluting among all industries, not only in terms of the volume of effluent generated, but also in terms of its characteristics as well. It generates about 0.2–10 liters of effluent per liter of processed milk (Vourch *et al.,* 2008) with an average generation of about 2.5 liters of wastewater per liter of the milk processed (Ramasamy *et al.,* 2004). According to the Ministry of Agriculture, Government of India, the average rate of milk production in India increased at the rate of 3.56 per cent annually in the period from 2001–2007 (http ://www. indiastat. com/table/agriculture/2/milkanddairyproducts/167/3858/data.aspx). It was estimated that about 110 million tonnes of the milk and about 275 million tonnes of wastewater was generated annually from the Indian dairy industries during the year 2010.

Operations of Milk Based Food Industry and Effluent Generation

Dairy industry involves processing raw milk into products such as market milk, butter, cheese, yogurt, condensed milk, dried milk (milk powder) and ice-cream using processes such as pasteurization, bottling, filling in cans, condensing, drying etc. Figure 12.1 is a composite flow diagram showing the major operations for the processing of the more common milk products. The following is a brief description of these processes.

Milk is received at the plant or receiving station in cans. It is dumped to a weigh vat and the cans are washed in a can washer and returned to the producer. From the weigh vat milk is pumped to a silo or, if the plant is a only chilling station, the milk is cooled and pumped to a tank or truck for transporting it to processing plant. About 50 per cent of the milk produced in this country is used as liquid milk. A small amount of this is bottled as raw milk, but the major portion is pasteurized prior to further handling. The milk may then be packed in pouch for distribution, condensed to produce evaporated milk, or dried to milk powder. A small amount of whole milk is used in the manufacture of ice-cream mixes and in some type of cheese. About 41 per cent of the milk supply is separated into cream and skim milk. Some of cream is bottled for distribution or is used for ice-cream mix. A considerable portion however is used, in the manufacture of butter. In some cases the producer may separate the cream and deliver it to the plant where it is cooled and processed. Butter milk is by-product of butter manufacture and may be condensed in the vacuum pan or may be dried on heated rolls with or without pre-condensing. Powdered butter milk is used mainly in the preparation of stock and poultry feed. Skim milk from the separator may be condensed in the vacuum pan and/or dried to produce skim milk powder.

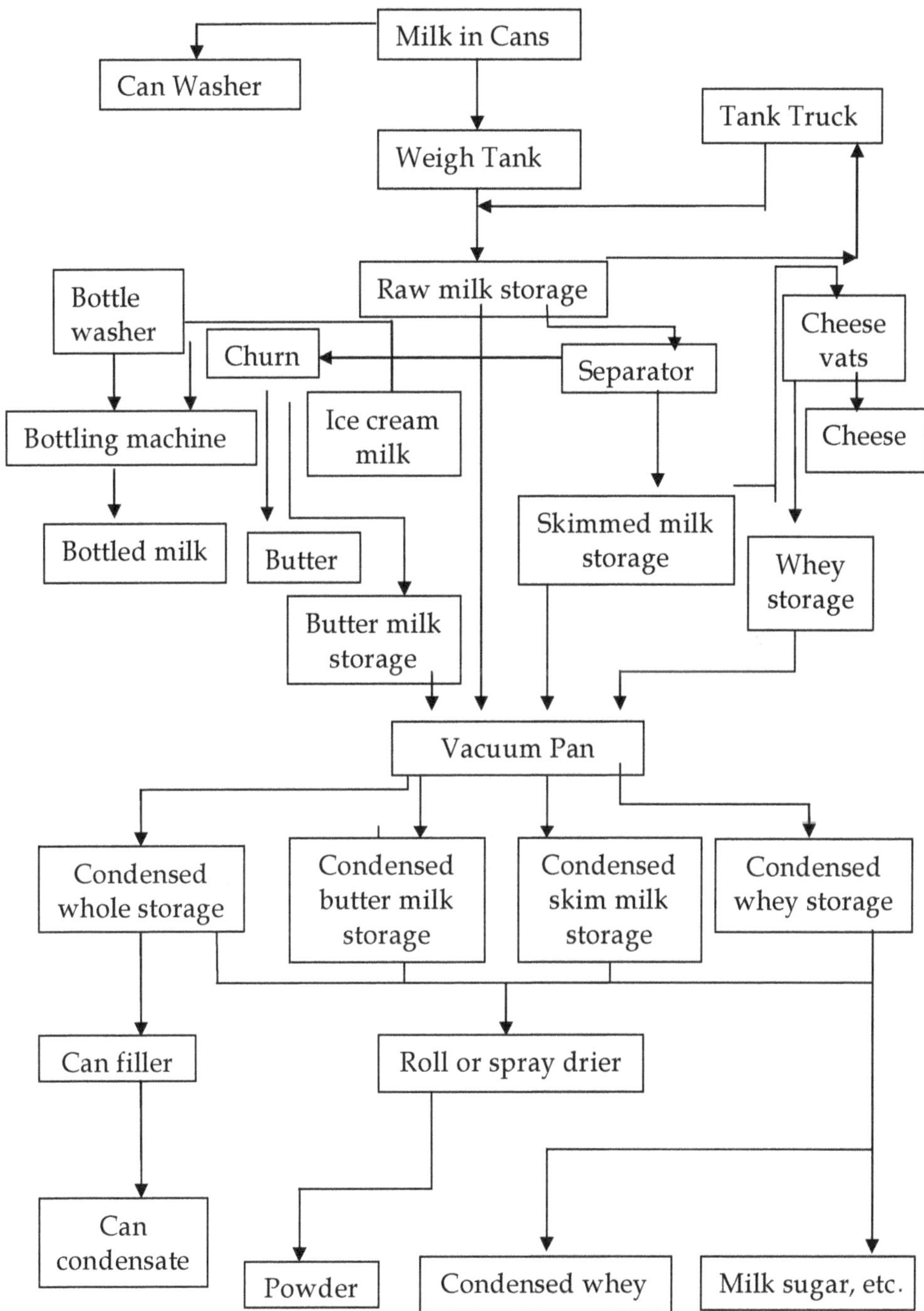

Figure 12.1: Composite Flow Diagram of Milk Product Operations.

Condensed and powdered whey are also used in food products and animal feeds. Some of the skim milk may be used for the manufacture of cottage cheese and casein. Whey is a by-product of cheese manufacture and is used in small plants for hog feeding. If the operations are large enough to warrant, it may be condensed in the vacuum pan or dried in the spray drier. Condensed and powdered whey are also used in food products and in animal feeds.

Wastewater and its Sources in Dairy Industry

Dairy industry produces different products, such as milk, butter, yogurt, ice-cream, and various types of desserts and cheese, the characteristics of these effluents also vary widely both in quantity and quality, depending on the type of system and the methods of operation used (Vidal *et al.,* 2000; Rico *et al.,* 1991). Wastes from milk product manufacture contain milk solids in a more or less dilute condition, but in varying concentration. These solids enter the waste from almost all of the operations. Approximately 65 per cent of dairy factory losses enter wastewater discharge streams and these can have a major impact on the environment. Sources of wastewater generated by milk based food industry are shown in Table 12.1. Contaminants expected in the wastewater from milk based food industry are shown in Table 12.2.

The sources of waste generated from milk based food industry are explained below (Verheijen *et al., 1996*):

Raw Milk Receiving Dock (RMRD)

Washing and cleaning out of product remaining in the tank, trucks, cans, piping, silos and other equipment is performed routinely after every processing cycle. It contains milk solids, detergents, sanitizers and milk wastes.

Milk Processing

Spillage is produced by leaks, overflow, freezing-on, boiling over, equipment malfunction or careless handling.

Processing losses include:

i. Discharge of sludge from clarifiers and cream separator
ii. Product wasted during pasteurized start-up, shut-down and product changeover
iii. Evaporator entrainment
iv. Discharges from bottles and washers
v. Splashing and container breakage in automatic packaging equipment
vi. Product change-over in filling machines.

Milk Packaging

Spoiled products cut opened milk packets of product, leakage of milk pouch during filling crates

Table 12.1: Sources of Wastewater from Milk Based Food Industry

Operations	*Processes*	*Sources of Waste*
Preparation Stages	Raw milk receiving dock	✰ Poor drainage of tankers ✰ Spills and leaks from hoses and pipes ✰ Spills from storage silos/tanks ✰ Foaming ✰ Cleaning operations
	Pasteuriztion/U.H.T	Liquid losses/leaks ✰ Recovery of downgradedproduct ✰ Cleaning operations ✰ Foaming ✰ Deposits on the surfaces ofpasteurization and heatingequipment
	Homogenization	✰ Liquid losses/leaks ✰ Cleaning operations
	Separation/Clarification (Centrifuge, reverse osmosis)	✰ Foaming ✰ Cleaning operations ✰ Pipe leaks
Product Processing stages	Market milk	✰ Foaming ✰ Product washing ✰ Cleaning operations ✰ Overfilling ✰ Poor drainage ✰ Sludge removal from ✰ Clarifier/Separators ✰ Leaks ✰ Damaged milk packages ✰ Cleaning of filling machinery
	Cheese making	✰ Overfilling vats ✰ Incomplete separation of wheyfrom curd ✰ Using salt in cheese making ✰ Spills and leaks ✰ Cleaning operations
	Butter and ghee making	✰ Vacreation and salt use ✰ Product washing ✰ Cleaning operations
	Powder manufacture Condensing and drying milk	✰ Spills of powder handling ✰ Start-up and shut-down losses ✰ Plant malfunction ✰ Stack losses ✰ Cleaning of evaporators ✰ Bagging losses

Source: E.P.A Guidelines, 1997.

Cleaning in Place

Wastewater is mainly produced during cleaning operations. Especially when different types of product are produced in a specific production unit, clean-up operations between product changes are necessary. In developing countries, the main problem is pollution through spoilage of milk. After every batch of product equipment like silos, pasteurizer, H.T.S.T., milk packaging machine etc are cleaned with detergent, hot water and acid also some of the sanitizing solution which is used for sanitization of ice-cream freezer, cheese vat are discharged as waste. Entrainment of lubricants from conveyors, stackers and other equipment appear in the wastewater from cleaning operations.

Table 12.2: Contaminants Present in Wastewater from Milk Based Food Industry

Plant Process	*Waste Generating Process*	*Nature of Waste*
Milk receiving	Tank truck washing	Milk solids + detergents
Clarifying and/or standardizing	Sludge from centrifugal machine	Milk solids high inprotein and cells
Storage of rawMilk	Tank washing and sanitizing	Milk solids + detergents + sanitizer
HTST Pasteurization, Homogenization	HTST start-up and product change over cleaning	Milk solids + detergent + sanitizer
Standardizing, separating, cooling	Sludge from separator	Milk solids high in proteinsand cells
Cream vat processing	Cleaning and sanitizing	Milk solids+ detergent+ sanitizer
Storage of milk + cream	Cleaning and sanitizing	Milk solids+ detergent+ sanitizer
Filling and transfer to cooler	Drips, broken package convey or lubrication cleaning	Milk solids + detergents + sanitizer + lubricant
Cold storage and distributors	Broken packages convey or lubrication returns	Milk solids + detergents + sanitizer + Lubricant

Source: Sharma, 2008.

Cheese/Whey/Curd

Waste results mainly from the production of whey, wash water, curd particles etc. Cottage cheese curd for example is more fragile than rennet curd which is used for other types of cheese. Thus the whey and wash water from cottage cheese may contain appreciably more fine curd particles than that from other cheeses. The amount of fine particles in the wash water increases if mechanical washing processes are used.

Butter/Ghee

Butter washing steps produce wash water containing buttermilk.

Routine operation of toilets, washrooms and restaurant facilities at the plant contribute waste.

Waste constituents may be contained in the raw water which ultimately goes to waste.

Non-dairy ingredients (such as sugar, fruits, flavors, nuts and fruit juices) utilized in certain manufactured products (include ice-cream, flavoured milk, frozen desserts, yoghurt and others) when drained contributes to waste.

Milk by-products that are deliberately wasted, significantly whey and sometimes buttermilk.

Uncontaminated water from coolers and refrigeration systems, which does not come in contact with the product, is not considered process wastewater. Such water is recycled in many plants. Sanitary sewage from plant employees and domestic sewage from washrooms and kitchens is usually disposed of separately from the process wastes and represent a very minor load of the plant.

Characterization of Wastewater Generated from Milk Based Food Industry

Dairy factory wastewaters commonly contain milk, byproducts of processing operations, cleaning products and various additives that may be used in the factory. Bovine milk typically contains water (87 per cent), fat (4 per cent), protein (3.5 per cent), lactose (4.7 per cent) and ash (0.8 per cent) (Bylund, 1995). Characteristics of dairy wastewater can be explained based on;

1. Physical Characteristics
2. Chemical Characteristics

Physical Characteristics

The principal physical characteristics of wastewater are its *solids content, colour, odour* and *temperature.*

Solids Content

The *total solids* in wastewater consist of the insoluble or suspended solids and the soluble compounds dissolved in water. The *suspended solids* content is found by drying and weighing the residue removed by the filtering of the sample. When this residue is ignited the *volatile solids* are burned off. Volatile solids are presumed to be organic matter, although some organic matter will not burn and some inorganic salts break down at high temperatures. In dairy waste effluent, total solids, total dissolved solids, total supended solids are composed mainly of carbonates, bicarbonates, chlorides, sulphate, phosphate, nitrate, Ca, Mg, Na, K, Mn and organic matter. Usually about 60 per cent of the suspended solids in dairy wastewater are settleable. Solids may be classified in another way as well: those that are volatilized at a high temperature (600°C) and those that are not. The former are known as volatile solids, the latter as fixed solids. Usually, volatile solids are organic.

Devi (1980) reported total plankton, which showed a sterkling parallelism with suspended solids from the different industries may have the different amount of solid particulate matter. When the effluent flows through the open drainage system particulate matter is expected to show greater degree of variance. If the effluent is highly acidic then the solid may dissolved in it, therefore it is necessary to evaluate effluent for the particulate matter.

Colour

Colour is very important factor of the aquatic life for making food from sunlight. Thus, photosynthetic activity reduced due to dark colouration and aquatic ecosystem is totally changed. Colour also affects the other parameters like temperature, DO, BOD etc. (Kolhe and Pawar, 2011). It is a qualitative characteristic that can be used to assess the general condition of wastewater.

Odour

The determination of odour has become increasingly important, as the general public has become more concerned with the proper operation of wastewater treatment facilities. The odour of fresh wastewater is usually not offensive, but a variety of odorous compounds are released when wastewater is decomposed biologically under anaerobic conditions.

Temperature

Temperature is important for its effects on certain chemical and biological reactions taking place in water and in organisms inhabiting aquatic media and will depend upon seasons and time of sampling. No specific limit for temperature is prescribed by WHO or ISI for the water quality use for the domestic purpose. Measurement of temperature is an important parameter required to get an idea of self purification capacity of river, reservoir and control of treatment plant. Water temperature is also important parameter for aquatic life. It is an important factor for calculating solubility of oxygen and carbon dioxide, bicarbonates and carbonates (Kolhe and Pawar, 2011).

Chemical Characteristics

Organic Substances

The **organic substances** in dairy wastewaters are contributed primarily by the milk and milk products wasted, and to a much lesser degree, by cleaning products, sanitizing compounds, lubricants and domestic sewage that are discharged to the waste stream. Over the years, a number of different tests have been developed to determine the organic content of wastewaters. Laboratory methods commonly used today to measure gross amounts of organic matter (greater than 1 mg/l) in wastewater include (1) biochemical oxygen demand (BOD), (2) chemical oxygen demand (COD) and (3) total organic carbon (TOC). Specific organic compounds are determined to assess the presence of priority pollutants. The BOD, COD and TOC tests are gross measures of organic content and as such do not reflect the response of the wastewater to various types of biological treatment technologies. It is therefore desirable to divide the wastewater into several categories, as shown in Figure 12.2.

Volatile Organic Carbons (VOC)

Volatile organic compounds such as benzene, toluene, xylenes, trichloroethane, dichloromethane, and trichloroethylene, are common soil pollutants in industrialized and commercialized areas. One of the more common sources of these contaminants is leaking underground storage tanks. Improperly discarded solvents and landfills, built before the introduction of current stringent regulations, are also significant sources of soil VOCs. Many of organic substances are classified as priority pollutants such as polychlorinated biphenyls (PCBs), polycyclic aromatic, acetaldehyde,

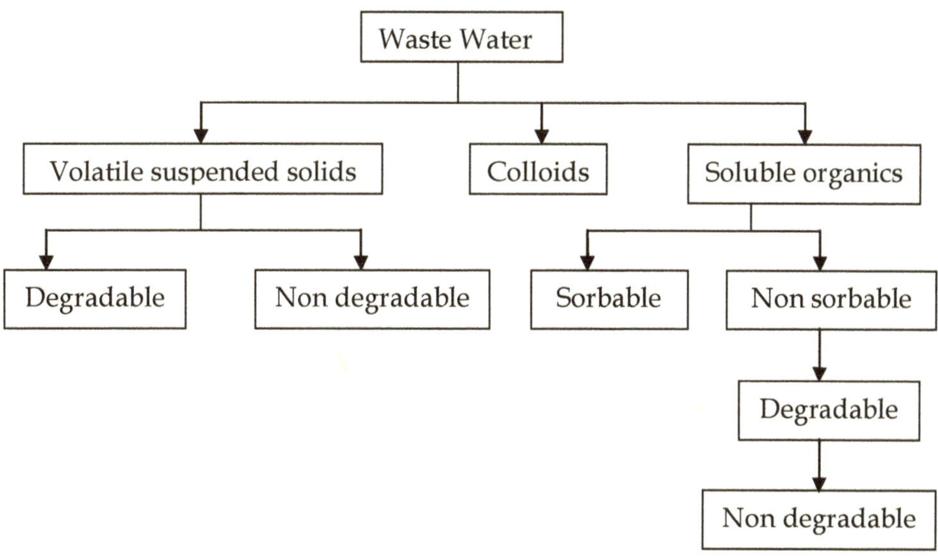

Figure 12.2: Partition of Organic Constituents of Wastewater (Eckenfelder, 1989).

formaldehyde, 1,3-butadiene, 1,2-dichloroethane, dichloromethane, hexachlorobenzene (HCB), etc,. In Table 12.3, a list of typical inorganic and organic substances present in industrial effluents is presented.

Table 12.3: Substances Present in Industrial Effluents

Substances	*Present in Wastewaters from*
Acetic acid	Acetate rayon, beet root manufacture
Acids	Chem. manufacture, mines, textiles manufacture
Alkalies	Cotton and straw kiering, wool scouring
Ammonia	Gas and coke and chemical manufacture
Arsenic	Sheep dipping
Cadmium	Plating
Chromium	Plating, chrome tanning, alum anodizing
Starch	Food processing, textile industries
Sugars	Dairies, breweries, sweet industry
Tartaric acid	Dyeing, wine, leather, chemical manufacture
Organic acids	Distilleries and fermentation plant

Source: Bond and Straub, 1974.

Inorganic Substances

The **inorganic substances** of dairy wastewaters have received much less attention as sources of pollution than the organic wastes because product manufactured is edible materials which do not contain hazardous quantities of inorganic substances. However, the non-edible materials used in the process, do contain inorganic substances which by themselves, or added to those of milk products and raw water, pose a potential pollution problem. Such inorganic constitute include phosphates (used as deflocculants and emulsifiers in cleaning compounds), chlorine (used in detergents and sanitizing products) and nitrogen (contained in wetting agents and sanitizers).

In dairy wastewaters, nitrogen originates mainly from milk proteins, and is either present in organic nitrogen form such as proteins, urea and nucleic acids, or as ions such as NH_4^+, NO_2^-, and NO_3^-. Phosphorus is found mainly in inorganic forms such as orthophosphate ($PO^{3''}_4$) and polyphosphate $(P_2O_4)_7^{--}$ as well as in organic forms also (Demirel *et al.,* 2005). The detergents and their additives are also present in small quantities in dairy wastewater. They may be alkaline or acidic, and very often contain additives like phosphates, sequestering agents, surfactants, etc. (Grasshoff, 1997).

Significant amount of Na, Cl, K, Ca, Mg, Fe, CO, Ni, Mn are also always present in dairy wastewater. The presence of high concentration of Na and Cl is due to the use of large amount of alkaline cleaners in dairy plant (Demirel *et al.,* 2005). Lens *et al.* (1994) observed that pine bark achieved a reduction of 72 per cent TSS, 63 per cent COD, 97 per cent BOD, 64 per cent NH^{4+}-N, and 35 per cent total N from wastewater. Bolan *et al.* (2004) studied a pilot-scale field experiment and showed that bark treatment achieved a considerable reduction in the concentration of nitrogen and phosphorus. The N and P concentrations approached the estimated maximum permissible level (MPL) value of 75 gm^{-3} and 3 gm^{-3}. The typical characteristics of wastewaters from various types of dairy facilities are presented in Table 12.4 and characterization of overall effluent generated from dairy wastewater is shown in Table 12.5.

Table 12.4: Characteristics of Untreated Wastewater from Dairy Plants

Effluent	*pH*	*BOD_5 (g/m³)*	*SAR*	*N (g/m³)*	*P (g/m³)*	*Electrical Conductivity (µS/cm)*	*Ref**
Whey	4.6	35000	3	1400	640	N/A	a
Condensate	8.3	N/A	N/A	0.6	0.1	7700	b
Cheese/evaporated milk manufacturing, clean effluent stream	N/A	12	N/A	N/A	N/A	880	b
Cheese/evaporated milk manufacturing,dirty effluent stream	8-12	700-1700	N/A	50-70	10	2600	b
Cheese/milk powder manu facturing, effluent	10.6	1500	N/A	0.01	35	2600	c
Cheese manufacture effluent	6.9	2800	21	150	42	3500	c

N/A: Not Available; BOD: Biochemical Oxygen Demand; SAR; Sodium Adsorption Ratio.

* a: Barnett *et al.,* 1994; b: Watkins and Nash, 2010; c: Carroll, 2008.

Table 12.5.: Characterization of the Overall Effluents Generated in Dairy Factories (composition in mg/L, expect for pH)

DF	*COD*	*BOD*	*Fat*	*N*	*P*	*pH*	*TS*	*VS*	*TSS*	*VSS*	*Ref**
1.	4000	2600	400	55	35	9	–	–	675	635	i.
2.	4000	2160	–	200	60	5-9	5100	4300	–	500	ii.
3.	2926	1580	294	36	21	6.7	2750	1880	–	–	iii.
4.	633	260	–	106	–	8.9	710	447	240	–	iv.
5.	2209	1112	60	–	–	7.2	–	–	278	–	v.
6.	4500	2300	–	56	33	7.2	2540	1093	816	–	vi.
7.	–	285	–	296	–	8.1	–	–	943	–	vii.
8.	2125	1250	–	70	100	9.8	1500	–	280	250	viii.
9.	–	241	–	191	50.9	8.5	–	–	–	804	viii.
10.	4500	2300	350	60	50	–	–	–	800	–	ix.
11.	4000	2000	–	60	–	–	–	–	800	–	x.
12.	1750	–	–	75	9.1	–	–	–	400	355	x.

DF: Dairy Factory; COD: Chemical oxygen in demand; BOD: Biochemical oxygen in demand; N: Nitrogen; P: Phosphorus; VS: Volatile solids; TS: Total solids; TSS: Total soluble solids; VSS: Volatile suspended solids.

* (i) Kasapgil *et al.*, 1994; (ii) Berg *et al.*, 1983; (iii) Rico *et al.*, 1991; (iv) Lo *et al.*, 1985; (v) Timofeyeva, 1992; (vi) Harper, 1974; (vi.) Shamir *et al.*, 2001; (viii) Monroy *et al.*, 1995; (ix) Craggs *et al.*, 2000; (x) Koyuncu *et al.*, 2000.

Condensate collected from the evaporation of milk or whey is one of the cleanest wastewaters although it may contain volatile organic components (Verheyen *et al.*, 2009) and possibly liquid droplets of milk or whey entrained into the vapour stream from the evaporators. Sodium hydroxide is often used for removal of fats and proteins from milk lines and other surfaces. It contributes sodium to wastewater and increases wastewater pH. Nitric, phosphoric, hydrochloric, acetic and citric acids may also be used to remove remaining deposits, especially mineral scale. These acids can decrease the wastewater pH significantly which can necessitate neutralization of the excess acid before treatment. Nitric and phosphoric acids contribute to the nutrient load of the wastewater and, as these nutrients can be difficult to remove from the wastewater, it can lead to accelerated eutrophication when discharged to the environment. From an environmental standpoint, phosphoric acid is the least desirable acid to be used for cleaning so factories have moved away from using phosphoric acid to minimise discharge concentrations of phosphorus. Strong oxidants or bleaches such as peroxyacetic acid, sodium hypochlorite and chlorine dioxide are used for sanitizing equipment and chlorine bleaches may produce toxic organochlorine compounds that contaminate the wastewater. Other chemicals that are used for specific applications include enzymes and detergents that are particularly useful for cleaning cool surfaces and have less adverse downstream consequences. In most plants caustic chemicals are generally preferred for higher temperature surfaces which are more difficult to clean. Where wastewaters are ultimately applied to land, some factories use blends of sodium hydroxide and potassium hydroxide to minimise the concentration of sodium in the effluent. Considering that dairy factories produce a range of products and use a range of chemicals, it is not surprising that dairy processing wastewaters are highly variable in their composition.

Decomposition of Organic Matter

Organic material present in the dairy wastewater is not stable and is decomposed (oxidized) by microorganisms and chemical reaction. In both the processes of reduction oxygen is consumed.

Chemical Oxygen Demand (COD)

The amount of oxygen needed to oxidize the organic and inorganic materials is called the Chemical Oxygen Demand (COD). It determines, the oxygen required for chemical oxidation of organic matter with the help of strong chemical oxidant. The COD is a test which is used to measure pollution of domestic and industrial waste. The waste is measured in terms of quantity of oxygen required for oxidation of organic matter to produce CO_2 and water. It is a fact that all organic compounds with a few exceptions can be oxidizing agents under the acidic condition. COD test is useful in pinpointing toxic condition and presence of biological resistant substances. It is expressed in milligrams per liter (mg/l) also referred to as ppm (parts per million), which indicates the mass of oxygen consumed per liter of solution. Dairy wastewater contains milk solids, detergents, sanitizers, milk wastes, and cleaning water. It is characterized by high concentrations of nutrients, and organic and inorganic contents (USDA-SCS, 1992).

Biochemical Oxygen Demand (BOD)

In any water system, microorganism will consume any organic and inorganic matter added to it and will produce biomass using oxygen present in the water. The oxygen required for the degradation of the organic matter biologically is called the Biochemical Oxygen Demand (BOD). While stabilizing organic matter under the aerobic conditions, it is necessary to provide standard conditions of nutrient, temperature and pH. Absence of microbes because of the low solubility of oxygen in water, strong wastes is always diluted to ensure that the demand of available oxygen does not increase. The BOD value is most commonly expressed in milligrams of oxygen consumed per litre of sample during 5 days of incubation at 20°C and is often used as a robust surrogate of the degree of organic pollution of water.

Wastewater Treatment Technologies for Milk Based Food IKndustry

Wastewater Treatment Unit Operations and Processes

Wastwater treatment methods are broadly classifiable into physical, chemical and biological processes:

1. Physical unit operations
   - ✰ Screening

   - ☆ Comminution
   - ☆ Flow equalization
   - ☆ Sedimentation
   - ☆ Flotation
   - ☆ Granular-medium filtration
2. Chemical unit operations
   - ☆ Chemical precipitation
   - ☆ Adsorption
   - ☆ Disinfection
   - ☆ Dechlorination
   - ☆ Other chemical applications
3. Biological unit operations
   - ☆ Activated sludge process
   - ☆ Aerated lagoon
   - ☆ Trickling filters
   - ☆ Rotating biological contactors
   - ☆ Pond stabilization
   - ☆ Anaerobic digestion
   - ☆ Biological nutrient removal

Options for Wastewater of Milk Based Food Industry

- ☆ Treatment to suitable standards for reuse or recycling
- ☆ Discharge to local authority sewers under a trade waste agreement (with pretreatment as necessary)
- ☆ Appropriate treatment and land discharge wherever practicable and environmentally beneficial

Best practice for wastewater systems is shown in Figure 12.3.

Physical Process

Segregation

Waste streams from the plant should also be segregated *e.g.*, whey can be reused to produce whey powder or stock feed. Spent cleaning solutions should be separated from other wastewater streams as they can be treated to recover cleaning agents. Highly saline wastewater should also be discharged separately to an evaporation pond where the salts can be removed and recycled.

Pre-treatment

Pretreatment of wastewater from milk based food industry consists of screening, grit separation, flow balancing, pH control and removal of coarse solids. The settleable

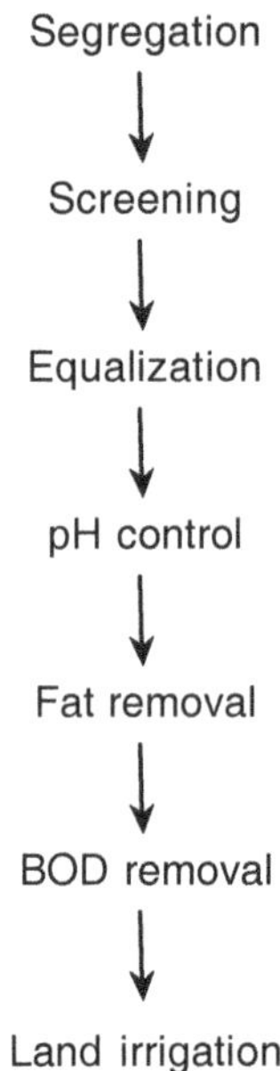

Figure 12.3: Practice for Wastewater Systems (adapted from EPA, 1997).

60 per cent of the suspended solids and 35 per cent of the BOD can be eliminated during primary treatment. Primary treatment is essential activity that needs to be undertaken for a proper application of various secondary treatment systems.

Screening

Wire mesh screens are advisable to prevent blockage of drains, particularly in bottling dairies to retain broken bottles, caps, labels and other solid material. Screens are also necessary at the cheese factories to remove cheese curd. Each screen is generally installed in the wastewater channel or at a wastewater storage tank inlet so that wastewater flows through it due to gravity. Trapped matter reduces the flow rate by friction loss and raises the upstream water level if the matter is left on the screens. Therefore, the matter must be raked up occasionally. Raking is done mechanically or manually. A bar screen, vibration screen, or rotating drum is often used. The bar screen is the most popular because of its easy installation, easy maintenance, reasonable cost, and other attractive features.

Fat Removal

Coarse milk solids should be removed by screening. Fats can constitute up to 50 per cent of the organic load. Its recovery is therefore significant in any treatment process. Dissolved air flotation is very effective method. Usually oil in wastewater floats and is separated from the water by using the difference of the densities between oil and water (Onishi, 2002).

Balancing of Flow

The daily time table in the dairy industry causes wide fluctuations in effluent volume and strength. To maximize treatment plant efficiency it is necessary to operate it at constant flow rates with relatively consistent untreated wastewater composition. To achieve this objective a balance or equalization tank is essential, the size being governed by local operating conditions and the necessity of accommodating peak flows. Aeration has been included in balance tank to prevent the possibility of septic conditions and subsequent protein coagulation developing. Aeration can be achieved by passing air through diffuser pipes, by pumping the incoming effluent against a splash plate or, by a surface aerator.

Settlement

The amount of settleable material in the waste depends upon its origin; waste from a bottling dairy or collecting depot contains very little settleable solids, while the waste from a milk products manufacturing creamery can contain sufficient settleable material to warrant the inclusion of this type of treatment scheme. A detention time of 4-6 h was sufficient for the removal of easily settleable material, thereby reducing the BOD of the waste by some 10-20 per cent and protecting the biological filter beds from overloading. In more recent schemes, employing activated sludge treatment and biofilters, initial settlement has been eliminated to ensure that aerobic conditions are maintain all times (Metcalf and Eddy, 2003).

Chemical Process

Chemical treatment processes like coagulation/flocculation, adsorption, and membrane process are required to remove suspended, colloidal, and dissolved constituents. Coagulation and flocculation is a frequently applied process in the primary purification of industrial wastewater. Coagulation using chemical coagulants consists of combining insoluble particles and/or dissolved organic matter present in dairy wastewater into large aggregates, thereby facilitating their removal in subsequent sedimentation, floatation, and filtration stages. Only few studies have been reported in literature for the coagulation treatment of dairy wastewater. Rusten *et al.* (1993) studied the different combinations of coagulants for the treatment of dairy wastewater. They observed that $FeClSO_4$ removed 2–3 per cent more COD than H_2SO_4 combined with CMC, and 4–6 per cent more COD than lactic acid combined with CMC. Feofanov and Litmanova (2000; 2001) studied the removal of organic impurities from contaminated wastewater of a dairy plant by coagulation with aluminum oxychloride and aluminum oxochloride at pH 2-12. Hamdani *et al.* (2004) found that the treatment of dairy wastewater by coagulation-decantation with iron chloride ($FeCl_3.6\ H_2O$), aluminum sulfate ($Al_2\ (SO_4)_3.18\ H_2O$), and calcium hydroxide ($Ca\ (OH)_2$) removed only 40 per cent of organic matter and nitrogen content. However, this treatment considerably reduced the suspended matter (94 per cent) and total phosphorus (89 per cent) with calcium hydroxide. Selmer-Olsen *et al.* (1996) used various types of chitosan as coagulants for the treatment of dairy wastewater. Mukhopadhyay *et al.* (2003) isolated lactose from whey by treatment with chitosan followed by alcohol precipitation. They reported BOD removal efficiency of 87 per cent. Among the various physico-chemical treatment methods, adsorption has been found to be attractive for the removal of organic compounds in wastewaters. Activated carbon (AC) is generally used as an adsorbent for the treatment of various types of wastewaters. However, many investigators have utilized several low-cost adsorbents like coal fly ash, rice husk ash, and bagasse fly ash (BFA), etc., for the treatment of a wide variety of wastewaters. Rao and Bhole (2002) used some low-cost adsorbents along with powdered activated carbon (PAC) for the treatment of dairy wastewater. PAC was found to be better in lowering the total dissolved solids (TDS) than other pretreated adsorbents like bagasse, straw dust, saw dust, coconut coir, and fly ash.

Sarkar *et al.* (2006) employed coagulation by chitosan and various inorganic coagulants followed by adsorption with powdered activated carbon (PAC) as pretreatment steps before treating the dairy wastewater by the membrane separation method. Chitosan was found to be a better coagulant, giving 57 per cent COD removal at a coagulant dosage of 10-50 mg/l. The electro coagulation (EC) process can be another alternative process for treating dairy wastewaters. An EC unit consists of one anode and one cathode. When electric power is applied from a power source, the anode material gets oxidized and the cathode is subjected to reductive deposition of elemental metals, and due to further reactions, metal hydroxides and/or polyhydroxides are produced, which remove the waste matter from wastewater by electrostatic attraction followed by coagulation. Sengil and Ozacar (2006) studied the EC process for the treatment of dairy wastewater and reported COD and oil-grease removal efficiencies of 98 and 99 per cent at electrolysis time of 7 and 1 min, respectively. The membrane-treatment process includes microfiltration (MF), ultrafiltration (UF), naofiltration (NF), reverse osmosis (RO), dialysis, and electrodialysis. These methods are very promising where product recovery is feasible and produce high quality effluent suitable for direct reuse. NF membrane treatment is a viable alternative to the conventional treatment by RO because it can operate at lower pressures, lower-energy consumption, and higher permeate recoveries than RO (Owen *et al.*, 1995). Frappart *et al.* (2006) reported the recovery of lactose and milk proteins as well as COD and ionic concentration reduction of dairy process waters having initial COD of 36,000 mg/l by NF using high shear rate dynamic filtration systems. Vourch *et al.* (2005) studied the performance of one-stage and two-stage (NF + RO and RO + RO) membrane treatment process. They evaluated the performances in terms of permeate flux, milk components rejection, and purified water characteristics. They concluded that both total organic carbon (TOC) and conductivity of water treated by a single RO or NF + RO operations were suitable for reuse as heating, cooling, cleaning, and boiler feed water.

Secondary Treatment or Biological Treatment

There are basically two types of biological wastewater treatment systems; aerobic and anaerobic systems. In view of high BOD load in the wastewater from milk based food industry, aerobic processes (for low organic load) and anaerobic processes (for high organic loads) are adopted for the treatment of wastewater from milk based food industry. The selection of processes for any particular plant plant will depend upon the size of the problem, location of the plant and the necessary degree of treatment (Sharma, 2008).

Aerobic Treatment

Aerobic biological treatment involves microbial degradation and oxidation of waste in the presence of oxygen. Conventional treatment of dairy wastewater by aerobic processes includes processes such as activated sludge, trickling filters, aerated lagoons, or a combination of these (Carta-Escobar *et al.,* 2004). Except protein and fat all other compounds of dairy effluents are biodegradable (Omil *et al.,* 2003). The presence of high organic matter, dairy wastewaters demands for biological treatment, especially anaerobic treatment (Rico *et al.,* 1991). However, the presence of fats shows the inhibitory action during anaerobic treatment of dairy wastewaters (Vidal *et al.,* 2000). This inhibition is due to the presence of long-chain fatty acids formed during the hydrolysis of lipids, which causes retardation in methane production (Hanaki *et al.,* 1981). Long-chain fatty acids were reported to be inhibitory to methanogenic bacteria (Koster, 1987) but lipids do not cause serious problems in aerobic processes (Komatsu *et al.,* 1991).

Sequential batch reactor (SBR) seems to be the most promising technology for treatment of dairy wastewater amongst the various aerobic technologies. It is a fill-and draw-activated sludge system. In this system, wastewater is added to a single batch reactor, treated to remove undesirable components, and then discharged. Equalization, aeration, and clarification can all be achieved using a single-batch reactor. Hence, savings on the total cost are obtained by elimination of clarifiers and other equipment (U.S. Environmental Protection Agency, 1999). Mohseni-Bandpi and Bazari (2004) used a bench scale aerobic SBR to treat the industrial milk factory wastewater. More than 90 per cent COD removal efficiency was achieved when COD concentration was varied from 400 to 2500 mg/l.

The optimum dissolved oxygen in the reactor was 2 to 3mg/l and the mixed-liquor volatile-suspended solids (MLVSS) were around 3000 mg/l. Sometimes a membrane is coupled to SBR and the reactor is named as membrane-sequencing batch reactor (MSBR). Bae *et al.* (2003) used MSBR for BNR. Nitrogen and phosphorus removal were found to be 96 and 80 per cent, respectively; whereas BOD removal was found to be in the range of 97-98 per cent. A stable SS free effluent was obtained by membrane separation. Compared to SBR, MSBR performed better even at higher organic loadings. Sirianuntapiboon *et al.* (2005) reported that high organic loading of 1.34 kg BOD_5/m^3 d; the COD, BOD_5, total kjeldahl nitrogen (TKN), and oil and grease removal efficiencies were found to be 89.3 ± 0.1, 83.0 ± 0.2, 59.4 ± 0.8, and 82.4 ± 0.4 per cent, respectively. The respective removal efficiencies in the conventional SBR system were only 87.0 ± 0.2, 79.9 ± 0.3, 48.7 ± 1.7, and 79.3 ± 10 per cent, respectively. The bio-sludge generated in MSBR was also three times lower as compared to conventional SBR. Neczaj *et al.* (2008) studied the dependency of the removal efficiency on the operating parameters using two SBR for the treatment of dairy wastewaters having initial COD concentration in the range of 400-7500 mg/l. The aeration time of 19 h and an anoxic phase of 2 h gave 98.6 per cent COD and 80.1 per cent TKN removal. The removal efficiencies of the SBRs decreased with increasing organic loading or decreasing HRT.

Under organic loading of 0.8 kg BOD_5/m^3 d and HRT of 10 d showed best quality effluent. Nitrogen is the main source of eutrophication. In this regard, the complete oxidation of ammonia during the treatment is favorable. Li and Zhang (2002) studied biological nitrogen removal using single-stage and two-stage SBR systems. In the single-stage SBR system, the removal efficiencies for TKN, total nitrogen (TN), COD, total solids, and volatile solids were found to be 75, 38.3, 80.2, 63.4, and 66.2 per cent, respectively. For complete ammonia oxidation in the single-stage SBR system, 4 d HRT was required. However, 1/3 HRT was required in a two-stage system (SBR and a complete-mix biofilm reactor) for complete ammonia oxidation as compared to the single SBR system. In recent trends, aerobic granular activated sludge SBR (GAS-SBR) have been reported to give a very high-rate of aerobic treatment and better settling (Wichern *et al.*, 2008). Schwarzenbeck *et al.* (2005) reported removal efficiencies of 90 per cent COD, 80 per cent TN, and 67 per cent total phosphorus in a GAS-SBR. Although the mixed culture activated sludge is widely used by the researchers for the treatment of dairy wastewaters but the bioaugmentation (addition of external microorganisms with high degradation capacity for specific wastewater) has been successfully used to give better performance by the various authors (Fantroussi and Agathos, 2005). Loperena *et al.* (2007) reported that although the commercial and mixed activated-sludge inocula gave similar COD removal in batch experiments for the treatment of a dairy industrial effluent; however, the COD degradation rate was greater for the commercial inoculum.

Anaerobic Treatment

High energy requirement by aerobic treatment methods is the primary drawback of these processes. Dairy effluents have high COD and organic content and are warm, enabling them to be ideal for anaerobic treatment (Wheatley, 1990). Involves no requirement for aeration, low amount of excess sludge production, and low area demand. Advantages of anaerobic treatment processes (Table 12.6). Consequently, a number of studies have been reported in open literature for the treatment of dairy wastewater by anaerobic methods. UASB reactors have been widely used for the dairy wastewater treatment in full-scale applications (Rico *et al.*, 1991; Gavala *et al.*, 1999; Yan *et al.*, 1989; Hwang and Hansen, 1992). The basic elements of a typical UASB reactor are a sludge blanket, influent-distribution system, gas-solid separator, and the effluent-withdrawal system. In the UASB reactor, the influent is distributed at the bottom and travels in up-flow mode (Tchobanoglous *et al.*, 2003).

In one of the studies, COD reduction of 90 per cent at organic loading rate of 0.031 kg COD/m^3d (t = 0.07 d) was achieved operating in steady-state conditions using a wastewater with a COD influent of 2050 mg/l. Dairy wastewater contains fats and the inhibitory action of the fat to the anaerobic treatment does not allow fast and increased removal efficiency (Vidal *et al.*, 2000). Various authors reported that the enzymatic hydrolysis of fats as pre-treatment may remove this problem. Cammarota *et al.* (2001) found that treatment of dairy wastewaters containing elevated fat and grease levels (868 mg/l) in a UASB reactor resulted in effluents of high turbidity (757 NTU), volatile suspended solids (VSS) up to 944 mg/l, and COD removal below 50 per cent. However, the same dairy wastewater pre-treated with 0.1 per cent (w/v)

Table 12.6: Comparison of Advantages and Disadvantages of Aerobic and Anaerobic Treatment of Dairy Industry Wastewaters (Kushwaha *et al.,* 2011)

Factors	*Aerobic Process*	*Anaerobic Process*
Reactors	Aerated lagoons, oxidation ditches, Stabilization ponds, Trickling filters and Biological discs	UASB, Anaerobic filter, Upflow packed bed reactor, CSTR, Down flow fixed-film reactor, Buoyant Filter Bioreactor
Reactor Size Size of reactor	Aerated lagoons, oxidation ditches, Stabilization ponds, Trickling filters and Biological discs requires larger land area but SBR needs comparatively lower area.	Smaller reactor size is required.
Effluent Quality Quality of waste water	Excellent effluent quality in terms of COD, BOD and nutrient removal is achieved.	Effluent quality in terms of COD is fair but further treatment is required. Nutrient removal is very poor.
Energy	High energy is required.	These processes produce energy in the form of methane.
Biomass yield	In comparison to anaerobic process, 6-8 times greater biomass is produced (Tchobanoglous *et al.,* 2003).	Lower biomass is produced
Loading Rate Rode of loading	Maximum 9000 g COD/m^3 d (Wichern *et al.,* 2008) is reported in literature.	Very high Loading rate of 31 kg COD/m^3 d has been reported (Rico Gutierrez *et al.,* 1991). This is the reason for smaller reactor volume and lesser area.
Oil and Grease Removal	These do not cause serious problems in aerobic processes (Komatsu *et al.,* 1991).	Fats in wastewater shows the inhibitory action during anaerobic treatment of dairy wastewaters (Vidal *et al.,* 2000)
Shock Load	Excellent performance in this regard.	Anaerobic processes showed not good responses to this shock loading.
Alkalinity Addition	No need	There is need for alkalinity addition to maintain the pH because pH changes during the digestion of lactose.

of fermented babassu cake containing lipases showed higher COD removal efficiency of 90 per cent when treated in the same UASB reactor.

In another study, Leal *et al.* (2006) used synthetic dairy wastewater containing high levels of oil and grease (200, 600, and 1000 mg/l) using two identical UASB reactors, one fed with wastewater from enzymatic hydrolysis step and the other with raw wastewater. 90 and 82 per cent COD removal efficiencies were achieved in the reactors fed with the hydrolyzed effluent and with raw wastewater, respectively.

Many authors have reported reduced efficiency of continuous UASB reactors due to the buildup of organic matter in the reactor (Morgan *et al.*, 1991; Motta-Marques *et al.*, 1990). Nadais *et al.* (2005) studied the intermittent operation of mesophilic UASB reactors to overcome the buildup of organic matter and the influence of cycle duration in the reactor. They reported that a 96 h cycle (48 h feed + 48 h feed less) reflected the highest conversion to methane. Anaerobic filter reactors are generally suitable for dairy effluents containing low concentrations of SS. The anaerobic filter reactors capture SS and provide sufficient retention time for bio-solids. Hence, the SS retention time and the HRT necessarily become satisfying. A laboratory-scale plastic medium anaerobic filter reactor provided average COD removal rates between 78 and 92 per cent at a HRT of 4 d (Viraraghavan and Kikkeri, 1990, 1991).

Performance of an upflow anaerobic filter reactor treating a high solid-containing dairy waste at 0-18 kg COD/m^{3}d organic loading was investigated by Monroy *et al.* (1994). The mean COD removal was found to be 70 per cent and the type of support media (porous or non-porous) greatly affected the reactor's performance in anaerobic filter reactors. The ideal packing media for the anaerobic filter is that which maximizes both the surface area and the porosity. A large surface area of the media enhances the attachment of the biomass and increased porosity.

The performances of porous and non-porous support media in anaerobic filter reactors on mesophilic anaerobic treatment of milk-bottling plant wastewater were investigated and it was concluded that the reactor with non-porous packing showed instability above an organic loading of 4 kg COD/m^{3}d, whereas the reactor with the porous packing was stable at very high organic loading of 21 kg COD/m^{3}d (Anderson *et al.*, 1994).

In an another study, the strength and the performance of a porous media in an upflow anaerobic filter (UFAF) were investigated up to an organic loading rate of 21 kg COD/m^{3}d with a HRT of 0.5 d. The highest quantity of the attached biomass (103 mg VSS/media ring) was found at the bottom and the lowest (23 mg VSS/media ring) at the top. After 8 months of operation, a reduction in compressive strength was 50 per cent of the media, and during this time, and 80 per cent COD removal efficiency was achieved (Ince *et al.*, 2000). Due to the gas generation, all types of anaerobic reactors are well mixed but the reactor performance gets affected by the degree of mixing and the solid content in the waste. The fluidized bed reactors removed the problem of mixing and gave better mass transfer characteristics; however, they require a long time for stable biofilm establishment (Rockey and Forster, 1985). The "two-stage" reactor concept improves process stability and efficiency and is particularly suitable for dairy wastewaters containing high concentrations of organic SS (Demirel *et al.,* 2005). In a two-stage system, acidogenic and methanogenic phases can be operated separately under respective optimum conditions. In the first stage, hydrolysis and degradation of lipids and proteins into amino acids and fatty acids take place (Schroder and Haast, 1989; Malaspina *et al.,* 1995). This degradation gives a high-energy yield to the microorganisms. In the second stage, the end-products of the fermentation process (acetate, butyrate, propionate, formic acid, carbon dioxide, and hydrogen) are produced by methanogenic microorganisms into methane and carbon dioxide.

Yu and Fang (2002b) studied acidogenesis of a dairy wastewater in batch reactors at pH 5.5 and 55°C, with reactors fed COD of 8000 mg/l. At the start, the concentration of acetate and butyrate increased rapidly and reached maxima at some points, then declined. Increase in substrate concentration, increased the butanol and propanol fractions. Beyond 8000 mg/l of COD, the metabolism shifted to the alcohol-producing pathways. Temperature and pH are important parameters in treatment with two-stage anaerobic reactors. Degradation efficiency is generally more at higher temperatures (Ghosh, 1991). Thus, the thermophilic operation facilitates the treatment at higher loading rates, and therefore requires smaller treatment plants than the mesophilic systems. Yu and Fang (2002a) reported that degradation of dairy wastewater pollutants increased with an increase in pH from pH 4.0 to 5.5. 48.4 per cent of COD was converted into volatile fatty acids and alcohols, 6.1 per cent into hydrogen and methane, and the remaining 4.9 per cent into biomass. On increasing the pH up to 6.5, there was very slight increase in degradation. HRT is also an important variable in two-stage anaerobic treatment processes. The effects of HRT on anaerobic acidogenesis have been investigated earlier (Yilmazer and Yenigun, 1999; Horiuchi *et al.,* 2002). In a recent study, the performance of a mesophilic acidogenic reactor was evaluated in terms of volatile fatty-acid production, with respect to HRT variations between 1-0.5 d with the organic loading rate of about 9.3 kg COD/m^3 d, in a continuous flow-completely mixed anaerobic reactor, coupled with a conventional gravity settling tank, and a continuous recycling system. It was shown that, acid production increased proportionally to the organic loading rate with decrease in HRT. The highest degree of acidification and the rate of acid production were 56 per cent and 3.1 kg/m^3 d, respectively, at 0.5 d of HRT. Variations in HRT affected volatile fatty acid production (Demirel and Yenigun, 2004).

Anaerobic processes treat only the carbonaceous pollutant with subsequent production of methane, while nitrogenous organic material is partly converted to ammonia which is also not desirable. The removal of phosphorus is also low. So a complete treatment, compatible with a good quality discharge to the natural environment, needs an additional process (Montuelle *et al.,* 1992). A hybrid system comprising of both anaerobic and aerobic reactor are used to treat dairy wastewaters.

The combination of a UASB and activated sludge (AS) gave an overall removal efficiency of 98.9 per cent for COD, 99.6 per cent for BOD$_5$, and 98.9 per cent for oil and grease at a total HRT of 1.08 d and 1.9–4.4 kg COD/m^{3}d of organic loading. TSS and VSS removal averaged above 72 and 75 per cent, respectively (Tawfik *et al.,* 2008). In another study, dairy wastewater was fed to anaerobic and anoxic reactors with an aerobic reactor in series. The overall system was operated at an HRT of 7 d. Overall

COD and total nitrogen removal was more than 90 and 65 per cent (Donkin and Russell, 1997). Successful anaerobic treatment of wastewaters discharged by raw milk-quality control laboratories was achieved in SBR with an anaerobic filter (AF) without fat removal prior to the anaerobic reactor, with the COD removal being higher than 90 per cent in AF, and most of the fat was successfully degraded (Omil *et al.*, 2003).

Biotreatment of Dairy Waste Effluent

Biotreatment leading to bioconversion of the waste materials is probably the most cost-effective technique for managing and utilizing waste (Vida *et al.*, 2007) The removal of organic matter from the wastewater using chemicals are used in many industrial wastes, and these methods are more expensive than the biotreatment method using microflora from the same area of the tests. Also the chemical methods may cause further contaminations to the environment. Dairy waste is basically biodegradable that produces an undesirable odour and contains an appreciable quantity of oil. Also dairy waste contains sufficient nutrients for biological growth. Biological treatment methods are considered more ideal and economical (Warner, 1976). Due to capital draining expense, attention has been drawn to the use of microbial culture preparations for waste treatment. The microorganisms of choices should have a strong degradative capacity and high toxic resistance. The process of seeding inoculation of microorganisms for degrading waste materials on streams, rivers and treatment tanks has been rapidly increasing practice in many countries because it is economical and the application is uncomplicated (Zamora and Lit 1995). In Table 12.7 below microflora associated with wastewater from dairy industry is shown.

Rajeshkumar and Jayachandran (2004) found that dairy wastewater had a high organic load (chemical oxygen demand [COD]: 5095 mg/L), an acidic pH (6.4), and a high probability of coliforms (most probable number [MPN] >1100/ml). The various bacterial strains isolated and purified were identified as *Sporolactobacillus* sp., *Citrobacter* sp., *Pseudomonas* sp., *Alcaligenes* sp., *Bacillus* sp., *Staphylococcus* sp., and *Proteus* sp. *Alcaligenes* sp. MMRR7 was found to give a maximum reduction upto 62 per cent in COD in 5 d of incubation. Chemical coagulation using alum at a concentration of 0.5 g/100 ml was found to be effective in the primary treatment of the effluent.

Vida *et al.* (2007) studied the microflora of the effluents from a dairy factory in Tehran (Pegah Dairy Processing Plant) for their ability to reduce the organic matter content and COD of the effluents and isolated 10 bacteria, which showed reduction in COD content from the 4th to 6th day of incubation at 30°C, shaking rate 150rpm and pH = 11. When condition for test organism was optimized, higher reductions in COD, carbohydrate, fat and protein content of the effluents were observed by an isolate number BP3 up to 84.70 per cent, 98 per cent, 45.30 per cent and 53 per cent, respectively.

Table 12.7: Microflora of Wastewater Generated from Dairy

Microrganisms	*Reference*
Sporolactobacillus sp., *Citrobacter* sp., *Pseudomonas* sp., *Alcaligenes* sp., *Bacillus* sp., *Staphylococcus* sp., and *Proteus* sp.	Rajeshkumar and Jayachandran, 2004 and Loperena *et al.*, 2009
Bacillus amyloliquefaciens, *Bacillus subtilis* and *Bacillus pumilus*	Loperena *et al.*, 2009
Pseudomonas fluorescens and *Bacillus* sp.	Kabbout *et al.*, 2011
Saccharomyces cerevisiae and *Humicola* sp.	Singh *et al.*, 2011
B. subtilis, B. licheniformis, B. amyloliquifaciens, S. marsescens, S. aureus, P. Aeruginosa	Prasad and Manjunath, 2011
Streptococcus strain, lactic yeast; *Bacillus* strain	Kosseva *et al.*, 2003
Bacillus sp.	Gowland *et al.*, 1987; Becker *et al.*, 1997
Acinetobacter sp., *Rhodococcus* sp.	Wakelin and Forster, 1997; Keenan and Sabelnikov, 2000
Pseudomonas sp.	Watanabe *et al.*, 1977; Pabai *et al.*, 1996
Pithophora sp.	Silambarasan *et al.*, 2012

Loperena *et al.* (2009) isolated milk fat/protein degrading microorganisms from different locations of a dairy wastewater treatment system with the goal of developing an inoculum for bioaugmentation strategies. Eight isolates, identified by 16S rRNA gene sequence which belongs to the genera *Bacillus, Pseudomonas,* and *Acinetobacter* and these were tested for their ability to remove COD and protein from a milk-based medium (3000 mg/l COD) compared to a commercial bioaugmentation inoculum. Based on the individual degradation capacity and growth behaviour of the isolates, three microorganisms *Bacillus amyloliquefaciens* strain NBRC 14141 (AB325582), *Bacillus subtilis* strain CICC10075 (AY881647) and *Bacillus pumilus* strain FO-036b (AF234854) were further selected and tested together. This consortium exhibited a COD removal similar to the commercial inoculum (44 per cent, 34 per cent and 42 per cent and 63 per cent, respectively). Djelal and Poignant (2010) studied the commercial mycelium consortium and showed that this consortium was able to grow in the presence of lactose and on dairy effluent. The maximum yield of the COD biodegradation was 76 per cent and 72 per cent respectively for the synthetic and industrial wastewater. It was also observed that the reduction of COD by the total flora (fungus and endogenous flora) was effective from 24 hours of treatment. Kabbout *et al.* (2011) showed that physicochemical treatment by coagulation-flocculation has reduced 33 per cent of the chemical oxygen demand, 45 per cent of the turbidity, 72 per cent of suspended matter and 20 per cent of total phosphorus while the biological treatment carried out using *Pseudomonas fluorescens* for 20 days resulted in very high reduction (upto 90 per cent) in COD while the effluent inoculated with *Bacillus* sp. showed reduction of only 54 per cent in COD under the similar conditions.

Singh *et al.* (2011) evaluated that *Saccharomyces cerevisiae* and *Humicola species* had ability to grow and produce biomass and reduce the organic load of the production of dairy wastewater. These species were able to reduce COD by about 62 per cent and 93 per cent respectively, with continuous biomass production, the decrease in lactose *i.e.* Organic load and increase in biomass (SCP) occurred in parallel and growth rate also increased simultaneously with increasing lactose consumption rate.

Prasad and Manjunath (2011) formulated the bacterial consortia of *B. subtilis, B. licheniformis, B. amyloliquifaciens, S. marsescens, S. aureus, P. aeruginosa* for use in treatment of wastewater or industrial effluent rich in lipid content and also showed that the role of lipid degradation capacity for *P. aeruginosa* is high compared to other bacteria. The formulated mix culture was found to be effective in treatment of lipid rich wastewater and BOD value was reduced from 3200 mg/L to less than 40 mg/L within 12 days of incubation.

Silambarasan *et al.* (2012) studied that when the effluent was treated with *Pithophora* sp. the colour changed to yellowish green on 10th day. On the 20th day the colour changed to green. When the effluent was treated with alga the colour changed into green and odourless also there was reduction in turbidity (Total Suspended Solids), COD, BOD, electrical conductivity, iron and free ammonia which was 302 mg/L, 350 mg/lit, 992 mg/lit, 1970 µS/cm, 2.38 mg/L, 70.56 mg/lit on 0th day and was reduced to 12mg/lit, 130 mg/lit, 399 mg/lit, 1465 µS/cm, 1.86 mg/lit, 1.34 mg/lit on 20th day respectively.

Biological Basis of Wastewater Treatment (Davies, 2005)

Bio augmentation can be defined as the controlled application of specially selected natural microorganisms to enhance the performance of an operating biological wastewater treatment plant in order to improve final effluent quality or reduce operating costs. In the aeration basin of a typical industrial wastewater treatment plant a biomass develops based on numerous strains of degradative and floc forming bacteria. This diversity of bacteria is necessary because different strains of bacteria are more effective in degrading different organic compounds. The biomass becomes adapted to the influent and will provide the required results provided a steady state of operation can be maintained. Over time the desirable characteristics of the plant biomass may be reduced due to shocks and undesirable characteristics become more dominant. The use of bioaugmentation can boost the desirable population and reduce the undesirable population The purpose of bio augmentation is to provide a controlled way of shifting the microbial population to one which is more effective in terms of degradation and settlement of the floc (http://www.biofuture.ie/news.html).

Bacterial Flocs

In a well-maintained aeration tank, the bacteria are concentrated in the flocculent material of the activated sludge, although some always occur free in the wastewater. The flocs are formed from aggregates of non-living organic polymers that are probably secreted by bacteria. They have an open porous structure, and are sufficiently robust to withstand the shear forces created by water movement, during aeration of the tanks. They vary in size from less than 10 µm up to 1mm (1000 µm) (Davies, 2005).

The bacteria are adsorbed on to the internal and external surfaces of the floc, and a medium sized floc may harbor several million bacteria. Immediately after the wastewater enters the aeration tank, the fine particulates, colloidal particles and large molecules, become entangled with, and adsorbed to, the floc material. This has the advantage that the enzymes that are secreted by the bacteria into the water, will tend to be confined in the vicinity of the substrate, thereby facilitating their digestion. However, for the bacteria living on the inside of the floc, oxygen availability may be a problem. This is because oxygen has to diffuse along a concentration gradient from the wastewater through the floc material to the inside. As in all ecosystems, the constituent organisms are in a dynamic steady state. Thus the dominant bacterial species may change, sometimes on a daily basis, in response to changes in the composition of the wastewater. Those species of bacteria that have the ability to secrete the enzymes to break down a novel food source will grow more rapidly, thereby increasing in relative number. This process is known as *adaptation* or *acclimation* (Davies, 2005).

Metabolism of Bacteria

Treatment of sewage in the aeration tank involves the removal of organic carbon from the mixed liquor by ingestion by the bacteria. Once inside, the carbon compounds are metabolised. Metabolism comprises the thousands of simultaneous chemical reactions that are going on at any one time inside the bacterium. In each of these reactions, a substrate, in the presence of an enzyme (which acts as a catalyst), is converted into a product.

$$\text{Substrate} \xrightarrow{\text{Enzyme}} \text{Product}$$

The product then becomes the substrate for the next step in the chain, and is almost immediately converted, in the presence of another specific enzyme, into a different product - and so on. For some of these reactions to take place, chemical energy needs to be provided (endergonic reactions). In other reactions (exergonic reactions), energy is given off, usually in the form of heat (Davies, 2005).

Growth of Bacteria

Bacteria show prodigious feats of growth. Some bacteria may double their biomass in as little as 20 minutes, provided they have the right conditions of temperature, pH and an abundance of organic carbon, other nutrients, trace elements etc. Note that an individual bacterium has limited capacity for growth, only growing from the size of a daughter cell produced at the time of division to the normal cell size. Growth rate is therefore measured as the increase in number of cells with time (Davies, 2005).

Legal Requirements

Treatment of domestic and industrial effluent is statutory requirement. Each state has a state level 'Pollution Control Board' which ensures that the effluents are treated by the industries and the municipal bodies before disposal. This applies to the dairy industry as well.

The IS 2490 part – I (1974) provides in detail the prescribed limits for various parameters of pollution. An abstract of the standard for parameters applicable to dairy industry is presented in Table 12.8.

Table 12.8: Standards for Disposal of Treated Effluent (Ref: IS 2490 Part – I)

Sl.No.	Parameters	Limits for Discharge of Effluents			
		Into Inland Surfaces Water	Into Public Sewers	On Land for Irrigation	Marine Disposal
1.	pH	5.5 – 9			
2.	Temperature maximum ºC	40	45	40	45
3.	Total Suspended Solids, mg/lit	100	600	200	100
4.	Total Dissolved Solids, mg/lit	2100	2100	2100	—
5.	BOD (5 days), mg/lit	30	350	100	100
6.	COD, mg/lit	100	—	250	250
7.	Oil and Grease, mg/lit	10	20	10	20
8.	Percent Sodium	—	—	60	—
9.	Ammonium nitrogen, (as N), mg/lit	50	50	—	50
10.	Total Kjeldahl nitrogen (as N), mg/lit	100	—	—	100
11.	Phosphate, mg/lit (dissolved as P)	5	—	—	5

Table 12.9: Minimum National Standards for Dairy Industry (Ref: Environment Protection Third Amendment Rules, 1992, GSR. 475 E, item 56)

Parameters	Standards	
Effluents	mg/lit except pH	Quantum per Product Processed
pH	6.5-8.5	—
*BOD_5 at 20 ºC	100	—
**Suspended Solids	150	—
Oil and Grease	10	—
Wastewater generation	—	3 cu.m./kl of milk

* BOD may be stringent upto 30 mg/lit if the recipient freshwater body is a source for drinking water supply. BOD shall be upto 350 mg/lit for the chilling plant effluent for applying on land provided the land is designed and operated as a secondary treatment system with suitable monitoring facilities. The drainage water from land after secondary treatment has to satisfy a limit of 30 mg/lit of BOD and 10 mg/lit of nitrate expressed as 'N'. The net addition to the groundwater quality should not be more than 3mg/lit of BOD and 3 mg/lit of nitrate expressed as 'N'. This limit for applying on land is allowed to subject to the availability of adequate land for discharge under the control of the industry. BOD value is relaxable upto 350 mg/lit, provided the wastewater is discharged into a town sewer leading to the secondary treatment of the sewage.

** Suspended solids limit is relaxable upto 450 mg/lit, provided the wastewater is discharged into town sewer leading to secondary treatment of the sewage.

In May, 1992, a notification of the 'Minimum National Standards (MINAS) for the dairy industry has been issued by Ministry of Environment and Forests of Government of India. These standards are given in the Table 12.9 Both the statutory requirements are applicable to dairy and chilling plants. The 'Consent' (acceptable) condition specified by the state pollution control board would govern the statutory requirements for a given ETP.

References

Anderson, G. K., Kasapgil, B., and Ince, O. (1994). Comparison of porous and nonporous media in up flow anaerobic filters when treating dairy wastewater. *Water Res*, 28(7): 1619–1624.

Bae, T., Han, S., and Tak T. (2003). Membrane sequencing batch reactor system for the treatment of dairy industrywastewater. *Process Biochem.* 39: 221–231.

Barnett, J. W., Kerridge, J.M. and Russell, J. M. (1994). Effluent treatment systems for dairy industry, *Australas Biotechnol*, 4, 26-30.

Berg, V. L. and Kennedy, K. J. (1983), "Dairy waste treatment with anaerobic stationary fixed film reactors". *Water Sci. Tech.* 15, 359-68.

Bolan N.S., Wong, L. and Adriano, D.C. (2004). Nutrient removal from farm effluents, *Bioresource Technology*, 94, 251–260.

Bond. R. G and Straub. C. P. (1974). *Wastewater Treatment and Disposal, Handbook of Environmental Control*, CRC Press, Volume 4.

Bylund G. (1995). *Dairy processing handbook.* Lund: Tetra Pak Processing Systems, 436.

Cammarota, M. C., Teixeira, G. A, and Freire, D. M. G. (2001). Enzymatic prehydrolysis and anaerobic degradation of wastewaters with high fat contents. *Biotechnology Lett.* 23: 1591–1595.

Carroll, M. (2008). Environmental Report 2008. Brunswick: Murray Goulburn Co-operative Co. Ltd., 39.

Carta-Escobar, F., Pereda-Marin, J., Alvarez-Mateos, P., Romero-Guzman, F., Duran-Barrantes, M. M., and Barriga-Mateos, F. (2004).Aerobic purification of dairy wastewater in continuous regime. Part I: Analysis of the biodegradation process in two reactor configurations. *Biochem. Eng. J.* 21: 183– 191.

Craggs, R. J., Tanner, C. C., Sukias, J. P. S. and Davies-Colley, R. J. (2000), Nitrification potential of attached biofilms in dairy farm waste stabilization ponds, *Water Sci. Tech.*, 42, 195-202.

Davies, P. S. (2005). The biological basis of wastewater treatment, Strathkelvin Instruments Ltd, 4-5.

Demirel, B. and Yenigun, O. (2004). Anaerobic acidogenesis of dairy wastewater: the effects of variations in hydraulic retention time with no pH control. *J Chem. Technol. Biotechnol.* 79: 755–760.

Demirel, B., Yenigun, O., and Onay, T. T. (2005). Anaerobic treatment of dairy wastewaters: A review. *Process Biochem.* 40: 2583–2595.

Devi (1980). Ecological studies of limon plankton of three freshwater body, Hyderabad. *Ph.D. thesis Osmania University, Hyderabad*.

Djelal, H. and Poignant, E. (2010). A study the behaviour of a commercial mycelium consortium for the biodegradation of dairy effluent, *J Biotechnol*, 150, 281-281.

Donkin, M. J. and Russell, J. M. (1997). Treatment of a milk powder/butter wastewater using the activated sludge configuration. *Water Sci. Technol.* 36(10): 79–86.

Eckenfelder, W.W. (1989). Industrial Water Pollution Control, Pemberton Press, McGraw-Hill, 2nd ed.

Eldridge, E. F. (1953). Milk products waste in *Industrial Wastes: Their Disposal and Treatment* (1996): W. Rudolfs (Ed.), New York: Reinhold Publishing Corp.

Environmental guidelines for the Dairy processing industry (1997). Environment Protection Authority State Government of Victoria, 14.

Fantroussi, E. L. and Agathos, S.N. (2005). Is bioaugmentation a feasible strategy or pollutant removal and site remediation? *Curr. Opin. Microbiol.* 8: 268–275.

Feofanov, Y. A., and Litmanova, N. L. (2000). Influence of solution pH and of coagulation agent dose on removal of organic impurities from wastewater of dairy plants by treatment with aluminum oxychloride. *Russian J. Appl. Chem.* 73(8): 1465–1466.

Feofanov, Y. A., and Litmanova, N. L. (2001). Mechanism of coagulation purification of wastewaterwith aluminium oxochloride. *Russian J. Appl. Chem.* 73(2): 344–346.

Frappart, M., Akoum, O., Ding, L. H., and Jaffrin, M. Y. (2006). Treatment of dairy process waters modelled by diluted milk using dynamic nanofiltration with a rotating disk module. *J. Membrane Sci.* 282: 465–472.

Gavala, N., Kopsinis, H., Skiadas, I. V., Stamatelatou, K., and Lyberatos, G. (1999). Treatment of dairy wastewater using an upflow anaerobic sludge blanket reactor. *J. Agri. Eng. Res.* 73: 59–63.

Ghosh, S. (1991). Pilot-scale demonstration of two-phase anaerobic digestion of activated sludge. *Water Sci. Technol.* 23(7–9): 1179–1188.

Government of India (2012). Basic Animal Husbandry Statistics. Department of Animal Husbandry, Dairying and Fisheries.

Grasshoff, A. (1997). Fouling and cleaning of heat exchanges. In: Bulletin of the IDF, No. 328, IDF, 32–44. Brussels.

Hamdani, A., Chennaoui,M., Assobhei, O., and Mountadar,M. (2004). Dairy effluent characterization and treatment by coagulation decantation. *LAIT*, 84(3): 317–328.

Hanaki, K., Matsuo, T., and Nagase,M. (1981). Mechanism of inhibition caused by long-chain fatty acids in anaerobic digestion process. *Biotechnol. Bioeng*, 23: 1591–1610.

Harper, W. J. (1974). Implant control of dairy wastes. *Food Tech*, 28, 50-5.

Horiuchi, J. I., Shimizu, T., Tada, K., Kanno, T., and Kobayashi, M. (2002). Selective production of organic acids in anaerobic acid reactor. *Bioresource Technol,* 82, 209–213.

http: //www. indiastat. com/table/agriculture/2/milkanddairyproducts/167/ 3858/data.aspx

http: //www.biofuture.ie/news.htmls

Hwang, S. H. and Hansen, C. L. (1992). Performance of upflow anaerobic sludge blanket (UASB) reactor treating whey permeate. *Transation ASAE* 35: 1665–1671.

Ince, O., Ince, B. K., and Donnelly, T. (2000). Attachment, strength and performance of a porous media in an upflow anaerobic filter treating dairy wastewater, *Water Sci. Technol,* 41(4–5), 261–270.

Indian Standards (1974) IS: 2490. Tolerance limits for industrial effluents discharged into inland surface waters: Part I General limits.

Kabbout, R., Baroudi, M., Dabboussi, F., Halwani, J. and Taha, S. (2011). Characterization, Physicochemical and Biological Treatment of Sweet Whey (Major Pollutant in Dairy Effluent), *International Conference on Biology, Environment and Chemistry,* 24, 123-127.

Kasapgil, B., Anderson, G.K. and Ince, O. (1994). "An investigation into pretreatment of dairy wastewater prior to aerobic biological treatment", *Water Sci. Tech.* 29, 205-12.

Kolhe, A. S. and Pawar, V. P. (2011). Physico-chemical analysis of effluents from dairy industry, *Recent Research in Science and Technology,* 3(5): 29-32.

Komatsu, T., Hanaki, K., and Matsuo, T. (1991). Prevention of lipid inhibition in anaerobic processes by introducing a two–phase system, *Water Sci. Technol,* 23, 1189–1200.

Koster, I. (1987). Abatement of long–chain fatty acid inhibition of methanogenic by calcium addition. *Biological Wastes,* 25: 51–59.

Koyuncu, I., Turan, M., Topacik, D. and Ates, A. (2000), "Application of low pressure nanofiltration membranes for the recovery and reuse of dairy industry effluents", *Water Sci. Tech,* 6, 213-21.

Kushwaha, J., Srivastava, V. C., Mall, I. D. (2011), An Overview of Various Technologies for the Treatment of Dairy Wastewaters, *Critical Reviews in Food Science and Nutrition,* 51, 442–452.

Leal, M. C. M. R., Freire, D. M. G., Cammarota, M. C., Anna, G. L., and Sant, Jr. (2006). Effect of enzymatic hydrolysis on anaerobic treatment of dairy wastewater, *Process Biochem,* 41: 1173–1178.

Lens, P.N., Vochten, P.M., Speleers, L., Verstraete,W.H. (1994). Direct treatment of domestic wastwater by percolation over peat, bark and woodchips, *Water Res.,* 28, 17–26.

Li, X. and Zhang, R. (2002). Aerobic treatment of dairy wastewater with sequencing batch reactor systems. *Bioprocess Biosystems Eng.* 25(2): 103–109.

Lo, K.V., Bulley, N.R. and Kwong, E., 1985. Sequencing aerobic batch reactor treatment of milking parlour wastewater, *Agric wastes,* 13, 131-6.

Lo, K.V., Bulley, N.R. and Kwong, E., 1985. Sequencing aerobic batch reactor treatment of milking parlour wastewater, *Agric wastes,* 13, 131-6.

Loperena, L., Ferrari, M. D., Díaz, A. L., Ingold, G., Pérez, L. V., Carvallo, F., Travers, D., Menes, R. J. and Lareo, C. (2009). Isolation and selection of native microorganisms for the aerobic treatment of simulated dairy wastewaters, *Bioresource Technology*, 100, 1762–1766.

Loperena, L., Ferrari, M. D., Saravia, V., Murro, D., Lima, C., Ferrando, L., Fernandez, A., and Lareo, C. (2007). Performance of commercial inoculums for the aerobic biodegradation of a high fat content dairy wastewater, *Bioresource Technol.,* 98, 1045–1051.

Malaspina, F., Stante, L., Cellamare, C. M., and Tilche, A. (1995). Cheese whey and cheese factory wastewater treatment with a biological anaerobic–aerobic process. *Water Sci. Technol.* 32: 59–72.

Metcalf and Eddy (2003), "Wastewater Engineering: Treatment and Reuse". Tata McGraw- Hill Edition.

Mohseni-Bandpi, A. and Bazari, H. (2004). Biological treatment of dairy wastewater by sequencing batch reactor. *Iranian J. Environ. Health Sci. Eng.* 1(2): 65–69.

Monroy, O., Johnson, K. A., Wheatley, A. D., Hawkes, F., and Caine, M. (1994). The anaerobic filtration of dairy waste: Results of a pilot trial. *Bioresource Technol.* 50: 243–251.

Monroy, O., Vazquez, F., Derramadero, J. C. and Guyot, J. P. (1995), "Anaerobicaerobic treatment of cheese wastewater with national technology in Mexico: thc case of "el sauz". *Water Sci. Tech.* 32, 149-56.

Montuelle, B., Coillard, J., and Lehy, J. B. (1992). A combined anaerobicaerobic process for the co-treatment of effluents from a piggery and a cheese factory. *J. Agri. Eng. Res.* 51: 91–100.

Morgan, J. W., Evison, L. M., and Forster, C. F. (1991). Changes to the microbial ecology in anaerobic digesters treating ice cream wastewater during start-up. *Water Res.* 25: 639–653.

Motta-Marques, D. M. L., Cayless, S. M., and Lester, J. N. (1990). Start-up regimes for anaerobic fluidized systems treating dairy wastewater. *Biological Wastes* 34: 191–202.

Mukhopadhyay, R., Talukdar, D., Chatterjee, B. P. and Guha, A. K. (2003). Whey processing with chitosan and isolation of lactose. *Process Biochem.* 39: 381–385.

Nadais, H., Capela, I., Arroja, L., and Duarte, A. (2005). Optimum cycle time for intermittent UASB reactors treating dairy wastewater. *Water Res.* 39: 1511–1518.

Neczaj, E., Kacprzak, M., Kamizela, T., Lach, J., and Okoniewska, E. (2008). Sequencing batch reactor system for the co-treatment of landfill leachate and dairy wastewater. *Desalination* 222: 404–409.

Omil, F., Garrido, J. M., Arrojo, B. and Mendez, R. (2003). Anaerobic filter reactor performance for the treatment of complex dairy wastewater at industrial scale. *Water Res.* 37: 4099–4108.

Onishi, M. C. (2002), The Best Treatment of Food Processing Wastewater Handbook, 351 (Science Forum).

Owen, J., Bandi, M., Howell, J. A., and Churchhouse, S. J. (1995). Economic assessment of membrane processes for water and wastewater treatment. *J. Membrane Sci.* 102: 77–91.

Prasad, M. P., Manjunath, K. (2011). Comparative study of biodegradation of lipid rich wastewater using lipase producing bacterial species, *Indian Journal of Biotechnology,* 10, 121-124.

Rajeshkumar, K. and Jayachandran, K. (2004). Treatment of dairy wastewater using a selected bacterial isolate, Alcaligenes sp. MMRR7. *Appl Biochem Biotechnol,* 118(1-3): 65-72.

Ramasamy, E. V., Gajalakshmi, S., Sanjeevi, R., Jithesh, M. N., Abbasi, S. A. (2004). Feasibility studies on the treatment of dairy wastewaters with upflow anaerobic sludge blanket reactors. *Bioresource Technol.* 93: 209 – 212.

Ramjeawon T. (2000). Cleaner Production in Mauritian Cane-sugar factories, Journal of Cleaner Production, 8 (6), 503-510(8).

Rao, M. and Bhole, A. G. (2002). Removal of organic matter from dairy industry wastewater using low-cost adsorbents, *J. Indian Chem. Eng. Section A,* 44(1): 25–28.

Rico, J.L., Garcia, P.A. and Fernandez-Polanco, F. (1991). "Anaerobic treatment of cheese production wastewater using UASB reactor". *Bioresource Tech.* 37, 271-276.

Rockey, J. S. and Forster, C. F. (1985). Microbial attachment in anaerobic expanded bed reactors. *Environ. Technol. Lett.* 6: 115–122.

Rusten, B., Lundar, A., Eide, O., and Odegaard, H. (1993). Chemical pretreatment of dairy wastwater. *Water Sci. Technol,* 28 (2): 67–76.

Sarkar, B., Chakrabarti, P.P., Vijaykumar, A., and Kale, V. (2006). Wastewater treatment in dairy industries — possibility of reuse. *Desalination* 195: 141–152.

Schroder, E. W. and De-Haast, J. (1989). Anaerobic digestion of deproteinated cheese whey in an up flow sludge-blanket reactor. *J. Dairy Res.* 56: 129–139.

Schwarzenbeck, N., Borges, J. M., and Wilderer, P. A. (2005). Treatment of dairy effluents in an aerobic granular sludge sequencing batch reactor. *Appl. Microbiol. Biotechnol.* 66: 711–718.

Selmer-Olsen, E., Ratanweera, H. C., and Pehrson, R. (1996). A novel treatment process for dairy wastewater with chitson produced from shrimp-shell waste. *Water Sci. Technol.* 11: 33–40.

Sengil, A. and Ozacar, M. (2006). Treatment of dairy wastewaters by electrocoagulation using mild steel electrodes. *J. Hazard. Mat.* B137: 1197–1205.

Shamir, E., Thompson, T. L., Karpiscal, M. M., Freitas, R. J. and Zauderer, J. (2001), "Nitogen accumulation in a constructed wetland for dairy wastewater treatment". *J. Am Water Resources Assoc,* 37, 315-25.

Sharma, P. (2008). Performance Evaluation of Wastewater Treatment Plant for Milk Based Food Industry, M. Tech Thesis submitted to Thapar University, Patiala.

Silambarasan, T., Vikramathithan, M., Dhandapani, R., Mukesh D. J. and Kalaichelvan, P.T. (2012). Biological treatment of dairy effluent by microalgae, *World Journal of Science and Technology,* 2(7): 132-134.

Singh, J. K., Meshram, R. L. and Ramteke, D. S. (2011). Production of Single cell protein and removal of 'COD' from dairy wastewater, *European Journal of Experimental Biology, 2011,* 1 (3): 209-215.

Sirianuntapiboon, S., Jeeyachok N., and Larplai, R. (2005). Sequencing batch reactor biofilm system for treatment of milk industry wastewater. *J. Environ. Management* 76: 177–183.

Tawfik, A., Sobhey, M., and Badawy,M. (2008). Treatment of a combined dairy and domestic wastewater in an up-flow anaerobic sludge blanket (UASB) reactor followed by activated sludge (AS system). *Desalination* 227: 167–177.

Tchobanoglous, G., Burton, F.L., and Stensel, H.D. (2003). Metcalf and Eddy Inc.- Wastewater Engineering Treatment and Reuse. Fourth Edition, Tata McGraw-Hill Publishing Company Limited, New Delhi, India. Torrijos, M., Sousbie, Ph., Moletta, R., and Delgenes J. P. (2004).

Timofeyeva, S.S. (1992). "Wastewater from dairy industry enterprises and modern methods of their decontamination", *J. Water Chem. Tech.,* 14, 43-9.

Timofeyeva, S.S. (1992). "Wastewater from dairy industry enterprises and modern methods of their decontamination", *J. Water Chem. Tech.,* 14, 43-9.

U.S. Department of Agriculture-Soil Conservation Service (USDA-SCS) (1992). Agricultural Waste Management Field Handbook. Washington, DC.

U.S. Environmental Protection Agency (USEPA). (1999). Wastewater technology fact sheet, Sequencing Batch Reactors. EPA 832–F-99-073.

Verheijen, L. A. H. M, Wiersema, D. and Pol, L. W. H. (1996), Management of Waste from Animal Product Processing, FAO Corporate Document Repository.

Verheyen, V., Cruickshank, A. and Wild, K. (2009). Soluble, semi-volatile phenol and nitrogen compounds in milk processing wastewaters, *J Dairy Sci* (In Press).

Vida, M., Akbar, S. and Zahra, G. (2007). Biodegradation of Effluents from Dairy Plant by Bacterial Isolates, *Iran. J. Chem. Chem. Eng., 26(1), 55-59.*

Vidal, G., Carvalho, A., Mendez, R., and Lema, J. M. (2000). Influence of the content in fats and proteins on the anaerobic biodegradability of dairy wastewaters, *Bioresource Technol,* 74, 231–239.

Viraraghavan, T. and Kikkeri, S. R. (1990). Effect of temperature on anaerobic filter treatment of dairy wastewater, *Water Sci. Technol,* 22: 191–198.

Viraraghavan, T. and Kikkeri, S. R. (1991). Dairy wastewater treatment using anaerobic filters, *Canadian Agri. Eng,* 33: 143–149.

Vourch, M. Balannec, B., Chaufer, B., and Dorange, G. (2005). Nanofiltration and reverse osmosis of model process waters from the dairy industry to produce water for reuse, *Desalination,* 172, 245–256.

Vourch, M. Balannec, B., Chaufer, B., and Dorange, G. (2008). Treatment of dairy industry wastewater by reverse osmosis for water reuse, *Desalination,* 219: 190–202.

Warner, J. N. (1976). Principles of dairy processing. Weley Eastern, New Delhi.

Watkins, M. and Nash, D. (2010), Dairy Factory Wastewaters, Their Use on Land and Possible Environmental Impacts – A Mini Review, *The Open Agriculture Journal, 4,* 1-9.

Wheatley, A. (1990). Anaerobic digestion: A waste treatment technology. Elsevier Applied Science, New York.

Wichern, M., Lubken, M., and Horn, H. (2008). Optimizing sequencing batch reactor (SBR) reactor operation for treatment of dairy wastewater with aerobic granular sludge. *Water Sci. Technol.* 58(6): 1199–1206.

Yan, J. Q., Lo, K. V., and Liao, P. H. (1989). Anaerobic digestion of cheese whey using up–flow anaerobic sludge blanket reactor. *Biological Wastes* 27: 289–305.

Yilmazer, G. and Yenigun, O. (1999). Two-phase anaerobic treatment of cheese whey. *Water Sci. Technol.* 40: 289–295.

Yu, H. Q. and Fang, H. H. P. (2002a). Acidogenesis of dairy wastewater at various pH levels. *Water Sci. Technol.* 45(10): 201–206.

Yu, H. Q. and Fang, H. H. P. (2002b). Anaerobic acidification of a synthetic wastewater in batch reactors at 55%C. *Water Sci. Technol.* 46(11–12): 153–157.

Zamora, A. Z. and Lit, M. A. L. (1995). "Biodegradation of Effluents from Dairy Products Processing Plant", Environmental Biotechnology: Principles and Applications, Kluwer, Academic Publishers, Printed in the Netherlands, 481-490.

2015, Dairy Product Technology: Recent Advances *Pages 251–282*
Editors: **Subrota Hati, Surajit Mandal and Birendra Kumar Mishra**
Published by: **DAYA PUBLISHING HOUSE, NEW DELHI**

Chapter 13

History of Fermented Functional Milk Products and their Biofunctionalities

Salma and Subrota Hati

Introduction

Humans began consuming cultured dairy products thousands of years ago. They likely present the earliest form of the unintentional food preservation. At some time, it was theorized, fermented milks were also regarded as having therapeutic value, even though there was no scientific basis for this notion (Shortt, 1999). After all, the existence of bacteria and their role in fermentation weren't even recognized until Pasteur's research in the 1860s.Then, at the beginning of the twentieth century, the Russian scientist Elie Metchnikoff (who was working at the Pasteur Institute in Paris) suggested that the health benefits of fermented milk were due to the bacteria involved in the fermentation. This hypothesis proposed that people consuming fermented milk regularly live longer, as lactic acid bacteria (LAB) ingested in the fermented milk colonize the intestine and inhibit putrefaction caused by harmful bacteria, thus slowing the ageing process. The longevity of peasants from the Balkans who regularly consumed fermented milk was cited as the evidence. The modern-day interest in the health effects of fermented milks is said to have been stimulated by this theory of longevity formulated by Metchnikoff. At around the same time and shortly thereafter, other scientists (as recounted by Shortt, 1999) reported that consumption of other lactobacilli (including *Lactobacillus acidophilus*) and bifidobacteria also had positive health effects, including reducing the rate of infant diarrhoea. These early reports, by highly regarded scientists and research institutes, attracted the attention of the medical community, and by the 1920s, studies using bacteria therapy (with milk as the carrier vehicle) had begun. Unfortunately, many of these subsequent studies suffered from the absence of established measurement criteria, the use of misidentified strains, and other design flaws (Shortt, 1999).

These Fermented products were easy to produce, had good shelf-lives, were free of harmful substances and had a pleasant sensory appeal. Milk was particularly suitable as a fermentation substrate owing to its carbohydrate-rich, nutrient-dense composition. Fresh bovine milk contains 5 per cent lactose and 3.3 per cent protein and has a water activity near 1.0 and a pH of 6.6 to 6.7, perfect conditions for most microorganisms.

Table 13.1: Milestones in the History of Fermented Dairy Foods

10,000 B.C. to Middle Ages	Evolution of fermentation from salvaging the surplus, probably by pre-Aryans
7000 B.C.	Cheese and bread making practiced
5000 B.C.	Nutritional and health value of fermented milk and beverages described
2000 B.C.–1200 A.D.	Different types of fermented milks from different regions
1907	Publication of book *Prolongation of Life* by Eli Metchnikoff describing therapeutic benefits of fermented milks
1900–1930	Application of microbiology to fermentation, use of defined cultures
1970–present	Development of products containing probiotic cultures or friendly intestinal bacteria

Fermented Milk — Definition

Campbell-Platt has defined fermented foods as those foods which have been subjected to the action of micro-organisms or enzymes so that desirable biochemical changes cause significant modification to the food.

The International Dairy Federation published general standards of identity for fermented milks that could be briefly defined as follows: Fermented milks are prepared from milk and/or milk products (*e.g.*, any one or combinations of whole, partially or fully skimmed, concentrated or powdered milk, buttermilk powder, concentrated or powdered whey, milk protein (such as whey proteins, whey protein concentrates, soluble milk proteins, edible casein and caseinates), cream, butter or milk fat—all of which have been manufactured from raw materials that have been at least pasteurized) by the action of specific microorganisms, which results in a reduction of the pH and coagulation.

Diversity and Classification of Fermented Dairy Products

Robinson and Tamime proposed a scheme for the classification of fermented milks into

- ✰ Lactic fermentations that include
  - (a) Mesophilic type, *e.g.*, cultured buttermilk;
  - (b) Thermophilic type, *e.g.*, yoghurt, Bulgarian buttermilk, dahi and

(c) Therapeutic or probiotic type, *e.g.*, acidophilus milk, Yakult;

products within this group constitute by far the largest number known worldwide;

- Yeast – lactic fermentations (kefir, koumiss, acidophilus yeast milk); and
- Mould – lactic fermentations (villi).

Certain, closely related products are manufactured from fermented milks by de-wheying; examples include labneh, skyr, ymer and shrikhand.

Table 13.2

Product	*Country of Origin*	*Period*	*Characteristics and Use*
Dahi	India	6000–4000 B.C.	Coagulated sour milk eaten as a food item; an intermediate product formaking butter and ghee
Chhash (Butter milk)	India	6000–4000 B.C	Diluted dahi or the butter milk left after churning of dahi into butter; used as beverage with meal
Laban zeer/ Khad	Egypt	5000–3000 B.C.	Sour milk, traditionally coagulated in earthenware vessels
Leben	Iraq	ca. 3000 B.C.	Traditional fermented milk containing yogurt bacteria; whey partially drainedby hanging the curd
Zabady	Egypt and Sudan	2000 B.C.	Natural type yogurt; firm consistency and cooked flavour
Cultured cream	Mesopotamia	1300 B.C.	Naturally soured cream
Shrikhand	India	400 B.C.	Concentrated sour milk, sweetened and spiced; semi-solid mass eaten with meals as sweet dish
Kishk	Egypt and Arab World	–	Dry fermented product made from Laban zeer and par boiled wheat; yellowish brown in colour with hard texture; highly nutritious with high amino acids and vitamin content
Kumys, Kumiss	Central Asia (Mongol, Russia)	400 B.C. (probably known around 2000 B.C.)	Traditionally mare's milk fermented by lactobacilli and yeast; sparkling beverage containing lactic acid, alcohol, and carbon dioxide
Mast	Iran	–	Natural type yogurt; firm consistency and cooked flavour
Villi	Finland	–	High viscosity fermented milk with lactic acid bacteria and mold
Taette	Norway	–	Viscous fermented milk also known as Cellarmilk
Langfil, Tattemjolk	Sweden	–	Milk fermented with slime-producing culture of lactococci
Ymer	Denmark	–	Protein fortified milk fermented by Leuconostocs and lactococci; whey is separated
Skyr	Iceland	870 A.D.	Made from ewes' milk by addition of rennet and starter; today concentrated by membrane technology
Prostok-vasha	Soviet Union	–	Fermented milk made from ancient times by fermenting raw milk with mesophilic lactic bacteria

Contd...

Table 13.2–*Contd...*

Product	*Country of Origin*	*Period*	*Characteristics and Use*
Kefir	Caucasu-sian China	–	Milk fermented with kefir grains; foamy effervescent product with acid andalcoholic taste
Yogurt (Kisle mliako)	Bulgaria	–	Cow's or ewe's milk fermented by *Str. thermophilus* and *Lb. bulgaricus*
Yogurt	Turkey	800 A.D.	Custard like sour fermented milk
Bulgarian milk	Bulgaria	500 A.D.	Very sour milk fermented by *Lb. bulgaricus* alone or with *Str. thermophilus*
Trahana	Greece	–	Traditional Balkan fermented milk; fermented ewe's milk mixed with wheat flour and then dried
Churpi	Nepal	–	Fermented milk is churned and the buttermilk remaining is heated to forma solid curd; may be further dried
Airan	Central Asia, Bulgaria	1253–1255 A.D.	Cow's milk soured by *Lb. bulgaricus*, used as refreshing beverage
Yakult	Japan	1935 A.D.	Highly heat treated milk fermented by *Lb. casei* strain Shirota; beverage and health supplement

Potential Benefits of Fermented Milk Products

Principal health effects of consuming fermented dairy foods are usually divided into two groups. One group refers to 'nutritional function', which is expressed as the function of supplying more nutrition efficiently. The other is 'physiological function', which includes the prophylactic and therapeutic functions beyond nutritional effects.

Nutritional Functions

Improvement of Protein Digestibility

Lactic acid bacteria require several amino acids, but there are not enough free amino acid molecules in milk to support their growth. They produce proteases to break down milk proteins and utilize the degradation products. The unused protein degradation products remain in the fermented milk as amino acids and peptides. As the pH decreases, caused by lactic acid formation from lactose, non degraded or partially degraded proteins become insoluble and form a gel. With these changes accompanying fermentation, the following observations have been reported regarding digestion and absorption of milk proteins. When yogurt is digested with artificial gastric juice, the size of the protein particles decreases and there are increases in nonprotein nitrogen and amino acids compared to the levels in non-fermented milk, suggesting that the digestibility will be increased by fermentation. It is an excellent source of essential amino acids such as tryptophan and cysteine. Tryptophan regulates appetite, sleep-waking rhythm, and pain perception. Cysteine is important in functions of –SH compounds.

Production of Bioactive Peptides

Functional peptides are generated during digestive processes in the body and during the fermentation processes used in fermented dairy foods. They arise from casein as well as from whey proteins. These peptides are inactive in the native proteins but assume activity after they are released from them. They contain 3 to 64 amino acids and largely display a hydrophobic character and are resistant to hydrolysis in the gastrointestinal tract. They can be absorbed in their intact form to exert various physiological effects locally in the gut or may have a systemic effect after entry into circulatory system. Casokinins are antihypertensive (lower blood pressure), casoplatelins are antithrombotic (reduce blood clotting), immunopeptides are immunostimulants (enhance immune properties), and phosphopeptides are mineral carriers.

The proteolytic system of milk bacteria consists of proteinase bound to the cell wall, and several intracellular peptidases. Lactic acid bacteria proteinases, such as those of *Lactococcus lactis* release biologically active oligopeptides from α- and β-caseins, which contain amino acid sequences that are present in casomorphines, lactorphines, casokinines and immunopeptides. They are peptides with two or more biological activities. Casomorphines and lactorphines have pharmacological characteristics similar to morphine (opiate-like effects), whereas lactoferroxins and casooxins act as opoid antagonists. The opoids have analgesic properties similar to aspirin. They act as analgetics; they stimulate excretion of some hormones, especially insulin and somatostatin; they prolong gastrointestinal resorption of nutrients; they modulate transport of amino acids in the intestine; and they act also as antidiarrhea agents (Brandsch *et al.*, 1994; Meisel 1997, 2005). These typical opiate peptides differ from endogenous opiates, such as enkephalines and endorphines, only in their N-end sequences (Teschemacher and Brantl, 1994).

The casein peptides also offer a promising role in regulating blood pressure. Conversion of angiotensin-I to angiotensin-II is inhibited by certain hydrolyzates of casein and whey proteins. Since Angiotensin-II raises blood pressure by constricting blood vessels, its inhibition causes lowering of blood pressure. This ACE inhibitory activity would therefore make dairy foods a natural functional food for controlling hypertension.

The glycomacropeptide released from casein as result of proteolysis may be involved in regulating digestion, as well as in modulating platelet function and thrombosis in a beneficial way. It is reported to suppress appetite by stimulating CCK hormone. Consequently, it may be a significant ingredient of satiety diets designed for weight reduction.

Source of Vitamins

Fermented dairy foods are an important source of complex B vitamins (particularly B_1, B_2, B_6 and B_{12}) as well as vitamin D. Fermented milks are also a good source of vitamin A and pantothenic acid. The level of fat-soluble vitamins, particularly vitamin A, is dependent on the fat content of the product. Some lactic bacteria are able to synthesize the B vitamin and folic acid and because of this reason vitamin content of yogurt in general is higher as starter bacteria especially *S. thermophilus* synthesize certain B group vitamins and folic acid during fermentation. Levels of some B vitamins, particularly vitamin B_{12}, are reduced due to requirement of this vitamin by some lactic acid bacteria for their growth.

- ☆ **B_2** (Riboflavin) - essential for releasing energy from nutrients, forms red blood cells, needed for tissue formation and required for maintaining healthy eyes and skin
- ☆ **B_{12}** - helps to form red blood cells and DNA, needed for the nervous system
- ☆ **B_6** - required for protein and red cell metabolism, immune and nervous systems functioning.

Enhancement of Mineral Absorption and Bioavailability of Calcium

One of the primary functions of calcium is to provide strength and structural properties to bone and teeth. The major source of dietary calcium is dairy products, supplying 75 per cent of the intake. Milk and dairy products are excellent sources of bioavailable calcium. Not only does milk contain more calcium than other foods, but also its absorption from milk has been considered to be superior to when the same amount of calcium is given to humans in other forms. Several possible explanations have been proposed for this, including the hypothesis that lactose or phosphopeptides released by the hydrolysis of casein act as absorption accelerators. Recent animal studies on the amount of calcium in bone, as well as bone weight and strength, showed that lactic acid is involved in the utilization of calcium. Also, increases in bone density and bone strength of femur were observed when administering *Bifidobacterium longum* bacterial powder to the osteoporotic model rat. These observations suggest that fermented milk can enhance calcium absorption.

Addition of lactic acid to unfermented yogurt, as well as regular fermented yogurt displays an improved bone mineralization as compared to the unfermented yogurt. It is postulated that the acidic pH due to added lactic acid or naturally contained in fermented yogurt converts colloidal calcium to its ionic form and allows its transport to the mucosal cells of the intestine.

Milk Fat

The digestibility of fat is also improved during fermentation. Milk fat is known for its high proportion of saturated fatty acids; advice is frequently given to avoid it because it contributes to an atherogenic blood profile and increased risk of coronary heart disease. However, one look at the composition of milk fat reveals that of the many different saturated fatty acids in milk only three (lauric, myristic and palmitic) have the property of raising blood cholesterol, and that at least one-third of the fatty acids are unsaturated, with a cholesterol-lowering tendency (Gurr, 1992). Furthermore, fermented milks contain components with at least protective if not hypoholesterolemic effects; these include calcium, linoleic acid, conjugated linoleic acid (CLA), antioxidants and lactic acid bacteria or probiotic bacteria (Rogelj, 2000). Milk fat contains a number of components such as CLA, sphingomyelin, butyric acid, ether lipids, carotene and vitamins A and D, with anti-carcinogenic potential (Jahreis *et al.*,

1999, Parodi, 1999a). Numerous *in vitro* and animal studies have confirmed the anti-carcinogenic activity of CLA, as well as its role in preventing atherosclerosis and in modulating certain aspects of the immune system (Cook and Pariza, 1998; MacDonald, 2000). It also supplies essential fatty acids including arachidonic acid, linolenic acid, omega 3-linoleic, eicosopentaenoic acid, and docosahexaenoic acid. The essential fatty acids cannot be synthesized by the body and must be supplied by our diet. The omega-3 fatty acids have a role in memory development and maintenance.

CLA is a strong antioxidant constituent of milk fat, and may prevent colon cancer and breast cancer. CLA has been shown to enhance immune response. Prostaglandin PGE-2 promotes inflammation, artery constriction and blood clotting. CLA may reduce the risk of heart disease by reducing the levels of prostaglandin PGE-2.

Another constituent of milk fat is sphingolipids. They occur at a level of only 160 g/kg. Studies have shown that they are hydrolyzed in the gastrointestinal tract to ceramides and sphingoid bases, which help in cell regulation and function. Studies on experimental animal show that sphingolipids inhibit colon cancer, reduce serum cholesterol, and elevate the good cholesterol HDL. They could protect against bacterial toxins and infections as well.

Physiological Functions

Alleviation of Lactose Intolerance

Lactose intolerance has a genetic basis, affecting African, Asian, American Indian and other non- Caucasian populations far more frequently than Caucasian groups. These individuals could typically tolerate lactose (*i.e.*, milk) when young, but lose this ability during adulthood. It is a condition that is characterized by the inability of certain individuals to digest lactose. Its specific cause is due to the absence of the enzyme β- galactosidase, which is ordinarily produced and secreted by the cells that line the small intestine. In individuals expressing β -galactosidase, lactose is hydrolyzed and glucose and galactose are absorbed across the epithelial cells and eventually enter into the blood stream (in the case of galactose, only after conversion to glucose in the liver). This beneficial action results from presence of β-galactosidase in the bacterial cells. Apparently being inside the bacterial cells protects the enzyme during passage through the stomach so that it is present and active when yogurt reaches the small intestine. Once the yogurt culture reaches the small intestine, it interacts with bile, which increases permeability of the cells of these bacteria and enables the substrate to enter and be hydrolyzed (Noh and Gilliland, 1993). If β-galactosidase is not produced in sufficient levels, however, the lactose remains undigested and is not absorbed. Instead, it passes to the large intestine, where it either causes an increase in water adsorption into the colon (via osmotic forces) or is fermented by colonic anaerobes. The resulting symptoms can include diarrhoea, gas, and bloating, leading many lactose intolerant individuals to omit milk and dairy products from their diet. As mentioned previously, the starter cultures used for yogurt manufacture (*Lb. delbrueckii* subsp. *bulgaricus* and *S. thermophilus*) are not bile resistant and thus are not expected to survive and grow in the intestinal tract. Despite this limitation, consumption of these bacteria provides a means of transferring β-galactosidase into the small intestine where it can improve lactose utilization in lactose maldigestors. Thus, the yogurt culture serves as an exogenous source of this enzyme, substituting for the β-galactosidase not produced by the host. In contrast, lactococci produce phospho-β galactosidase.The substrate of this enzyme is lactose-6-phosphate (the product of the lactose PTS), which can only be formed by intact cells.This enzyme does not hydrolyze free lactose. Thus, when lactococci are ingested and lysed in the small intestine, there will be little, if any impact on lactose digestion and lactose-intolerant individuals will unfortunately get no relief.

Probiotic Effect and Intestinal Health

One of the reasons for the increasing interest in fermented foods is its ability to promote the functions of the human digestive system in a number of positive ways. This particular contribution is called probiotic effect. Probiotics have been defined as living microorganisms, which upon ingestion in certain numbers have beneficial effects on human health beyond inherent general nutrition. These effects are attributed to the restoration of increased intestinal permeability and unbalanced gut microflora, improvement of the intestine's immunological barrier functions and alleviation of the intestinal inflammatory response. The human intestinal microbial flora is estimated to weigh about 1000 grams and may contain 10^{16} - 10^{17} colony forming units representing more than 500 strains. For physiological purposes, it can be considered to be a specialised organ of the body with a wide variety of functions in nutrition, immunology and metabolism.

Studies on mice have shown that the indigenous microorganisms in the stomach are *Lactobacillus, Streptococcus* and *Torulopsis* while in the small intestine, ceacum and colon several different species (*Bacteroides, Fusobacterium, Eubacterium, Clostridium,* etc.) coexist (Savage, 1983). Even though there is a wide variation among individuals, the number of species and size of the population are usually kept stable in normal healthy subjects. However, in patients with systemic insult like starvation, shock, injury and infection or specific insult of the gastrointestinal canal through inflammation, chemotherapy or radiation, the gut mucosal permeability will be increased leading to translocation of microbes (Carrico and Meakin, 1986; Alexander *et al.,* 1990; Wells, 1990; Kasravi *et al.,* 1997). A fermented food product or live microbial food supplement which has beneficial effects on the host by improving intestinal microbial balance is generally understood to have probiotic effect (Fuller, 1989).

Antioxidant Effects

Free radicals are produced in the body in the course of regular metabolism but when exposed to xenobiotic agents from foods and environment the risk of radical production significantly increases. The most important are the free radicals derived from oxygen. Provided the antioxidant system of the organism does not manage to neutralize them rapidly enough, they can cause destructive or lethal changes (such as apoptosis) through oxidation of membrane lipids, proteins, enzymes and DNA. The classical yoghurt bacteria *Lactobacillus delbrueckii* ssp. *bulgaricus* and *Streptococcus thermophilus* inhibit peroxidation of lipids through scavenging the reactive oxygen radicals, such as hydroxyl radical, or hydrogen peroxide (Ling and Yen, 1999).

The antioxidant activity of several species and strains of milk bacteria contained in fermented milk can significantly affect human health. This has been confirmed also by clinical studies of goat milk fermented with a starter culture *Lactobacillus fermentum* ME-3 (Kullisaar *et al.*, 2003). The healthy volunteers have consumed for 21 days each day 150 g of milk, either sour or non fermented. Sour milk compared to non fermented milk has shown important improvement of the overall antioxidant activity of blood, as well as antioxidant status, prolonged resistance of lipoprotein fraction to oxidation, reduced level of peroxide lipoproteins and oxidized LDL cholesterol, reduced level of glutathione redox ratio, and increased overall antioxidant activity.

Some lactobacilli produce antioxidant factors also in the human GI tract (Ljungh *et al.*, 2002). The majority of milk bacteria show antioxidant behavior (eliminating the excess oxygen free radicals) producing superoxide dismutase, or glutathione. Perhaps the very first of such targeted products is the Estonian cheese Picante, in the manufacturing of which also *Lactobacillus fermentum* ME-3 with significant antioxidant and antimicrobial effects is added to the starting culture mix (Songisepp *et al.*, 2004).The antioxidant activity is also exerted by various peptides derived from α-lactalbumin, β-lactoglobulin and α-casein (Fitzgerald and Murray, 2006).

Lowering Serum Cholesterol

Cholesterol is an essential component of cell membranes and is used to produce some hormones and bile acids. It is synthesized in the liver and skeletal muscle and also supplied by absorption through the alimentary tract from ingested foods. Blood cholesterol level is controlled by a complex mechanism. Cholesterol is supplied to each organ through the bloodstream as lipoprotein and also transported to the liver from each organ, to be finally excreted as bile acids.

High serum cholesterol is a risk factor in heart disease, which is one of the main causes of death in developed countries. It is believed that the arterial sclerosis that proceeds to cardiac infarction is caused by cholesterol accumulation in the blood vessel wall. Investigations of the cholesterol-reducing potential of fermented milk are triggered by the observation that the Masai people of East Africa have low blood cholesterol levels although they consume much milk and meat. Fermented milk with surfactant, which accelerates fat absorption, was given to them. Although their weight increased from consuming fermented milk, their blood serum cholesterol decreased. As a possible mechanism for lowering serum cholesterol level by fermented milks, a substance hydroxymethylglutarate by probiotic bacteria, which inhibits hydroxymethyl-glutaryl CoA reductase, an important enzyme in cholesterol synthesis in the body is known. It has also been reported that in rats serum cholesterol was lowered and cholesterol synthesis was inhibited by a methanol extract of milk fermented by *Streptococcus thermophilus*.

Deconjugation of bile acids by lactobacilli can occur in the small intestine. *Lb. acidophilus* more actively deconjugates glycocholic acid than it does taurocholic acid (Corzo and Gilliland, 1999) this becomes significant because the dominant conjugated bile acid in the human intestine is glycocholic acid. Free bile acids are less well absorbed in the small intestine than are conjugated bile acids and thus more are excreted through faeces (Chickai *et al.*, 1987). Excretion of bile acids through faeces represents one of the major mechanisms whereby the body eliminates cholesterol. This is because cholesterol is a precursor for synthesis of bile acids and many bile acids that are excreted from the body are replaced by synthesis of new ones. Thus, there is a potential for reducing the cholesterol pool in the body. Furthermore, free bile acids do not support absorption of cholesterol from the intestinal tract as well as do conjugate ones (Eyssen, 1973). Thus, deconjugation of bile acids in the intestinal tract may reduce the efficiency by which cholesterol is absorbed from the intestinal tract, attempts are made to select a strain with an ability to deconjugate bile acid or a strain that directly decreases absorption of cholesterol in the intestine, by assimilating or adsorbing cholesterol to the bacterial cells. In addition, viscous exopolysaccharides produced by *Lactococcus lactis* subsp. *cremoris* in fermented milks have been suggested to interfere with absorption in a similar manner to dietary fibre. Studies on supplementation of infant formula with *Lb. acidophilus* showed that the serum cholesterol in infants was reduced from 147 mg/ml to 119mg/100 ml (Harrison and Peat, 1975).

Antihypertensive Effects

High blood pressure is another risk factor in cardiovascular diseases. Blood pressure is controlled by complicated neural and humoral factors, which are interrelated and form an auto-control system. Among them, the renin–angiotensin system plays a particularly important role. Renin, a proteolytic enzyme secreted from the kidney, acts on angiotensinogen to generate angiotensin I. The two C-terminal amino acid residues of angiotensin I are cleaved off by the angiotensin converting enzyme (ACE) to generate an octapeptide, angiotensin II, which raises blood pressure.

ACE inhibitors are widely used clinically and have been proven to work effectively in hypertensive patients. Therapeutic effects on cardiovascular diseases have been attributed to fermented milks since ancient times. They were used for high blood pressure, diabetes and heart disease. Yogurt is included in the group of foods that exhibit relatively strong ACE inhibitory activity. Also, some cheese or peptides obtained by proteolytic breakdown of milk proteins have been reported to show ACE inhibitory activity.

When hypotensive effects of several fermented milks made with various LAB were compared by feeding these milks to spontaneously hypertensive rats (SHR), only milk fermented by *Lb. helveticus* was observed to decrease blood pressure, along with a strong ACE inhibitory activity. The stronger proteolytic activity of *Lb. helveticus*, in comparison to other LAB, was thought to be related to the hypotensive effect of milk fermented by *Lb. helveticus.*

When a starter culture with *Lb. helveticus* was added to skim milk, ACE inhibition increased during the process of fermentation, suggesting that an ACE inhibitor(s) was/were generated. The active compounds were purified by using high performance liquid chromatography and identified as two tripeptides; Val-Pro-Pro (VPP) and Ile-Pro-Pro (IPP). Most of the ACE inhibitory activity in the fermented milk has been attributed to these two tripeptides, which were generated from β-casein by the action of proteinase and several peptidases of *Lb. helveticus*. Furthermore, there was lowered ACE activity in the aorta of SHR rats that were orally given fermented milk, and these tripeptides were detected in the aorta. It has thus been proved that peptides taken orally are absorbed in the intestine and produce hypotensive effects by inhibiting ACE activity in the body.

In the clinical studies, ingesting 95 ml of fermented milk daily for 8 weeks significantly lowered the systolic and diastolic blood pressure of hypertensive patients. Based on this evidence, a beverage using fermented milk as its main ingredient has been approved as a 'Food for specified health use' (FOSHU) in Japan.

Anticancer Effects

Research in Netherlands comparing breast cancer patients and healthy people showed that patients with breast cancer consume less fermented milk, suggesting that fermented milk may prevent cancer. As an experimental animal model of human cancer, rats were given dimethylhydrazine (DMH) to induce colon cancer and fed with fermented milk or selected LAB. It has been reported that feeding *Lb. acidophilus* bacteria delayed the onset of cancer, and that milk soured by *Lb. helveticus* or yogurt decreased cancer incidence. The following possible mechanisms for the inhibitory effect on cancer onset have been proposed:

1. Decrease of Mutagenic Activity

The active element was either casein, which is a major component of milk, or the cell wall of *E. faecalis*, and further research is needed to elucidate the mechanisms of action. Furthermore, oral administration of mutagens such as DMH trigger DNA damage in intestinal cells of rats, but feeding yogurt or LAB prevents the DNA damage.

2. Modification of Intestinal Microflora

In rats and humans who are on a predominantly meat diet, considered to be a high-risk diet for colon cancer, high activities of such enzymes as β-glucuronidase, nitroreductase, azoreductase and steroid-7-dehydroxylase, that are all linked to the generation of carcinogenic compounds by converting procarcinogens into carcinogens have been observed. However, the activity of these enzymes was lowered by supplementing *Lb. acidophilus* or acidophilous milk. Together with results showing inhibitory effects on chemical carcinogenesis in rats, administration of live *Lb. acidophilus* is suggested to be effective in inhibiting colon cancer. Short-chain fatty acids produced by *L. acidophilus* and bifidobacteria were also reported to inhibit the generation of carcinogenic products by reducing enzyme activities.

3. Suppression of Cancer Cell Growth

In an early anticancer study with use of LAB, the inhibitory effect of parenterally administered *Lb. bulgaricus* cell wall glycopeptide was reported. Later, in oral administration studies, several types of fermented milks, such as yogurt, cow colostrum fermented by *Lb. acidophilus* or *Lb. bulgaricus* plus *S. thermophilus*, milk fermented with *Lb. helveticus*, etc., have been found to have an inhibitory effect on cancer cell growth. Several factors have been reported as effective in suppressing the growth of cancer cells. These include the anion fraction of yogurt dialysate separated by ion exchange chromatography, a fraction separated by ion chromatography of the supernatant of milk fermented by *Lb. bulgaricus*, LAB cells in yogurt such as *Lb. delbrueckii* subsp.

bulgaricus, *S. thermophilus* and *Lb. helveticus* subsp. *jugurti*, polysaccharides made by *Lb. helveticus* subsp. *jugurti*, and kefir. These fermented milks or effective compounds in them are thought to work primarily on the immune system. Cancer cells can be considered foreign to the organism, and their growth is usually suppressed by the immune surveillance system. Those that escape this system proliferate and develop cancer.

Prevention and Reduction of Diarrhoea Symptoms

One of the main applications of probiotics has been the treatment and prevention of antibiotic-associated diarrhoea, which is often caused by occurrence of *C. difficile* after an antibiotic treatment. *C. difficile* is an indigenous gastrointestinal organism usually encountered in low numbers in the healthy intestine; however, the antibiotic treatment may lead to a disruption of indigenous microflora and subsequently to an increase in the concentration of this organism and toxin production, which causes symptoms of diarrhoea. The administration of an exogenous probiotic preparation is required to restore the balance of the intestinal microflora (Sazawal *et al.*, 2006). The strongest evidence of a beneficial effect of defined strains of probiotics has been established for *L. rhamnosus* GG and *B. animalis* Bb-12. Administration of oral rehydration solution containing *Lactobacillus* GG to children with acute diarrhoea resulted in a reduction of the duration of diarrhoea, lower chance of a protracted course and faster discharge from the hospital (Guandalini *et al.*, 2000).

A competitive exclusion is the mechanism by which probiotics inhibit the adhesion of rotavirus by modifying the glycosylation state of the receptor in epithelial cells via excreted soluble factors (Freitas *et al.*, 2003). The presence of probiotics also prevents the disruption of the cytoskeletal proteins in the epithelial cells caused by the pathogen, which leads to the improved mucosal barrier function and prevention of the failure in the secretion of electrolytes.

Effect on Immunological Function

The immune system not only defends the body against bacterial and viral infection, but also plays a role in many diseases such as cancer, allergies and autoimmune diseases. In addition to the effect on cancer discussed above, consumption of fermented milks has been reported to inhibit infection. Studies on mice administered Salmonellae to develop intestinal infections or on mice infected in the nasal cavity with *Klebsiella pneumoniae* to cause pneumonia demonstrate that mice given fermented milk lived longer. The translocation of *Candida albicans* to the liver in immunosuppressed mice was also suppressed. Alleviation of allergies by fermented milks has also been reported.

LAB can affect functions of immune cells; for example, activation of macrophages and 'natural killer' (NK) cells have been observed. Activation of these cells was reported in relation to anticancer properties of parenterally administered *Lb. casei*. Effect of oral administration was shown by administering *Lb. delbrueckii* subsp. *bulgaricus* or *Lb. Casei* and observing the increase in phagocytosis activity and lysozome release by peritoneal macrophages. In the study in which an increase in the numbers of surviving mice by administering fermented milk after nasal infection with *Kl. pneumoniae* was demonstrated, a concomitant increase in phagocytosis activity by pulmonary macrophages was observed.

The effects of lactic acid bacteria and fermented milk on cytokines, which regulate immune responses by mediating information between cells, have also been reported. Parenteral administration of *Lb. casei* to mice increased the serum levels of colony growth stimulating factor (CSA) which takes part in macrophage differentiation as well as those of -interferon which activates NK cells and macrophages. An increase in interferon production *in vitro* by human peripheral lymphocytes has been observed by the addition of yogurt. In a human clinical study, an increase in the serum level of -interferon as well as in the NK cell count was observed after the ingestion of yogurt or LAB used in yogurt, compared to unfermented skim milk.

Inhibition of *Helicobacter pylori* and Other Intestinal Pathogens

H. pylori is an intestinal pathogen, which causes peptic ulcers, type B gastritis, and chronic gastritis. It resides in the stomach as an opportunistic pathogen without causing any symptoms. An increased density of *H. pylori* on the gastric mucosa is associated with more severe gastritis and an increased incidence of peptic ulcers. One of the measures which may help reduce rate of infection is a diet modulation with the inclusion of probiotics (Khulusi *et al.,* 1995). Antibiotic treatments are successfully used to eradicate *H. pylori.* However, some side effects are usually encountered including antibiotic-associated diarrhoea and likelihood of induction of the antibiotic resistance in the intestinal pathogens. Probiotic organisms do not appear to eradicate *H. pylori,* but they are able to reduce the bacterial load in patients infected with *H. pylori.*

L. casei Shirota and *L. acidophilus* were able to inhibit the growth of *H. pylori.* In an intervention study, 14 patients infected with *H. pylori* received *L. casei* Shirota (2×10^{10} CFU per day) fermented milk for 6 weeks. Ureolytic activity was reduced in 64 per cent of the patients who consumed fermented products containing probiotics, compared to 33 per cent of the control group (Cats *et al.,* 2003). Several mechanisms in regard to the effect of probiotics on *H. pylori* have been suggested including production of antimicrobial substances, enhanced gut barrier function and competition for adhesion sites.

Prevention of Oral Disease

Dental caries and periodontal disease are major public health problems that bother all countries in the world. Dental caries is an infectious, communicable disease that acid-forming bacteria of dental plaque can destroy tooth structure in the presence of fermentable carbohydrates such as sucrose, fructose and glucose. The mineral content of teeth is sensitive to increases in acidity by the production of lactic acid. So, the infection results in loss of tooth minerals from the outer surface of the tooth and can progress through the dentin to the pulp, finally compromising the tooth vitality or in short caries lesions result from interactions of odontopathogenic bacteria that colonize the tooth surface. These bacteria utilize dietary sugars to produce mutans and organic acids, which in turn demineralize calcium and other cations from the tooth's enamel. The body, however, counteracts demineralization by the salivary protein statherin binding calcium to remineralize the tooth's surface. Simply, dental caries worsen if odontopathogenic bacteria overcome the body's ability to remineralize the tooth.

Milk bioactive peptides such as caseinophosphopeptides and glycomacropeptides have shown to exert an inhibitory effect on cariogenic bacteria such as *Streptococcus mutans* and other species and are used in common personal hygiene products to prevent dental caries. Also to have a beneficial effect in oral cavity, a probiotic should have a tendency to form a biofilm that acts as a protective lining for oral tissues against oral diseases. Probiotics strains have been shown to vary broadly in their adhesiveness to saliva-coated HA and so in biofilm formation ability. Among probiotics strains *L. rhamnosus* GG exhibited the maximum values of adhesion. Extra attention should be paid in the selection of strain if a prebiotic product should be administered since some strains such as *Lactobacillus salivarius* may have cariogenic properties.

Periodontal Diseases

The first studies of the use of probiotics for enhancing oral health were for the treatment of periodontal inflammation. Patients with various periodontal diseases, gingivitis, periodontitis and pregnancy gingivitis, were locally treated with a culture supernatant of a *L. acidophilus* strain (Kragen H, 1954). Sig-nificant recovery was reported for almost every patient. The probiotic strains used in these studies include *L. reuteri* strains, *L. brevis* (CD2), *L. casei* Shirota, *L. salivarius* WB21, and *Bacillus subtilis*. *L. reuteri* and *L. brevis* have improved gingival health, as measured by decreased gum bleed-ing (Krasse *et al.*, 2006).

Oral Candida

Only two studies have investigated the effects of probiotic bacte-ria on oral *Candida* infection in humans. When a test group of elderly people consumed cheese containing *L. rhamnosus* strains GG and LC705 and *Propionibacterium freudenreichii ssp. shermanii* JS for 16 weeks, the number of high oral yeast counts decreased, but no changes were observed in mu-cosal lesions (Hatakka *et al.*, 2007).

One promising example is the generation of an *S. mutans* strain with a complete deletion of the open reading frame of lactate hydrogenase and thus significantly reduced cariogenicity (Hillman *et al.*, 2007).

Weight Management and Obesity

A variety of epidemiological, clinical, animal, and *in vitro* investigations provide evidence that dairy products play a role in controlling body weight and enhancing fat loss. This beneficial effect is mainly observed with the consumption of low-fat dairy products such as milk and yogurt and is specifically attributed to their calcium content.

A significant inverse relationship between dietary calcium and body fat was reported in a 5-year longitudinal study of preschool children. There are two mechanisms by which calcium facilitates weight and fat loss. The calcitrophic hormone 1, 25 dihydroxy-vitamin D is increased when dietary calcium is low. The elevated levels of 1, 25 dihydroxy-vitamin D favor the entrance of calcium in adipocytes and this increases lipogenesis, reduces lipolysis, and promotes fat storage. Conversely, when dietary calcium is high, calcium levels in adipocytes decline, lipogenesis is reduced, lipolysis is facilitated, and fat loss is promoted (Zemel, 2002). Another mechanism whereby dietary calcium intake may reduce body adiposity is by inhibiting fat absorption from the gastrointestinal tract and increasing faecal loss of fatty acids and energy through the formation and excretion of calcium–fatty acid soaps.

Recent Trends in Fermented Milk Product Development

Ever-growing consumer demand for convenience, combined with a healthy diet and preference for natural ingredients has led to a growth in functional food markets. Current trends and changing consumer needs indicate a great opportunity for innovations and developments in fermented milks. Scientific and clinical evidence is also mounting to corroborate the consumer perception of health from fermented milks. Probiotics, prebiotics, synbiotics and associated ingredients also add an attractive dimension to cultured dairy products. Also, owing to expanding market share and size of dairy companies, there has been a reduction of clearly structured markets *i.e.* merging of dairy products and fruit beverage markets with introduction of 'juiceceuticals' like fruit-yogurt beverages that are typical example of hybrid dairy products offering health, flavour and convenience. Another potential growth area for fermented milks includes added-value products such as low calorie, reduced-fat varieties and those fortified with physiologically active ingredients including fibers, phytosterols, omega-3-fatty acids, whey based ingredients, antioxidant vitamins, isoflavones that provide specific health benefits beyond basic nutrition. World over efforts have been devoted to develop fermented milks containing certain nonconventional food sources like soybeans and millets and convert them to more acceptable and palatable form thus producing low cost, nutritious fermented foods especially for developing and underdeveloped nations where malnutrition exists. Furthermore, use of biopreservatives and certain innovative technologies like membrane processing, high pressure processing and carbonation lead to milk fermentation under predictable, controllable and precise conditions to yield hygienic fermented milks of high nutritive value.

1. Product Development Strategies and Potential

Nutritionally improved foods with at least one nutritional improvement over the conventional counterpart have been successful in the marketplace (Duncan, 1998). In addition to basic technologies, modern processes lead to milk fermentation under predictable, controllable and precise conditions to yield hygienic fermented dairy products of high nutritive value (Kurmann, 1984). It appears that accentuating the positive attributes of inherent milk constituents, incorporating health-promoting cultures and offering a variety of flavours and textures to the consumer could enhance fermented milk consumption (Rudrello, 2004). Product modification strategies include removal or reduction of fat, cholesterol, sodium and calories and fortification with vitamins, calcium, fiber, active cultures and other physiologically active ingredients to align with health perceptions of consumers (Chandan, 1999).

Probiotics, Prebiotics and Synbiotics

Probiotics, prebiotics and associated ingredients might add an attractive dimension to cultured dairy products for augmenting current demand for functional foods. Probiotic fermented milks, is one major segment amongst fermented milks that has tremendous potential for growth and development (Rudrello F, 2004). The definition of probiotics proposed by Guarner and Schaafsma, 1998 is largely adopted, which is as follows: *Oral probiotics are living microorganisms, which upon ingestion in certain numbers exert health benefits beyond inherent basic nutrition.* Although, probiotics have been with us for as long as people have eaten fermented milk, but their association with health benefits dates only from the turn of the last century, when Metchnikoff drew attention to the adverse effects of some gut microflora on the host, and suggested that ingestion of fermented milk ameliorated this so-called autointoxication (Anuradha *et al.,* 2005).

The importance of these probiotic-containing products, commonly regarded as functional foods, in the maintenance of health and well-being is becoming a key factor affecting consumer choice (Maragkoudakisa *et al.,* 2006). This has resulted in rapid growth and expansion of the market for such products, in addition to increased commercial interest in exploiting their proposed healthful attributes. Fermented milks, such as yogurt and buttermilk have received the most attention in this regard (Gardiner *et al.,* 1998).

Prebiotics is another important aspect linking gut health and probiotics. This is a relatively new concept in the dairy products market. A prebiotic is a non digestible food ingredient that beneficially affects the host by selectively stimulating the growth, activity or both. The health effect of a prebiotic therefore resembles that of a probiotic. For a food ingredient to be classified as a prebiotic, it must:

- ☆ Neither be hydrolyzed nor absorbed in the upper part of the gastrointestinal tract
- ☆ Be a selective substrate for one or a limited number of potentially beneficial commensal bacteria in the colon, thus stimulating the bacteria to grow, become metabolically activated or both, and
- ☆ Be able as a consequence to alter the colonic microflora toward a healthier composition (Chandan, 1999).

Fructooligosaccharides (FOS) are the only products presently recognized and used as food ingredients that meet these criteria (Hartemink *et al.,* 1999). Experimental evidence suggests that certain other carbohydrate based components such as transgalactosylated dissaccharides and soybean oligosaccharides may also fit this classification.

Another possibility in microflora management procedures is the use of synbiotics, in which probiotics and prebiotics are used in combination (Gibson *et al.,* 1995). The living microbial additions (probiotics) may be used in conjunction with specific substrates (prebiotics) for growth (*e.g.,* an FOS in conjunction with a bifidobacterial strain or lactitol in conjunction with a *Lactobacillus* organism). This combination could improve the survival of the probiotic organism, because its specific substrate is readily available for its fermentation, and result in advantages to the host that the live microorganism and prebiotic offer. Probiotics, prebiotics and synbiotics that may be suitable for human consumption can be incorporated into various fermented milks.

Use of Biothickeners — EPS Cultures

Exopolysaccharides (EPS) synthesized LAB play a major role in the manufacturing of fermented dairy products such as yogurt, cheese, fermented cream, milk based desserts. EPS are high molecular weight carbohydrates composed of a backbone of repeated subunits of monosaccharides (Cerning, 1990). D-galactose, D-glucose and L-rhamnose are almost always present, but the ratios vary considerably. Interest in EPS producing LAB has increased recently because these food grade organisms produce polymers important in determining the rheological properties of dairy products (Cerning, 1990; Duboc *et al.,* 2001). When added to food products, polysaccharides function as thickeners, stabilizers, emulsifiers, gelling agents and water binding agents (De Vuyst and Degeest, 1999). EPS may act both as texturizers and stabilizers, firstly increasing the viscosity of a final product, and secondly by binding hydration water and interacting with other milk constituents, such as proteins and micelles, to strengthen the rigidity of the casein network. As a consequence EPS can decrease syneresis and improve product stability (Rimada *et al.,* 2003).

Furthermore, it has been reported that some of these EPS materials contain gluco- and/or fructooligosaccharides and may generate short-chain fatty acids upon hydrolysis in the intestinal tract by the colonic microflora, and may have potential health (*e.g.,* anti-tumor, cholesterollowering or immunomodulatory effects) and nutritional benefits as a prebiotic to the intestinal microflora (Nagaoka *et al.,* 1994; Oda *et al.,* 1983). At present isolated strains of *Lactococcus* species and thermophilic LAB are used extensively in the manufacture of fermented milks and many EPS producing LAB have been studied extensively (Cerning *et al.,* 1992; Cerning, 1995; Sikkema *et al.,* 1998). They are of commercial interest as they act as biothickeners and aid in enhancing texture, mouthfeel and stability of the product. EPS producing cultures have been successfully used for the manufacture of Nordic ropy milk (Neve *et al.,* 1988). Scandinavian fermented milk drinks display a firm thick, slimy consistency and these rely on the souring capacity of mesophilic ropy strains of *Lactococcus lactis* subsp. *lactis* and subsp. *cremoris* and concomitant production of heterotype EPS for texture (Gamer *et al.,* 1997). Kumar P, 2000 determined that *dahi* prepared using EPS culture had better body and texture and exhibited little syneresis. Pandya, 2002 reported that EPS producing cultures significantly improved the rheological properties of *dahi* and reduced syneresis. Folkenberg *et al.,* 2005 produced set yoghurts with seven different exopolysaccharide-producing starter cultures and observed that yoghurts in which the EPS were associated with protein had high ropiness, low serum separation and appeared more resistant to stirring. A better understanding of the structure-function relationship of EPS in a dairy food matrix remains a challenge to further improve applications of EPS to better satisfy the consumer demand for appealing, tasty and even healthier products.

Low Calories/Low Fat Fermented Milks

With billions of dollars spent on dieting each year, consumer's desire for nutritious low calorie dairy products continues to grow (Kantor, 1990) and consumption of low - or non fat dairy products has increased in recognition of their health benefits, and consumer's health problems (Haque *et al.,* 2003). Consumers of low-cal foods no longer accept compromise and with upto- date sweetening methods, the food industry can manufacture low-cal cultured milk products, which satisfy highest organoleptic demands. Fat free and low-fat formulations for yogurt have earned highly acceptable place in consumer lifestyles that are seeking fat and cholesterol reduction.

Although, the manufacture of low or nonfat dairy products has been possible for many years, the use of fat replacers in the manufacture of dairy products is still novel. Fat replacers, which decrease the calorific value of food, can be used to solve some physical and organoleptic problems originating from low-fat levels in the final products (Guven *et al.,* 2005). Even though yoghurt does not have a high fat content compared with cheese and ice cream, fat replacers have been used to reduce the fat content of yoghurt (Tamime *et al.,* 1994; Barrantes *et al.,* 1994; Tamime *et al.,* 1996).Various fat replacers and replacer blends that have been used to produce low fat/low calorie cultured dairy products and the technically developed fat substitutes are divided into two main types: modified starches or proteins which have good emulsifying or gel properties along with low calorie values; and modified fat/oil based products that contain bonds resistant to digestion, *e.g.,* glycerol ethers and complex carbohydrates or fatty acid esters. In one study, low calorie yoghurts were made from reconstituted Skimmed milk powder (SMP) and seven types of starch based commercial fat substitutes (LitesseTM – improved polydextrose, N – OilR II, LycadexR 100 and 200 – maltodextrin, PaselliRSA2, and P-fibre 150C and 285F) added at the rate of 1.5 per cent, and these were compared with the control made with anhydrous milk fat (AMF) (Barrantes *et al.,* 1994a; 1994b). Guven *et al.,* 2005 concluded that yogurt samples containing 1 per cent of inulin showed similar characteristics to the control yogurt containing 3 per cent of milk fat. However, increased use of inulin in fat-free yogurt negatively influenced some physical properties of yogurt, *i.e.* whey separation, consistency and organoleptic scores.

Non-nutritive sweeteners can also be used to impart an attractive calorie reduction in fermented milks. Aspartame and Acesulfame-K individually and in combination have been used as sweeteners for the formulation of numerous low-cal and sugar free yoghurts, milk beverages, whey based beverages and cultured milk products (Botma, 1988; Best, 1989; Lotz *et al.,* 1992). Ten different plain and fruit yoghurts were prepared using sucrose, aspartame, sorbitol, calcium saccharine, sodium saccharine, fructose, acesulfame-K, dihydrochalcone, sucrose plus monoammonium glycyrrhizinate (MAG) and fructose plus MAG as sweeteners wherein aspartame yoghurt was reported to be the most preferred on the basis of consumer panel (Keating *et al.,* 1990). Lotz *et al.,* 1992 determined that acesulfame-K yoghurt permits stable sweetening during fermentation phase however aspartame is degraded to some extent. The recommended level of each type of sweetener in strawberry yoghurt was 0.016g/ 100gm. Farooq and Haque, 1992 assessed the influence of sugar esters (.05 per cent) of various hydrophilic-lipophilic balances on the textural properties of nonfat, low calorie yogurts over 14 day storage at 4°C. Aspartame was used (200 ppm) to sweeten a skim milk based yogurt that was stabilized with starch (5 per cent). It was observed that sugar esters improved the overall quality of the yogurt. Kumar M, 2000 advocated that 0.08 per cent Aspartame on curd basis was most acceptable level to prepare low-cal *Lassi*. Also no effect on pH and acidity of *Lassi* was seen however there was a remarkable decrease in viscosity after Aspartame addition.

Fruits and Fermented Milks – Product Diversification

Recently there has been an increased trend to fortify cultured milk products with fruit juices/pulps. Owing to expanding market share and size of dairy companies, there has been a reduction of clearly structured markets *i.e.* merging of dairy products and fruit beverage markets with introduction of 'juiceceuticals' that include products like fruit-yogurt beverages (Litcher, 2001). Addition of fruit preparations fruit flavors, and fruit purees has enhanced versatility of flavor, texture, colour, variety to fermented milks and additionally fruits also have a healthy image. The association of fruits with cultured dairy products has endorsed healthy perception even more in the consumer mind. Categories of fruits and cultured milk products are typical example of hybrid dairy products offering health, flavor and convenience that will drive growth in coming years (Veeneman, 1999). Keeping in view the market trends, incorporation of fruits in traditional fermented milk products not only aids in value addition and product diversification but also helps in checking the post harvest losses and hence economic loss. It may also enhance the profitability of milk and fruit producers as well as processors. Fruits are rich sources of various important phytonutrients namely, vitamins, minerals, antioxidants and dietary fibers. Processed fruits are more widely employed; they may be added to cultured milk in various forms namely fruit purees, fruit pieces, fruit syrup/juices, crushed fruit, frozen/osmodehydrofrozen fruits, fruit preserves and other miscellaneous fruit products.

Suitability of different fruits *i.e.* mango, sapota, papaya, pineapple, kokum @ 10, 15, 20 per cent levels each was studied for preparation of fruit yoghurt and it was concluded that that mango pulp and pineapple juice could be used satisfactorily up to 20 per cent level. However, sapota pulp, papaya pulp and kokum juice produced inferior quality yoghurt (Desai *et al.*, 1994). Fruit *dahi* was prepared using mango, banana, pineapple and strawberry @ 6. 8, 6 and 4 percent levels each and mango fortified *dahi* was found to be most acceptable on basis of organoleptic quality (Pandya, 2002). Fruit based *shrikhand* has also been prepared using fruits like apple, papaya, mango (Bardale *et al.*, 1986). Coconuts have also been employed in yoghurt production wherein four types of yoghurts were made from mixtures of cow milk and coconut milk in different combinations. Using coconut milk in yoghurt production could be an interesting alternative option in the regions with high coconut production (Imele and Atemnkeng, 2001).

Whey Based Fermented Milks

Whey is a byproduct from cheese and casein production. It is an important source of lactose, calcium, milk proteins and soluble vitamins, which make this product to be considered as a functional food and a source of valuable nutrients.

Whey products have certain essential amino acids, good digestibility, and protein efficiency index higher than 3.0. Vitamins such as thiamin, riboflavin, pantothenic acid, vitamin B_6 and B_{12} are also present. Functional properties of whey proteins, such as emulsifying, water/fat holding, foaming, thickening and gelling properties, also make them interesting to be used as a food ingredient (Hall and Iglesias, 1997). Due to their functional properties, whey solids/whey as such could be used in conjunction with fermented milks (Huffmann, 1996).

Several studies have focused on the use of milk whey in yoghurt making and use of whey powder or whey–milk powder mixtures. This process leads to the increase of milk total solid content in order to provide better consistency, texture and creaminess to the product. In other studies, replacement of skimmed milk by whey protein concentrates (WPC) and milk protein concentrates (MPC) was studied. Thus, yoghurts with different mineral and protein composition were obtained. It was observed that these components are of decisive importance in the fermentation and gelling process and also in the type of gel obtained. However, yogurt microorganisms should be plenty and alive in the final product.

Good quality whey based fermented milk drink containing 2.5 per cent fat and 10 per cent sugar was prepared by Otero *et al.,* 1995. Macedo *et al.,* 1999 prepared low cost, probiotic whey milk beverage using buffalo milk cheese whey, cow skim milk and soymilk. Lassi like cultured milk-whey beverages have been developed using paneer whey and cheese whey. In another study whey powder was used to substitute partially the milk powder in yoghurt, which led to a slower acidification rate in yoghurts that become a little yellowish. Also, better sensory, flow properties and greater syneresis was obtained for products prepared with whey powder as compared to those prepared using skim milk powder. Augustin *et al.,* 2003 made set and stirred yogurts using 80:20 blends of skim milk solids and sweet WPC. The resulting yogurts were reported to have higher gel strength, viscosity and lesser whey separation.

Use of Non-Conventional Food Sources

Soy-based foods may provide additional benefits for the consumer for example due to their hypolipidemic, anti- cholesterolemic and antiatherogenic properties and also to their reduced allergenicity (Öner *et al.,* 1993; Park *et al.,* 2005). Thus, incorporation of these nonconventional food sources like soybeans and different kinds of millets with fermented milks may help in increasing utilization of these non conventional food sources and producing low cost, nutritious fermented foods apart from extending the variety of fermented milks. Soymilk is low in fat, carbohydrate, calcium, phosphorus, and riboflavin, but high in iron, thiamine, and niacin in comparison with cow's milk. Soymilk contains higher amount of protein than buffalo milk and is deficient in sulfur containing amino acids (Kumar and Mishra, 2003). Replacing a part of milk used in making yogurts with soymilk enriches nutritional value of the product. Soymilk is characterized by beany or soy flavor which can be modified by lactic acid fermentation (Nsofor and Chukwu, 1992).

Kumar and Mishra (2003) used response surface methodology (RSM) to optimize the formulation of mango soy fortified yogurt (MSFY). The independent variables were proportions of mango pulp, soymilk, and fat content of buffalo milk yogurt-like products were prepared from a combination of skim milk and soymilk (100:0, 75:25, 50:50, 25:75, and 0:100) containing saccharified-rice solution by lactic fermentation of four different cultures. The ratio of skim milk and soymilk had no significant effect on titratable acidity, also there was no significant difference in texture and overall acceptability among yogurts produced from mixed substrates and skim milk-based yogurt.

2. Fotification with Physiologically Active Ingredients

Various popular ingredients of functional significance are being incorporated into cultured dairy products to enhance their market value (Chandan, 1999). Since consumption of functional foods containing nutraceuticals is being highly encouraged, thus fermented milks are produced with incorporation of these. In the presence of such new components, the gel structure and other properties of fermented milks change. Some of these ingredients designed to enhance consumer appeal, which may be incorporated into fermented milks, include:

Essential Minerals and Vitamins

Certain minerals like Calcium claimed to prevent osteoporosis, cancer and control hypertension can be fortified in cultured milks. An attempt to fortify yoghurt with calcium salts revealed that yoghurt is a suitable vehicle for fortification with calcium salts and Calcium content of the fortified yoghurts could be increased with about 34.3, 37.6, and 39.4 per cent by addition of Ca Lactate, Ca Gluconate and Ca Lactate + Ca Gluconate, respectively. Antioxidant vitamins (C and E) to prevent cancer, cardiovascular disease, and cataracts as well as multivitamin-mineral mixes are being incorporated in fat free cultured milks to provide meal replacements for consumers within a targeted niche (Chandan, 1999).

Dietary Fibers

The beneficial role of dietary fibre in human nutrition has lead to a growing demand for incorporation of novel fibres into foods. There is little information about fiber fortification in cultured dairy products however various fibers like psyllium, guar gum, gum acacia, oat fiber and soy components can be used. In one experiment, pectin and raspberry concentrate was incorporated in commercial stirred yogurt samples, increasing the consistency and it was found that yogurt with pectin was more shear stable in comparison with yogurt with raspberry concentrate (Ramaswamy *et al.*, 1992).

In another study, seven types of insoluble dietary fibers from five different sources (soy, rice, oat, corn and sugar beet) were used to fortify sweetened plain yogurt. Fiber addition caused acceleration in the acidification rate of the experimental group yogurts, and most of the fortified yogurts also showed increases in their apparent viscosity. However, soy and sugar beet fibers caused a significant decrease in viscosity due to partial syneresis. In general, fiber addition led to lower overall flavor and texture scores as a grainy flavor and a gritty texture were intense in all fiber fortified yogurts, except in those made with oat fiber, which gave the best results. Similarly, β-glucan was used to prepare low fat yoghurt and as the amount of β -glucan increased a corresponding increase in yogurt consistency and firmness as well as a decrease in syneresis was reported. Palacios *et al.,* 2005 prepared yogurt systems from whole milk, with Calcium (50 mg of calcium/100 mL of yogurt) and three levels of fiber from two wheat-bran sources. In comparison with a plain yogurt, the presence of fiber and calcium augmented the consistency, diminished the syneresis and the pH was higher.

Omega-3-Fatty Acids

Milk fat composition in dairy products can be altered by reducing the ratio of saturated to unsaturated fatty acids and increasing the contents of fatty acids that are more desirable for human nutrition, such as the omega-3 polyunsaturated fatty acids (PUFAs). The importance of omega- 3 - fatty acids like - linolenic has been widely publicized because they are precursors of important long-chain fatty acids, such as eicosapentaenoic acid (EPA) and docosahexaenoic acid (DHA), which cannot be synthesized in the human body. Yet they are vital for the normal functioning and development of the brain, and are believed to reduce plaque formation in the arteries (Milner *et al.,* 1999). They are also claimed to exert cancer inhibition, anti-allergy effects and improvement in learning ability (Wolfram, 2003). Increased levels of healthy fatty acids in dairy products can be efficiently achieved by the use of selected bacteria during fermented milk manufacture (Kim *et al.,* 2002) or the substitution of milk fat by oils with high levels of PUFAs. However, replacement of milk fat by oils with high levels of PUFAs yielded yoghurts with less firmness and higher syneresis (Barrantes *et al.,* 1996). A different possibility for increasing omega-3 PUFA content in milk is to include fish and vegetable oils or marine algae in animal diets (Dave *et al.,* 2002). In one study, a modified milk where fat had been replaced by oils enriched in omega-3 polyunsaturated fatty acids was used for the manufacture of a set-type fermented product and no effect was found on yoghurt flavour however, product texture was adversely affected (Diana *et al.,* 2004). In another experiment yogurt rich in poly- and mono- unsaturated fatty acids were prepared by replacing milk fat. Fortification with oils did not have any effect on microbial growth however texture and flavor were adversely affected (Barrantes *et al.,* 1994a, 1994b).

Phytosterols and Phytostanols

Phytosterols are plant derived sterols that have similar structure as cholesterol thus these interfere with the uptake of cholesterol from the intestinal tract and are one natural way of achieving low level cholesterol in the blood stream (Jones *et al.,* 2002).

Awaisheh *et al.,* 2005 prepared yoghurts from modified milk base containing three important nutraceuticals, namely omega-3-fatty acids, isoflavones and phytosterols. The cultures employed to make the yoghurts were single probiotic strains of *Lactobacillus gasseri* or *Bifidobacterium infantis* and, to achieve a short production time, a two-stage fermentation procedure was used with *Streptococcus thermophilus* and *Lactobacillus delbrueckii* subsp. *bulgaricus.* The nutraceuticals appeared to have no adverse effect on flavour and storage trials at 50°C showed that the viability of the probiotic cultures was retained over 15 days (Khurana and Kanawjia, 2007).

Gamma-Aminobutyric Acid (GABA)

It is an amino acid that has long been reported to lower blood pressure by intravenous administration in experimental animals (Takahashi *et al.,* 1955; Stanton

HC, 1963; Lacerda *et al.,* 2003) and in human subjects (Elliott *et al.,* 1959). GABA is present in many vegetables and fruits but not in dairy products. However, the effect of dietary GABA has attracted little attention as a factor that may influence blood pressure. A novel fermented milk product containing GABA was reported to lower blood pressure in people with mild hypertension (Inoue *et al.,* 2003). Hayakawa *et al.,* 2004 investigated the blood-pressure-lowering effects of GABA and a GABA enriched fermented milk product (FMG) by low-dose oral administration to spontaneously hypertensive and normotensive Wistar–Kyoto rats and it was suggested that low-dose oral GABA has a hypotensive effect in spontaneously hypertensive and finally concluded that the hypotensive effect of FMG was due to GABA.

Recent Developed Novel Fermented Milk Products

Herbal Probiotic Lassi

First two indigenous probiotic cultures from India- *Lactobacillus helveticus* MTCC 5463 and *Lactobacillus rhamnosus* MTCC 5462 were deposited as Indian patent at the Institute of Microbial Technology in Chandigarh by Sheth M.C. College of Dairy Science (Dairy Microbiology Department). Using these cultures, an Indian patent has been filed for a process to manufacture herbal probiotic fermented milk product 'lassi' having herbal content as Safed Musli. Lassi was further tested to check its potential in reducing cholesterol level in humans. Volunteers have been recruited to conduct clinical trials of the probiotic lassi with a view to check its potential to reduce cholesterol levels in humans. Its unparalleled therapeutic and medicinal properties have made it a key ingredient in the preparation of a number of Ayuvedic formulations.

Synbiotic Dahi

Heat treated cows' whole milk (95 °C/5 min. and subsequently cooled to 37°C) was inoculated with probiotic culture *L. acidophilus* $LBKV_3$ @ 2.0 per cent v/v, and prebiotic inulin @ 2.0 per cent as well as other functional food ingredients like WPC @ 3.0 per cent, NMCP @ 200 mg/100ml as a plain synbiotic *dahi* (blend A), or with non nutritive sweetener sucralose @ 19.5 mg/100 ml (blend B), or sucrose @ 9.0 per cent w/v (blend C), which were decided on the basis of sensory profile during preliminary studies carried out. These three blends were compared with control *dahi* (M) made by using cows' milk, without any additives and fermented with the probiotic culture. All the four *dahi* prepared were stored at 3 ± 1 °C for the period of 28 days, and microbiological and chemical parameters as well as sensory attributes were evaluated at an interval of seven days. From the sensory analysis on 9-point hedonic scale; it was clear that all the parameters (flavour, colour and appearance, body and texture and overall acceptability) for the three blends of synbiotic *dahi* were superior from control sample (M) when they were fresh and even at the end of storage. Flavour scores (around 8.50) of fresh as well as refrigerated synbiotic *dahi* samples (B and C) were superior to sample A (7.97) and control *dahi* M (7.17). It was observed that body and texture scores of synbiotic *dahi* made with low calorie sweetener sucralose (B) and with 9.0 per cent natural sugar (C) could not be differentiated by judges throughout the entire storage period of 0 to 28 days, and the samples were statistically at par (Gawai, 2006).

Synbiotic Lassi with Honey as Prebiotic

Lassi (stirred yoghurt) is a popular fermented milk product of Indian sub-continent. It can be one of the most appropriate vehicles to carry scientifically proven probiotic bacteria and prebiotic ingredients. On the other hand, there are many natural plant materials, which have proven health benefits and presence of oligosaccharides, which can act as prebiotic. Honey being one of them.

Level of addition of honey was decided by incorporating honey at 0, 3, 5 and 7 per cent and comparing the quality of lassi based on chemical, microbiological and sensory attributes. The lassi was prepared from toned milk fermented by isolates of *Streptococcus thermophilus* MTCC 5460 and a probiotic culture *Lactobacillus helveticus* MTCC 5463 as per the protocol standardized earlier. Overall there were no significant difference in microbiological counts and sensory scores, but product with 5 per cent honey showed relatively better acceptance, better viable count of lactobacilli and optimum level of acid production.

The product could provide optimum dose of probiotic lactobacilli count of 65 x 10^7 CFU/g and is organoleptically acceptable till 21 days of storage at 5±20C. Such product is ready for commercial manufacture (Sharma, 2010).

Artificially Carbonated Fermented Milk

Carbonation is a treatment, which has sold billions of dollars worth of water and flavoring thus using this process with products of genuine nutritive value like yogurt/cultured dairy products; new products with tremendous potential could be created. Carbonated fermented milk occupies a prime position in present era since they have been nutritionally proven and time tested for safety and acceptability. Soft drinks have a huge market all over the world. However, as people are becoming more conscious about health and nutrition, and the society has been aware of the hazardous effects of soft drinks, a special niche for self carbonated nutritious fermented milk based drink is emerging. Typical carbonated milks are Kefir and Koumiss, which are known since ages in the regions between Eastern Europe and Mongolia for their sensory characteristics including sparkling taste and alcoholic flavour. However, control of carbonation and yeast fermentation are major problem in such products. Hence, artificially carbonated fermented milk without alcohol in the product shall be more appreciated.

Streptococci thermophilus MD2 (MTCC 5460) and a probiotic isolate of *Lactobacillus acidophilus* V3 (MTCC 5463) were used to prepare fermented milk from double tonned milk. The fermented milk was carbonated at three different CO_2 pressure levels *viz.* 10,15, and 20 kg/cm^2 and a pressure of 15 kg/cm^2 was found to be optimum based on sensory evaluation and viable counts. During heat treatment of milk; 10, 12 and 15 per cent sugar was incorporated and it was found that the carbonated milk with 15 per cent sugar gives highest scores for overall acceptability (8.22) as compared to 12 per cent (7.52) and 10 per cent (7.11). Three levels of salt were tested for improving the acceptability of the product. It was observed that addition of salt at 0.8 per cent level has better sensory profile as compared to 0.5 per cent or 1.0 per cent. Salt addition has no effect on acidity, pH and viable count in the product. The last phase of optimization was done by preparing a product with a combination of 15 per cent sugar and 0.8 per cent salt and comparing it with either sugar or salt added product. Overall acceptability of the salt and sugar added product was 8.08 as compared to 7.51 for only sugar and 6.88 for only salt added product (Shah, 2009).

Conclusion

It is evident that the market for fermented milks is booming specially probiotics and those with special added ingredients. Modern consumers are increasingly interested in their personal health, and expect the food that they eat to be healthy or even capable of preventing illness. Producers and marketers of cultured milks are making every effort to keep them growing through product development and packaging innovations while delivering a 'good for you' flavourful products suited for all occasions of gastronomic indulgence.

There are now products with complete supplementation offered as medical foods, as well as healthy products for people who have problems obtaining all the nutrients they need. It is clear from the literature that new kinds of fermented milks containing various nutrients are being tested as curatives for specific diseases and are approaching medical food effectiveness in conventional food format and will continue to be introduced to the food supply. The occurrence of diet-related diseases of deficiency and excess, points to the importance of the development of functional foods (science). Functional food science must be viewed world over beyond the short-term commercial prospects and should be considered for long-term research and development.

References

Alexander JW, Boyce ST and Babcock GF (1990). The process of microbial translocation. *Ann. Surg.* 212, 496-512.

Anuradha S, Rajeshwari K. (2005). Probiotics in Health and Disease. *JIACM*, 6: 67-72.

Augustin MA, Cheng LJ, Glagovskaia O, Clarke PT, Lawrence A. (2003). Use of blends of skim milk and sweet whey protein concentrates in reconstituted yogurt. A*ust J Dairy Technol*, 58: 3-10.

Awaisheh SS, Haddadin MSY, Robinson RK. (2005). Incorporation of selected nutraceuticals and probiotic bacteria into a fermented milk. *Int Dairy Journal*, 15: 1184-1190.

Bardale PS, Waghmare PS, Zanzad DM, Khedkar DM. (1986). The preparation of Shrikhand like product from skim milk chakka by fortifying with fruit pulps. Ind *J Dairy Sci*, 39: 480-483.

Barrantes E, Tamime AY, Davies G, Barclay M (1994).Production of low-calorieyoghurt using skim milk powder and fat substitutes. 2. Compositional quality. *Milchwissenschaft*, 49: 135-139.

Barrantes E, Tamime AY, Sword AM, Muir DD, Kalab M. (1996).The manufacture of set-type natural yoghurt containing different oils—2. Rheological properties and microstructure. *Int Dairy Journal*, 6: 827-837.

Barrantes E, Tamime AY, Sword AM. (1994a). Production of low-calorie yoghurt using skim milk powder and fat substitutes. 3. Microbiological and organoleptic qualities. *Milchwissenschaft*, 49: 205-208.

Barrantes E, Tamime AY, Sword AM.(1994b). Production of low-calorie yoghurt using skim milk powder and fat substitutes. 4. Rheological properties. *Milchwissenschaft*, 49: 263-266.

Best D. (1989). High-intensity sweeteners lead low-calorie stampede. *Prepared Foods*, 158: 97-98.

Botma Y. (1988). Aspartame in refreshing beverages and milk products. *Fluessiges Obst*, 55: 80-82.

Brandsch M., Brust P., Neubert K., Ermisch A.(1994). β-Casomorphins –chemical signals of intestinal transport systems, pp. 207–219 in V. Brantl, H. Teschemacher (Eds). *β-Casomorphins and Related Peptides: Recent Developments.* VCH, Weinheim (Germany).

Carrico J and Meakin JI (1986). Multiple organ failure syndrome. *Arch. Surg.* 121, 196-208.

Cats A, Kuipers EJ, Bosschaert MA, Pot RG, Vandenbroucke-Grauls CM, Kusters JG. (2003). Effect of frequent consumption of a *Lactobacillus casei*-containing milk drink in *Helicobacter pylori*-colonized subjects. *Aliment. Pharmacol. Ther*. 17: 429–435.

Cerning J, Bouillanne C, Landon M, Desmazeaud M.(1992). Isolation and characterization of exopolysaccharides from slime-forming mesophilic lactic acid bacteria. *J Dairy Sci*, 75: 692-699.

Cerning J. (1990). Exocellular polysaccharides produced by lactic acid bacteria. *FEMS Microbiol Rev*, 87: 113-130.

Cerning J. (1995). Production of exopolysaccharides by lactic acid bacteria and dairy propionibacteria. *Lait*, 75: 463-472.

Chandan RC. (1999). Enhancing market value of milk by Adding Cultures. *J Dairy Sci*, 82: 2245-2256.

Chickai T, Nakao H, Uchida K. (1987). Deconjugation of bile acids by human intestinal bacteria implanted in germ-free rats. *Lipids* 22: 669.

Cook M. E., Pariza, M. (1998). The Role of Conjugated Linoleic Acid (CLA). in Health. *Int. Dairy Journal*, 8(5/6). 459-462.

Corzo G, Gilliland SE. (1999). Bile salt hydrolase activity of three strains of *Lactobacillus acidophilus. J Dairy Sci* 82: 472.

Dave RI, Ramaswamy N, Baer RJ. (2002). Changes in fatty acid composition during yogurt processing and their effects on yogurt and probiotic bacteria in milk procured from cows fed different diets. *Aust J Dairy Technol*, 57: 197-202.

De Vuyst L, Degeest B. (1999). Heteropolysaccharides from lactic acid bacteria. *FEMS Microbiol Rev*, 23: 153-177.

Desai SR, Toro VA, Joshi SV. (1994). Utilization of different fruits in the manufacture of yoghurt. *Ind J Dairy Sci*, 47: 870-874.

Diana ABM, Janer Carolina, Pelaez Carmen, Requena, Teresa. (2004). Effect of milk fat replacement by polyunsaturated fatty acids on the microbiological, rheological and sensorial properties of fermented milks. *J Sci Food Agric*, 84: 1599-1605.

Duboc P, Mollet B. (2001). Applications of exopolysaccharides in the dairy industry. *Int Dairy J*, 11: 759-768.

Duncan SE. (1998). Dairy products: the next generation. Altering the image of dairy products through technology. *J Dairy Sci*, 81: 877- 883.

Elliott CAK, Hobbiger F. (1959). Gamma aminobutyric acid: circulatory and respiratory effects in different species: re-investigation of the anti-strychnine action in mice. *J Physiol*, 146: 70-84.

Eyssen H. (1973). Role of the gut microflora in metabolism of lipids and sterols. *Proc Nutr Soc* 32: 59.

Farooq K, Haque ZU. (1992). Effect of Sugar Esters on the Textural Properties of Nonfat Low Calorie Yogurt. *J Dairy Sci*, 75: 2676-2680.

Fitzgerald R.K., Murray B.A. (2006). Bioactive peptides and lactic farmentations. *Internat. J. Dairy Technol.* **59**, 118–125.

Folkenberg DM, Dejmek P, Skriver A, Ipsen R. (2005). Relation Between Sensory,Texture Properties And Exopolysaccharide Distribution In Set and In Stirred yoghurts Produced with Different Starter Cultures. *J Texture Studies*, 36: 174-189.

Freitas M, Tavan E, Cayuela C, Diop L, Sapin C, Trugnan G.(2003). Host-pathogens cross-talk. Indigenous bacteria and probiotics also play the game. *Biol. Cell.* 95: 503–506.

Fuller R (1989). Probiotica in man and animals. *Journal of Applied Bacteriology* 66.

Gamer L, Blondeau K, Simonet JM. (1997). Physiological approach to extracellular polysaccharide production by *Lactobacillus rhamnosus* strain C83. *J App Microbiol*, 83: 281.

Gardiner SG, Ross RP, Collins JK, Fitzgerald G, Stanton C. (1998). Development of a Probiotic Cheddar Cheese Containing Human- Derived *Lactobacillus paracasei*. *App Env Micro*, 64: 2192- 2199.

Gawai KM (2006). Formulation and properties of new functional dairy products-Synbiotic Dahi. *M.Tech Thesis*. Anand, India: SMC College of Dairy science.

Gibson GR, Roberfroid MB. (1995). Dietary modulation of the human colonic microbiota: introducing the concept of prebiotics. *J Nutr*, 125: 1401-12.

Guandalini S, Pensabene L, Zikri MA, Dias JA, Casali LG, Hoekstra H, Kolacek S, Massar K, Micetic-Turk D, Papadopoulou A, de Sousa JS, Sandhu B, Szajewska H, Weizman Z. (2000). *Lactobacillus* GG administered in oral rehydration solution to children with acute diarrhea: A multicenter European trial. *J. Pediatr. Gastroenterol. Nutr.* 30: 214–216.

Guarner F, Shaafsma GJ. (1998). Probiotics. *Int J Food Microbiol*, 30: 237-238.

Gurr M. I. (1992). Milk products: contribution to nutrition and health. J. *Soc. Dairy Technol.* 45: 61-67.

Guven M, Yasar K, Karaca, OB, Hayaloglu, AA. (2005). The effect of inulin as a fat replacer on the quality of set-type low-fat yogurt manufacture. *Int J Dairy Technol*, 58: 180-184.

Hall GM, Iglesias O. (1997). functional properties of dried milk whey. *Food Sci Technol Int*, 3: 381-383.

Haque ZU (2003). Cheddar whey processing and source: II. Effect on non-fat ice cream and yogurt. *Int J Food Sci Technol*, 38: 463- 473.

Harrison VC, Peat G. (1975). Serum cholesterol and bowel flora in the newborn. *Am J Clin Nutr*, 28: 1351.

Hartemink R. (1999). Prebiotic effects of Non-Digestible Oligo- and Poly- Saccharides. *Ph.D. Thesis*, Wageningen, Ponsen and Looijen: Landbouwuniversiteit Wageningen.

Hatakka K, Ahola AJ, Richardson M *et al.* (2007). Probiotics reduce the prevalence of oral candida in the elderly-a randomized controlled trial. *J Dent Res*, 86: 125-130.

Hayakava K, Kimura M, Kasaha K, Matsumoto K, Sansawa H, Yamori Y. (2004). Effect of a g-aminobutyric acid-enriched dairy product on the blood pressure of spontaneously hypertensive and normotensive Wistar-Kyoto rats. *British J Nutr*, 92: 411-417.

Hillman JD, Mo J, McDonell E, Cvitkovitch D, Hillman CH. (2007). Modification of an effector strain for replacement therapy of dental caries to enable clinical safety trials. *J Appl Micro-biol*, 102: 1209-1219.

Huffmann LM. (1996). Processing whey protein for use as a food ingredient. *Food Technol*, 50: 49-52.

Imele H, Atemnkeng A. (2001). Preliminary study of the utilisation of coconut in yoghurt production. *J Food Technol Africa*, 6: 11-12.

Inoue K, Shirai T, Ochiai H, Kasao M, Hayakawa K, Kimura M, Sansawa H.(2003). Blood-pressure-lowering effect of a novel fermented milk containing g-aminobutyric acid (GABA). in mild hypertensives. *Eur J Clin Nutr*, 57: 490-495.

Jahreis G., Fritsche J., Möckel P., Schöne F., Möller U., Steinhart H. (1999). The potential anti-carcinogenic conjugated linoleic acid, *cis-9, trans-11* C18: 2, in milk of different species: cattle, goats, sheep, pigs, horses, human beings. *Nutrition Research*, 19(10). 1541-1549.

Jones PJ. (2002). Clinical nutrition: functional foods–more than just nutrition. *Can Med Assoc J*, 166: 1555-1563.

Kantor MA. (1990). Light dairy products: the need and consequences. *Food Technol*, 44: 81.

Kasravi FB, Adawi D, Molin G, Bengmark S and Jeppson B (1997). Effect of oral supplementation of lactobicilli on bacteria translocation in acute liver injury induced by D-galactosamine. *J. Hepatology* 26, 417-424.

Keating KR, White CH. (1990). Effect of alternative sweeteners in plain and fruit flavored yoghurts. *J Dairy Sci*, 73: 54-62.

Khulusi S, Mendall MA, Patel P, Levy J, Badve S, Northfield TC. (1995). *Helicobacter pylori* infection density and gastric inflammation in duodenal ulcer and non-ulcer subjects. *Gut* 37: 319–324.

Khurana HK and Kanawjia SK (2007), Recent Trends in Development of Fermented Milks, *Current Nutrition and Food Science* (*3*). 91-108

Kim YJ, Liu RH. (2002). Increase of conjugated linoleic acid content in milk by fermentation with lactic acid bacteria. *J Food Sci*, 67: 1731-1737.

Kragen H. (1954). The treatment of inflammatory affections of the oral mucosa with a lactic acid bacterial culture prepara-tion. *Zahnarztl Welt*,9: 306-308.

Krasse P, Carlsson B, Dahl C, Paulsson A, Nilsson A, Sinkiewicz G. (2006). Decreased gum bleeding and reduced gin-givitis by the probiotic Lactobacillus reuteri. *Swed Dent J*, 30: 55-60.

Kullisaar T., songisepp e., Mikelsaar M., Zilmer K., Vihalemm T., Zilmer M. (2003). Antioxidative probiotic fermented goats' milk decreases oxidative stress-mediated atherogenicity in human subjects. *Brit. J. Nutr*. 90, 449–456.

Kumar M. (2000). Physicochemical characteristics of low calorie Lassi and flavored dairy drink using fat replacers and artificial sweeteners. *M.Sc. Thesis*. Karnal, India: National Dairy Research Institute.

Kumar P, Mishra HN. (2003). Optimization of mango soy fortified yogurt formulation using response surface methodology. *Intl J Food Prop*, 6: 499-517.

Kumar P. (2000). Physico-chemical and micro structural properties of dahi using EPS producing strains. *M.Sc Thesis*. Karnal, India: National Dairy Research Institute.

Kurmann JA. (1984). Aspects of the production of fermented milks. In: *Fermented Milks*, Bulletin no.179. Brussels, International Dairy Federation, 16- 26.

Lacerda CEJ, Campos RR, Araujo CG, Andreatta-Van LS, Lopes OU, Guertzenstein PG. (2003). Cardiovascular responses to microinjections of GABA or anesthetics into the rostral ventrolateral medulla of conscious and anesthetized rats. *Braz J Med Biol Res*, 36: 1269-1277.

Ling M.Y., Yen C.L. (1999). Antioxidative ability of lactic acid bacteria. *J. Agric. Food Chem*. 47, 1460–1466.

Litcher A. Progress through change. *Eur Dairy Mag* 2001, 3: 26-28.

Ljungh A., Lan J., Yanagisawa N. (2002). Isolation, selection and characteristics of *Lactobacillus paracasei* subsp. *paracasei* F16. *Microb. Health Dis*. **3** (Suppl.), 4–6.

Lotz A, Klug C, Kreuder K. (1992). Sweetener stability. *Dairy Ind Int*, 57: 27-28.

MacDonald H. B. (2000). Conjugated Linoleic Acid and Disease Prevention: A Review of Current Knowledge. *Journal of the American College of Nutrition*, 19(2). 111-118.

Macedo RF, Renato J, Freitas S, Pandey A, Soccol CR. (1999). Production and shelf-life studies of low cost beverage with soymilk, buffalo cheese whey and cow milk fermented by mixed cultures of *Lactobacillus casei* ssp. *shirota* and Bi-fidobacterium adolescentis. *J Basic Micro*, 39: 243-251.

Maragkoudakisa PA, Miarisa C, Rojeza P, *et al.* (2006). Production of traditional Greek yoghurt using Lactobacillus strains with probiotic potential as starter adjuncts. *Int Dairy J*, 16: 52-60.

Meisel H. (1997). Biochemical properties of regulatory peptides derived from milk proteins. *Biopolymers* 43, 118–128.

Meisel H. (2005). Biochemical properties of peptides encrypted in bovine milk proteins. *Curr.Med.Chem.* 12, 1905–1919.

Mensink RP, Ebbing S, Lindhout M, Plat J, Marjolien MA. (2002). Effects of plant stanol esters supplied in low-fat yoghurt on serum lipids and lipoproteins, non-cholesterol sterols and fat-soluble antioxidant concentrations. *Atherosclerosis*, 160: 205-213.

Milner JA, Alison RG. (1999). The role of dietary fat in child nutrition and development. *J Nutr*, 129: 2094-2105.

Nagaoka M, Hashimoto S, Watanabe T, Yokokura T, Mori Y. (1994). Antiulcer effects of lactic acid bacteria and their cell-wall polysaccharides. *Biol Pharm Bull*, 17: 1012-1017.

Neve H, Geis A, Teuber M. (1988). Plasmid encoded functions of ropy lactic acid streptococcal strains from Scandinavian fermented milks. *Biochemie*, 70: 437.

Noh DO, Gilliland SE. (1993). Influence of bile on cellular integrity and β-galactosidase of *Lactobacillus acidophilus*. *J Dairy Sci* 76: 1253.

Nsofor LM, Chukwu EU. (1992). Sensory evaluation of soy milk-based yoghurt. *J Food Sci Technol Mysore*, 29: 301-304.

Oda M, Hasegawa H, Komatsu S, Kambe M, Tsuchiya F. (1983). Antitumor polysaccharide from *Lactobacillus* sp. *Agric Biol Chem*, 47: 1623-1625.

Öner, MD, Tekin AR, Tanzer E. (1993). The Use of Soybeans in the Traditional Fermented Food– Tarhana. *Lebensmittel-Wissenschaft und-Technologie*, 26: 371-372.

Otero M, Rodriguez T, Camejo J, Cardoso F. (1995). A fermented milk beverage. *Alimentaria*, 260: 93-95.

Palacios AA, Morales ME, Vélez RJF. (2005). Rheological and Physicochemical Behavior of Fortified Yogurt, With Fiber and Calcium. *J Texture Studies*, 36: 333-349.

Pandya CN. (2002). Development of technology for fruit dahi. *M.Sc Thesis*. Karnal, India: National Dairy Research Institute.

Park DJ, Sejong O, Hyung K, Mok C, KimSae H, Imm JY. (2005). Characteristics of yogurt-like products prepared from the combination of skim milk and soymilk containing saccharified-rice solution. *Int J Food Sci Nutr*, 56: 23-34.

Parodi, P. W. (1999a). Symposium: A Bold New Look at Milk Fat. Conjugated Linoleic Acid and Other Anticarcinogenic Agents of Bovine Milk Fat. *J. Dairy Sci.* 82(6). 1339-1349.

Pirkul T, Temiz A, Erdem YK. (1997). Fortification of yoghurt with calcium salts and its effect on starter microorganisms and yoghurt quality. *Int Dairy J*, 7: 547-552.

Ramaswamy HS, Basak S. (1992). Pectin and raspberry concentrate effects on the rheology of stirred commercial yoghurt. *J Food Sci*, 57: 357-60.

Resta-Lenert S, Barrett KE. (2003). Live probiotics protect intestinal epithelial cells from the effects of infection with enteroinvasive *Escherichia coli* (EIEC). *Gut* 52: 988–997.

Rimada PS, Abraham AG. (2003). Comparative study of different methodologies to determine the exopolysaccharide produced by kefir grains in milk and whey. *Lait*, 83: 79-87.

Rogelj, I. (2000). Milk, Dairy Products, Nutrition and Health. *Food Technol. Biotechnol.* 38(2). 143-147.

Rudrello F. Health trends shape innovation for dairy products (online). Euromonitor international archive, 2004 Oct 5.http: //www.euromonitor.com/ article.asp?id=4011.

Savage DC (1983). Microbial ecology of the gastrointestinal tract. *Nutrition and the Intestinal Flora*. ed. B. Hallgren ISBN 91 22 00593 5.

Sazawal S, Hiremath G, Dhinga U, Malik P, Deb S, Black RE.(2006). Efficacy of probiotics in prevention of acute diarrhoea: A meta-analysis of masked, randomised, placebo-controlled trials. *Lancet Infect. Dis.* 6: 374 382.

Shah NP (2009). Development of Artificially Carbonated Fermented Milk. *M.Sc. Thesis.* Anand, India: SMC College of Dairy Science.

Sharma S (2010). Development of Synbiotic Lassi with Honey as prebiotic. *M.Tech Thesis*. Anand, India: SMC College of Dairy Science.

Shortt C. (1999). The probiotic century: historical and current perspectives. *Trends Food Sci Technol*, 10: 411-417.

Sikkema J, Oba T. (1998). Extracellular polysaccharides of lactic acid bacteria. *Snow Brand R&D Rep*, 107: 1-31.

Songisepp E., Kullisaar T., Hutt P., Elias P., Brilene T., Zilmer M., Mikelsaar M.: (2004). A new probiotic cheese with antioxidative and antimicrobial activity. *J. Dairy Sci.* 87, 2013–2017.

Stanton HC. (1963). Mode of action of gamma amino butyric acid on the cardiovascular system. *Arch Int Pharmacodyn Ther*, 143: 195- 204.

Takahashi H, Tiba M, Iino M, Takayasu T. (1955). The effect of gaminobutyric acid on blood pressure. *Jpn J Physiol*, 5: 334- 341.

Tamime AY, Barclay, MNI, Davies G, Barrantes E. (1994). Productions of low-calorie yoghurt using skim milk powder and fat substitutes. 1. A review. *Milchwissenschaft*, 49: 85-87.

Tamime AY, Barrantes E, Sword AM. (1996). The effect of starch based fat substitutes on the microstructure of set-style yogurt made from reconstituted skimmed milk powder. *J Soc Dairy Technol*, 49: 1-10.

Teschemacher H. and Brantl V. (1994). Milk proteins derived atypical opioid peptides and related compounds with opioid antagonist activity pp. 3–17 in V. Brantl, T. Teschemacher (Eds). β-*Casomorphins and Related Peptides: Recent Developments.* VCH Publishers, Weinheim (Germany).

Veeneman MW. (1999). The European non alcoholic beverage market. *Fruit Processing*, 9: 484-489.

Wells CL (1990). Relationships between intestinal microecology and the translocation of intestinal bacteria. *Antoine van Leeuwenhoek* 58, 87-93.

Wolfram G. (2003). Dietary fatty acids and coronary heart disease. *Eur J Med Res*, 8: 321-324.

Zemel, M.B. (2002). Regulation of adiposity and obesity risk by dietary calcium: Mechanisms and implications, *J. Am. College Nutr.*, 21, 146S–151S.

2015, Dairy Product Technology: Recent Advances *Pages 283–307*
Editors: **Subrota Hati, Surajit Mandal and Birendra Kumar Mishra**
Published by: **DAYA PUBLISHING HOUSE, NEW DELHI**

Chapter 14

Functional Biomolecules and Food Ingredients Elaborated by LAB and their Potential Food Applications

Sreeja Mudgal

Versatility of the microbes to synthesize as well as breakdown complex substances in to various metabolites are finding application in the food industry for a number of nutritional and health reasons. Lactic acid bacteria (LAB) are among the most important groups of microorganisms which has a long tradition of use in the food industry and the number and diversity of their applications has increased considerably over the years. They have been considered for the production of functional biomolecules and food ingredients such as biothickeners, bacteriocins, vitamins, bioactive peptides, enzymes, flavouring compounds, organic acids and amino acids. These metabolites have found application in the food industry for product manufacturing, in the improvement of product characteristics, in reducing allergenicity, for value addition and preservation. Some of these metabolites and their potential applications especially in the dairy field are discussed in this chapter.

LAB – An Update

Lactic acid bacteria are a group of Gram-positive, non-spore forming, cocci or rod shaped organisms which produce lactic acid as the major end product during the fermentation of carbohydrates (Rattanachaikunsopon and Phumkhachorn, 2010; Khalid, 2011). They are catalase negative, aerotolerant anaerobes (Michaela *et al.,* 2009) and are fastidious in their nutritional requirements. They are among the most important group of microorganisms typically associated with the human gastrointestinal tract. They can be also found in the gastrointestinal tract of various animals and habitats that are rich in nutrients, such as dairy and meat products, seafood products, soil, green plants and fermenting vegetables (Holzapfel *et al.*, 2001; Khalid, 2011). Phylogenetically, LAB belong to the Clostridium branch of Gram positive bacteria with a DNA base composition of less than 53 mol per cent G+C (Stiles and Holzapfel, 1997; García-Fruitós, 2012). Based on their carbohydrate metabolism LAB are divided into two distinct groups. The homo-fermentative group utilizes the Embden-Meyerhof-Parnas (glycolytic) pathway to transform a carbon source mainly into lactic acid. Hetero-fermentative bacteria produce equimolar amounts of lactate, CO_2, ethanol or acetate from glucose exploiting phosphoketolase pathway. Homo-fermentative group consist of *Lactococcus, Pediococcus, Enterococcus, Streptococcus* and certain species of Lactobacilli. Hetero-fermentative group include certain species of Lactobacilli, *Leuconostoc* and *Weisella*. As per the major taxonomic revisions of LAB, the group now include genera such as *Aerococcus, Carnobacterium, Enterococcus, Lactobacillus, Lactococcus, Leuconostoc, Pediococcus, Streptococcus, Tetragenococcus, Vagococcus*, and *Weissella* (Sangoyomi *et al.*, 2010).

Selection of an ideal starter culture for any particular food application and improving the functionality of the exisiting starter cultures are two very important aspects, as far as the fermented food industry is concerned. Both these aspects have been advanced considerably through scientific achievements during the last few years. Until recently, the isolation and selection of a desirable culture was time consuming and cumbersome involving screening of large number of isolates. However, research carried out in the last two decades lead to the development of tools which allow us to specifically target the individual genes and metabolic pathways responsible for desired performance parameters of a starter culture. The use of high throughput screening (HTS) methods can overcome the problems associated with traditional screening techniques. In HTS repetitive laboratory manipulations are carried out using the power of robotics. It provides rapidity and better reproducibilty. This technology has seen rapid growth and development in the last decade, with many equipment suppliers having developed dedicated robots for specific laboratory operations. HTS differs from manual screening in the sense that in HTS all assays are performed in (96- or 384-well) microtitre plates, which greatly facilitates automation and high sample throughput. The analysis robot can be used for measuring enzymatic activities, growth rates or various metabolites. The robotic screening allows faster selection of a small number of strains (10–50) with the requisite attributes. These are then further analysed to confirm that they do indeed perform as required. Only those strains, which pass the rigorous retesting and reanalysis, are evaluated in product trials (Ercolini, 2013).

Genomic characterization of LAB has brought a revolution in the understanding of LAB and thus opened the doors for ways to manipulate their genes for various applications. Molecular approaches have been used to introduce phage resistance mechanisms in starter cultures and for improving the robustness of LAB strains. Even though such methods possess huge potential for introducing protection against phage, all use recombinant DNA approaches, which restrict their industrial use at present (Mills *et al.,* 2010). A food-grade protocol for the generation of bacteriophage-insensitive mutants (BIMs) of industrial *S. thermophilus* starters was developed by Mills *et al.* (2007). The resultened strains were completely resistant to phage attack. Studies done on this revealed that the genome of *S. thermophilus* harbour a few hypervariable regions, including three clustered regularly interspaced short palindromic repeats (CRISPR) (Deveau *et al.,* 2008; Horvath *et al.,* 2008). *S. thermophilus* can integrate novel spacers into its CRISPR loci in response to phage attack (Horvath *et al.,* 2008), where they function as small interfering RNAs, base pairing with target mRNAs and promoting their degradation or translation shutdown (Sorek *et al.,* 2008). Efforts are made to improve robustness of LAB strains through manipulation of gene expression levels. Improved salt tolerance, heat resistance, oxidative stress tolerance, enhanced survival in the presence of ethanol and H_2O_2 is reported for different species of LAB through such manipulations (Bron and Kleerebezem, 2011). Additionally, concepts such as patho-biotechnology and receptor mimic therapy has been tried for improving probiotic cultures (Sleator and Hill, 2008). The availability of genome sequences of LAB species further advanced understanding of these beneficial microbes. It facilitate their study as an integrated and interacting network of genes, proteins, and biochemical reactions and thus allows deeper understanding of individual organism's functions. About more than 140 draft (gapped) or finished LAB genome sequences are publicly available, representing more than 50 LAB species (excluding enterococci and pathogenic streptococci) (Teusink *et al.,* 2011; De Vos , 2011; Steele *et al.,* 2013). Inspite of all the advancement and advantages, the genetic manipulation techniques have its own limitations. LAB which has been enhanced through genetic modifications comes under genetically modified organisms and with the exception of the United States and Canada, there is still uncertainty in the public arena towards the use of genetic manipulation (Mills *et al.,* 2011). Moreover, the use of patho-biotechnology for genetic modification of probiotics may be a hard to accept concept for the consumers owing to the use of pathogen derived genes. Yet despite this, genetically modified LAB especially, designer probiotics offer huge potential for both technological and clinical applications. Also, advanced knowledge can open interesting perspectives to improve the performances of LAB.

Lactic Acid Bacterial Metabolites

Metabolic activities of LAB which are necessary for their survival and growth are also important for their industrial application. LAB are normally used to ferment milk, meat, vegetables, cereals and wine. Among food fermentations, milk-based fermentations are considered very common and traditional in many societies. Metabolic activities of LAB and the resultant metabolites have been used in the dairy industry for product manufacturing, preservation and for health benefits (probiotics). The GRAS status of LAB provides an additional impetus to their industrial use. *Lactococcus, Streptococcus, Pediococcus, Leuconostoc* and *Lactobacillus* are the commonly used genera for milk fermentations. The species belonging to these genera are increasingly being used as starter cultures in the manufacture of various fermented dairy products (Table 14.1). During lactic fermentations, they produce variety of metabolites such as organic acids, diacetyl, hydrogen peroxide, bacteriocins, etc. (Table 14.2) which provide fermented foods distinctive flavours, textures, and aromas while preventing spoilage, extending shelf-life, and inhibiting pathogenic organisms (Rattanachaikunsopon and Phumkhachorn, 2010).

Table 14.1: Lactic Acid Bacteria Associated with Fermented Milk Products

LAB	*Fermented Dairy Products*
Lb. delbrueckii subsp. *bulgaricus*, *S. thermophilus*	Yoghurt
L. lactis subsp. lactis, *L. lactis* subsp. *lactis* var. *diacetylactis*, *L. lactis* subsp. *cremoris*,*Leu. menesteroides* subsp. *cremoris*	Butter and buttermilk
Lb. kefir, *Lb. kefiranofaciens*, *Lb. brevis*	Kefir
L. lactis subsp. *lactis, L. lactis* subsp. *cremoris*	– Hard cheeses without eyes
L. lactis subsp. *lactis*, *L. lactis* subsp. *lactis* var. *diacetylactis*, L. lactis subsp. *cremoris*, *Leu. menesteroides* subsp. *cremoris*	– Cheeses with small eyes
Lb. delbrueckii subsp. *lactis*, *Lb. helveticus*, *Lb. casei*, *Lb. delbrueckii* subsp. *bulgaricus*, *S. thermophilus*	– Swiss-and Italian-type cheeses
Lb. casei, *Lb. acidophilus*, *Lb. rhamnosus*,*Lb. johnsonii*, *B. lactis*, *B. bifidum*, *B. breve*	Fermented, probiotic milk
L. acidophilus	Acidophilus Milk
L. delbreukii ssp *bulgaricus*, *L. acidophilus*	Kumiss
L. lactis ssp *lactis*, *S. thermophilus*, *L. delbreukii* ssp *bulgaricus*, *L. plantarum*	Dahi
S. thermophilus, *L. delbreukii* ssp *bulgaricus*	Shrikhand
L. lactis ssp *lactis*, *S. thermophilus*, *L. bulgaricus*	Lassi

Compiled from Rattanachaikunsopon and Phumkhachorn (2010); Khurana and Kanawjia (2007).

Table 14.2: Functional Biomolecules and Food Ingredients Elaborated by LAB

Category	*Uses in the Food Industry*
Organic acids	
Lactic, acetic, propionic	Acidulant, pH regulator, flavor enhancer and preservative
Hydrogen peroxide	Antimicrobial effect, activates the lactoperoxidase system,
Carbon dioxide	Antimicrobial effect, creates anaerobic conditions, contribute to product characteristics such as in kefir, Koumiss, Swiss cheese
Bacteriocins	Biopreservative, bioactive packaging
Flavouring compounds	
Diacetyl	Buttery aroma of dairy products such as butter, cultured butter milk and cottage cheese, antimicrobial effect
Acetaldehyde	Flavour of yoghurt, some cheeses
Enzymes	
β-galactosidase	Low lactose products
Proteinases, proteases, peptidases, lipases	Product characteristics, accelerated cheese ripening
Exopolysaccharides	Emulsifiers, stabilizers, gelling agents, thickeners and encapsulants
Therapeutic peptides	Health benefits

Contd...

Table 14.2–*Contd...*

Category	*Uses in the Food Industry*
Vitamins	
Folate, Vitamin B12, Vitamin K, Riboflavin	Health benefits
Low calorie sweeteners	Health benefits, product characteristics
Fatty acids	Antimicrobial effect, product characteristics
Conjugated linoleic acid	Health benefits

Source: Compiled from various sources.

Organic Acids

Milk fermentation by LAB is characterized by the accumulation of organic acids such as lactic, acetic, formic and propionic acid. These organic acids bring reduction in pH of milk and influence the subsequent microbial activity in the fermented milk. The levels and types of organic acids produced during the fermentation process depend on the species of organisms, culture composition and growth conditions (Lindgren and Dobrogosz, 1990). The inhibitory effect of organic acids is mainly caused by undissociated form of the molecule, which diffuses across the cell membrane towards the more alkaline cytosol and interferes with essential metabolic functions. The toxic effects of lactic and acetic acid include the reduction of intracellular pH and dissipation of the membrane potential (Suskovic *et al.*, 2010). Lactic acid is the major metabolite of LAB fermentation where it is in equilibrium with its undissociated and dissociated forms, and the extent of the dissociation depends on pH. At low pH, a large amount of lactic acid is in the undissociated form, and it is toxic to many bacteria, fungi and yeasts. However, different microorganisms vary considerably in their sensitivity to lactic acid. In addition, the stereoisomers of lactic acid also differ in antimicrobial activity, L-lactic acid being more inhibitory than the D-isomer (Benthin and Villadsen, 1995). Acetic and propionic acids produced by LAB strains through heterofermentative pathways may interact with cell membranes, and cause intracellular acidification and protein denaturation (Huang *et al.*, 1986). They are more antimicrobially effective than lactic acid due to their higher pKa values (lactic acid 3.08, acetic acid 4.75, and propionic acid 4.87), and higher percent of undissociated acids than lactic acid at a given pH (Earnshaw, 1992). Acetic acid was more inhibitory than lactic towards yeasts (Richards *et al.*, 1995), and towards the growth and germination of *Bacillus cereus* (Wong and Chen, 1988). Acetic acid also acted synergistically with lactic acid; lactic acid decreases the pH of the medium, thereby increasing the toxicity of acetic acid (Adams and Hall, 1988). In addition to their preservative effect due to antimicrobial action, the organic acids also contribute to sensory attributes of the fermented foods.

Hydrogen Peroxide and Carbon Dioxide

Hydrogen peroxide is produced by LAB in the presence of oxygen as a result of the action of flavoprotein oxidases or nicotinamide adenine hydroxy dinucleotide

(NADH) peroxidase. The antimicrobial effect of H_2O_2 may result from the oxidation of sulfhydryl groups causing denaturation of a number of enzymes and from the peroxidation of membrane lipids thus the increased membrane permeability (Suskovic *et al.,* 2010). H_2O_2 may also act as a precursor for the production of bactericidal free radicals such as superoxide (O_2 -) and hydroxyl (OH.) radicals which can damage DNA. It has been reported that the production of H_2O_2 by *Lactobacillus* and *Lactococcus* strains inhibited *Staphylococcus aureus, Pseudomonas* sp. and various psychotrophic microorganisms in foods (Davidson *et al.,* 1983). In raw milk, H_2O_2 activates the lactoperoxidase system, producing hypothiocyanate (OSCN-), higher oxyacids (O_2SCN- and O_3SCN-) and intermediate oxidation products that are inhibitory to a wide spectrum of Gram-positive and Gram-negative bacteria (Daechel 1989, Conner, 1993).

Carbon dioxide is mainly produced by heterofermentative LAB. Its antimicrobial action may be due to creation of an anaerobic environment which inhibits enzymatic decarboxylations and the accumulation of CO_2 in the membrane lipid bilayer which may cause a dysfunction in permeability (Suskovic *et al.,* 2010; Eklund, 1984). Many of the food spoilage microorganisms, especially Gram-negative psychrotrophic bacteria are effectively inhibited by CO_2. (Farber, 1991; Hotchkiss, 1999). The antimicrobial activity of CO_2 varies considerably between the organisms. CO_2 at 10 per cent could lower the total bacterial counts by 50 per cent (Wagner and Moberg 1989), and at 20-50 per cent it had a strong antifungal activity (Lindgren and Dobrogosz, 1990).

Flavouring Compounds: Diacetyl and Acetaldehyde

In addition to the calabolism of milk sugar lactose, LAB also has the capability to metabolize other substrates, such as citrate. Citrate is present in small amounts in milk and is also added as a preservative to foods. Citrate fermentation by LAB leads to the production of 4-carbon compounds, mainly diacetyl, acetoin and butanediol, which have aromatic properties. Diacetyl is responsible for the buttery aroma of dairy products such as butter, cultured butter milk and cottage cheese. In addition, it is an important component of the flavour of different kinds of cheeses. Moreover, the CO_2 produced as a consequence of citrate metabolism contributes to the formation of "eyes" (holes) in Swiss variety cheeses. Thus, the utilization of citrate in milk by LAB has a very positive effect on the quality of the end products. Strains of *Lactococcus lactis* ssp. *lactis* biovariety *diacetylactis* and some species belonging to *Leuconostoc* and *Weissella* genera are used as diacetyl producers by dairy industry. In addition, in traditional cheeses *Enterococcus* species contribute to the aroma by fermenting citrate (Quintans *et al.,* 2008).

Diacetyl also possess antimicrobial effect. It is found to be more inhibitory towards Gram-negative bacteria than Gram positive bacteria. Diacetyl concentration of 344 µg/ml inhibited strains of *Listeria, Salmonella, Yersinia, Escherichia coli,* and *Aeromonas* (Jay, 1992). But the use of diacetyl as a food preservative is limited due to low quantity, *e.g.* 4 µg/ml, of its production by producer organisms such as *L. lactis* ssp. *diacetylactis* during lactic fermentation. Also the acceptable sensory levels of diacetyl are at 2-

7µg/mL (Earnshaw, 1992). However, synergistic action of diacetyl along with other antimicrobial factors can be exploited for combined preservation systems in fermented foods (Jay 1992).

Acetaldehyde is another important aroma compound produced especially by yoghurt starter cultures. LAB can produce acetaldehyde through several pathways including from glucose through the pyruvate and acetyl-CoA intermediates of glycolysis, from amino acids and other metabolites that are converted to pyruvate or through the conversion of threonine (Papagianni *et al.,* 2011). Acetaldehyde at a concentration of 10-100 ppm is inhibitory towards the growth of *Staphylococcus aureus, Salmonella typhimurium* and *E. coli* in dairy products (Piard and Desmazeaud 1991).

Enzymes

The biological catalysts of LAB influence the processing, composition, organoleptic properties, overall quality and shelf life of fermented foods. LAB can be used as a source for the preparation of enzyme extracts that are able to function under the environmental conditions of fermentation (Tamang, 2011). They also release various enzymes into the gastrointestinal tract and exert potential synergistic effects on digestion and alleviate symptoms of intestinal malabsorption (Naidu *et al.,* 1999). Among the enzymes elaborated by LAB, most commonly used for dairy applications include â- galactosidase for low lactose products, proteinases for accelerated cheese ripening, proteases for reducing allergic properties of cow milk products for infants, and lipases for the development of lipolytic flavours in speciality cheeses. An important microbial enzyme which has found its place in the dairy industry is â-galactosidase or lactase. Treatment of milk and milk products with lactase reduce their lactose content and hence help in dealing with the problems of lactose insolubility and lack of sweetness. Furthermore, this treatment could make milk suitable for consumption by a large number of adults and children who are lactose intolerant. Whey hydrolysis by â- galactosidase converts lactose into a very useful product like sweet syrup, which can be used in various processes of dairy, confectionary, baking, and soft drink industries. It can also solve the environmental problems linked with whey disposal. The enzyme â-galactosidase can also be used in transglycosylation of whey lactose to synthesize prebiotic, galacto-oligosaccharides (GOSs). These are nondigestible oligosaccharides, not hydrolyzed or absorbed in the upper intestinal tract, and hence pass onto the colon where they are fermented selectively by beneûcial intestinal bacteria. The role of proteolytic enzymes such as certain peptidases and lipolytic enzymes produced by starter cultures in improving the sensory quality of cheese varieties has been reported by research workers (Guldfeldt *et al.,* 2001; Gonzalez *et al.,* 2010).

Gums/Stabilizers

Exposolysaccharides (EPS) released by LAB have received considerable attention as natural thickener and are finding application in the food industry as emulsifiers, stabilizers, gelling agents, thickeners and encapsulants. Many strains of LAB are capable of elaborating EPS which can be used to improve rheology and texture of fermented dairy products. Most LAB producing EPS belong to the genera *Streptococcus,*

Lactobacillus, Lactococcus, Leuconostoc, and *Pediococcus* (Ruas-Madiedo and de los Reyes-Gavila´n, 2004). EPS can either be attached to the bacterial cells as capsules or found as unattached material in the growth medium (Khurana and Kanawjia, 2007). The physiological role of EPS is to provide protection against desiccation, bacteriophage attack, antimicrobial compounds, and phagocytosis. They may also help in the bacterial adhesion to surfaces such as the human intestinal mucosa (Ruas-Madiedo *et al.,* 2008). LAB starter cultures producing EPS *in situ* during milk fermentation have found application as a natural source of food biothickeners (*e.g.,* yogurt and Scandinavian fermented milk viili). In addition, certain EPS produced by LAB may have benecial effects on human health (Table 14.3), although more *in vivo* studies are needed to demonstrate their efficacy. Approximately 30 species of LAB produce EPS and the best known are *L. casei, L. acidophilus, L. brevis, L. curvatus, L. delbrueckii bulgaricus, L. helveticus, L. rhamnosus, L. plantarum* (Badel *et al.,* 2011, Mollea *et al.,* 2013). The production of EPS is dependent on the nutrients available in the medium (sugars, amino acids and vitamins) and other factors that support the growth of bacteria such as temperature, pH, oxygen tension, and incubation time. Milk derivatives (lactose based media) and MRS (glucose based medium) are the widely used media for EPS production by lactobacilli (Badel *et al.,* 2011; Werning *et al.,* 2012). Presence of EPS confer beneficial physical properties to dairy products, such as the improvement of rheological properties and reduction of syneresis in yoghurts, better consistency of curd in low-fat ripened cheeses (Ruas-Madiedo *et al.,* 2008) and enhanced yield and functionality to low fat (6 per cent) Mozzarella cheese (Broadbent *et al.,* 2003). Additionally, the production of EPS does not add any flavors to the product (Badel *et al.,* 2011).

Exposolysaccharides (EPS) released by LAB have received considerable attention as natural thickener and are finding application in the food industry as emulsifiers, stabilizers, gelling agents, thickeners and encapsulants. Many strains of LAB are capable of elaborating EPS which can be used to improve rheology and texture of fermented dairy products. Most LAB producing EPS belong to the genera *Streptococcus,* *Lactobacillus, Lactococcus, Leuconostoc,* and *Pediococcus* (Ruas-Madiedo and de los Reyes-Gavila´n, 2004). EPS can either be attached to the bacterial cells as capsules or found as unattached material in the growth medium (Khurana and Kanawjia, 2007). The physiological role of EPS is to provide protection against desiccation, bacteriophage attack, antimicrobial compounds, and phagocytosis. They may also help in the bacterial adhesion to surfaces such as the human intestinal mucosa (Ruas-Madiedo *et al.,* 2008). LAB starter cultures producing EPS *in situ* during milk fermentation have found application as a natural source of food biothickeners (*e.g.,* yogurt and Scandinavian fermented milk viili). In addition, certain EPS produced by LAB may have benecial effects on human health (Table 14.3), although more *in vivo* studies are needed to demonstrate their efficacy. Approximately 30 species of LAB produce EPS and the best known are *L. casei, L. acidophilus, L. brevis, L. curvatus, L. delbrueckii bulgaricus, L. helveticus, L. rhamnosus, L. plantarum* (Badel *et al.,* 2011, Mollea *et al.,* 2013). The production of EPS is dependent on the nutrients available in the medium (sugars, amino acids and vitamins) and other factors that support the growth of bacteria such as temperature, pH, oxygen tension, and incubation time. Milk derivatives (lactose based media) and MRS (glucose based medium) are the widely used media for EPS production by lactobacilli (Badel *et al.*, 2011; Werning *et al.*, 2012). Presence of EPS confer beneficial physical properties to dairy products, such as the improvement of rheological properties and reduction of syneresis in yoghurts, better consistency of curd in low-fat ripened cheeses (Ruas-Madiedo *et al.*, 2008) and enhanced yield and functionality to low fat (6 per cent) Mozzarella cheese (Broadbent *et al.*, 2003). Additionally, the production of EPS does not add any flavors to the product (Badel *et al.*, 2011).

Table 14.3: Some Selected EPS Alongwith their Producer Organism and Application

EPS	*Producer strain*	*Application/Functional properties*	*Reference*
Dextran	*Leuconostoc mesenteroides*	Gel filtration compound (Sephadex) and as a blood plasma substitutes (Dextran 70)	Monsan *et al.*, 2001; Badel *et al.*, 2011
		immunomodulator	Gombocz *et al.*, 2007
Levan	*Lactobacillus sanfranciscensis*	Prebiotic	Dal Bello *et al.*, 2001
Kefiran	Kefir grain	Reduces blood pressure, cholesterol and blood glucose rates anti-inflammatory, anti-tumoural and stimulation of immunoglobulins secretion	Maeda *et al.*, 2004a,b; Rodrigues *et al.*, 2005; Vinderola *et al.*, 2006
HePS	*L. delbrueckii bulgaricus* and *L. acidophilus*	Antitumor activity and an enhancement of macrophage function	De Vuyst *et al.*, 2007
β-(2,1) fructans (inulin like polysaccharide)	*L. reuteri* 121	Prebiotic	
curdlan	*Agrobacterium* sp.	Authorized by Food and Drugs Administration as food additive (under the name Pureglucan) for its particular gelifying properties. Anti-tumour agents activating macrophages and white blood cells	Badel *et al.*, 2011
Reuteran and levan	*L. reuteri* 121	Used in baking industry. Contribute to bread flavour, texture and shelf life	Tieking *et al.*, 2005

Exposolysaccharides (EPS) released by LAB have received considerable attention as natural thickener and are finding application in the food industry as emulsifiers, stabilizers, gelling agents, thickeners and encapsulants. Many strains of LAB are capable of elaborating EPS which can be used to improve rheology and texture of fermented dairy products. Most LAB producing EPS belong to the genera *Streptococcus*, *Lactobacillus*, *Lactococcus*, *Leuconostoc*, and *Pediococcus* (Ruas-Madiedo and de los Reyes-Gavila´n, 2004). EPS can either be attached to the bacterial cells as capsules or found as unattached material in the growth medium (Khurana and Kanawjia, 2007). The physiological role of EPS is to provide protection against desiccation, bacteriophage attack, antimicrobial compounds, and phagocytosis. They may also help in the bacterial adhesion to surfaces such as the human intestinal mucosa (Ruas-Madiedo *et al.,* 2008). LAB starter cultures producing EPS *in situ* during milk fermentation have found application as a natural source of food biothickeners (*e.g.,* yogurt and Scandinavian fermented milk viili). In addition, certain EPS produced by LAB may have benecial effects on human health (Table 3), although more *in vivo* studies are needed to demonstrate their efficacy. Approximately 30 species of LAB produce EPS and the best known are *L. casei, L. acidophilus, L. brevis, L. curvatus, L. delbrueckii bulgaricus, L. helveticus, L. rhamnosus, L. plantarum* (Badel *et al.,* 2011, Mollea *et al.,* 2013). The production of EPS is dependent on the nutrients available in the medium (sugars, amino acids and vitamins) and other factors that support the growth of bacteria such as temperature, pH, oxygen tension, and incubation time. Milk derivatives (lactose based media) and MRS (glucose based medium) are the widely used media for EPS production by lactobacilli (Badel *et al.,* 2011; Werning *et al.,* 2012). Presence of EPS confer beneficial physical properties to dairy products, such as the improvement of rheological properties and reduction of syneresis in yoghurts, better consistency of curd in low-fat ripened cheeses (Ruas-Madiedo *et al.,* 2008) and enhanced yield and functionality to low fat (6 per cent) Mozzarella cheese (Broadbent *et al.,* 2003). Additionally, the production of EPS does not add any flavors to the product (Badel *et al.,* 2011).

Therapeutic Peptides

Lactic acid bacteria are known to liberate certain specific protein fragments that have a positive impact on human body functions and conditions and may ultimately influence human health. These peptides are called bioactive peptides. Bioactive

Table 14.4: Bioactive Peptides Identified in Fermented Milk Products

Product	*Identified Bioactive Peptide*	*Activity*
Cheese type Parmigiano- Reggiano	β-CN (8–16), β-CN (58–77), αs2-CN(83–33)	Phosphopeptides, precursor of β-casomorphin
Cheddar	αs1-CN fragments, β-CN fragments	Several phosphopeptides
Italian varieties: Mozzarella, Crescenza, Gogonzola, Italico	β-CN (58–72)	ACE-inhibitory
Gouda	αs1-CN (1–9), β-CN (60–68)	ACE-inhibitory
Festivo	αs1-CN (1–9), αs1-CN (1–7), αs1-CN (1–6)	ACE-inhibitory
Emmental	αs1-CN fragments β-CN fragments	Immunostimulatory, several phosphopeptides, antimicrobial
Manchengo	Ovine αs1-CN, αs2-CN, β-CN fragments	ACE-inhibitory
Fermented milks Sour milk	β-CN (74–76), β-CN (84–86), •-CN (108–111)	Antihypertensive
Yogurt	Active peptides not identified	Weak ACEinhibitory
Dahi	Ser-Lys-Val-Tyr-Pro	ACE-inhibitory

Source: Korhonen, 2009; Haque *et al.,* 2009.

Table 14.5: Some Commercially Available Dairy Products Containing Bioactive Peptides

Brand Name	*Dairy Product*	*Made by*	*Claimed Functional Bioactive Peptide*	*Health Claims*	*Producer*
Calpis/Ameal STM	Sour milk	Skim milk inoculated with starter cultures containing *Lactobacillus helveticus* CP790 and *Saccharomyces cerevisiae*	VPP, IPP from β-CN and •-CN	Blood pressure reduction	Calpis Co., Japan
Evolus	Calcium enriched fermented milk drink	*Lactobacillus helveticus* LBK-16Hstrain as starter	VPP, IPP from β-CN and •-CN	Blood pressure reduction	Valio Oy, Finland
Festivo	Fermented low-fat hard cheese		α_s1-CN (1–9), α_s1-CN (1–7), α_s1-CN (1–6)	No health claim as yet	MTT Agrifood Research, Finland
PRODIETF200/ Lactium	Flavored milk drink, confectio-nery, capsules		α_s1-CN (91–100)	Reduction of stress effects	Ingredia,France

Source: Korhonen, 2009; Haque *et al.,* 2009; Tidona *et al.,* 2009; Sharma *et al.,* 2011.

peptides can be generated by digestive enzymes and during milk fermentation with the starter cultures (Table 14.4) traditionally employed by the dairy industry (Korhonen and Pihlanto, 2006; Sharma *et al.*, 2011). Peptides with various bioactivities have been identified in several dairy-products such as milk protein hydrolysates, fermented milks and many cheese varieties (Gobbetti *et al.*, 2002; Korhonen and Pihlanto-Leppälä, 2004; Sieber *et al.*, 2010). Some commercially available dairy products with bioactive peptides are listed in Table 14.5.

Bacteriocins

Increasing consumer awareness of the risks derived not only from food borne pathogens, but also from the articial chemical preservatives used to control them have forced the food industry to look for alternatives which are free from such risks. Bacteriocins produced by LAB can be a suitable option and are finding application as biopreservatives to enhance safety and extend shelf life of food (De Vuyst and Leroy, 2007). The term "biopreservative" includes the antimicrobial compounds that are of plant, animal and microbial origin and have been used in human food for long time without any adverse effect on human health. Fermented foods are good examples of biopreserved foods in which the starter cultures are allowed to grow so that they can produce anti microbial metabolites. The small heat-stable bacteriocins of lactic acid bacteria have been recognized as perhaps the most promising entities for use in applications of food preservation for a number of reasons. They are widely distributed and established as food-grade bacteria and often have a suitable spectrum of bacterial targets, for example, strains of *Listeria, Clostridium* and other Gram-positive bacteria including LAB. Furthermore, these bacteriocins can endure harsh treatments, such as boiling, without losing much of their activity. Among the bacteriocins of LAB, Nisin is the single bacteriocin which has found commercial application as biopreservative. It is considered safe by Food and Drug Administration (FDA) and also by World Health Organization (WHO) and has received the denomination of Generally Recognized as Safe (GRAS). Nisin is produced by *Lactococcus lactis* subsp. *lactis*. It is used in pasteurized, processed cheese products to prevent outgrowth of spores such as those of *Clostridium tyrobutyricum* that may survive heat treatments as high as 85–105°C. Use of nisin allows these products to be formulated with high moisture levels and low NaCl and phosphate contents and also allows them to be stored outside chill cabinets without risk of spoilage. The level of nisin used depends on food composition, likely spore load, required shelf life and temperatures likely to be encountered during storage. Nisin is also used to extend the shelf life of dairy desserts which cannot be fully sterilized without damaging appearance, taste or texture. Nisin can signicantly increase the limited shelf life of such pasteurized products. Nisin is added to milk in the Middle East where shelf-life problems occur owing to warm climate, necessity to transport milk over long distances and poor refrigeration facilities. It can double the shelf life at chilled, ambient and elevated temperatures and prevent outgrowth of thermophilic heat resistant spores that can survive pasteurization. It can also be used in canned evaporated milk. Nisaplin, the commercial nisin, contains 2.5 per cent pure nisin A. It is legally used in more than 50 countries for specific food applications (Khurana and Kanawjia, 2007).

Bacteriocin, lacticin 3147 is similar to that of nisin with respect to its broad host range and effectively inhibits many Gram-positive food pathogens and spoilage microorganisms. It is produced by *Lactococcus lactis.* Such starters may provide a very useful means of controlling the proliferation of undesirable microorganisms during Cheddar cheese manufacture. Another bacteriocin, Pediocin PA-1, has been observed to inhibit Listeria in dairy products such as cottage cheese, ice cream, and reconstituted dry milk (Rodríguez *et al.,* 2002). Bacteriocinogenic strains of LAB can also be used for *in situ* production of bacteriocins in dairy products (Khalid, 2011). The use of milk fermented by a bacteriocin producer as an ingredient in milk-based foods may be a useful approach for introducing bacteriocins into foods at low cost. Microgard is such a preparation commercially produced from grade A skim milk fermented by a strain of *Propionibacterium shermanii.* It has a wide antimicrobial spectrum including some Gram-negative bacteria, yeasts and fungi. It is added to a variety of dairy products such as cottage cheese and yoghurt. In cottage cheese, it inhibits psychrotrophic spoilage bacteria. The inhibitory activity of microgard depends primarily on the presence of propionic acid, but there has also been a role proposed for a bacteriocin-like protein produced during the fermentation. Another commercial product is BioProfit, which contains *Lactobacillus rhamnosus* LC 705 and *Propionibacterium freudenreichii* JS. It is used as a protective culture to inhibit yeasts in dairy products and Bacillus spp. in sourdough bread (Khurana and Kanawjia, 2007). Other example of a secondary metabolite produced by lactic acid bacteria which has antagonistic activity include an antibiotic called Reuterocyclin, produced by *Lactobacillus reuteri.* The spectrum of inhibition of this antibiotic is limited to gram-positive bacteria including *Lactobacillus* spp., *Bacillus subtilis, Bacillus cereus, Enterococcus faecalis, Staphylococcus aureus* and *Listeria innocua* (Rattanachaikunsopon and Phumkhachorn, 2010).

Another potential application of bacteriocins is in bioactive packaging, where bacteriocins from LAB can be incorporated into packaging destined to be in contact with food. This system combines the preservation function of bacteriocins with conventional packaging materials, which protects the food from external contaminants. Bioactive packaging can be prepared by directly immobilizing bacteriocin to the food packaging, or by addition of a sachet containing the bacteriocin into the packaged food, which will be released during storage of the food product. Studies investigating the effectiveness of bio-active cellulose based packaging inserts and a vacuum packaging pouch made with polyethylene/polyamide to improve shelf life and safety aspects have proved promising. When considering bio-active packaging, the stability and the ability to retain activity while immobilised to the packaging film is of vital importance. While many LAB bacteriocins possess signicant antimicrobial qualities that could greatly enhance the safety of a food, it may yet emerge that industrially they will be most frequently applied as a 'final hurdle' in a food system where another hurdle(s) already exists to eliminate pathogens and spoilage agents that survive only in adventitious circumstances. To decide the optimal conditions for application of bacteriocins in foods certain important aspects such as the most effective conditions for application of each particular bacteriocin, antimicrobial effects of bacteriocins and bacteriocinogenic cultures in food ecosystems especially, in terms of microbial interactions, and knowledge of the characteristics of bacteriocin resistant variants and the conditions that prevent their emergence should be studied in detail.

Vitamins

The use of vitamin-producing micro-organisms is considered a more natural and economically viable alternative than fortification with chemically synthesized pseudo-vitamins as the former would allow the production of foods with elevated concentrations of vitamins that are less likely to cause undesirable side effects (LeBlanc *et al.,* 2011). Certain strains of LAB have the capability to synthesize water-soluble vitamins such as those included in the B-group. LAB used in the dairy industry such as *Lactococcus lactis, Lactobacillus acidophilus, L. delbrueckii* ssp. *bulgaricus,* and *Streptococcus thermophilus* have been shown to synthesize folate and secreate into the medium during fermentation and thus improving the folate content of products. Vitamin B12, an important cofactor for the metabolism of fatty acids, amino acids, carbohydrates and nucleic acids is synthesized by propionibacteria. Genera *Lactococcus, Lactobacillus, Leuconostoc, Bifidobacterium* and *Enterococcus* synthesize the vitamin K. The vitamin producing strains may be useful as dairy starters to increase the vitamin content in fermented dairy products (Wouters *et al.,* 2002; Crittenden *et al.,* 2003; Burgess *et al.,* 2004; Beitane and Ciprovica, 2012). Vitamin production by LAB is considered a strain-dependent trait; hence the selection of adequate combination of strains is essential to develop fermented foods with increased vitamin concentrations. Besides riboflavin, folate and vitamin B12; increased levels of other B-group vitamins such as niacin and pyridoxine have also been reported as a result of the LAB fermentation in yoghurt, cheeses and other fermented products (LeBlanc *et al.,* 2011).

Fatty Acids

Under certain conditions, some lactobacilli and lactococci possessing lipolytic activities may produce significant amounts of fatty acids in fermented milk (Rao and Reddy, 1984). The unsaturated fatty acids are active against Gram-positive bacteria, and the antifungal activity of fatty acids is dependent on chain length, concentration, and pH of the medium (Gould, 1991). The antimicrobial action of fatty acids has been thought to be due to the undissociated molecule, not the anion, since pH had profound effects on their activity, with a more rapid killing effect at lower pH (Kabara, 1993).

Low-Calorie Sugars

Low-calorie sugars are one more addition to functional biomolecules and food ingredients elaborated by LAB and include both sugars and sugar alcohols (polyols) such as mannitol, sorbitol, tagatose, and trehalose. These are said to exert a number of technological and health benefits (Table 14.6). Mannitol and sorbitol are low-calorie sugars that could replace sucrose, lactose, glucose or fructose in food products as they display equivalent sweetness and taste (Hugenholtz *et al.,* 2002). Mannitol can also serve as anti-oxidant in biological cells (Shen *et al.,* 1997). Tagatose is considered as a potential sucrose replacement as it has almost equal sweetening power as sucrose and higher than similar components such as mannitol and sorbitol,

Table 14.6: Low Calorie Sweeteners, Producer Strains and Potential Applications

Sweetener	*Producer LAB/Strains*	*Technological Application*	*Functional Application*
Mannitol	Heterofermentative LAB.*Ln. pseudomesenteroides*, *Lactobacillus* spp. B001, *Lactobacillus* spp. KY-107, *Lactobacillus* spp.Y-107 and *Leuconostoc* spp. Y-002, *Ln. pseudomesenteroides* ATCC 12291, *Lb. sanfranciscensis*, and *Lb. intermedius* NRRL B-3693	Sweetener in sugar-free products, bodying and texturizing agent, anticaking agent, humectant, preservative	Osmotic diuretic, constituent in chewable tablets and granulated powders, antioxidant
Tagatose	*Lb. gayonii*, *Lb. plantarum*, and *Bifidobacterium longum*.	Low calorie sweetener, Flavour enhancer, as a partial replacer for sucrose	Promotion of weight loss, control of diabetes, prebiotic, antiplaque, noncariogenic, antihalitosis
Sorbitol	*Lb. casei*, *Lb. plantarum*	Sweetener, humectant, texturizer, and softener	Antioxidant, bacterio-static agent, cryo-protectant, stabilizer, excipient in pharmaceutical fields
Trehalose	*P. acidipropionici* and *P. freudenreichii* subsp. *Shermanii*, *P. freudenreichii* subsp. *shermanii* strain NIZO B365	Preservative, flavour and texture enhancer	Anticariogenic, dietetic, preservation and protection of biologic materials, antioxidant

Compiled from Hugenholtz *et al.*, 2002; Patra *et al.*, 2009; Monedero *et al.*, 2010.

but much lower caloric value since it is poorly degraded by the human body (Hugenholtz *et al.,* 2002). Looking to the technological and functional benefits of these safe sweeteners, their producer LAB strains can be exploited for natural enrichment of fermented products with these sweeteners for development of novel low-calorie foods with added functional values especially for diabetic patients and weight watchers (Patra *et al.,* 2009).

Conjugated Linoleic Acid (CLA)

LAB belonging to the genera *Lactobacillus* and *Streptococcus* are capable of biosynthesizing CLA which may have positive effects on inflammation, cancer (apoptosis induction), metabolic disorders (insulin resistance, body weight control), and cardio-vascular diseases. Different CLA isomers have differential effects on health. Some isomers of CLA reduce carcinogenesis, atherosclerosis, and body fat (Ogawa *et al.,* 2005). Also CLA concentrations play a role in the balance toward beneficial or detrimental effects. Anti-carcinogenic effects are observed at CLA dosages of 0.5–1 per cent (w/w) of the total diet and studies showed that humans generally excrete 20 mg of linoleic acid per day. This substrate can then be available for CLA producing bacteria in the intestine. But higher dietary intake of linoleic acid may imply risks (Ewaschuk *et al.,* 2006). All these considerations suggest that further studies are needed to better characterize CLA producing LAB and their metabolic products (Pessione, 2012). Such LAB strains can be exploited for functional food applications.

Others

Strains of LAB are found to elaborate metal-fixing enzymes. Several Lactobacillus species are able to intracellularly fix sodium selenite into selenocysteins and selenomethionines, thus providing a more bio-available form of this metal, which is generally poorly adsorbed by human cells in its inorganic form. *Lactobacillus reuteri*, for instance, express a selenocysteine-lyase, an enzyme having a key role for new seleno-proteins biosynthesis (Pessione, 2012). LAB can also function as chelators for other nutraceutically important metals at the colon level. Zinc-bearing and zinc-extruding LAB can thus function as immunomodulators useful to control viral gastroenteritis (Salvatore *et al.,* 2007).

As Probiotics

Certain strains of LAB are gaining popularity for their specific metabolic activities which are of great importance in human health. Known popularly as probiotics, such strains may exert their effects through modification of gut pH, antagonization of pathogens through production of antimicrobial compounds, competition for pathogen binding and receptor sites as well as for available nutrients and growth factors, stimulation of immunomodulatory cells and production of lactase (Parvez *et al.,* 2006). The metabolic end products of probiotic strains of LAB such as organic acids tend to lower the pH of the intestinal contents, creating conditions less desirable for harmful bacteria. It also improves intestinal mobility and relieves constipation possibly through a reduction in gut pH. Probiotics may also influence other protective functions of the intestinal mucosa including synthesis and secretion of antibacterial peptides and mucins (Parvez *et al.,* 2006). Bacteriocins, hydrogen peroxide, and biosurfactants produced by the probiotics aid their survival in the gastrointestinal tract and competitively inhibit the adherence of more pathogenic bacteria to the intestinal epithelium (Boyle *et al.,* 2006; Soccol *et al.,* 2010). Hydrogen peroxide produced by LAB provide protection against urogenital infections, especially in case of bacterial vaginosis (Reid, 2008). The enzymes and vitamins which are released into the intestinal lumen by LAB exert synergistic effects on digestion and alleviating symptoms of intestinal malabsorption. Folate-producing probiotics have been proposed to efficiently confer protection against inflammation and cancer, as this strategy combines the beneficial effects of probiotics and prevention of folate deficiency that is associated with premalignant changes in the colonic epithelia (Rossi *et al.,* 2011). Enzymatic hydrolysis of milk constituents such as lactose, proteins and fat enhance their bioavailability and helps in the release of free amino acids, organic acids and short chain fatty acids which exert health benefits in the host. Short chain fatty acids may provide protection against pathological changes in the colonic mucosa, helps to maintain an appropriate pH in the colonic lumen, which is critical in the expression of many bacterial enzymes and in foreign compound and carcinogen metabolism in the gut. Probiotics might suppress the growth of bacteria that convert procarcinogens into carcinogens, thereby reducing the amount of carcinogens in the intestine (Parvez *et al.,* 2006). EPS production by probiotics increase the colonization of probiotic bacteria by cell to cell interactions in gastrointestinal tract (Kanmani *et al.,* 2013). The antitumor activity of some of these EPSs has been reported. Additionally, the enzyme lactase alleviate symptoms of lactose intolerance.

Conclusion

The functional biomolecules and food ingredients of LAB are increasingly finding application in the food industry for reasons such as increased demand for health foods, novel foods and natural products. Also the use of antimicrobial biomolecules as a natural preservative in food and as a novel way of food preservation has gained much importance in recent years. Since LAB are food-grade microorganisms with the GRAS status, use of their biomolecules as natural alternatives to produce all-natural food products without additives are receiving much attention from food manufacturers. The health benefits exerted by the functional biomolecules of probiotics can be exploited for application in probiotic therapies. Meeting such demands require careful selection of LAB strains, process optimization to have maximum productivity and genetic engineering strategies to improve upon the strains and to design novel strains possessing desirable functional attributes. Being a treasure trove of functional biomolecules, an indepth understanding of metabolic pathways of LAB is of prime importance to food scientists and researchers. The availability of modern genetic manipulation tools and systems biology of LAB where in an organism is studied as an integrated and interacting network of genes, proteins, and biochemical reactions will provide the means for understanding and engineering complex metabolic pathways of LAB for efficient production of their biotechnologically important molecules.

References

Adams MR and Hall CJ (1988). Growth inhibition of food borne pathogens by lactic and acetic acids and their mixtures. *Int. J. Food Sci. Technol.* 23: 287-292.

Aureli P, Capurso L, Castellazzi AM, Clerici M, Giovannini M, Morelli L, Poli A, Pregliasco F, Salvini F, Zuccotti (2011). Probiotics And Health: An Evidence-based Review. *Pharmacol. res.* 63: 366-376.

Badel S, Bernardi T, Michaud P (2011). New perspectives for Lactobacilli Exopolysaccharides. *Biotechnology Advances*, 29: 54-66.

Barrangou R and Horvath P (2012). CRISPR: New Horizons in Phage Resistance and Strain Identification. *Annu. Rev. Food Sci. Technol* 3:143–62.

Barinov A, Bolotin A, Langella P, Maguin E, Van De Guchte M (2011). Genomics Of The Genus Lactobacillus. In: Sonomoto K, Yokota A, editors. Lactic Acid Bacteria and Bifidobacteria: Current Progress in Advanced Research. Caister Academic Press, Portland, USA.

Beitane I, Ciprovica I (2012). The study of added prebiotics on b group vitamins concentration during milk fermentation. *Romanian Biotechnol Letters* 16: 92-96.

Benthin S and Villadsen J (1995). Different inhibition of *Lactobacillus delbrueckii* subsp. *bulgaricus* by D- and L-lactic acid: effects on lag phase, growth rate and cell yield. *J. Appl. Bacteriol.* 78: 647-654.

Boyle RJ , Robins-Browne RM and Tang MLK (2006). Probiotic use in clinical practice: what are the risks? *Am J Clin Nutr* 83:1256–64.

Broadbent J, McMahon DJ, Welker DL, Oberg CJ, and Moineau S (2003). Biochemistry, Genetics, and Applications of Exopolysaccharide Production in *Streptococcus thermophilus*: A Review. *J. Dairy Sci.* 86:407–423.

Bron PA and Kleerebezem M (2011). Engineering lactic acid bacteria for increased industrial functionality. *Bioengineered Bugs* 2(2): 80-87.

Burgess C, O'Connell-Motherway M, Sybesma, W, Hugenholtz J, van Sinderen D. (2004). Riboflavin Production in *Lactococcus lactis*: Potential for *in situ* production of vitamin-enriched foods, *Appl Environ Microbiol.* 70: 5769-5777.

Conner, D.E. 1993. Naturally occuring compounds. In: Antimicrobials in Foods, 2nd edition. eds. Davidson, P.M. and Branen, A.L. pp. 441-468. Marcel Dekker Inc., New York.

Crittenden RG, Martinez NR, Playne MJ (2003). Synthesis and utilisation of folate by yoghurt starter cultures and probiotic bacteria. *Int. J. Food Microbiol.* 80: 217-222.

Dal Bello FD, Walter J, Hertel C, and Hammes WP (2001). *In vitro* study of prebiotic properties of levan-type exopolysaccharides from lactobacilli and non-digestible carbohydrates using denaturing gradient gel electrophoresis. *Syst. Appl. Microbiol.* 24: 232–237.

Davidson, P.M., Post, L.S., Braner, A.L., and McCurdy, A.R. 1983. Naturally occuring and miscellaneous food antimicrobials. In: *Antimicrobials in Foods.* eds. Davidson, P.M. and Branen, A.L. pp. 385-392. Marcel Dekker Inc., New York.

Deveau H, Barrangou R, Garneau JE, et al (2008). Phage response to CRISPR-encoded resistance in *Streptococcus thermophilus. J Bacteriol* 190: 1390–1400.

De Vos WM (2011). Systems solutions by lactic acid bacteria: from paradigms to practice. *Microbial Cell Factories.* 10(Suppl 1):S2- S13.

De Vuyst L and Leroy F (2007). Bacteriocins from Lactic Acid Bacteria: Production, Purification, and Food Applications. *J Mol Microbiol Biotechnol.* 13:194–199.

De Vuyst L, De Vin F, Kamerling JP (2007). Exopolysaccharides from lactic acid bacteria. In: Comprehensive Glycoscience, Kamerling JP (ed) Vol. 2. Oxford: Elsevier. p. 477–518.

Earnshaw RG (1992). The antimicrobial action of lactic acid bacteria: natural food preservation systems. In: The Lactic Acid Bacteria in Health and Disease. ed. Wood, B.J.B. pp. 211-232. Elsevier Applied Science, London and New York.

Eklund T (1984). The effect of carbon dioxide on bacterial growth and on uptake processes in the bacterial membrane vesicles. *Int. J. Food Microbiol.* 1: 179-185.

Ercolini D (2013). High-Throughput Sequencing and Metagenomics: Moving Forward in the Culture-Independent Analysis of Food Microbial Ecology. *Appl. Environ. Microbiol.* 79:10 3148-3155.

Ewaschuk J B, Walker J W, Diaz H, Madsen K L (2006). Bioproduction of conjugated linoleic acid by probiotic bacteria occurs *in vitro* and *in vivo* in mice. *J Nutr* 136: 1483–1487.

Farber JM (1991). Microbiological aspects of modified-atmosphere packaging technology - a review. *J. Food Prot.* 54: 58-70.

García-Fruitós E (2012). Lactic acid bacteria: a promising alternative for recombinant protein production. *Microbial Cell Factories.* 11:157

Gobbetti M, Stepaniak L, De Angelis M, Corsetti A, Di Cagno R (2002). Latent Bioactive Peptides in Milk Proteins: Proteolytic Activation and Significance in Dairy Processing. Crit. Rev. *Food Sci. Nutr.* 42: 223-239.

Gombocz K, Beledi A, Alotti N, et al (2007). Influence of dextran-70 on systemic inflammatory response and myocardial ischaemia-reperfusion following cardiac operations. *Critical Care.* 11 (4): R87.

Gould GW (1991). Antimicrobial compound. In: Biotechnology and Food Ingredients. eds. Goldberg, I. and Williams, R. pp. 461-483. Van Nostrand Reinhold, New York.

Guo T, Kong J, Zhang L, Zhang C, Hu S (2012). Fine tuning of the lactate and diacetyl production through promoter engineering in *Lactococcus lactis.* PLoS ONE 7:e36296. http://dx.doi.org/10.1371/journal.pone.0036296.PMid:22558426 PMCid:3338672.

Guldfeldt LU, Sorensen KI, Stroman P, Behrndt H, Williams D, Johansen E (2001). Effect of Starter Cultures with a Genetically Modified Peptidolytic Or Lytic System on Cheddar Cheese Ripening. *Int. dairy J.* 11: 373–382.

Gonzalez L, Sacristan N, Arenas R, Fresno JM, Tornadijo ME (2010). Enzymatic Activity of Lactic Acid Bacteria (with Antimicrobial Properties) Isolated from a Traditional Spanish Cheese. *Food microbiol.* 27: 592–597.

Haque E , Chand R and Kapila S (2009). 'Biofunctional Properties of Bioactive Peptides of Milk Origin'. *Food Reviews International.* 25(1): 28 – 43.

Hoefnagel MHN, Starrenburg MJC, Martens DE, Hugenholtz J, Kleerebezem M, van Swam II, Bongers R, Westerhoff HV, Snoep JL (2002) Metabolic engineering of lactic acid bacteria, the combined approach: kinetic modeling, metabolic control and experimental analysis. *Microbiology.* 148:1003-1013.

Holzapfel W H, Haberer P, Geisen R, Björkroth J, and Schillinger U (2001). Taxonomy and important features of probiotic microorganisms in food and nutrition. *Am J Clin Nutr.* 73(suppl):365S–73S.

Horvath P, Romero DA, Coute-Monvoisin AC, et al (2008). Diversity, activity, and evolution of CRISPR loci in *Streptococcus thermophilus*. *J Bacteriol.* 190: 1401-1412.

Hotchkiss JH, Chen JH and Lawless HT (1999). Combined effects of carbon dioxide addition and barrier films on microbial and sensory changes in pasteurized milk. *J. Dairy Sci.* 82: 690-695.

Huang L, Forsberg CW and Gibbins LN (1986). Influence of external pH and fermentation products on *Clostridium acetobutylicum* intracellular pH and cellular distribution of fermented products. *Appl. Environ. Microbiol.* 51: 1230-1234.

Hugenholtz J (1993). Citrate metabolism in lactic acid bacteria. *FEMS Microbiol Rev.* 12:165-178.

Hugenholtz J, Sybesma W, Groot MN et al (2002). Metabolic engineering of lactic acid bacteria for the production of Nutraceuticals. *Antonie van Leeuwenhoek* 82: 217–235.

Jay JM (1982). Antimicrobial properties of diacetyl. *Appl. Environ. Microbiol.* 44: 525-532.

Kabara JJ (1993). Medium-chain fatty acids and esters. In: Antimicrobials in Foods, 2nd edition. eds. Davidson, P.M. and Branen, A.L. pp. 307-342. Marcel Dekker Inc., New York.

Kanmani P, Satish kumar R, Yuvaraj N, Paari K A, Pattukumar V, and Venkatesan Arul (2013). Probiotics and Its Functionally Valuable Products—A Review. *Crit rev food sci nutr.* 53:641–658.

Khalid K (2011). An overview of lactic acid bacteria. *International Journal of Biosciences* (IJB). 1(3): 1-13.

Khurana H K and Kanawjia SK (2007). Recent Trends in Development of Fermented Milks. *Current Nutrition and Food Science.* 3: 91-108.

Korhonen H (2009). Milk-derived bioactive peptides: From science to applications. *Journal of Functional Foods.* 1: 177-187.

Korhonen H and Pihlanto A (2004). Milk-derived bioactive peptides: formation and prospects for health promotion. In: Handbook of functional dairy products. Functional foods and nutraceuticals series 6, Shortt C and O'Brien J (eds.). CRC Press, Boca Raton, FL, USA, pp. 109-124.

Korhonen H and Pihlanto A (2006). Bioactive peptides: production and functionality. *Int Dairy J*. 16: 945-960.

Kumar A, Grover S, Sharma J, and Batish VK (2010). Chymosin and other milk coagulants: sources and biotechnological interventions. Critical Reviews in Biotechnology. DOI: 10.3109/07388551.2010.483459.

LeBlanc JG, Lain˜ o1JE, Juarez del Valle M, et al (2011). B-Group vitamin production by lactic acid bacteria – current knowledge and potential applications. *Journal of Applied Microbiology* 111: 1297–1309.

Lindgren SE and Dobrogosz WJ (1990). Antagonistic activities of lactic acid bacteria in food and feed fermentations. *FEMS Microbiol. Rev.* 87: 149-164.

Lopez de Felipe F, Kleerebezem M, de Vos WM, Hugenholtz J (1998) Cofactor engineering: a novel approach to metabolic engineering in *Lactococcus lactis* by controlled expression of NADH oxidase. *J Bacteriol.* 180: 3804-3808.

Maeda H, Zhu X, Omura K, Suzuki S, Kitamura S (2004a). Effects of an exopolysaccharide (kefiran) on lipids, blood pressure, blood glucose, and constipation. *Biofactors.* 22: 197–200.

Maeda H, Zhu X, Suzuki S, Suzuki K, Kitamura S (2004b). Structural characterization and biological activities of an exopolysaccharide kefiran produced by *Lactobacillus kefiranofaciens* WB-2B. *J Agric Food Chem.* 52: 5533–8.

Michaela S, Reinhard W, Gerhard K, Christine ME (2009). Cultivation of anaerobic and facultatively anaerobic bacteria from spacecraft-associated clean rooms. *Applied and Environmental Microbiology.* 11(75): 3484-3491.

Mlichová Z and Rosenberg M (2006). Current trends of â-galactosidase application in food technology. *Journal of Food and Nutrition Research.* 45 (2): 47-54.

Mills S, Coffey A, McAuliffe OE, Meijer WC, Hafkamp B and Ross RP (2007). Efficient method for generation of bacteriophage insensitive mutants of Streptococcus thermophilus yoghurt and mozzarella strains. *J Microbiol Methods.* 70: 159-164.

Mills S, Griffin C, Coffey A, Meijer WC, Hafkamp B and Ross RP (2010). CRISPR analysis of bacteriophage-insensitive mutants (BIMs) of industrial *Streptococcus thermophilus* – implications for starter design. *J Appl Microbiol.* 108 : 945–955.

Mollea C, Marmo L and Bosco F (2013). Valorisation of Cheese Whey, a By-Product from the Dairy Industry. Chapter 24. In : Food Industry. Muzzalupo I (ed). pp 556-557.

Monsan P, Bozonnet S, Albenne C, Joucla G, Willemot RM, Simeon MR (2001). Homopolysaccharides from lactic acid bacteria. *Int Dairy J.* 11: 675–85.

Monedero V, Perez-Martínez G, Yebra MJ (2010). Perspectives of Engineering Lactic Acid Bacteria For Biotechnological Polyol Production. *Appl microbial biotechnol.* 86: 1003–1015.

Naidu AS, Bidlack WR, Clemens RA (1999). Probiotic Spectra of Lactic Acid Bacteria (LAB). *Crit rev food sci nutr.* 38: 13-126.

Ogawa J, Shigenobu Kishino , Akinori Ando, Satoshi Sugimoto,Kousuke Mihara, Sakayu Shimizu (2005). Production of conjugated fatty acids by lactic acid bacteria. *Journal of Bioscience and Bioengineering*.100 (4): 355–364.

Olempska-Beer ZS, Merker RI, Ditto MD, DiNovi MJ (2006). Food-processing enzymes from recombinant microorganisms—a review. *Regulatory Toxicology and Pharmacology*. 45: 144–158.

Oliveira AP, Nielsen J, Forster J (2005). Modeling *Lactococcus lactis* using a genome-scale flux model. *BMC Microbiology* 5:39. http://dx.doi.org/10.1186/1471-2180-5-39 PMid:15982422 PMCid:1185544

Panesar PS, Kumari S, and Panesar R (2010). Potential Applications of Immobilized α-Galactosidase in Food Processing Industries. *Enzyme Research* Article ID 473137, 16 pages doi:10.4061/2010/473137.

Papagianni M, Avramidis N (2011). *Lactococcus lactis* as a cell factory: A twofold increase in phosphofructokinase activity results in a proportional increase in specific rates of glucose uptake and lactate formation. *Enzyme Microb Technol.* 49:197-202.

Papagianni M (2012). Metabolic engineering of lactic acid bacteria for the production of industrially important compounds. *Computational and Structural Biotechnology Journal.* 3(4): e201210003, http://dx.doi.org/10.5936/csbj.201210003

Parvez S, Malik KA, Kang S AH and Kim HY (2006). Probiotics and their fermented food products are beneficial for health. *Journal of Applied Microbiology* 100: 1171–1185.

Patra F, Tomar SK and Arora S. (2009). Technological and Functional Applications of Low-Calorie Sweeteners from Lactic Acid Bacteria. *Journal of Food Science.* 74 (1): R16-R23.

Pessione E (2012). Lactic acid bacteria contribution to gut microbiota complexity: lights and shadows. *Front Cell Infect Microbiol.* 2: 86. doi: 10.3389/fcimb.2012.00086.

Piard JC and Desmazeaud M (1991). Inhibiting factors produced by lactic acid bacteria: 1. Oxygen metabolites and catabolism end-products. *Lait* 71: 525-541.

Platteeuw C, Hugenholtz J, Starrenburg MJC, van Alen-Boerrigter I, de Vos WM (1995). Metabolic engineering of *Lactococcus lactis*: influence of the overproduction of á-acetolactate synthase in strains deficient in lactate dehydrogenase as a function of culture conditions. *Appl Environ Microbiol* 61:3967-3971.

Prajapati JB (1995). Microbiology of starter cultures. In: Fundamentals of Dairy Microbiology. Akta Prakashan, Nadiad, India. pp.54.

Quintans NG , Blancato V , Repizo G, Magni C and López P (2008). Citrate metabolism and aroma compound production in lactic acid bacteria, Chapter 3. In: Molecular Aspects of Lactic Acid Bacteria for Traditional and New Applications. Editors: Baltasar Mayo, Paloma López and Gaspar Pérez-Martínez pp 65-88. ISBN: 978-81-308-0250-3

Rao DR and Reddy JC (1984). Effect of lactic fermentation of milk on milk lipids. *J. Food Sci.* 49: 748-750.

Rattanachaikunsopon P and Phumkhachorn P (2010). Lactic acid bacteria: their antimicrobial compounds and their uses in food production. *Annals of Biological Research.* 1 (4) : 218-228.

Reid G (2008). Probiotic lactobacilli for urogenital health in women. *J. Clin. Gastroenterol.* (Suppl. 3), 42: 234–236.

Richards RME, Xing DKL and King TP (1995). Activity of *p*-aminobenzoic acid compared with other organic acids against selected bacteria. *J. Appl. Bacteriol.* 78: 209-215.

Rodrigues KL, Carvalho JCT, Schneedorf JM (2005). Anti-inflammatory properties of kefir and its polysaccharide extract. *Inflammopharmacol.* 13: 485–92.

Rodríguez JM, Martínez MI, Kok J (2002). Pediocin PA-1, a wide-spectrum bacteriocin from lactic acid bacteria. *Crit Rev Food Sci Nutr.* 42: 91–121.

Rossi M, Amaretti A and Raimondi S (2011). Folate production by probiotic bacteria. *Nutrients.* 3: 118–134.

Ruas-Madiedo R and Reyes-Gavila´n CG de (2004). Invited Review: Methods for the Screening, Isolation, and Characterization of Exopolysaccharides Produced by Lactic Acid Bacteria. *J. Dairy Sci.* 88: 843–856.

Ruas-Madiedo P, Abraham A, Mozzi F, Los Reyes-Gavilán C G de (2008). Functionality of exopolysaccharides produced by lactic acid bacteria.. In: Molecular aspects of lactic acid bacteria for traditional and new applications. Mayo B, López P and Pérez-Martínez G (Eds). pp. 137-166.

Salvatore S, Hauser B, Devreker T, Vieira M C, Luini C, Arrigo S, Nespoli L, Vandenplas Y (2007). Probiotics and zinc in acute infectious gastroenteritis in children: are they effective? *Nutrition.* 23: 498–506.

Sangoyomi TE, Owoseni AA, Okerokun O (2010). Prevalence of enteropathogenic and lactic acid bacteria species in wara: A local cheese from Nigeria. *African Journal of Microbiology Research.* 4(15): 1624-1630.

Sharma S, Singh R, Rana S (2011). Bioactive Peptides: A Review. *Int. J. Bioautomation.* 15(4): 223-250.

Shen B, Jensen RG and Bohnert HJ (1997). Mannitol protects against oxidation by hydroxyl radicals. *Plant Phys.* 115: 527–532.

Sieber R, Bütikofer U, Egger C, Portmann R, Walther B, Wechsler D (2010). ACE-inhibitory activity and ACE-inhibiting peptides in different cheese varieties. *Dairy Sci. Technol.* 90: 47–73.

Sleator RD and C. Hill C (2008). New frontiers in probiotic research. Letters in Applied Microbiology 46: 143–147.

Soccol CR , Vandenberghe LPS, Spier MR et al (2010). The Potential of Probiotics: A Review. *Food Technol. Biotechnol.* 48 (4): 413–434.

Sorek R, Kunin V and Hugenholtz P (2008). CRISPR – a widespread system that provides acquired resistance against phages in bacteria and archaea. *Nat Rev Microbiol.* 6: 181–186.

Starovoitova VV, Velichko TI, Baratova LA, Filippova IY, Lavrenova GI (2006). A comparative study of functional properties of calf chymosin and its recombinant forms. *Biochemistry Mosc.* 71: 320–324.

Steele J, Broadbent J and Kok J (2013). Perspectives on the contribution of lactic acid bacteria to cheese avor development. *Current Opinion in Biotechnology.* 24:135–141.

Stiles ME, Holzapfel WH (1997). Lactic Acid Bacteria Of Foods And Their Current Taxonomy. *Int. J. food Microb.* 36: 1–29.

Suskovic J, Kos B, Beganovic J, Pavunc A L, Habjanic K and Matosic S (2010). Antimicrobial Activity – The Most Important Property of Probiotic and Starter Lactic Acid Bacteria. *Food Technol. Biotechnol.* 48 (3): 296–307.

Tamang JP (2011). Prospects Of Asian Fermented Foods In Global Markets. 11th ASEAN Food Conference, Bangkok, Thailand.

Teusink B, Bachmann H, Molenaar D (2011). Systems biology of lactic acid bacteria: a critical review. *Microb Cell Fact.* 10:S11.

Tidona F, Criscione A, Guastella A M, Zuccaro A, Bordonaro S, Marletta D (2009). Bioactive peptides in dairy products. *Ital.J.Anim.Sci.* 8: 315-340.

Tieking M, Kaditzky S, Valcheva R, Korakli M, Vogel RF, Gänzle MG (2005). Extracellular

homopolysaccharides and oligosaccharides from intestinal lactobacilli. *J Appl Microbiol.* 99: 692–702.

Vinderola G, Perdigón G, Duarte J, Farnworth E, Matar C (2006). Effects of the oral administration of the exopolysaccharide produced by Lactobacillus kefiranofaciens on the gut mucosal immunity. *Cytokine.* 36: 254–60.

Wagner MK and Moberg LJ (1989). Present and future use of traditional antimicrobials. *Food Technol.* 1: 143-147.

Werning ML, Sara Notararigo, Montserrat Nácher, Pilar Fernández de Palencia, Rosa Aznar and Paloma López (2012). Biosynthesis, Purification and Biotechnological Use of exopolysaccharides Produced by Lactic Acid Bacteria, Food Additive, Prof. Yehia El-Samragy (Ed.), ISBN: 978-953-51-0067-6, InTech, Available from: http://www.intechopen.com/books/food-additive/biosynthesis-purification-and-biotechnological-use-of exopolysaccharides- produced-by-lactic-acid-bac.

Wong,HC and Chen YL (1988). Effects of lactic acid bacteria and organic acids on growth and germination of *Bacillus cereus*. *Appl. Environ. Microbiol.* 54: 2179-2184.

Wouters JTM, Ayad EHE, Hugenholtz J, Smit G (2002). Microbes from raw milk for fermented dairy products. *Int. Dairy J.* 12: 91-109.

2015, Dairy Product Technology: Recent Advances ***Pages 309–324***
Editors: **Subrota Hati, Surajit Mandal and Birendra Kumar Mishra**
Published by: **DAYA PUBLISHING HOUSE, NEW DELHI**

Chapter 15

Trends in the Packaging of Traditional Indian Dairy Products

Kunal K. Ahuja, Sumit Goyal,
Kiran Bala and G.K. Goyal

India is the largest milk producing country in world with annual production of 133 million tons during 2012 -13 (DAHD, 2013). About half of milk produced in India is converted into various traditional Indian dairy products, whereas 46 per cent of total milk produced is consumed as fluid milk. Traditional dairy products of India are the products which evolved over ages utilizing locally available fuels and cooking wares. In India, no celebration is considered complete without serving the traditional sweets. Traditional dairy products of India can be classified under six broad categories namely, heat desiccated, heat acid coagulated, fermented, fat rich, cereal based and frozen dairy products.

In recent years rapid developments are taking place in the process of production and packaging of dairy products. The wide range of traditional dairy products, *viz.*, gulabjamun, rasogolla, shrikhand, paneer have already taken strong industrial footing in the country with the advancement in manufacturing technologies. However, a large proportion of traditional dairy products are still produced by small scale sweet makers popularly known as '*halwais*', which results into batch to batch large variation in the quality of products. Most of the *halwais* do not pay much attention to hygiene during production process, which leads to serious contamination of the products during manufacture. On the other hand, high water activity of traditional dairy products assists deterioration process and limits their shelf-life. Processing methods like heating, freezing, radiation, addition of preservatives either from chemical sources (*e.g.* potassium sorbate, hydrogen peroxide, sodium benzoate etc.) or biological sources (*e.g.* bactiriocins like nisin, reutrin etc.), or in selected combination, have been attempted by several researchers to inhibit deteriorative changes during storage. However, increased costs for processing methods like heating, freezing and bio-preservation; and growing consumer concern towards chemical preservatives has prompted the food industry to look for alternative methods of preservation.

Packaging can be described as a coordinated system of preparing foods for transport, storage, sale and end use. It provides gateway to know the product and protects it from the adverse effects of the environment. Generally, a package performs three basic functions: containment, protection, and attracts the customer, thus helps in selling of the product. Many food products must be protected from atmospheric influences such as water vapour, oxygen and odours. Since, most of the traditional Indian dairy products are rich in fat content; they are susceptible to oxidative and hydrolytic rancidity. Improperly packaged dairy products may easily go under bio-deterioration as getting contaminated with microbes or attacked by insects and rodents during storage and distribution. Changes in body and textural properties of traditional dairy products may also occur during stacking and transportation making them unacceptable to the consumers, which can be avoided by proper packaging techniques.

Many efforts have been made for mechanized production of traditional Indian dairy products, thereby reducing batch to batch variation in the quality of product and reduce the risk of microbial contamination. Today most of the dairy products in Indian market are being sold in packages like plastic pouches, laminates, moulded containers, paper cartons, etc. Such packages are light in weight, thus reduce the cost of transportation and can be used for single-service. Rapid developments have already taken place for inline packaging system for western dairy products such as cheese, butter etc. packed in packages like paper cartons, plastic pouches, laminates, etc. However, appropriate attention has not been given for designing and developing of new packaging machinery suitable for traditional dairy products.

Increasing inland demands as well as export potential for traditional dairy products have indicated an urge to extend shelf life by adopting good manufacturing practices and innovative packaging techniques. Lack of proper understanding about the ideal characteristics of traditional dairy products and their compatibility with packaging material counter with the selection or development of new packages. A good package being used in packaging of traditional dairy products should preserve its colour, flavour, odour, shape, body and texture, and prevent physico-chemical changes during storage.

Now-a-days, modified atmosphere packaging (MAP) technique is being extensively used for shelf life extension of food products. As the name implies, a modified atmosphere is one in which the normal composition of air is changed or modified within a high barrier package. Barrier properties of packaging materials provide resistance to passage of gases, vapours and odours, thereby prevent early spoilage. This method of packaging replaces the air headspace of package with a gas or mixture of gases like N_2 and CO_2 or a mixture thereof (Addington, 1991). MAP is capable of substantially extending the shelf life of many perishable foods, and has accordingly undergone developments to meet the consumer demand for fresh, high quality processed convenience foods. The spoilage of processed dairy foods occurs mainly due to microbial growth and oxidative changes. MAP can significantly extend the shelf life of dairy foods by retarding the microbial growth and reducing the oxidative changes. It helps in increasing the distribution radius, retains appeal to consumers, and retards the growth of molds and bacteria. Besides, MAP also prevents mechanical damage of the product by adding a cushioning effect of gas around the product.

Role of Packaging

Three basic functions of packaging are to contain the product, to protect the product and help in boosting the sale of the product. Other requirements such as to preserve, measure, communicate, display and glamorize may also act as additional functions of package. However, a package should be free from any deception like expanded and over packaging. It is essential to know the nature and composition of the product, its desired shelf-life under specified conditions of storage. The ideal packaging material should have the following characteristics (Punjirath, 1995):

- It should not impart its own odour/flavour to the product.
- It should be inert to food and must be non- toxic.
- It should protect the product from environmental influences such as moisture, oxygen, and light.
- It should be convenient, tamper proof, machinable and economic.
- It should indicate difference among different packaged products.

The package not only protects the product but also gives information about the contents, storage conditions, methods of use, date of manufacture, expiry date, price and nutritional facts.

Choice of Packaging Material

Choice of an appropriate packaging material depends on several factors including sensitiveness of the product towards moisture, oxygen and light; storage conditions; design of package; intrinsic factors: pH, acidity, redox potential, ionic strength and other compositional attributes of food materials; required speed of packaging and machine capability; migration of package constituent to foodstuff; shelf life; environmental concern regarding biodegradability and recycling potential; mode of transportation; convenience to end users, *viz.*, boil in bag, dual ovenable, re-closable, cut-open and mix etc.; and compatibility with the processing techniques such as aseptic packaging, retort processing, radiation processing, post-packaging heat treatment etc.

Packaging of Traditional Dairy Products

At present, *halwais* used to keep milk-based sweets in open trays made up of aluminum or tin or stainless steel in refrigerated or non-refrigerated counter show cases (Figure 15.1).

Figure 15.1: Traditional Dairy Based Sweets kept in Stainless Steel Trays at Local Market.

On demand by the consumers, items are weighed and placed either in ordinary paper box or paper box lined with butter paper or laminated paper box. Of late, plastic boxes have also appeared in market scene for packaging of sweets. However, in much more traditional way some shopkeepers place sweets on dhak leaves and give to the consumers. Some *halwais* wrap sweets in glassine or greaseproof paper and sell them in duplex board boxes. No better packaging are available for traditional dairy products like gulabjamun and rasogulla for local consumption, but it is canned for export purposes and long term storage. Present trend for packaging of khoa, some khoa based sweets namely peda, kalakand, burfi, gulabjamun; paneer; chhana and chhana based sweets are discussed hereunder:

Khoa

Khoa is concentrated whole milk product obtained by open pan condensing of milk. It is a partially heat desiccated milk product obtained by continuous boiling of cow, buffalo or mixed milk or concentrated milk till desired flavour, texture and consistency is achieved. As stipulated by Food Safety and Standard Rules (2011), the fat content of khoa should not be less than 30 per cent on dry matter basis. Total solids content in khoa may vary with the variety, which remains in the range of 55-70 per cent. Khoa can also be prepared by using steam jacketed kettle, and by continuous methods using scraped surface heat exchanger. The quality and shelf life of khoa largely depends on the type and quality of milk and manufacturing technique employed for its production. The product coming out from the processing equipments retains its temperature around 90°C and therefore, remains almost free from any spoilage organisms. Post processing handling and storage conditions are responsible for its microbial contamination resulting into its limited shelf-life of less than 5 days at room temperature. In traditional marketing, blocks of khoa are shaped hemispherical and packed in bamboo basket lined with green leaves or used newspaper sheets and covered with a gunny cloth or dhak leaves. Khoa in large quantity is also transported to far off places packed in gunny bags. Packaging of khoa in this manner is improper and unsuitable from hygienic point of view.

The most common defects of khoa during storage are hardening of the product due to loss of moisture, fat oxidation, yeast and mould growth etc. Khoa if stored at higher temperature will develop acidity, whereas khoa stored at very low temperature for longer period will develop stale flavour due to obsolescence of sulfhydryl group and sandiness defect may occur due to crystallization of supersaturated lactose. Selection of proper packaging systems minimizes these defects and extends the shelf life of khoa for substantial longer period. By using 4-ply laminated pouches made of PP/LDPE/foil/LDPE, the shelf life of khoa can be increased to 14 days at 30°C and 75 days in the cold storage (Kumar *et al.,* 1975). According to Goyal and Rajorhia (1991) hot filling of khoa at 80 to 90° C in pre-sterilized tin cans enhances its shelf life up to 14 days at 37° C, while the use of 3-ply laminate made of paper/aluminum foil/ LDPE or 2-ply laminate of MST cellulose/LDPE keeps khoa in good condition for 10 days at 37°C and for 60 days under refrigerated storage. Hot filling of khoa at such a high temperature helps extending its shelf life, but causes undesirable browning in khoa making it unsuitable for manufacturing of some sweets like burfi. However, brown colour remains desirable in sweets like kalakand (product prepared from danedar variety of khoa), dharwad peda (a product in Dharwad region of Karnataka State, India), *lal peda* (prepared in some parts Eastern Uttar Pradesh, India) and thabdi (famous product of western parts of Gujarat, India) etc. High barrier laminates based on polyester/ethylene vinyl alcohol (EVOH)/polythene were suggested for products like khoa and milk desserts (Punjrath, 1995). Such laminates can be used for bulk packaging of khoa in cold stores for longer duration. Tin cans and rigid plastic containers of 15 kg can also be advantageously used for bulk packaging of khoa under vacuum and storage at room temperature.

Danedar khoa is highly perishable. Its shelf life could be extended up to 60 days at 11°C by packaging under nitrogen/vacuum in a flexible pouch of poster paper/ aluminum foil/LDPE (Sharma *et al.*, 2001). The antimicrobial agent, potassium sorbate (0.03 per cent) coupled with vacuum packaging enhanced the shelf-like of khoa in cellophane/LDPE laminates to 21 days. Vacuum packaging enhanced the shelf life of khoa up to 9 days in the conventional laminates (NDRI, 1993).

Khoa Based Sweets: Burfi, Peda and Kalakand

Burfi, peda and kalakand occupy dominating place in the several khoa-based sweets in terms of popularity and market demand. Burfi is prepared using dhap khoa, while pindi variety of khoa is used for the preparation of peda. Base material for the preparation of kalakand is danedar khoa. However, good quality of kalakand is only obtained when prepared in one step from milk instead of two -steps method using danedar khoa as a base material. About 30 per cent granulated sugar is blended vigorously with hot khoa for making these sweets, but the shape, body and texture and chemical composition of such products differ from each other.

The present trend for packaging of such products is (Alam *et al.*, 2005): peda: paperboard boxes (Figure 15.2a) with wax paper lining, paper bags, dhak leaves; kalakand: dhak leaves, paperboard cartons lined with was paper, plastic boxes; burfi: dhak leaves, paperboard cartons with paper lining, duplex board box containing plastic tray (Figure 15.2b), paper bags and plastic boxes (Figure 15.2c). These sweets are mostly packaged in paper cartons or duplex board boxes with or without butter paper lining. In a more attractive package, more than one traditional dairy sweet can be arranged in columns of a plastic tray retained in duplex board boxes (Figure 15.3). Such packages are very popular for gift purposes during festive occasions.

The traditional packages do not provide sufficient protection to milk sweets from atmospheric hazards and unhygienic handling, and thus susceptible to become dry, hard, mouldy and develop off flavours. Also, the product packed in these wrappers/packages are not suitable for distant transportation and outstation retail sale or sale through super markets, because they lack necessary mechanical and protective properties.

Tin containers can be used but their cost is excessive. Recently, some of the reputed manufacturers of these sweets have started packaging burfi and peda in HDPE (high density polyethylene)/polypropylene boxes (Figure 15.2c) and cartons of 500g and 1 kg size. The modern flexible polyfilms (Figure 15.2d) and laminates (Figure 15.2e) offer alternate choices. The chemical composition of the sweet, the transportation hazards, and the period of storage under specified conditions of temperature and humidity are the major factors, which should largely decide the type of packaging materials. Various types of post manufacture defects in burfi, peda

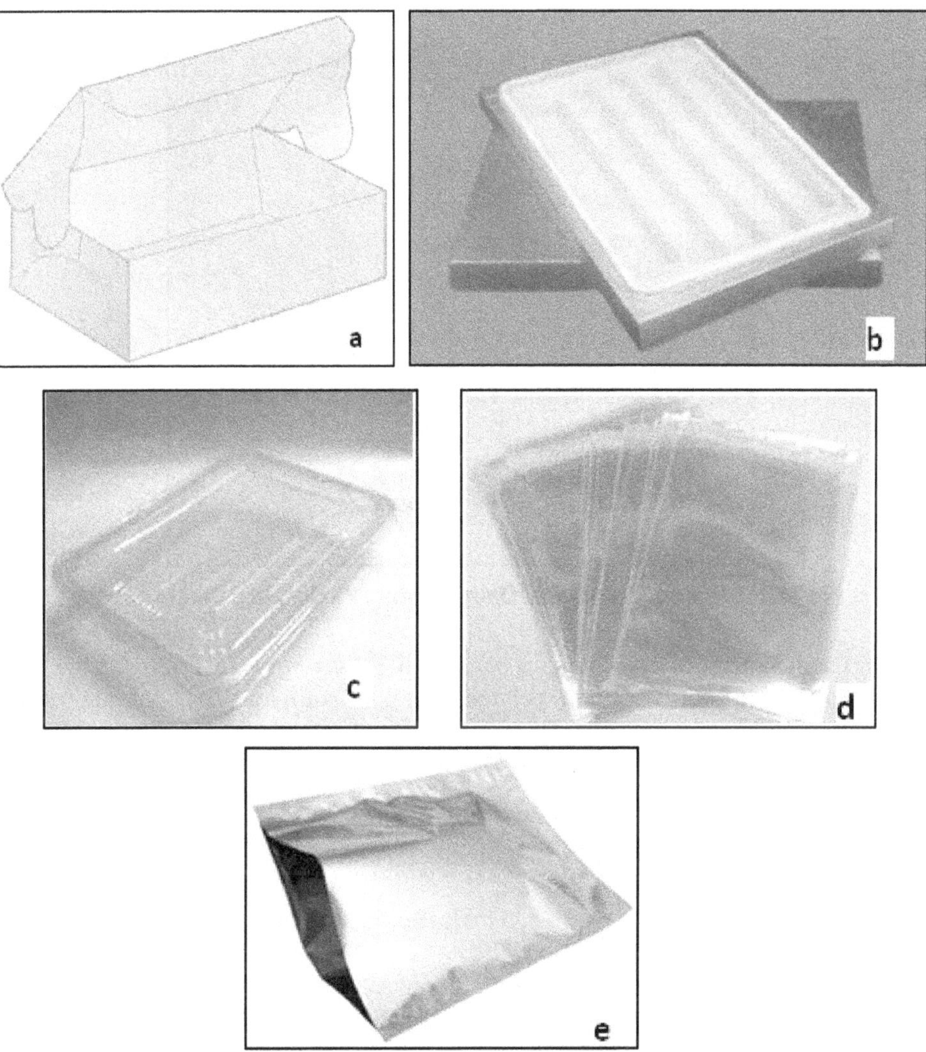

Figure 15.2: a: Paperboard Boxes; b: Duplex Board Box with Plastic Tray; c: Polypropylene Box; d: LDPE Pouches; e: Metalized Polyester Laminate.

and kalakand and their prevention with appropriate packaging techniques are discussed hereunder.

Defects in Body and Texture

Moisture content of burfi, peda and kalakand varies widely with amount of sugar added and extent of desiccation achieved during their manufacture. In general, lab made burfi, peda and kalakand contain approximately 15 per cent moisture. At particular moisture level, the sweets have unique texture and typical chewing properties. For storage of burfi, an optimum relative humidity (RH) of 70 per cent is recommended. Higher RH makes the product moist and pasty; whereas, low RH

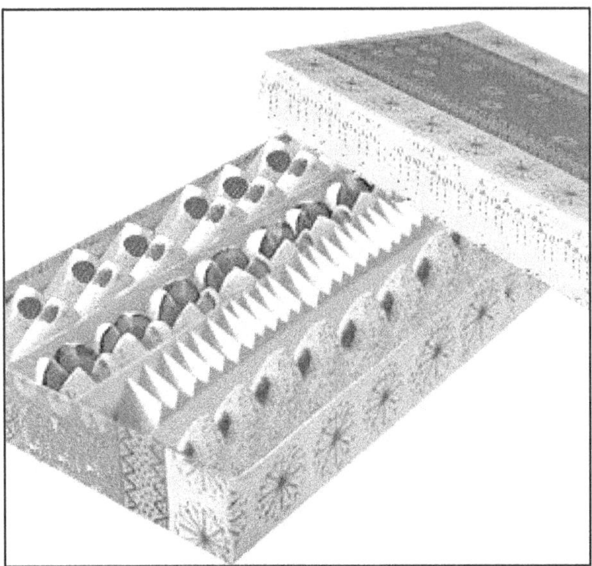

Figure 15.3: Different Traditional Dairy Product Packed in a Single Package.

results in hard product. The choice of package to prevent moisture loss may be from HDPE, PP, MXXT (moisture proof, polymer coated on both sides, transparent), polycel or other suitable combinations. This will prevent the ingress of moisture into the product and also prevent the product becoming pasty under humid conditions.

Prevention of Rancid and Oxidized Flavours

Khoa based sweets are quite rich in milk fat and hence susceptible to rancidity and oxidative changes during storage. Rancidity resembles the sour and pungent odour of butyric acid, which is caused by the hydrolysis of milk fat, while oxidized flavour develops due to the action of oxygen with the fatty acid radicals of the unsaturated fats. Light is among the major factors next to oxygen, responsible for initiating oxidation reaction in the presence of free radicals. Proper packaging plays a key role in preventing the rancidity. Exposure of direct light to the product can be prevented by using packaging materials having reflecting pigments and denser films. Packaging materials which have very good oxygen barrier properties such as MST cellulose, MXXT, metallized polyester/poly, 5-layer co- extruded films, laminates having Al – foil are recommended for preventing oxidative deterioration. Vacuum packaging of traditional products also enhances the shelf life to a great extent (Goyal and Rajorhia, 1991).

Prevention from Discolouration and Absorption of Foreign Odours

Burfi, peda and kalakand often lose their original colour and appearance during storage due to light induced oxidation leading to loss of colour intensity. Maillard type browning – a common storage defect of milk sweets, is also accelerated by exposure to light and moisture. These fat rich dairy products quickly absorb foreign odours and rapidly lose their inherent delicate colour and flavour. It is extremely important that these products are packed in such materials which can stop the two-way traffic of odours/gases in the products in order to preserve their original colour and flavour. Packaging material should also be grease resistant in order to avoid seepage of fat.

Modified Atmospheric Packaging of Khoa Based Sweets

The common types of spoilage in burfi, peda and kalakand can be significantly delayed or prevented by using modified atmospheric packaging. Kumar *et al.* (1997) packed peda under modified atmospheric conditions (80 per cent nitrogen and 20 per cent CO_2) in high barrier multilayer (EVA/EVA/PVdC/EVA) film, and at atmospheric conditions containing oxygen scavenger. Samples with modified atmospheric conditions exhibited shelf life of 15 days at 37°C and 30 days at 20°C, whereas peda packed with oxygen scavenger in high barrier films exhibited an extended shelf life up to 2 months at 37°C and 6 months at 20°C. Peda samples packed in conventional paperboard boxes were reported to indicate signs of mold growth within 7 days. Londhe *et al.* (2012) attempted to improve the shelf-life of brown peda by modifying the atmosphere within package of nylon based co-extruded film and storage up to 40 days at 30±1°C. Brown peda packaged under vacuum was reported to have a shelf of 40 days. However, samples packed under modified atmosphere and paperboard boxes lined with wax paper deteriorated rapidly when compared with vacuum packed brown peda samples.

Chhana

Chhana is a heat-acid coagulated product popular in Eastern and Northern regions of India. It is used as a base material for the preparation of large number of sweets including rasogolla, rajbhog, sandesh, cham-cham, rasomalai, chhana murki, chhana podo pantooa etc. In traditional way, chhana is packed in bamboo baskets lined with leaves. For transportation purposes, chhana is dipped in whey kept in earthen pots. The product has 50-70 per cent moisture and therefore, it has very limited keeping quality. It requires protection from heat, light, O_2, microbial contamination, moisture loss, and odour. Fresh chhana is preferred for the preparation of good quality sweets. Therefore, not much attempts have been taken for its shelf-life extension.

Chhana Based Sweets

Chhana based sweets like sandesh, rasogolla, etc. are extremely popular in the eastern and north eastern regions of the country. Sandesh is generally packaged in paperboard cartons with a wax paper lining, ordinary paper bags and dhak leaves. The rosogolla is packaged in tinplate cans, kulhads etc. Canning of rosogolla is expensive and the other methods of packaging are unhygienic, inconvenient and unsuitable for outstation retail sales.

Gulabjamun and Rosogolla

Gulabjamun is a khoa based sweet popular all over India. Dhap variety of khoa having 40-45 per cent moisture is preferred for making of gulabjamun. Gulabjamun should have brown colour, smooth and spherical shape, soft and slightly spongy body, free from both lumps and hard central core, uniform granular texture, with cooked flavour and free from doughy feel and the sweet should be fully succulent with sugar syrup with optimum sweetness. Rasogolla is popularly known as king of Bengal sweets, first prepared by Nobin Chandra Das in 1868. It resembles ping-pong ball in shape, snow-white in colour and possesses a spongy, slightly chewy and juicy body. Rasogolla balls are stored and served in sugar syrup. The product may be flavoured with kewara, pistachio and rose flavour. These sweets need to be saved from mechanical hazards, light, oxygen, ingress or egress of moisture, yeasts and moulds. The similarity between the two is based on their shape, texture and method of storage. Both are spherical in shape, spongy, porous and kept in sugar syrup. Their shape and porosity attributes are very critical and have to be maintained till the product reaches to the consumer. On an average, they contain about 40 per cent moisture and 50 per cent sugar. Yeast and mould growth is a common problem associated with yeasty/fruity flavour defects during storage of both the sweets. Since the body and texture of gulabjamun and rosogolla is very delicate and has to be preserved in sugar syrup, they are invariably packaged in lacquered tin cans (Figures 15.3a and 15.3b) of 500g and 1kg capacity, respectively. The proportion of rosogolla and syrup is kept 40:60. Using hot filling technique product can be stored in good condition for more than 6 months at ambient conditions.

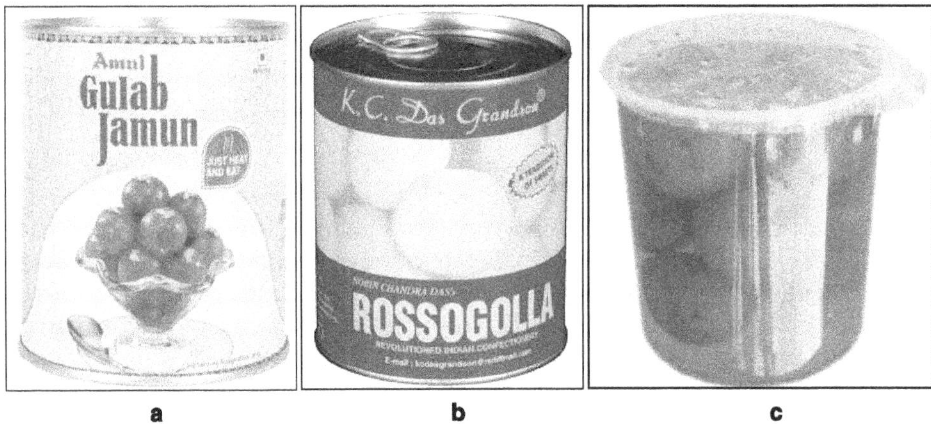

Figure 15.3: a: Gulabjamun Packed in Tin Can; b: Rasogolla Packed in Tin Can; c: Gulabjamun Packed in Plastic Container with Heat Sealed Lid.

Tin cans with 'S' resistant lacquer should be used, if sulphur containing compound is being used as preservative. Gulabjamun is largely packaged without syrup in paper cartons or plastic boxes lined with wax paper. Though lacquered tin can is the most suitable packaging material for rosogolla and gulabjamun, but it is very expensive. Hence, there is a need to pack these products in composite cans made of plastic and laminated with polypropylene – aluminum foil material. The material is heat resistant and suitable for hot filling. Recent days, polypropylene cups (Figure 153.3c) and high impact polystyrene (HIPS) cups along with heat sealable caps are also being used for the packaging of rasogolla.

Paneer

Paneer, a popular traditional Indian dairy product, is obtained by acid coagulation of hot milk. Paneer is used as a base material for the preparation of large number of culinary dishes including paneer curry, paneer tikka, chilli paneer etc. Since the product is manufactured at elevated temperatures, its spoilage microflora may come either from chilled water and post manufacturing handling practices. Paneer is generally packaged in polyethylene bags and has a short shelf life of about 7 days at refrigeration storage, and less than a day at room temperature. High moisture content, post-manufacturing contamination and improper packaging are major contributors for its limited shelf life. The spoilage in paneer occurs due to the surface growth of microorganisms. A greenish yellow slime forms on the surface and the discolouration which is accompanied by off-flavour. Hence, it is invariably the surface that gets spoiled earlier while the interior remains good for a longer time at refrigeration storage. Use of an appropriate packaging material and creating air free environment significantly enhances the shelf life of paneer. Sachdeva *et al.* (1991) vacuum packaged paneer blocks (10 x 4 x 6 cm^3) in cryovac/PE pouches and no deterioration was reported up to 30 days at 6 ± 10 °C. Vacuum packaging improves the body and texture of paneer as it becomes more compact and better sliceable.

Punjrath *et al.* (1997) reported that vacuum packaging of paneer in laminated film (EVA/EVA/PVdC/EVA) followed by heat treatment at 90°C for one minute could help extending the shelf life up to 90 days at refrigeration temperature. Recently, some organizations have already started vacuum packaging in high barrier films (Figures 15.4a and 4b). According to Srivastava (2007), the shelf-life of paneer packaged under vacuum was 20 days as compared to 10 days in conventional air packed samples stored at 3±1°C, representing 200 per cent increase in the shelf life of paneer. Storage of vacuum packed paneer samples at -10 to-15 °C was reported to further enhance the shelf life up to 45 days. The study of Rai *et al.* (2008) reveals that, when paneer was packaged in high barrier bags (LLD/BA/Nylon-6/BA/LDPE) under 4 atmospheres, namely air (atm 1), vacuum (atm 2), 100 per cent CO_2 (atm 3) and 100 per cent N-$_2$ (atm 4) and stored at 7 ±1 °C, atm 3 was the best, followed by atm

Figure 15.4: a: Vacuum Packed Paneer Block;
b: Frozen Paneer Cubes Packaged in Metalized Laminate.

4, atm 2, and atm 1 in terms to retard different physico-chemical changes occurring during storage of paneer. The Cryovac system using shrink film has been successfully used for the packaging of paneer. Long life of paneer can be achieved by packaging in Metallized PET or Nylon – PET or Nylon or LDPE/LLD. Shrivastava and Goyal, (2009) studied the suitability of MAP paneer stored at 3±1 °C by preparing paneer-curry and subjecting it for sensory evaluation by trained sensory panelists. Paneer curry prepared from 30 days stored modified atmosphere packaged paneer samples was found acceptable by the panelists, while it was acceptable when only 20 and 10 days stored vacuum packed and air packed samples, respectively used for the preparation. The preservative effect of 100 per cent CO_2 were reported maximum followed by 50 per cent CO_2/50 per cent N_2 and 100 per cent N_2 respectively, in descending order. Frozen samples of paneer packed in high barrier films such as metalized laminates can be kept for longer period of 6 months and above.

Paneer Tikka

Paneer tikka, a very popular dish for vegetarians, is an exotic kebab of Indian cottage cheese popularly known as paneer (Kalra and Gupta, 2001). It is highly rich

Figure 15.4: Vacuum Packed Paneer Tikka in LLD/BA*/Nylon-6/BA*/LDPE.

in proteins, fat, vitamins and minerals content; and is laxative in nature. It is extensively used as fast- food during get-togethers, marriage parties, birthday parties, and also in restaurants. The shelf-life of unpacked paneer tikka at room temperature is hardly one day.

Verma *et al.* (2007) successfully attempted to increase the shelf life of stored paneer tikka up to 21 days by packaging in high barrier films under modified atmosphere (100 per cent CO_2) as compared to 7 days for air packed samples. Vacuum packaging of paneer tikka in high barrier package (LLD/BA*/Nylon-6/BA*/LDPE) has been successfully demonstrated to retard chemical changes and extending the shelf life of the product at 3 ± 1°C (Ahuja and Goyal, 2013).

Ghee

Approximately, 35 per cent of the total milk produced in India is converted into ghee. Majority of the dairies are packing ghee in lacquered or unlacquered tin cans of various capacities ranging from 500 g to 15 kg. Tin cans protect the product well against tampering and during transportation to far off places without significant wastage. The most common and serious deterioration in ghee is the development of rancid flavour, caused by the formation of volatile compounds. The modern packaging plays a vital role in delaying the onset of this defect. The packaging material should possess good water vapour barrier properties. High-density polyethylene (HDPE) and polypropylene (PP) are known to have low water vapour transmission rates (WVTR), and easily available at low cost. If such films are laminated to other suitable basic packaging materials, one can get almost negligible value for WVTR, which would be ideal. The package to be selected should show sufficient tensile strength, elongation, tear resistance and burst strength, besides overall good mechanical strength. The packaging of ghee can also be done in polymer coated cellophane, polyester, nylon – 6, or food grade PVC and their laminates.

Fermented Indigenous Product

Dahi, misti dahi and shrikhand are the popular indigenous fermented milk products. Dahi is popular all over India, whereas misti dahi is popular in eastern part of India. Shrikhand is a semisolid, sweetish- sour fermented milk product prepared from chakka, which is obtained by draining off whey from dahi using muslin cloth or any other mechanical means like quarg separator or basket centrifuge. Shrikhand forms part of meals on festive occasions in the states of Gujarat and Maharashtra in India. Now-a-days, their production is not only confined to small scale level but many organized dairies have started commercial production.

Dahi and misti dahi are traditionally sold in earthen pots, which provide fine body and texture but they are heavy in weight, breakable, expensive and cannot be covered properly. Shrinkage of product is major defect of dahi or misti dahi which takes place due to loss of moisture through the porous body of pot. Attempts have been made to pack dahi in the shellac coated earthen pots, which were recommended to minimize the problem of wheying off during transportation. Shellac coating also prevents shrinkage defect of dahi when compared with the dahi samples packaged in uncoated earthen pots. Earthen pots are eco-friendly with the environment. Set type dahi, misti dahi and shrikhand is currently packaged in polystyrene cups, high impact polystyrene cups (HIPS), polypropylene cups of 100, 200 and 500 g size by organized dairies. The drawbacks with plastic cups are wheying off from dahi during transportation due to mechanical shear, and environmental concerns related to very low degradability of plastic. Stirred dahi and lassi (stirred sweetened dahi based product) are commonly packed in LD/LLDPE pouches.

Kulfi

Kulfi is a frozen dairy product made from desiccated milk, sugar, and flavourings. Traditionally kulfi is packaged in aluminium/plastic cones, or tin cans. Kulfi is delivered to the customer by removing from the metal/plastic cones. Such cones are reusable by the kulfi makers. Recently the use of earthen pots has been started for the packaging of kulfi. These containers are eco-friendly and their use should be encouraged.

Instant Formulations for Traditional Dairy Product

Consumer packages for gulabjamun mix, kheer mix and kulfi mix include sachets and flexibles like metalized polyester kept in cartons. Jha (2000) prepared instant dry kheer mix having good reconstitution properties by two stage drying of the admixture of milk concentrate and rice flour obtained from partially pre-gelatinized along with sugar. Spray dried kheer powder with instant rice packed in metallized polyester laminates was reported to have shelf-life up to 6 months at room temperature. Kulfi mix powder (Ghosh,1991) packed in tin cans was found to be stable up to 7 months at 30°C. Dalia is milk and wheat-based particulate containing dairy dessert, which is popular as a breakfast food in India. Jha (2006) formulated instant dalia mix packed in PE-Paper board cartons.

Conclusion

The shelf life of high moisture and high fat food products is very limited mainly due to the chemical effect of atmospheric oxygen and growth of spoilage microorganisms. Both these factors, alone or in conjunction with one another, result in changes in colour, flavour, odour and overall deterioration of the product. Revolutionary changes are taking place at a very fast speed in packaging of food products. Product compression is unavoidable during vacuum packaging making it unsuitable for many products. Modified Atmosphere Packaging (MAP), a modern preservation technique, enhances the shelf life of food products including processed dairy foods many fold. MAP also saves the product from compression during storage and transportation. Since the demand for convenience foods in India is growing at a rapid rate, it is expected that new forms of packaging material such as roll wraps, pouches, cartons, PP – trays covered with transparent films of MXXT or such other films are likely to appear on the market place for packaging of Indian dairy products. However, due to increased environmental concern, it is essential to shift towards eco-friendly packaging of traditional dairy products.

References

Addington, P. (1991). MAP of dairy products. Leatherhead. Food RA MAP Training Course, T072 Notes.

Ahuja, K.K. and Goyal, G.K. (2013). Combined effect of vacuum packaging and refrigerated storage on the chemical quality of *paneer tikka*. *J. Food Sci. Technol.*, 50(3): 620-623.

Alam, T. (2013). Innovation in packaging systems for traditional dairy products. In: Souvenir, 8th Convention of Indian Dairy Engineers Association and National Seminar on Mechanised Production of Indian Dairy Products 2-3 September, 2013.

Alam, T. and Goyal, G.K. (2011). Effect of MAP on the microbiological quality of mozzarella cheese stored in different packages at 7±1°C. *J. Food Sci. Technol.*, 48 (1): 120-123.

Alam, T., Goyal, G.K. and Broadway, A.A. (2005). Packaging trends in dairy Industry. In: Indian Dairy Industry, volume I, Published by Dr. Chawla Dairy Information Centre (P). Ltd, New Delhi. pp: 180-185.

DAHD: Department of Animal Husbandry, Dairying and Fisheries (2013). Annual Report 2012-13.

Goyal, G. K. and Tanweer Alam (2004). MAP- A new frontier packaging in dairy and food industry. *Indian Dairyman*, 56 (8) 53-60.

Goyal, G.K. and Rajorhia, G.S. (1991). Role of modern packaging in marketing of indigenous dairy products. *Indian Food Industry* 10(4): 32-34.

Jha, A. (2000). Developments in the manufacture of cereal-based convenience foods. In: Compendium entitled "Advances in Formulated Foods", short course conducted by CAS-DT, NDRI Karnal, during June 19th - July 10, 2000, pp: 108-111.

Jha, A. (2006). Cereal-based convenience formulations for traditional dairy products, In short course on "Developments in Traditional dairy products" organized by Centre for Advanced Studies, Dairy Technology Division, NDRI, Karnal during 10-30th Dec, 2006, pp: 213-216.

Kumar, A., Rajorhia, G.S. and Srinivasan, M.R. (1975). Effect of modern packaging materials on the keeping quality of khoa *J. Food Sci. Technol.*, 12: 172.

Kumar, R., Bandyopadhyay, P. and Punjarath, J.S. (1997). Shelf-life extension of peda using different packaging techniques. *Indian J. Dairy Sci.*, 50(1): 40-49.

Londhe, G., Pal, D. and Raju, P.N. (2012). Effect of packaging techniques on shelf life of brown *peda*, a milk-based confection. *LWT-Food Science and Technology*, 47(1): 117-125.

NDRI (1993) Annual Report 1992-93.NDRI, Karnal, pp. 85.

Palit, C. and Pal, D. 2000. Studies on enhancement of shelf life of burfi. Modern trends and perspective in food packaging for 21st century. ICFOST Souvenir, CFTRI Mysore.

Punjrath, J.S. (1995). Trends in packaging of milk and milk products. *Indian Dairyman*. 47: 29–40.

Punjrath, J.S.; Kumar, R. and Bandyopadhyay, P. (1997). Tapping the potential of traditional dairy food. In Souvenir 28th Dairy Industry Conference, Bangalore 27-29 April, 1997.

Rai S, Rai, G.K., and Goyal, G. K. (2008). Effect of modified atmosphere packaging and storage on the chemical quality of paneer. *J. Dairying Food Home Sci.*, 27: 33-37.

Sachdeva,S., Prodopek,D. and Reuter, H. 1991. Technology of paneer from cow milk. *Jap. J.Dairy Food Sci.* 40(2): A85.

Sharma, H. K., Singhal, R. S. and Kulkarni, P. R. (2001). Effect of packaging under vacuum or under nitrogen on the keeping quality of danedar Khoa. *Int. J. Dairy Technol.*, 50 (3): 107-110.

Shrivastava, S. and Goyal, G.K. (2009). Effect of modified atmosphere packaging (MAP) on the chemical quality of paneer. *Indian J. Dairy Sci.*, 62(4): 255-261.

Verma, V., Goyal, G.K. and Bhatt, D.K. (2007) Influence of Modified Atmosphere Packaging (MAP) and Storage on the Chemical Quality of Paneer Tikka. *Int. J. Food Sci. Technol and Nutrition*, 1(2): 235-240.

2015, Dairy Product Technology: Recent Advances ***Pages 325–339***
Editors: **Subrota Hati, Surajit Mandal and Birendra Kumar Mishra**
Published by: **DAYA PUBLISHING HOUSE, NEW DELHI**

Chapter 16

Recent Biotechnological Approaches in Dairy and Food Industry

Pradip Behare, Vaibhao Lule,
S.K. Tomar and Surajit Mandal

Introduction

Biotechnology is the use of living systems and organisms to develop or make useful products, or "Any technological application that uses biological systems, living organisms or derivatives thereof, to make or modify products or processes for specific use" (CBD, 2013). It often overlaps with the related fields of Industrial biotechnology, bioengineering, biomedical engineering depending on the tool and application. Biotechnology as applied to food processing to make use of microbial inoculants to enhance properties such as the taste, aroma, shelf-life, texture and nutritional value of foods. The process in which microorganisms and their enzymes bring about these desirable changes in food materials is known as fermentation. Fermentation process is widely applied in the production of microbial cultures, enzymes, flavours, fragrances, food additives and a range of other high value-added products. These high value products are increasingly produced in more technologically advanced countries for use in their food and non-food processing applications. Many of these high value products are also imported by developing countries for use in their food-processing applications. Industrial biotechnology is known mainly in Europe as white biotechnology is the application of biotechnology for industrial purposes. This includes use of micro-organisms, or components of cells like enzymes, to generate industrially useful products in sectors such as chemicals, food and feed, detergents, paper and pulp, textiles and biofuels (Johnson 2008).

Biotechnological achievements of recent years have emerged as powerful tool to improve quality attributes of food products including milk and meat products. Biotechnological approaches can be employed for improving productivity, economy, physicochemical and nutritional attributes of a wide range of food products. Many of the biotechnological techniques can be explored in the area of quality assurance programs that help to produce food products of assured quality and public health safety. Traditional methods of food-safety monitoring such as the detection of pathogenic bacteria in food are generally based on the use of culture media. At least two to three days are required for the initial isolation of an organism, followed by the requirement for several days of additional confirmatory tests. Biotechnology-based methods can provide accurate results within a relatively short time frame. Biotechnological developments have resulted in the widespread availability of low-cost rapid methods of identification when compared with the significant cost/time requirements of conventional techniques.

All organisms are made up of cells that are programmed by the same basic genetic material, called DNA (deoxyribonucleic acid). Segments of the DNA called gene tell individual cells how to produce specific proteins. It is the presence or absence of the specific protein that gives an organism a trait or characteristic. More than 10,000 different genes are found in most plant and animal species. This total set of genes for an organism is organized into chromosomes within the cell nucleus. The process by which a multicellular organism develops from a single cell through an embryo stage into an adult is ultimately controlled by the genetic information of the cell as well as interaction of genes and gene products with environmental factors.

Genetic engineering (GE) is the technique of removing, modifying or adding genes to a DNA molecule in order to change the information it contains. By changing this information, genetic engineering changes the type or amount of proteins an organism is capable of producing. Genetic engineering is used in the production of drugs, human gene therapy, and the development of improved plants. For example, an "insect protection" gene (*Bt*) has been inserted into several crops- corn, cotton, and potatoes which give farmers new tools for integrated pest management (Nair 2008). GE is also being used to improve particular traits of beneficial organisms.

Applications of Biotechnology in Dairy Sector

Biotechnology has already made significant contributions in dairy industry which can be broadly categorized in to two areas like dairy production and processing (Table 16..1).

Out of these aforementioned areas, some have already been realized while others are still in the developmental stages and will take some time before reaching the end users. Some of these promising areas and their potential application in dairy industry are described in subsequent section.

Table 16.1: Application of Biotechnology in Dairy Sector

Dairy Production	*Dairy Processing*
☆ Recombinant Bovine/Buffalo Somatotropin - rbst	☆ Designing milk through genetic engineering
☆ Recombinant Vaccines	☆ Genetically Modified Organisms (GMOs)
☆ Rumen manipulation	☆ Starter Cultures
☆ DNA finger printing and RFLP	☆ Genetically modified foods
☆ PCR based diagnostics (Animal Health care)	☆ Food grade Biopreservatives
☆ Embryo Transfer Technology	☆ Recombinant dairy enzymes/proteins
☆ Animal Cloning	☆ Accelerated cheese ripening
☆ Gene Pharming and Transgenics	☆ Probiotics
	☆ Functional foods and neutraceuticals
	☆ Dairy waste management and pollution control
	☆ Gene probes and PCR based pathogen detection

Genetic Manipulation of Starter Cultures

Good quality starter cultures (Lactic Acid Bacteria-LAB) are the pre-requisite for preparing good quality fermented foods. The commercial value of the fermented foods is, therefore, mainly dependent upon the performance of the starter cultures. One of the most notable features of lactic starter cultures is that they generally harbour large complements of plasmids. It is very significant that many commercially important traits of these bacteria are in fact plasmid encoded. These include lactose and citrate utilization, phage resistance, proteinase production and bacteriocin production/ immunity. The location of such important genetic information of considerable industrial value on the plasmids could be detrimental to the host as these extra chromosomal elements may be lost from the cell population due to one or more reasons However, it has proved to be a blessing in disguise to the researchers since the plasmid encoded genes are more accessible for manipulation and more easily analyzed than the genes located on the chromosomes. Recent advances in gene transfer and cloning technologies provided numerous opportunities for the application of genetic approaches to strain development programme appropriate for their application in dairy industry. In addition, significant advances have already been made in the analysis of *Lactococcus* and *Lactobacillus* gene expression. Genetic manipulation of these cultures has now been developed to the point where it is now possible to clone and express homologous and heterologous genes. Cloning vectors have also been constructed to isolate regulatory signals such as promoters, ribosome blinding sites and terminators. This information could be extremely valuable in allowing the expression of foreign gene in lactic acid bacteria and enhancing the expression of desired phenotypic properties. The development of plasmid integration vectors for lactic acid bacteria offers the possibility of introducing and stabilizing genes of interest by incorporating them into the host chromosomes. The strategy could be extremely useful to tackle the problem of instability in the lactic cultures as a result of frequent loss of plasmid during their propagation. Plasmid integration vectors can also be used to explore and help characterize chromosomal DNA from lactic acid bacteria. Indeed, proteinase and phage resistant determinants have been successfully integrated and stabilized in lactococcal chromosomes where they are amenable to substantial amplification.

Food Grade Biopreservatives

Another potential area for commercial exploitation of lactic acid bacteria through their genetic engineering is the Biopreservative attributes of these organisms. Appropriate food-grade biopreservatives can be developed to control undesirable bacteria and molds in dairy foods by following these strategies for direct application in Dairy Industry

Bacterial Inhibition

Suppression of spoilage and food-borne pathogens by lactic acid bacteria could be extremely beneficial to human health and dairy industry as these attributes can considerably improve the shelf life and safety of fermented foods. The antimicrobial activity in lactic acid bacteria has been ascribed to a number of factors and the extent of their antibacterial spectrum varies from culture to culture and strain to strain. Among these factors, the bacteriocins appear to be the most promising candidate for exploring the preservative potentials of these bacteria on a large scale. This is precisely due to their proteinaceous nature since the production of these polypeptides could be conveniently manipulated at the molecular level.

The application of such food preservatives in dairy industry to combat undesirable food spoilage and pathogenic organisms could be more relevant to countries like India where the health of consumers is always at risk due to poor hygiene, tropical temperate conditions and inadequate refrigeration facilities (FAO 2010). Biotechnological techniques can now be applied to develop strains of lactic acid bacteria capable of enhanced production of these natural food grade preservative and also to combine within a single strain the ability to produce a number of such bacteriocins to extend their antibacterial spectrum. Much of the work being applied to bacteriocins is often based on approaches and protocols originally developed for the analysis of Nisin which has already been given the GRAS status. In this case, the structural genes have been sequenced; protein engineering has been employed to generate the variants of the molecules and engineered strains which can be used to study expression of the compound available.

Fungal Inhibition

Molds contamination in dairy foods and feeds is a serious problem especially in India where tropical humid conditions and unhygienic environment are extremely favorable for the growth and proliferation of these types of organisms. Apart from causing serious spoilage problems in these commodities, many of the incriminating molds could also pose serious health risks to the consumers on account of their ability to produce extremely toxic and carcinogenic metabolites such as aflatoxins. Although, mold contamination in foods can be effectively controlled by the application of some chemical additives, there are serious concerns regarding the use of these unnatural additives in food chain. Exploring the antifungal attributes of lactic acid bacteria could be a very useful strategy to tackle the mold problem in dairy foods.

Dairy Enzymes

Dairy industry requires large amount of rennet to produce cheese in bulk. Traditionally rennet is procured from calves. However due to global imbalance between production of cheese and calf slaughter and worldwide shortage of calf rennet, exploration of alternative source of rennet is required to be investigated. In certain countries like India, religious feeling has aggravated the need to rennet substitute. Advances in Biotechnology have also made a strong impact on production of several enzymes and proteins used in dairy industry in the processing of milk for the manufacture of some fermented products. Many of the enzymes are by and large of microbial origin. However, their production levels are not very high in the producer organisms and thus their production is not cost effective for commercial exploitation. Food processing has benefited from biotechnologically produced enzymes such as chymosin (rennet), proteases lipases, alpha amylase and lactase which can find lot of application in dairy industry. Many of these enzymes like alpha amylase, chymosin produced by genetically modified organisms have been granted GRAS status by FDA thereby allowing their use in place of conventional sources of these enzymes in the dairy industry. Genetically engineered enzymes are easier to produce than enzymes isolated from original sources and are favoured over chemically synthesized substances because they do not create by products or off flavours in foods.

Recombinant Chymosin and other Dairy Enzymes

Bovine chymosin traditionally known as calf rennet has been extensively used as milk coagulant during the manufacture of a variety of cheese all over the world. India's cheese industry is also expanding rapidly and hence the demand for calf rennet or its substitutes is also showing a phenomenal increase. However, in India, the cheese made with the help of calf rennet is not acceptable to a large sector of the consumers due to religious sentiments. This has stimulated lot of interest in search for rennet from alternate sources. Although, application of milk clotting enzymes from microbial sources have been in vogue in cheese manufacture, the quality of cheese manufacture, the quality of cheese made with such enzymes does not match with that made from calf rennet. In this context, buffalo rennet could be an excellent alternative to bovine calf rennet on account of its inherent compatibility with buffalo milk for producing high quality cheese.

With the rapid expansion of Fast Foods entering into the Indian market, the demand for specialized cheeses, such as Mozzarella cheese for Pizza etc. has increased tremendously. With this, demand for chymosin has also increased several folds. Hence, for large scale production of buffalo calf rennet, application of rDNA technology and Genetic Engineering techniques could be extremely valuable. By cloning the genes of interest from buffalo calves in *E. coli* and other hosts, it is now possible to produce recombinant Buffalo chymosin for commercial use in cheese industry. Pure chymosin produced from genetically engineered *E. coli* reduced coagulation time by 5 folds. Cheese prepared with use of bacterial chymosin and ripened at 6–7°C for 7 months showed similar characteristics and proteolysis to cheese made with calf chymosin (Singhal *et al.,* 1990). The DNA of calf chymosin has been successfully cloned into yeast (*Kluveromyces lactis*), bacteria (*E. coli*) and moulds (*Aspergillus niger*). These can be produced on large scale by fermentation methods and extracted by downstream processing. Technological performance of recombinant chymosin is excellent since the cheese produced using recombinant chymosin is essentially indistinguishable from traditional cheese (Mathur *et al.,* 2003). Similarly other dairy enzymes like proteinases lipases, beta-galactosidase, etc. can also be produced through rDNA technology. Treatment of milk with galactosidase results in hydrolysis of lactose, thereby making it digestible by lactose intolerant people. The enzyme galactosidase hydrolyses lactose into glucose and galactose. Since these enzymes are costly, biotechnology can help in its economic production as well application (Ramesh and Yadav 1990). Commercial galactosidase is produced from yeasts such as *Kluyveromyces lactis* and *Kluyveromyces marxianus* (formerly known as *Kluyveromyces fragilis* and *Saccharomyces fragilis*), and moulds such as *Aspergillus niger* and *Aspergillus oryzae*. The production and optimization of galactosidase enzyme using synthetic medium by *Kluyveromyces lactis* NRRL Y-8279 in shake flask cultures was found suitable (Dagbagli and Goksungur, 2008). Proteases and lipases can be used in cheese industry for flavour development whereas Betagalactosidase can find application in the production of lactose free milk intended for lactose intolerant patients.

Accelerated Cheese Ripening

Many cheese varieties require long ripening periods at low temperature for characteristic flavour and texture development. This process significantly increases the cost of the product. Controlled use of biotechnological products like genetically engineered proteolytic and lipolytic enzymes can accelerate cheese ripening. Modified / genetically tailored microorganisms and enzymes are now being used for enhancing flavour production in cheese. Enzyme addition is now one of the few preferred methods of accelerated ripening of cheese. The enzymatic reactions are specific and hence undesirable side effects caused by live microorganisms are avoided in the cheese. Enzymes may be immobilized or encapsulated for long term action on the production for quick action and homogenous distribution in the product. The molecular analysis of the lactococcal proteolytic system has created the exciting prospect of being able to use genetic strategies to manipulate these hosts to produce strains with new flavour characteristics. It is also likely that these approaches will allow cheeses to be ripened in a more controlled fashion and an accelerated rate, if desired. This will result in the production of fermented foods with new attractive properties as well as creating economic savings for the producers.

Genetically Modified Microorganisms (GMOs) in Accelerated Cheese Ripening: A Novel Cheese-Biotech Alliance

The production of cheese and a range of other fermented foods is one of the oldest manifestations of biotechnology. Many consumers consider cheese to be a delicacy with exquisite taste and aroma characteristics and would recoil at the thought of using Genetically Modified Microorganisms (GMOs) in the manufacturing process.

The use of recombinant DNA technology to produce GMOs for accelerated ripening of cheese is one of the most important scientific advances of the 20th century. It has great potential in research, because it allows the development of highly sensitive analytical procedures. It also has potential in industry, leading to processes and products that would be difficult to develop using conventional techniques. These include food and food processing. The use of GMOs in food and food processing is litigious due to a lack of acceptance by consumers; especially in Europe. Recent developments have opened many more possibilities for the use of gene technology. The next generation of bacterial cultures will probably contain strains with properties modified by gene technology, for example with new properties for cheese making created by changing the expression level of one or a few genes.

The considerable knowledge now available on the genetics of cell wall associated proteinase and many of the intracellular peptidases makes it possible to specifically modify the proteolytic system of starter *Lactococcus*. The gene for the neutral proteinase (Neutrase) of *B. subtilis* has been cloned in *Lc. lactis* UC317 by McGarry *et al.* (1994). Cheddar cheese manufactured with this engineered culture as the sole starter undergoes very extensive proteolysis and the texture became very soft within two weeks at 8°C. By using a blend of unmodified and Neutrase producing cells as starter, a more controlled rate of proteolysis was obtained and ripening was accelerated (McGarry *et al.,* 1994). The 80:20 blend of unmodified: modified cells gave best results. Since the genetically modified cells were not food grade, the cheese was not tasted but the results appear sufficiently interesting to warrant further investigation when a food-grade modified mutant becomes available. Since amino acids are widely believed to be major contributors, directly or indirectly, to flavour development in cheese, the use of a starter with increased aminopeptidase activity would appear to be attractive. Two studies (McGarry *et al.,* 1994; Christensen *et al.,* 1995) have been reported on the use of a starter genetically engineered to super-produce aminopeptidase N; although the release of amino acids was accelerated, the rate of flavour development and its intensity were not, suggesting that the release of amino acids is not rate limiting. In an on-going study, engineered *Lactococcus* starters harbouring PepG or PepI genes from *Lb. delbreuckii* have shown increased proteolysis, especially at the level of amino acids; PepG was the more effective. The cheeses have not yet been assessed organoleptically. Apart from legal aspects, the principal technical problems when engineering starters with improved cheese-ripening properties is the lack of knowledge on the key or limiting lactococcal enzymes involved in cheese ripening. A range of peptidase-deficient *Lactococcus* mutants (lacking 1 to 4 peptidases) has been developed (Mierau *et al.,* 1996) primarily with the objective of identifying the importance of the various peptidases, alone or in various combinations, to the growth of the organism in milk. Mutants deficient in 1 or 2 peptidases (in various combinations) were also used for the small-scale manufacture of Cheddar cheese (mutants deficient in 3 or 4 peptidases are unable to grow in milk at a rate sufficient for cheese manufacture). Perhaps surprisingly, cheeses made using the peptidases-deficient mutants, even those lacking both PepN and PepC, did not differ substantially from the control with respect to the level and type of proteolysis or flavour and texture. The results appear to suggest that there are alternative routes for amino acid production.

Rapid Detection of Food Pathogens

Although there has been dramatic increase in milk production in the country and more surplus milk is now available to dairy industry for processing, the microbiological quality of such milk and processed products has to be of acceptable quality. The safety of milk and milk products from public health point of view can be ensured by rigorous monitoring of dairy foods especially for high risk food pathogens such as *E. coli* 0157:H7, *Listeria monocytogenes, Salmonella, Shigella,* etc. which gain access into these foods due to mishandling during production and processing. Since, the conventional methods currently used in Quality Assurance labs for the detection of these potential pathogens are extremely laborious, lengthy and tedious, they do not serve any useful purpose to dairy industry as by the time results are available, the concerned food lot is already sold in the market and hence no follow up action is possible to avoid possible outbreak of diseases associated with such food pathogens. Hence, there is a need to develop simple and new innovative techniques for rapid detection of pathogens in dairy foods.

PCR Based Detection of Food Pathogens

Another extremely versatile, reliable, sensitive and practical method for rapid detection of food pathogens directly from foods without following lengthy culturing steps is polymerase chain reaction (PCR). Application of such methods in Food Quality Assurance and Public Health Labs can go a long way in protecting the health of the public. The crucial feature of sample preparation in the direct analysis of food without cultural enrichment is the isolation of target DNA with high reproducibility and efficiency. Construction of the actual PCR assay requires nucleotide sequence information and knowledge about the chosen target gene to guarantee maximum specificity and sensitivity. By using different formats of PCR, it is now possible to simultaneously detect more than one food pathogen in the food. PCR based detection of food pathogens has completely revolutionized the area of diagnostics as the results are available in less than 24 hrs to enable the food processing units to take appropriate follow up action in case a pathogen is detected in a food to avoid any risk to the consumers.

PCR Based Typing Methods

PCR (Polymerase Chain Reaction) automatically amplifies rapidly any particular stretch of DNA using a pair of primers each complementary to the flanking ends on both the strands, in a thermal cycler, within a short span of time. Thus primers can be designed to specifically amplify only a chosen region, specific to either a particular species or strain or various species simultaneously in a single reaction *i.e.*, (Multiplex PCR). Thus genus, species or even strains can be identified accurately using PCR. Among PCR based methods, RAPD (Randomly amplified Polymorphic DNA), ARDRA (restriction analysis of the amplified *r*DNA), species-specific PCR all are being used for differentiation of microorganism.

RAPD

In RAPD assay, also referred to as Arbitrary Primed PCR, patterns are generated by the amplification of random DNA segments with single short (typically 10 bp) primers of arbitrary nucleotide sequence. The primer is; not targeted to amplify any specific bacterial sequences and will hybridize under low stringency condition at multiple random chromosome locations and initiate DNA synthesis. After separation of the amplified DNA fragments by agarose gel electrophoresis, a pattern of bands results, which is characteristic of the particular bacterial strain (Williams *et al.,* 1990; Welsh and McCleland, 1992; Meunier and Grimont, 1993). The merits of this method are:

- ☆ Requirement of very short time
- ☆ Provides good level of discrimination
- ☆ Can be applied for large no. of strains

Species-specific PCR

As the name indicates, this type of PCR reaction makes use of species specific primers in for differentiation of different species. Primers sequences usually chosen from highly conserved 16S rRNA and 23S rRNA or 16S-23S intergenic spaces region. The advantage with these generic regions is that all these are part of highly conserved rRNA operon, which has remained well conserved during evolution and is common to all eubacteria moreover these also contain highly variable regions unique to each particular species. All these primers are being used for the identification of different microbial species. (Andrighetto *et al.,* 1998; Drake *et al.,* 1996; Song *et al.,* 2000; Fortina *et al.,* 2001). Although this is a highly specific technique for species level differentiation, but strain typing is not possible through this methodology. However, straight sequencing of these amplified 16s-23s intergenic spacer regions, in different isolates can easily differentiates even very much closely resembling strains and is readily becoming popular among researchers.

Amplification and Restriction Analysis of 16s rRNA Gene (ARDRA)

Among other 16s rRNA based techniques, ARDRA combines both PCR and restriction analysis and hence is a powerful technique and is capable of differentiating species as well as strains (Andrighetto *et al.,* 1998; Kullen *et al.,* 2000). However since it also has limitation of tedious handling steps, it cannot be applied for large number of sample simultaneously.

Genetically Engineered Milk – Humanization of Bovine Milk

Breast milk is nature's perfect food for human infants providing them all aspects of nutrition and protection against infections. However, a considerable number of infants are fed formulae based on bovine/buffalo milk. The composition of these infant formulations can be improved if the proteins contained therein more closely resemble those of human milk. It is now possible to add human lactoferrin or lysozyme to bovine milk by genetic engineering to produce new functional foods. The shelf life of such products is also expected to be very high due to antimicrobial activity of these proteins.

The goal of the dairy industry has been to create an efficient, healthy cow/ buffalo that can serve all the needs of the industry. Genetic engineering offers the opportunity for a paradigm shift, a reshaping of the industry from the producers to the processing plants. Dairy producers have the opportunity to choose to produce high protein milk, milk destined for cheese manufacture that has accelerated curd clotting time, milk containing neutraceuticals, orally administered biologicals that provide health benefits or a replacement for infant formulae. Such a scenario would be a radical change for the dairy industry. There is now little doubt that the products of genetic engineering will become a part of the dairy industry in the upcoming days.

Modified Milk in Transgenic Dairy Cattle

Bovine milk has been described as an almost perfect complete food, because it is a rich source of vitamins, calcium, and essential amino acids (Karatzas and Turner, 1997). Some of the vitamins found in milk include vitamin A, B, C, and D. Milk has greater calcium content than any other food source, and daily consumption of two servings of milk or other dairy products supplies all the calcium requirements of an adult person (Rinzler *et al.*, 1999). Caseins represent about 80 per cent of the total milk protein and have high nutritional value and functional property (Brophy *et al.*, 2003). The caseins have a strong affinity for cations such as calcium, magnesium, iron, and zinc. There are four types of naturally occurring caseins in milk, αS1, αS2, β, and κ (Brophy *et al.*, 2003). They are clumped in large micelles, which determine the physicochemical properties of milk. Even small variations in the ratio of the different caseins influence micelles structure, which in turn can change the milk's functional properties. The amount of caseins in milk is an important factor for cheese manufacturing, since greater casein content results in greater cheese yield and improved nutritional quality (McMahon and Brown 1984). It has been estimated that by enhancing the casein content in milk by 20 per cent would result in an increase in cheese production, generating an additional $190 million/year for the dairy industry (Wall *et al.*, 1997). Dairy cattle have only one copy of the genes that encode α (s1/s2), β, and κ-casein proteins, and out of the four caseins, κ and β are the most important (Bawden *et al.*, 1994). Increased milk κ-casein content reduces the size of the micelle resulting in improved heat stability. β-caseins are highly phosphorylated and bind to calcium phosphate, thus influencing milk calcium levels (Dalgleish *et al.*, 1989; Jimenez Flores and Richardson, 1988). Brophy *et al.* (2003), using nuclear transfer technology, produced transgenic cows carrying extra copies of the genes *CSN2* and *CSN3*, which encode bovine β- and κ-caseins, respectively. Genomic clones containing *CSN2* and *CSN3* were isolated from a bovine genomic library. Previous studies conducted with mice revealed that *CSN3* had very low expression levels (Persuy *et al.*, 1995). In order to enhance expression of *CSN3*, the researchers created a *CSN2/3*-fusion construct, in which the *CSN3* gene was fused with the *CSN2* promoter. The *CSN2* genomic clone and the *CSN2/3*-fusion construct were co-transfected into bovine fetal fibroblast (BFF) cells, where the two genes showed coordinated expression. The transgenic cells became the donor cells in the process of nuclear transfer, generating nine fully healthy and functional cows. Overexpression of *CSN2* and *CSN2/3* in the transgenic cows resulted in an 8-20 per cent increase in β- casein and 100 per cent increase in κ-casein levels (Brophy *et al.*, 2003; Jube and Borthakur, 2006).

Elimination of Carcinogenic Compounds from Food

Brewer's yeast (*Saccharomyces cerevisiae*) is one of the most important and widely used microorganisms in the food industry. This microorganism is cultured not only for the end products it synthesizes during fermentation, but also for the cells and the cell components (Aldhous 1990). Today, yeast is mainly used in the fermentation of bread and of alcoholic beverages. Recombinant DNA technologies has made possible to introduce new properties into yeast, as well as to eliminate undesirable by-products. One of the undesirable by-products formed during yeast fermentation of foods and beverages is ethylcarbamate or urethane, which is a potential carcinogenic substance (Ough 1976). For this reason, the alcoholic beverage industry has dedicated a large amount of its resources to funding research oriented to the reduction of ethylcarbamate in its products (Dequin 2001). Ethylcarbamate is synthesized by the spontaneous reaction between ethanol and urea, which is produced from the degradation of arginine, found in large amount in grapes. Yeasts, used in wine fermentation, possess the enzyme arginase that catalyzes degradation of arginine. If this enzyme can be blocked, arginine will no longer be degraded into urea, which in turn will not react with ethanol to form ethylcarbamate. In industrial yeast, the gene *CAR1* encodes the enzyme arginase (EC 3.5.3.1) (Dequin 2001). To reduce the formation of urea in sake, Kitamoto *et al.* (1991) developed a transgenic yeast strain in which the *CAR1* gene is inactivated. The researchers constructed the mutant yeast strain by introducing an ineffective *CAR1* gene, flanked by DNA sequence homologous to regions of the arginase gene. Through homologous recombination, the ineffective gene was integrated into the active *CAR1* gene in the yeast chromosome, interrupting its function. As a result, urea was eliminated and ethylcarbamate was no long formed during sake fermentation. This same procedure can be used to eliminate ethylcarbamate from other alcoholic beverages including wine (Kitamoto *et al.*, 1991).

Proteomics in Food and Dairy Sector

The use of proteomics for process development and validation in food technology and food biotechnology as well as corresponding quality control of raw materials and final products was at the beginning rather limited. In last years it has changed rapidly so proteomics technology is routinely used, and the terms 'Industrial process proteomics' (Incamps *et al.*, 2005) and 'Industrial proteomics' (Josic *et al.*, 2006) are now frequently used. Proteomics, through the application of gel electrophoresis and non-gel electrophoresis approaches, offers a powerful new way to characterise the protein component of foods. Whereas genomics provides information on the total genome of the organism, proteomics reveals which proteins are actually expressed in each tissue type. Furthermore the application of proteomic techniques offers a way to investigate differences in the protein composition of different tissues within a specific animal or vegetable food type, as well as between different varieties of it. In addition it has the power to follow changes in the protein component of various tissues during growth, maturation and post-mortem or post-harvest, as well as downstream treatments such as cooking. Milk contains a complex mixture of proteins that undergo several qualitative and quantitative changes during processing and storage. Elucidation of these changes is important to optimized milk processing and storage parameters. Proteomics has become widely used in the field of research on milk proteins due to improvements in the tools and diversified ways in which the methodologies of protein analysis have been developed. Proteomics is useful tool to study the milk proteins in relation to species–species differences, post-translational modification, effect of lactation, mastitis, changes in heated milk and during cheese ripening, and the detection of milk adulteration (Abd ElSalam, 2014).

Proteomic techniques *viz.* Mass spectroscopy (MS), FTIR-MS, LC-MS, SDS-PAGE, NMR and other are increasingly used for assessment of raw materials and final products as well as for control, optimization and development of new processes in food technology and biotechnology. However, most proteomic analyses are performed by the use of comparative 2D electrophoresis, and recently developed, faster and more effective methods such as quantitative isotope labeling (Clifton *et al.,* 2009; Martinez-Gomariz *et al.,* 2009) and label-free quantitative proteomics are scarcely used. The use of these methods combined with the already developed validation strategies (Ruebelt *et al.,* 2006; Natarajan *et al.,* 2006) will enable better in process control and characterization of batch-to-batch variations, as well as increasing use of proteomics for answering some key questions in food science detection of food contaminants and allergens, and further assessment of safety of GM foods. There are some papers discussing the potential of proteomics and its use to assess food quality (Carbonaro 2004) and technology (Han and Wang 2008). However, an overview about the use of this promising technique for the characterization of the complete production process in food manufacturing, biological and microbial safety and quality control of the final product is still missing.

References

Abd ElSalam MH 2014. Application of Proteomics to the areas of milk production, processing and quality control–A review. *Int J Dairy Technol* 67(2): 153–166.

Aldhous P 1990. Modified yeast fine for food. *Nature* 344: 186.

Andrighetto C, De Dea P, Lombardi A, Neviani E, Rossetti L, Giraffa G 1998. Molecular identification and cluster analysis of homofermentative thermophilic lactobacilli isolated from dairy products. *Res Microbiol* 149(9): 631-643.

Bawden WS, Passey RJ, Mackinlay AG 1994. The genes encoding the major milk-specific proteins and their use in transgenic studies and protein engineering. *Biotechnol Genet Eng* 12(1): 89-138.

Brophy B, Smolenski G, Wheeler T, Wells D, L'Huillier P, Laible G 2003. Cloned transgenic cattle produce milk with higher levels of β-casein and κ-casein. *Nat Biotechnol* 21(2): 157-162.

Carbonaro M 2004. Proteomics: present and future in food quality evaluation. *Trends Food Sci Tech* 15(3): 209-216.

CBD 2013. Text of the CBD. www.Cbd.int. Retrieved on March 20, 2013.

Christensen JE, Johnson ME, Steele JL 1995. Production of cheddar cheese using a Lactococcus lactis ssp. Cremoris SK11 derivative with enhanced aminopeptidase activity. *Int Dairy J* 5(4): 367-379.

Clifton JG, Huang F, Kovac S, Yang X, Hixson DC, Josic D 2009. Proteomic characterization of plasmaderived clotting factor VIII–von Willebrand factor concentrates. *Electrophoresis* 30(20): 3636-3646.

Dagbagli S, Goksungur Y (2008). Optimization of b-galactosidase production using Kluyveromyces lactis NRRL Y-8279 by response surface methodology. *Elect J Biotech* 11(4): 11-12.

Dalgleish DG, Horne DS, Law AJR (1989). Size-related differences in bovine casein micelles. *Bba-Gen Subjects* 991(3): 383-387.

Dequin S (2001). The potential of genetic engineering for improving brewing, wine-making and baking yeasts. *Appl Microbiol Biot* 56(5-6): 577-588.

Drake M, Small CL, Spence KD, Swanson BG (1996). Rapid detection and identification of Lactobacillus spp. in dairy products by using the polymerase chain reaction. *J Food Protect* 59(10): 1031-1036.

FAO 2010. Agricultural biotechnologies in developing countries: Options and opportunities in crops, forestry, livestock, fisheries and agro-industry to face the challenges of food insecurity and climate change (ABDC-10). In: FAO *International Technical Conference.*

Fortina MG, Ricci G, Mara D, Parini C, Manachini PL 2001. Specific identification of Lactobacillus helveticus by PCR with Pep C, pep N and htr A targeted Primers. *FEMS Microbiol Lett* 198(1): 85-89.

Han JZ, Wang YB 2008. Proteomics: present and future in food science and technology. *Trends Food Sci Tech* 19(1): 26-30.

Incamps A, HelyJoly F, Chagvardieff P, Rambourg JC, Dedieu A, Linares E, Quemeneur E 2005. Industrial process proteomics: alfalfa protein patterns during wet fractionation processing. *Biotechnol Bioeng* 91(4): 447-459.

Jimenez-Flores R, Richardson T 1988. Genetic engineering of the caseins to modify the behavior of milk during processing: A review. *J Dairy Sci* 71(10): 2640-2654.

Johnson FX 2008. Industrial Biotechnology and Biomass Utilisation. In: *Future Prospects for Industrial Biotechnology*, Stockholm,Austria, pp 196.

Josic D, Brown MK, Huang F, Lim YP, Rucevic M, Clifton JG, Hixson DC 2006. Proteomic characterization of interalpha inhibitor proteins from human plasma. *Proteomics* 6(9): 2874-2885.

Jube S, Borthakur D 2006. Recent advances in food biotechnology research. In: *Food Biochemistry and Food Processing*. Hui YH, Nip W-K, Nollet LML, Paliyath G, Sahlstrom S, Simpson BK (eds). Oxford, Blackwell Publishing, pp 35-70.

Karatzas CN, Turner JD 1997. Toward altering milk composition by genetic manipulation: current status and challenges. *J Dairy Sci* 80(9): 2225-2232.

Kitamoto KATSUHIKO, Oda K, Gomi K, Takahashi KOJIRO 1991. Genetic engineering of a sake yeast producing no urea by successive disruption of arginase gene. *Appl Environ Microb* 57(1): 301-306.

Kullen MJ, SanozkyDawes RB, Crowell DC, Klaenhammer TR 2000. Use of the DNA sequence of variable regions of the 16S rRNA gene for rapid and accurate identification of bacteria in the Lactobacillus acidophilus complex. *J Appl Microbiol* 89(3): 511-516.

Martinez-Gomariz M, Hernaez ML, Gutierrez D, Ximenez-Embun P, Prestamo G 2009. Proteomic analysis by two-dimensional differential gel electrophoresis (2D DIGE) of a high-pressure effect in Bacillus cereus. *J Agr Food Chem* 57(9): 3543-3549.

Mathur BN, Rook H, Shthi S 2003. Recent trends in processing of genetically modified dairy foods. *Ind Dairyman* 55: 29-35.

Meunier JR, Grimont PAD 1993. Factors affecting reproducibility of random amplified polymorphic DNA fingerprinting. *Res Microbiol* 144(5): 373-379.

McGarry A, Law J, Coffey A, Daly C, Fox PF, Fitzgerald GF 1994. Effect of genetically modifying the lactococcal proteolytic system on ripening and flavor development in Cheddar cheese. *Appl Environ Microb* 60(12): 4226-4233.

McMahon DJ, Brown RJ 1984. Composition, structure, and integrity of casein micelles: a review. *J Dairy Sci* 67(3): 499-512.

Mierau I, Kunji ER, Leenhouts KJ, Hellendoorn MA, Haandrikman AJ, Poolman B,. Kok J 1996. Multiple-peptidase mutants of Lactococcus lactis are severely impaired in their ability to grow in milk. *J Bacteriol* 178(10): 2794-2803.

Nair AJ 2008. *Principles of Biotechnology*. Laxmi Publications.

Natarajan SS, Xu C, Bae H, Caperna TJ, Garrett WM 2006. Characterization of storage proteins in wild (*Glycine soja*) and cultivated (*Glycine max*) soybean seeds using proteomic analysis. *J Agr Food Chem* 54(8): 3114-3120.

Ough CS 1976. Ethyl carbamate in fermented beverages and foods a naturally occurring ethylcarbamate. *J Agr Food Chem* 24(2): 323-328.

Persuy MA, Legrain S, Printz C, Stinnakre MG, Lepourry L, Brignon G, Mercier JC 1995. High-level, stage-and mammary-tissue-specific expression of a caprine κ-casein-encoding minigene driven by a β-casein promoter in transgenic mice. *Gene* 165(2): 291-296.

Ramesh K, Yadav JS 1990. Biotechnological applications in dairy processing. *Ind Dairyman* 42(6): 264-268.

Report 1992. *Convention on Biological Diversity*. United Nations. pp 1–30.

Rinzler CA, Jensen MD, Brody JE 1999. *The New Complete Book of Food: A Nutritional, Medical, and Culinary Guide*. New York, Infobase Publishing.

Ruebelt MC, Leimgruber NK, Lipp M, Reynolds TL, Nemeth MA, Astwood JD,. Jany KD 2006. Application of two-dimensional gel electrophoresis to interrogate alterations in the proteome of genetically modified crops. 1. Assessing analytical validation. *J Agr Food Chem* 54(6): 2154-2161.

Singhal RS, Samant SK, Gupte RK 1990. Biotechnology in dairy industry. *Ind Dairyman* 42(9): 372-380.

Song Y, Kato N, Liu C, Matsumiya Y, Kato H, Watanobe K 2000. Rapid identification of 11 human intestinal Lactobacillus species by multiplex PCR assays using group and species-specific primers derived from the 16S-23S rRNA intergenic spacer region and its flanking 23S rRNA. *FEMS Microbiol Lett* 187(2): 167-173.

Wall RJ, Kerr DE, Bondioli KR 1997. Transgenic dairy cattle: genetic engineering on a large scale. *J Dairy Sci* 80(9): 2213-2224.

Welsh J, McClelland M 1992. PCRamplified length polymorphisms in tRNA intergenic spacers for categorizing staphylococci. *Mol Microbiol* 6(12): 1673-1680.

Williams JG, Kubelik AR, Livak K J, Rafalski JA, Tingey SV 1990. DNA polymorphisms amplified by arbitrary primers are useful as genetic markers. *Nucleic Acids Res* 18(22): 6531-6535.

2015, Dairy Product Technology: Recent Advances ***Pages 341–359***
Editors: **Subrota Hati, Surajit Mandal and Birendra Kumar Mishra**
Published by: **DAYA PUBLISHING HOUSE, NEW DELHI**

Chapter 17

Bio-Functional Whey Based Beverages

Gopika Talwar and Santosh Kumar Mishra

Introduction

Whey is a nutritious by product from cheese, chhana and paneer industry containing valuable nutrients like lactose, proteins, minerals and vitamins etc. which have indispensable value as human food. Whey constitutes 45-50 per cent of total milk solids, 70 per cent of milk sugar (lactose), 20 per cent of milk proteins and 70-90 per cent of milk minerals and most importantly, almost all the water soluble vitamins originally present in milk (Horton, 1995). Whey proteins though present in small quantities has high protein efficiency ratio (3.6), biological value (104) and net protein utilization (95) and is next only to egg protein in terms of nutritive value. The continuing annual growth in the production and consumption of cheese and coagulated milk products represents the generation of large amounts of whey. In India, it is estimated that about 100 million kg of whey is annually derived as a by product which may cause substantial loss of about 70,000 tonnes of nutritious whey solids (Parekh, 2006). The main source of Indian whey is from production of channa and paneer. However, whey composition is very variable and significantly depends on the technology of whey production. Most compositional differences are in contents of calcium, phosphates, lactic acid and lactate which are present in much higher amounts in acid whey. Whey utilization is a great concern for the dairy industry due to techno-economic problems associated with it. Considerable efforts have been made over the past years to find new outlets for whey utilization and to reduce environmental pollution (González-Martínez *et al.*, 2002; Douaud, 2007; Jelièiæ *et al.*, 2008).Today modern industrial processing techniques such as ultrafiltration, reverse osmosis; hydrolysis, electro dialysis etc. have converted whey into a major source of functional ingredients. Still, large amount of whey is disposed off raw as cost of processing, handling and transportation exceeds the cost of whey product.

Therefore, conversion of whey into fermented or non-fermented beverages is one of the most attractive avenues for utilization of whey for human consumption. The development of any process for economical utilization of whey would be of great benefit to the dairy industry. Production of whey based beverages started in 1970's and until today a wide range of different whey beverages has been developed.

Table 17.1: Composition of Liquid Whey

Composition (per cent)	*Sweet Whey*	*Acid Whey*	*Casein Whey*
Total solids (per cent)	6.20	5.70	6.10
Lactose (per cent)	4.80	4.60	4.70
Proteins (per cent)	0.75	0.30	0.50
Fat (per cent)	0.05	<0.01	<0.01
Ash (minerals) (per cent)	0.60	0.80	0.90
pH	6.1	4.6	4.4

A variety of whey beverages such as plain, carbonated, alcoholic, soya and fruit types have been successfully developed and marketed all over the world, because they hold great potential for utilising whey solids. In India, a number of refreshing whey drinks and beverages have been developed that include whevit, acido-whey, whey-based fruit beverages, whey-based soups, whey-based lassi and whey-based sport beverage. These beverages are preferably prepared from paneer/chhana whey, which is acidic and has low protein content.

Different Types of Whey

Whey can be classified as

1. Sweet whey
2. Acid whey

Sweet whey is the liquid that is produced when making renneted hard cheese like cheddar or Swiss cheese. Acid whey (also known as "sour whey") is a by product of acid types of cultured dairy products such as cottage cheese or strained yogurt. Both types of whey contain mostly water, lactose (milk sugar), and some proteins, although sweet whey contains more proteins. Sweet whey and acid whey also contain vitamins and minerals such as vitamin A, riboflavin, vitamin B6, calcium, and magnesium, in varying amounts. Sweet whey is easy to use, but acid whey can't always be used in the same way that sweet whey is. Acid whey is more of an environmental hazard, because it deprives its surroundings of oxygen due to the bacteria it contains.

Uses for Acid Whey

a. To soak grain for making breads.

b. To lacto-ferment vegetables such as sauerkraut and other veggies, or to inoculate your next batch of yogurt.

c. To feed animals such as chickens, pigs, cows, cats, and dogs. Animals like sweet whey better, but it can upset their digestion if they consume too much.

d. As a cleaner for conditioner for hairs.

Uses for Sweet Whey

1. To reconstitute fruit juice to add nutritional value.
2. To make whey lemonade.
3. Can be added it to smoothes and shakes to provide more vitamins, minerals, and proteins.
4. It can be used to cook potatoes, rice, grits, pasta, and grains.
5. To make whey cheeses *e.g.* Ricotta
6. Make lacto-fermented drinks such as ginger ale and limeades.

Benefits of Whey-Based Fermented Drinks

Whey is an excellent growth medium for lactic acid bacteria to ferment lactose in whey to form lactic acid.

1. Whey as a drink can replace much of the lost organic and inorganic salts to the extracellular fluid.
2. Whey is rapidly adsorbed due to absence of fat emulsion.
3. Whey has been used to treat various ailments such as arthritis, liver complaints and dyspepsia.
4. It also possesses almost all the electrolytes of Oral Rehydration Solution (ORS), which is invariably used to control dehydration.
5. On fermentation with LAB, it becomes a suitable drink for lactose-intolerant people.
6. Fermentation of whey with LAB also masks the effect of curdy flavour of whey.
7. At industrial scale, large volumes of whey can be used directly from paneer/cheese vats, thus eliminating transportation and disposal problems.
8. Conversion of whey into beverages involves very simple processes.
9. Utilisation of whey generates additional revenue to the dairy plant.

Above all, its utilisation also solves the problems of environmental pollution. In this way fermented whey beverages with desirable nutritive and sensory properties is produced, without implementation of any complicated and expensive technologies like ultrafiltration and evaporation which are being used in case of processing of whey proteins isolates or concentrates or powdered whey to beverages.

Preparation of Whey Drink

To prepare whey drink good quality milk can standardized with 4.5 per cent fat and heated to boiling temperature with continuous stirring. Milk after this is allowed to cool and 2.0 per cent citric acid is added and it is stirred till coagulation takes place. Then whey is strained through a muslin cloth. The whey thus obtained is clear and greenish yellow in colour. The filtered whey can be centrifuged at 45°C to remove the fat. The acidity of whey is adjusted ranging from 0.5-0.7 per cent by addition of citric acid. Now sugar and fruit flavour or pulp is added according to different treatments. The mixture is then pasteurized and cooled to room temperature. The product is then bottled or pouched and stored at 5-7° C under refrigerated conditions.

Different Types of Whey Drinks

1. Fruit Based whey beverages
2. Fermented whey beverages
3. Alcoholic whey beverages
4. Whey soups
5. Whey based energy drink.

1. Fruit Based Whey Beverages

Many attempts have been made to incorporate fruit juices to whey in order to mask the typical whey flavour. Firstly it was tried with citrus juices which have high consumer acceptability. Whey drinks containing tropical fruits such as mango, banana and papaya have been tried by Duric *et al.,* 2004.

Some of the technological advantages associated with utilization of whey proteins in beverages include:

- ☆ Their solubility over a broad range of pH (3-8), even in their iso-electric pH.
- ☆ They have bland flavour so their inclusion in formulation does not cause cheesiness, saltiness, in developed products. Instead they are carrier of aroma compounds and help in developing flavour to their full potential.
- ☆ They possess excellent buffering capacity that is advantageous particularly in probiotic drinks where it help in survival of "live" bacteria in stomach.
- ☆ Their addition improve the viscosity of beverages, hence they can substitute stabilizers in beverages and therefore enhance the "mouth feel".

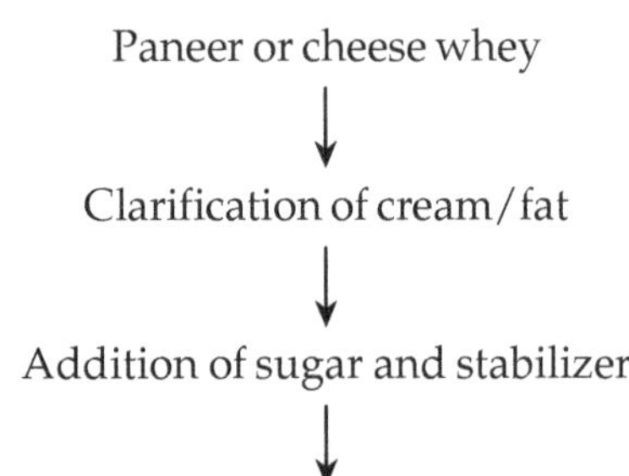

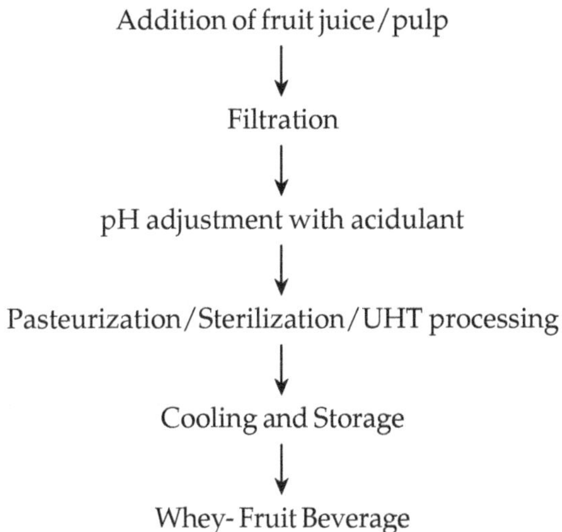

India has plenty of fruits that are a good source of vitamins and minerals. As production of fruits is seasonal, there is a glut in the market during a particular season. Fruits like mango, guava and banana are an excellent source of vitamin A and C, as well as a good source of potassium and beta-carotene. Besides being delicious, these beverages are highly nutritious. They may be particularly useful in places where there is inadequate nutrition, which could lead to nutritional deficiency diseases. In fruit producing countries, where malnutrition prevails, development of a nutritious, shelf-stable, whey-fruit beverage is a potential way to utilize surplus fruit.

Such attempts have been made by Dawdle *et al.* (2009) to develop a whey beverage with addition of mango pulp and sugar at different levels. The whey beverage was evaluated and optimized their level by sensory evaluation at 9- point hedonic scale. Mangos are high in fiber, but low in calories (approx. 110 per average sized mango), with traces amount of fat and sodium. The sensory score for flavour of mango drink ranged from 6.0-8.5. Maximum score (8.5) was given to product made from 5 per cent mango pulp and 7.5 per cent sugar. Sikder *et al.* (2001) formulated different blends of whey beverages by using various levels of mango pulp (8-12 per cent) with 0.04 per cent acidity. Singh *et al.* (1999) attempted to develop a soft beverage from paneer whey and guava. The pineapple flavoured chhana whey beverage has also been prepared with the addition of 5, 10 and 15 per cent of pineapple pulp in chhana whey. It was concluded that the good quality pineapple flavoured beverage can be prepared by addition of 5 parts of pineapple pulp and 95 parts of chhana whey with the addition of 8.0 per cent sugar.

The failure of whey-based beverages to perform well on the market is related to sedimentation problems. Ernest *et al.,* 2005 studied the selected characteristics of whey-fortified banana beverages stored at 4, 20, 30 and 40°C. Drinks were monitored at specific time intervals over a 60-day storage period. The sensory characteristics of the whey-banana beverage stored at 40°C were sour, sweet, smooth beverage, with distinctive banana flavour and minimum off-flavour However, when left standing;

this beverage separates immediately due to protein interactions between fruit constituents, such as the banana pectin and tannins.

To prevent the whey proteins' possible sedimentation during, whey can be mostly deproteinized either by heat treatment or by removing sediments through centrifugation or ultra filtration. To overcome sedimentation problems, high methoxyl pectin are used in acidified milk beverages. In such beverages pectin is added to increase the physical stability of the beverage. Such attempts have been carried out in case of development of whey based prickly pear beverage from prickly pear juice. The objective of this study was to test the impact of the addition of sugar, the treatment of whey and the addition of pectin and their potential interaction on the physical stability of beverages. The beverages obtained by mixing whey, prickly pear juice and other constituents (conservators (potassium sorbate and sugar), a colloidal stabilizer (HM-pectins) and an emulsifier (carboxy-methyl-cellulose)) were then homogenized, pasteurized at 80°C for 20 min and filled into glass bottles. The later were cooled at room temperature and then stored under refrigerated conditions (5 -1°C) for 40 days. (Baccouche *et al.*, 2013).

To provide whey drink with functional properties some drinks can be developed by addition of bottle gourd juice and herbs which posses' therapeutic, prophylactic, anti bacterial and organoleptic properties. Whey based beverage prepared from pineapple and bottle gourd juices in combination with edible herbal medicinal plant extract of *Mentha arvensis* had not only excellent nutritional and sensory properties but will also posses' therapeutic, prophylactic, antibacterial and organoleptic properties. It was prepared with the addition of mentha extract varying from 0 to 4 per cent concentration. For the preparation of 100 ml of herbal beverage, whey amount varying from 68 to 72 ml was added with 10 ml of pineapple juice and 10 ml of bottle gourd juice. The whey, juices and sugar (8 g/100 ml of beverage) were mixed in the given amount, preheated to 45°C before mixing mentha extract. Whey-based mango herbal beverage prepared with 2 per cent *Mentha* extract has been found to exhibit highest overall acceptability on the day of preparation as well as after 30 days of storage (Sirohi *et al.*, 2005). High-acid beverages (pH <4.4 and often <3.5) encounter very few microbial species that will grow. In addition, these products are generally pasteurized at >185°F and filled at approximately 182°F. The fill temperature—along with bottle inversion post-capping—essentially sterilizes the interior of the container and the cap. The resulting shelf life of these products is one year, but could effectively be longer than that. All whey sources being equal, and with adequate quality control of the ingredients, the choice of whey source has no bearing on shelf life.

Whey beverages can be compared to fruit drinks in terms of colour, protein content, anti oxidant activity and sensory score. In one such study orange whey drink was found to have 67 per cent more protein than orange drink. No significant differences were found between the beverages in respect of antioxidant activity against the ABTS* radical cation, remaining constant for the first 6 months of storage. Orange-whey beverages contained almost 3.5 times as much vitamin B_2 as orange beverages. Whey drink can also be prepared by reconstitution of sweet cheese whey powder by dissolving in distilled water to 10 per cent (w/v) final concentration in the ratio of 1:10 and the pH was 6.4. It can also be prepared by mixing reconstituted whey (RW) were with pasteurised semi-skimmed milk obtained from the market. All of these blends were heat treated at 80 °C for 15 min. No significant differences ($p>0.05$) in acidity was found between the samples which were fermented 3 or 4 hours under the conditions of the method. On the basis of the results it may be concluded that fermentation by yoghurt culture significantly affected the acidity of whey drink samples. Improvement of the whey-based beverages organoleptic properties because there were no significant differences in sensory properties between fermented and non fermented samples. The addition of milk was the most important factor influencing not only the total quality of the whey drinks but also their flavour, appearance, colour, viscosity and homogeneity.

Concentrated Fruit Based Whey Beverages

The purpose of developing concentrated fruit based beverage is to deliver the product in more convenient form, minimizing the transport and packaging requirement. It also improves the protein content. For preparing such beverages whey is concentrated in vacuum pans or evaporators, mixed with fruit juice concentrate, sugar and other additives, heat treated and packed. Singh *et al.* (2000) found that the whey mango concentrate obtained by mixing 15 per cent mango pulp (25° brix), 77 per cent paneer whey concentrate (37 per cent TS), 8 per cent sugar and a pH of 4.2 was most acceptable for developing whey-mango concentrate. Whey-apple and whey-kinnow concentrate is also available in the market.

The major problems in whey- fruit juice concentrates are as follow:

- ✰ Lactose crystallization during storage especially under refrigeration storage.
- ✰ High protein coagulation hinders their effective thermal processing.
- ✰ Whey protein coagulation during thermal processing.
- ✰ Low storage stability at elevated temperate.

These points should be investigated for developing acceptable shelf life nutritious whey-fruit juice concentrates.

2. Fermented Whey Beverages

The health giving and vitalizing properties of fermented milk and milk like beverages have been documented science ancient time. The nutritional and physiological benefits of fermented beverages are the promotion of growth and digestion, settling effects on gastrointestinal tract by decreasing harmful bacteria, improving bowel movement, ameliorating immunity and mineral absorption, suppression of cancer and lowering of blood cholesterol (Butriss, 1997).

Processing of whey to beverages began during the time of world war-2. Whey beverage called "Lactrone" was developed and patented by schulz. In the processing of lactrone the whey is first fermented with kefir culture and vacuum evaporated to remove whey waste. After that alcohol is distilled off the resulting stable product is concentrated. Vitamin-C can also be added to product to enhance the flavour of diluted beverage (Schulz, 1942).

Rivella

A sparkling, crystal clear drink is one if the oldest whey beverage from Switzerland. Rivella is prepared by fermenting deproteinized whey with lactic acid bacteria, filtering, condensing to a 7:1 concentrate and then adding sugar and flavouring. The product is re-filtered, diluted and carbonated. The last step is bottling and pasteurization. The finished product containes 9.7 per cent TS, 12.5 per cent total nitrogen and about 3.7 pH. (Anon, 1960). In some recent studies whey is fermented by using strains like *Lactobacilius acidophilus, Lactobacillus delbrueckii* sbsp. *bulgaricus, Streptococcus thermophillus, Lactobactcillus rhamnosus* and *Bifidobacterium animalis* subsp. *lactis.* There by the most successful ones proved to be fermentation with yogurt culture (*Streptococcus thermophillus and Lactobacillus delbrueckiisbsp. bulgaricus*) and co-culture *Streptococcus thermophillus, and Bifidobacterium animalis* subsp. *lactis* (Almeida *et al.,* 2008). Similar results was obtained by Pescuma *et al.* (2008) in which proposed co-culture *Streptococcus thermophillus, and Bifidobacterium animalis subsp. lactis* had high potential culture for whey fermentation.

There are some indications that fermentation of whey using yoghurt culture (*Lactobacillus delbrueckii* sbsp. *Bulgaricus and Streptococcus thermophillus*) produces a more intense yoghurt flavour as compared to the one obtained when skim milk is fermented. This suggest the possibility of producing beverages from whey with similar sensory to those of fermented milk drinks or with some flavour attributes of drinking yoghurt, following manufacturing procedures conventionally used for milk.(Gallardo-escamilla *et al.,* 2005). *Lactobacillus rhamnosus* belong to one of the frequently used strains. One of the most famous beverage obtained by whey fermentation with *Lactobacillus rhamnosus*is "**Gefilus**", which is produced in Finland using demineralized whey with prior lactose hydrolysis. It is necessary to hydrolyse lactose before starting fermentation with *Lactobacillus rhamnosus* because it does not have ability of fermenting lactose due to lack of enzyme *B*-galactosidase. This beverage is mostly being flavoured by addition of fruit juices or fruit aromas and fructose as sweeting agent (Tratnik, 1998).

A new technology was developed for manufacturing of whey beverage with kefir aroma. In this process heat treated whey (90°/30 min) was inoculated with 1-5 g kefir grains per 100 ml at room temperature with continuous shaking. Highest flavoured score was obtained using 3.5g kefir grains per 100 ml whey. The total nitrogen and non-protein nitrogen increased where as there was decrease in total solid and lactose with storage (Ismail *et al.,* 1992).

Functional fermented whey drink was formulated by mixing fermented whey concentrate, peach juice and 2 per cent calcium lactate. The product has low lactose and B-lactoglobulin contents and high essential amino acid concentration. In this process whey protein concentrate (WPC-35) was reconstituted in water (10 per cent w/v) and incubated at 37 degree for 24 h with *Lactobacillus delbrueckiisbsp. bulgaricus* CRL 656: *Streptococcus thermophillus* CRL 804:*Lactobacillus acidophilus* CRL636 at a 1:1.5:6.4 cfu/ml ratio. Then the fermented WPC 35 was mixed with peach juice and calcium lactate and stored at 10 degree for 28 days (Pescuma *et al.,* 2010).

Due to the low total solid content of the liquid whey the mouth feel of fermented whey beverage is poor and watery in comparison with fermented milk. Therefore it is required to use exopolysaccharide - producing probiotic strains or hydrocolloids. Hydrocolloids when added in low amount enhance viscosity of the product and prevent sedimentation of dispersed particles. Therefore the choice of proper type and level of hydrocolloid used is one of the most important factors in manufacturing of fermented dairy products. In fact, it is very important that the added hydrocolloids do not mask natural flavour of the product and effective at the typical product pH range. *i.e.* 4.0-4.6. Some most suitable and frequently used hydrocolloids are carboxy methyl cellulose (CMC), pectin, alginate and xanthan gum (XG). Addition of these significantly enhances mouth feel of the end product (Gallardo-Escamilla *et al.,* 2007).

Acido-whey

For the manufacture of acido-whey that is a non-alcoholic whey drink, deproteinised whey is fermented with a culture of *Lactobacillus acidophilus* and *Lactobacillus bulgaricus.* Sugar and flavour are then added and the product is heated at 75°C for 5 min, cooled to 5°C, packed in pouches and stored in refrigerated conditions.

Whey-Based Lassi

Whey-based lassi, in which up to 60 per cent of milk is replaced with whey, has also been developed at NDRI. The product formulation requires addition of pectin, CMC and trisodium citrate. The product has 2.0 per cent fat, 1.8 per cent protein, 4.6 per cent lactose and about 23.0 per cent total solids. This product can also be UHT-processed for long shelf-life.

Probiotic Whey Beverages

Probiotic whey beverages (with addition of sugar and pectin) is produced using probiotic cultures *Lactobacillus reuteri* and *Bifidobacterium bifidum.* The beverage is fermented by probiotic strain Bb-12 which has lower sensory score than beverages fermented by strains La-5 and Lc-1. *Lactobaciluus acidophilus,* YG culture, *Lactobacillus rhamnosus* and *Bifidobacterium animalis.* A fermented whey drink was prepared by using yoghurt cultures and co-cultures *Streptococcus thermophilus, Bifidobacterium animalis subsp. lactis. Streptococcus thermophillus* and *Lactobacillus delbrueckii* sbsp. *bulgaricus,* has a high potential culture for whey fermentation. Cultured whey was prepared from raw milk whey using *B. bifidum* having shelf life of 7 days at refrigeration temperature. Whey-based probiotic product was developed by using *L. reuteri* and *B. bifidum* with a shelf life of 14 days.

3. Alcoholic Whey Beverages

Alcoholic beverages can be manufactured by bioconversion of whey. A good beverage should be transparent, dear, and preferably sparkling. Various research efforts have been done on this theme, and yeasts like *Kluyveromyces fragilis* and *K. Marxianus* have been proposed as suitable biocatalysts for this bioprocess (Koutinas *et al.,* 2007). Distilled alcoholic beverages are characterised by the presence of volatile compounds (fusel alcohols, fatty acids, esters and others), which arise during fermentation, distillation and storage processes. Identification of these compounds is of major importance, not only to determine the flavour characteristics of the drink, but also to detect illicit spirits, and to identify anomalies that are indicative of inconsistent manufacturing practices (Fitzgerald *et al.,* 2000).

Although there are numerous literature reports about alcohol production from whey, most of them are based on the addition of fruit juices, such as mango, banana, pineapple, guava and strawberries, to whey (Kourkoutas *et al.,* 2002). In addition, the scale-up of the process has been little explored and the development of a suitable large-scale procedure for effective utilisation of lactose is still necessary. Moreover, information regarding the volatile compounds presents in the distilled drink is scarce, since most of the studies are only concerned in increasing the ethanol yield during fermentation. Givliano Drangone *et al.* (2008) revealed the presence of forty volatile compounds in the alcoholic distilled beverage produced by continuous fermentation of whey with *Kluyveromyces marxianus*. Most of these compounds are similar to those reported for other alcoholic beverages, although the concentration values are different. Higher alcohols (mainly isoamyl alcohol, isobutanol, and 1-propanol) and ethyl esters (mainly ethyl acetate) were the most dominant compounds present, contributing thus for the greatest proportion of the total aroma. Some short and long chain fatty acid esters that contribute to fruity and flowery aroma were also present, and the volatile compounds that can be harmful to the health (methanol, acetaldehyde and ethyl acetate) were found at low levels.

Several fermented whey beverages with varying alcohol contents have also been successfully produced. These drinks may be classified as:

- Beverages containing less than 1.5 per cent alcohol
- Whey beer
- Whey wine
- Whey champagne.
- Whevit

Beverages Containing less than 1.5 per cent Alcohol

These drinks are produced from whey permeate by fermenting the lactose with *Kluyveromyces fragilis* or *Saccharomyces lactis* to an alcohol content of 0.5-1.0 per cent, adding flavouring, sweetener and bottling. Acid whey permeate would lend a pleasantly tart flavour

to the product. One such product developed in Poland is produced by inoculation of acid whey permeate with kefir fungi (30 per cent addition, 5 hr incubation at 77°F). The fermented whey beverage contains 0.6-0.7 per cent lactic acid and 0.8-0.95 per cent alcohol. Koumiss-like drinks can be produced with mixtures of whey and buttermilk or by mixing whey and milk for fermentation. Guan and Brunner developed a koumiss-type product with a skim milk/whey blend in which the mixture has 2.5 per cent added saccharose, inoculated with 2.5-10 per cent of a culture containing lactobacilli and yeast and incubated at 78°F for 12-15 hr to a lactic acid content of 1 per cent. The fermented product is stirred, homogenized and bottled in glass bottles.

The bottles are stored at 68-77°F for 2 hr to produce some CO_2 and alcohol, and then cooled to less than 40°F. The koumiss-like product has a shelf-life of 4 weeks at 40°F.

Whey Beer

The production of beer from whey has been carried out since around the 1940s. Whey beer can be produced with or without addition of malt; it can be fortified with minerals or can contain starch hydrolysate and vitamins. Whey is suitable for the production of beer for the following reasons:

- The whey protein content and quantity of milk minerals in the colloidal state form the basis for a high degree of CO_2 binding.
- Whey, like wort, has a high mineral content.
- A caramel-like flavour develops, similar to that of kilned malt, mainly as a consequence of the browning reaction of lactose.
- Lactose in whey is only slightly sweet and does not alter the flavour of the final beer.

Some of problems that can occur here are presence of milk fat since which can cause loss of beer foam, undesirable odour and taste due to low solubility of whey proteins and inability of beer yeasts to ferment lactose and there can be high microbial load present in the whey.

Whey Wine

Whey wine contains relatively low alcohol amount (10-11 per cent) and is mostly flavoured with fruit aromas. Production of whey wine includes clearing, de-proteinization, lactose hydrolysis by ß-galactosidase, decanting and cooling, addition of yeasts and fermentation, decanting, aging, filtering and bottling (Popoviæ-Vranješ and Vujièiæ, 1997).Palmer and Marquardt also described a process for whey wine in which natural whey is converted to a clear fermented beverage base in a 5-step process of clarification, deproteinization, fermentation, de-ashing and polishing filtration. The process resulted in a wine with the following composition: 8.0 per cent alcohol by vol., 6-9 per cent invert sugar for sweetness, 2-4 per cent natural flavour extract, and 0.2-0.5 per cent organic acid. Panellists found no significant differences between a targeted flavoured wine and the fermented whey wine. However, economic projections indicated a significantly lower market price for whey wine compared to commercial flavoured wine. Yoo and Mattick indicated that total production of alcohol increases with lactose concentration to a maximum in whey of 12 per cent lactose. An ethanol concentration of 10 per cent was obtained from a 10 per cent acid whey solution containing 16 per cent added sucrose.

Whey Champagne

Polish workers have reported on a process for producing whey champagne. For production of 1000 litres of champagne, 200 litres of deproteinized acid whey, 90 kg of saccharose, 10 kg of caramel sugar, and 1.5 kg of dry yeast, 0.1 kg of fruit essence and 720 litres of water are used. The whey, water and sugar are blended together and pasteurized, followed by addition of the yeast and fruit essence and bottling. The whey champagne is fermented in the bottle at 65°F for 8-12 hr.

Whevit

Whevit, an orange, pineapple, lime or mango flavoured alcoholic drink from whey, was developed at National Dairy Research Institute (NDRI), Karnal. For its manufacture, fresh whey is efficiently separated in cream separator, deproteinised by steaming and cooled to room temperature. To the deproteinised and clarified whey, 22-23 per cent of 50 per cent sugar solution is added followed by 2-2.1 per cent of 10 per cent citric acid, colour and flavour. It is then fermented by incubation at 22°C for 14-16 h with a 1 per cent culture of *Saccharomyces cerevisae*. The product is bottled, pasteurised, cooled and stored at low temperature (5-10°C). The final product contains 0.5 to 1 per cent alcohol.

4. Whey-Based Soups

Soups are served as appetisers before meals as they stimulate the secretion of gastric enzymes that leads to feeling of hunger. In market a large number of ready-to-make soup mixes are available to suit the palate of consumers. But certain additives in such soups mixes are considered harmful particularly to children. Moreover apparently they do not seem to provide quality nutrients and utilisation of whey for soup preparation is an attractive possibility.

The common sequence of the operations in the production of a whey-based vegetable soup is: blending of the vegetable and corn flour in whey followed by heat processing. The time-temperature combination for cooking of vegetables, corn flour and seasoning is important for proper dispersion of vegetables, gelatinisation of starch and flavour perception of soups. The development of long shelf life soup involves proper sterilisation of soup. Whey based soup powders can be manufactured by cooking of vegetables in concentrated whey, mixing in it fried seasonings and gelatinised starch followed by spray drying. However, commercial soup powders available in the market comprise blending of dried vegetables in gelatinised starch. The process for the manufacture of whey-based soup involves blending of vegetables in whey and cooking of corn flour followed by heating. The time-temperature combination of cooking of vegetables, corn flour and seasoning is important for dispersion of vegetables, gelatinisation of starch and flavour perception of soup respectively (Singh and Kumar, 1997). The developed product could be stored for a week under refrigeration and UHT treatment can be adopted to improve the shelf-stability. Paneer and cheese whey were utilised for the potato-carrot-tomato and spinach soups. Cheese whey was preferred for the manufacture of vegetable soups than paneer whey (Singh *et al.*, 1994). The reason could be the low pH of paneer whey that resulted in acidic product not usually compatible with most vegetables. Whey-based soups have been reported to be more viscous as compared to water-based most probably gelation of whey proteins on heating. Whey-based soups require fewer amounts of salt, thickener and fat. Technology for manufacture of retort processed low fat tomato-whey soup has been developed recently at the NDRI. Alam *et al.* (2002) reported the technological aspects for the manufacture of tomato whey soup using paneer whey. Few years back Amul has introduced UHT processed tomato-whey soup in Tetra Pak and last year VITA has launched tomato-whey soup in polystyrene cup in Haryana.

Mushroom-whey soup powder was prepared by cooking mushroom with concentrated cheese whey and blending. The seasonings (onion, garlic and ginger) were fried separately in hydrogenated fat, and corn flour was added if thickening was desirable. After the addition of salt and a further quantity of concentrated whey, the soup mix was again blended. It was then spray-dried to produce a mushroom-whey soup powder. The physico-chemical properties of the soup powder, such as moisture content, loose and packed bulk density, wettability, insolubility index, thio-barbituric acid and hydroxy methyl furfural content, increased during storage. However, dispersibility and reflectance value decreased during storage. The soup powder reconstituted well when boiled in water for 2 min. There constituted soup was considered acceptable, with an overall acceptability score of 7.1 on a nine-point hedonic scale, after 8 months of storage of the soup powder at 30°C when packed in metalized polyester.

5. Whey Based Energy Drink

Proteins are macromolecules composed of one or more polypeptide chains, each with a characteristic sequence of amino acids linked by peptide bonds (Ohr, 2001).Whey proteins have exceptional nutritional value. Whey proteins are also a good source of sulfur-containing amino acids which are proven to maintain antioxidant levels in the body (Pasin and Miller, 2000). Whey proteins improve the host's antioxidant defences as well as lower the oxidant burden (Walzem, 1999). The ability of the proteins to inhibit harmful changes caused by lipid oxidation seems to be related to amino acid residues in the proteins. Such amino acids include tyrosine, methionine, histidine, lysine and tryptophan which are capable of chelating pro-oxidative metal ions. Antioxidant activity is also affected by the amino acid composition, sequence and configuration (Pena-Ramos and Xiong, 2001).

Studies done by pasin and miller in 2000 shows that Cysteine, methionine, lysine, leucine and glutamine are key amino acids that play vital roles in sports performance and health. The amino acids cysteine and methionine are thought to stabilize DNA during cell division, while arginine and lysine amino acids help to stimulate growth hormone release, thus resulting in muscle growth and increased muscle mass as well as a decrease in percent body fat. Whey proteins offer natural alternatives to anabolic-androgenous steroids. Glutamine helps to replenish muscle glycogen after exercise and also prevents decline in immune function caused by overtraining (Pasin and Miller, 2000). More over the biological value of whey protein is 1.14 higher than any other protein. Whey contains a number of valuable minerals as well. These minerals help to enhance the functionality of whey proteins. These include monovalent sodium, potassium and chloride ions, magnesium, citrate and phosphate (Anonymous, 2001).Whey is an excellent source of bio available calcium which functions to reduce the incidence of stress fractures during exercise and prevents bone loss in both hypoestrogenic female athletes as well as post-menopausal women (Pasin and Miller, 2000), and can enhance bone growth due to the presence of fibroblast growth factors (Walzem, 1999).

The ideal sports protein should have a good balance of essential and nonessential amino acids, an abundant supply of BCAAs, and should be low in fat and cholesterol

(Pasin and Miller, 2000). Whey protein is one of the best sources of BCAAs, which are the only amino acids that can be oxidized directly by the muscle, thereby providing immediate energy (Anonymous, 2002). The structure of the amino acids leucine, isoleucine, and valine allow them to be easily utilized (Anonymous, 2002). The iron status of athletes is very important, because if the loss of iron exceeds the intake of iron or the body's ability to recycle iron, the athlete is at risk for developing iron-deficiency anemia (Manore and Thompson, 2000). Most sports rely on the aerobic pathway which requires oxygen to produce ATP. The transportation and diffusibility of the oxygen molecules are limiting factors in aerobic performance. Shortage of iron can impair aerobic performance. Lactoferrin is a minor protein component of whey, comprising only 1-2 per cent of all whey protein. It is a precursor for the iron binding protein transferrin which is responsible for regulating iron absorption and regulating the bioavailability of iron as well (Whey Protein Institute, 2001). Lactoferrin in whey and whey beverages also acts as an iron supplement to boost blood oxygenation without any side effects (Pasin and Miller, 2000). Transferrin also acts as an antioxidant by binding iron and preventing it from participating in free radical formation. Free radicals have been identified as the cause of muscle injuries in athletes, resulting in decreased performance (Pasin and Miller, 2000; Brink, 2000).

Recently whey products have become very popular ingredients in sports nutrition. The past decade has seen increasing popular interest in healthy lifestyles based on regular exercise. This increase in the number of muscle and fitness enthusiasts has prompted a growing consumer demand for protein sports beverages, specialized nutritional drinks, nutritional snack bars and other products designed to optimize athletic performance. A growing body of scientific evidence indicates whey proteins deliver important physiological benefits for consumers seeking superior physical performance and recovery.

Benefits of Whey Proteins in Sports Nutrition

- ☆ Easily digestible high quality protein–provides additional energy, spares endogenous protein.
- ☆ Contains high levels of BCAAs: leucine, isoleucine, and valine
- ☆ Good source of sulfur-containing amino acids such as cysteine and methionine- maintains antioxidant levels in the body, and are thought to stabilize DNA during cell division
- ☆ Contains glutamine–helps muscle glycogen replenishment and prevents decline in immune function from overtraining
- ☆ Excellent source of bioavailable calcium–reduces stress fractures during exercise and prevents bone loss in hypoestrogenic female athletes.
- ☆ Contains high levels of arginine and lysine–may stimulate growth hormone release, and thus stimulates and increase in muscle mass and decline in body fat.

In addition, their bland flavour allows their formulation with products not normally associated with dairy products, like a fruit-bar or fruit-juice application.

Flow Chart for Manufacturing of Whey Energy Drink

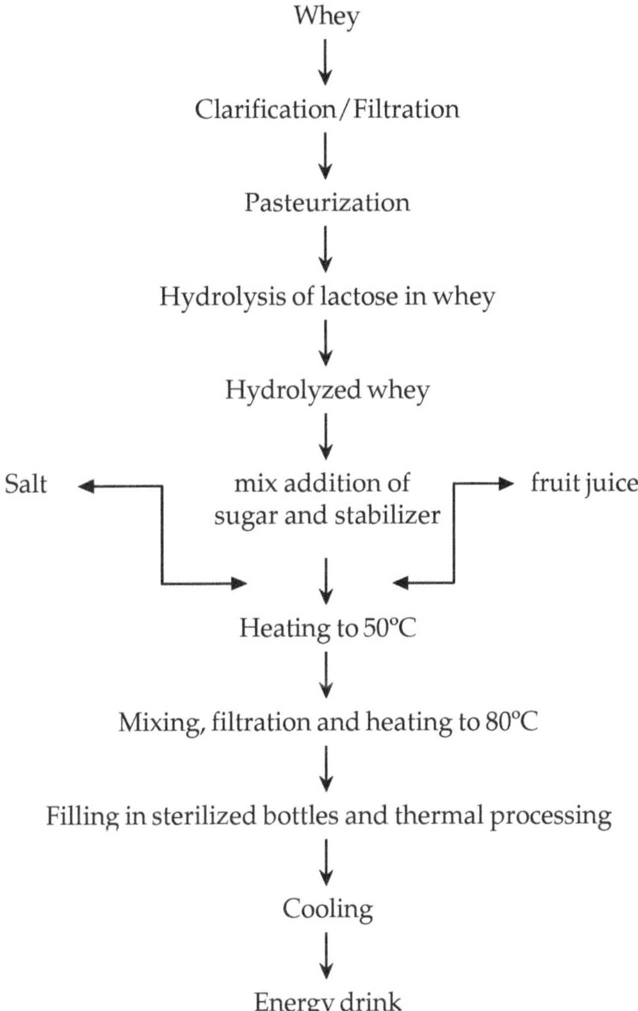

In sports drinks and bars WPIs are used:

- ☆ To provide quick energy
- ☆ To increase muscle weight 'gain'
- ☆ To repair muscle tissue

Whey proteins contain a high level of BCAAs (~26 per cent) that are taken up directly by skeletal muscles during extensive exercise, rather than first being metabolized through the liver like other amino acids. Since the body's demand for these amino acids increases during exercise, athletes who want to preserve muscle mass may benefit by increasing their consumption. Whey has 10g leucine, 6.5g isoleucine and 5.5g valine per100g of protein. There are some variations in amino acid composition depending on the origin and the processing of the protein. Different processes produce whey proteins with different BCAA contents. For example, an ion exchange process and microfiltration produce a profile with more-lacto globulin richer in BCAAs. In the US, dairy companies produce whey proteins with enhanced levels of BCAAs specifically tailored for sports beverages and sports bar applications.

Whey proteins are also used as a protein base for homemade sports products such as cookies and / or drinks. Sports nutritionists sometimes recommend homemade sports products as low-cost alternatives to commercial products, particularly for recreational athlete.

Lactose on hydrolysis gives glucose and galactose which are instant energy providing constituents. At present little work has been found in literature on hydrolyzed lactose beverages or drinks from whey. The manufacture of high energy, nutritious and mineral rich drink from whey to be used by hard working group of people, in order to meet instant energy requirement and compensate electrolytic loss, is an useful approach. The utilization of whey as a beverage production is more economical and value addition than other methods of treatment and disposal.

Following are some of the common whey energy drinks available in national and international market

Black Hole is an energy drink which contains Taurin. It´s based on whey and it is flavored with Bilberry

1. White Bull is a carbonated, whey-based energy drink, flavored with cacao.
2. Whey Fresh is energy drink based on skimmed milk whey. This stimulating drink is available in the flavors tonic and apple. Protein water
3. Protein water is pleasantly flavoured water that provides consumers the added benefit of high quality whey protein to help boost nutritional intake. Clear Protein™ 8855 is a unique and highly innovative whey protein isolate ingredient that is clear in solution. Amul pro whey protein malt beverage with malt is the best energy drinking which is been made from best healthy flours and natural grains. One can get this malt beverage in different varieties of flavors. Amul pro whey malt beverage is fastest and largest selling energy drink in the market.

Conclusions

Current consumer preference regarding food intake are for more protein and fibre with no significance change in carbohydrates and with less fat. Whey based beverages fit this preference pattern very well. Whey, a by-product of cheese making, is recognized as a value added ingredient in many food products including dairy, bakery, confections, food and beverages. Due to increasing awareness regarding potential health benefits of whey new technologies such as reverse osmosis (RO), ultrafiltration, demineralisation using ion exchange and nano filtration. Although these technological advances have created greater avenue for whey usage, simpler and cheap technologies are necessary for whey utilization. The beverage sector provides best answer to these questions because it utilizes whey with least processing and low cost.

References

Alam, M.T., Singh S., Broadway, A.A. 2002. Utilization of paneer whey for the preparation of tomato whey soup. *Egyptian J. Dairy Sci.*, **30:** 355-361.

Almeida, K.E., Tamime, A.Y., Oliveira, M.N. 2008. Acidification rates of probiotic bacteria in *Minas frescal* cheese whey, *LWT*, **41:** 311-316.

Anon, 1960. "Rivella" — a new form of whey utilisation. *Dairy Ind.*, **25(2):** 113.

Anonymous, 2001. Do it with dairy. DMI. Whey. p. 1-3. www.doitwithdairy.com/infolib/ingspecsheet/factwhey.htm

Anonymous, 2002. Do it with dairy. "Athletes get the most from their diets: whey protein provides high level of necessary protein and other nutrients".

www.extraordinarydairy.com/standard.asp?ContentPage1d=84

Anatovskiy, A., and V. Yaroshenko. 1950. Preparation of sparkling whey. *Mol-Prom.*, **11(9):** 31.

Arwa, B., Monia, E., Imene, F., Hamadi, A. 2013. A Physical stability of whey-based prickly pear beverages, *Food Hydrocolloids*, **33:** 234-244

Ðuriæ, M., Cariæ, M., Milanoviæ, S., Tekiæ, M., Paniæ, M. 2004. Development of whey based beverages, *European Food Research and Technology*, **219:** 321-328.

Douaud, C. 2007. Whey proteins sees demand from functional drinks. http: // www.nutraingredients-usa.com.

Ernest, K., Robert, S. and Louise, W. 2005. Storage stability and sensory analysis of UHT processed whey-banana beverages. *Journal of Food Quality*, **28:** 386-401

Fitzgerald, G., James, K. J., MacNamara, K., and Stack, M. A. (2000). Characterization of whiskeys using solid-phase microextraction with gas chromatography-mass spectrometry. *Journal of Chromatography*, **896:** 351–359.

Gallardo-Escamill, F.J., Kelly, A.L., Delahunty, C.M. (2005). Influence of starter culture on flavor and Headspace Volatile Profiles of Fermented Whey and Whey Produced from Fermented Milk, *Journal of Dairy Science*, **88:** 3745-3753.

Gallardo-Escamilla, F.J., Kelly, A.L., Delahunty, C.M. (2007). Mouthfeel and flavour of fermented whey with added hydrocolloids, *International Journal of Dairy Science*, **17:** 308-315.

Guan, J., and Brummer, J.R. 1987. Koumiss produced froma skimmilk/sweet whey blend. *Cultured Dairy Prod. J.*, **22(1):** 23.

González-Martínez, C., Becerra, M., Cháfer, M., Albros,A., Carot, J., Chiralt, A. 2002. Influence of substituting milk powder for whey powder on yogurt quality. *Trends in Food Science and Technology*, **13:** 334-340.

Gandhi, D.N. 1996. Fermented Whey Beverages, NDRI Bulletin, No. 278, pp. 1-7. Holsinger, V.H., L.P. Posati, and E.D. DeVilbiss. 1974. Whey beverages: a review. *J. Dairy Sci.*, **57:** 849-859.

Horton, B.S. 1995. Whey processing and utilization. *Bulletin of the International Dairy Federation,* **308:** 2-6.

Jelièiæ, R., Božaniæ, R., Tratnik, L. 2008. Whey-based beverages - a new generation of diary products. *Mljekarstvo,* **58(3):** 257-274.

Manore, M. and Thompson, J. 2000. Sport nutrition for health and human performance. Human Kinetics, Champaign, IL, 2000. p. 106, 111, 115, 121, 277-278,341.

Ohr, L. M. 2001. Proteins pack a one-two punch. *Prepared Foods.* Vol. August. p. 28-30, 32.

Pescuma, M., Hebert, E.M., Mozzi, F., Font De Valdez, G. (2008). Whey fermentation by thermophilic acid bacteria: Evolution of carbohydrates and protein content, *Food Microbiology,* **25:** 442-451.

Popoviæ-Vranješ I Vujièiæ, A., Vujièiæ, I. (1997): *Tehnologija surutke,* Poljoprivredni fakultet Novi Sad.

Pescuma, 2010. *International Journal of Food Microbiology,* **141:** 73–81.

Palmer, G.M., and R.F. Marquardt. 1978. Modern technology transforms whey into wine. *Food Prod. Development,* **12(1):** 31, 34

Peña-Ramos, E. A. and Y. L. Xiong. 2001. Antioxidative activity of whey protein hydrolysates in a liposomal system. *Journal of Dairy Science,* **84:** 2577-2583.

Pasin, G. and Miller, S.L. 2000. U.S. whey products and sports nutrition. U.S. Dairy Export Council. www.usdec.org/pdffiles/manuals/9SportsNut.pdf. p.1-8

Parekh, J.V. 2006. Emerging new technologies in the dairy industry in India. http: // www.fnbnews.com/article/detarchive.asp?articleid=19393§ionid=40

Sehulz, M. 1942. The stability of vitamin C in acid whey beverages. *Deut. Molkerei-Ztg.,* **63:** 492.

Sienkiewicz, T., and C.L. Riedel. 1990. Whey and Whey Utilization, 2nd ed., Verlag Th. Mann, Gelsenkirchen-Buer, Germany.

Singh, S., Kumar, A. 1997. Whey in soups and fruit beverages. In compendium of Short Course on "Technological advances in dairy by-products" organized under the aegis of CAS in Dairy Technology, pp. 64-68.

Singh, S., Ladkani, B.G., Kumar, A., Mathur, B.N. 1994. Whey utilization for the manufacture of ready to serve soups. *Indian J.Dairy Sci.,* **47:** 501- 504.

Sikder, B., Sarkar, K., Ray, P.R. and Ghatak, P.K. 2001. Studies on shelf-life of whey-based mango beverages. *Beverage Food World,* **28:** 53-54.

Singh, W., Kapoor, C.M. and Srivastava, D.N. 1999. Technology for the manufacture of guava- whey beverage. *International Journal of Dairy Science,* **52:** 268-270.

Sirohi, D., Patel, S., Choudhary, P.L. and Sahu, C. 2005. Studies on preparation and storage of whey-based mango herbal pudina (*Mentha arvensis*) beverage. *Journal of Food Science and Technology,* **42(2):** 157-161.

Singh, S., Singh, A.K.and Gandhi, D.N. 2000. Formulation of whey-mango concentrate. Paper presented in conference in biotechnological strategies in agro processing" on 9-11 Feb 2000. Organizes by Punjab State Council for Science and Technology. Chandigarh.

Tratnik L.J. 1998. Mlijeko-tehnologija, biokemija i mikrobiologija, Hrvatska mljekarska Udruga, Zagreb.

Walzem, R.D. and Rosemary L. 1999. Health enhancing properties of whey proteins and whey fractions. U.S. Dairy Export Council. www.usdec.org/pdffiles/manuals/7health.pdf. p.1-7.

Whey Protein Institute (WPI). 2001. Benefits of whey proteins. www.wheyoflife.org/benefits.html#3

Yoo, B.W., and Mattick, J.F. 1969. Utilization of acid andsweet cheese whey in wine production. *J. Dairy Sci.*, **52:** 900.

2015, Dairy Product Technology: Recent Advances ***Pages 361–387***
Editors: **Subrota Hati, Surajit Mandal and Birendra Kumar Mishra**
Published by: **DAYA PUBLISHING HOUSE, NEW DELHI**

Chapter 18

Recent Trends in Non Thermal Food Processing and Preservation

Kanchan Mogha and Subrota Hati

Introduction

Food preservation usually involves inhibiting the growth of bacteria, fungi (such as yeasts), and other micro-organisms as well as retarding the oxidation of fats that cause rancidity. It can be said that it is a continuous fight against microorganism spoiling the food and making it safe for consumption. Food preservation ensuring safety and quality has become a prime goal of the food processors looking towards increased demand of the consumers for more varied foods and more extension in shelf life during the last 25 years. Many processes designed to preserve food will involve a number of food preservation methods. Preserving fruit by turning it into jam, for example, involves boiling (to reduce the fruit's moisture content and to kill bacteria, yeasts, etc.), sugaring (to prevent their re-growth) and sealing within an airtight jar (to prevent recontamination). There are many traditional methods of preserving food that limit the energy inputs and reduce carbon footprint. Maintaining or creating nutritional value, texture and flavour is an important aspect of food preservation, although, historically, some methods drastically altered the character of the food being preserved. In many cases these changes are desirable qualities *i.e.* cheese, yoghurt and pickled onions being common examples. Innovations and technology developments in the field of food pasteurization and sterilization are continuously evolving which can replace traditional food preservation techniques (intense heat treatments, salting, acidification, drying and chemical preservation) by new preservation techniques. However, the most investigated non-thermal preservation technologies are high hydrostatic pressure (HHP), pulsed electric fields (PEF), ultrasounds (US), pulsed light treatment (PL) etc. In spite of the intensive research efforts and investments, very few of these new preservation methods are until now implemented by the food industry. So it is necessary to make a reflection on the possibilities and especially the advantages of the above mentioned recent technologies for the food industry and also its impact on environment.

Novel Non Thermal Processing Technologies

In the last decade, non-thermal inactivation techniques have been a major research area, driven by an increased consumer demand for nutritious, freshlike food products with a high organoleptical quality and an acceptable shelf life. Preservation of food was done since ancient times, right from the day fire was invented and cooking of food was started. In our day-to-day routine, we use number of food preservation techniques, including pasteurisation, boiling, cooling. The term 'non-thermal processing' is often used to designate technologies that are effective at ambient or sublethal temperatures. Recent advances in non-thermal processing of foods have created new approaches for preserving foods without comprising products quality. Investigated inactivation technologies are ionisation radiation, high pressure processing (HPP), pulsed electrical fields, high pressure homogenisation, UV decontamination, pulsed high intensity light inactivation, high intensity laser, pulsed white light, high power ultrasound, oscillating magnetic fields, high voltage arc discharge and streamer plasma have the ability to inactivate microorganisms to varying degrees (Butz and Tauscher, 2002). Some of these treatments may involve heat due to the generation of internal energy. They can eliminate the use of high temperatures to kill the microorganisms, avoiding the deleterious effects of heat on flavor, color and nutritive value of foods, mainly vitamins.

Pulsed Electric Field

A non-thermal inactivation technology gaining commercial interest and evolving more and more from laboratory and pilot scale to industrial scale is pulsed electric field technology (PEF). To change the behavior of microorganisms the concept of pulsed electric fields (PEF) was first proposed in 1967. The electric field phenomenon was identified as membrane rupture theory in the 1980s. Increasing the membrane permeability led to the application of PEF assisted extraction of cellular content and transfer of genetic material across cell membrane. The lethal effects of PEF to microorganisms were studied in 1990s when laboratory and pilot plant equipment were developed to evaluate the effect of PEF as a non thermal food process to provide consumers with microbiologically-safe and fresh-like quality foods. PEF offers the ability to inactivate microorganisms with minimal effects on the nutritional, flavor and functional characteristics of food products due to the absence of heat. For food quality attributes, PEF technology is considered superior to traditional thermal processing methods because it avoids or greatly reduces detrimental changes in the sensory and physical properties of foods (Quass, 1997).The first commercial PEF pasteurization on apple cider products took place in 2005 in the United States. Pulsed electric field (PEF) of high intensities is a viable alternative for sterilization of liquid foods such as fruit juices, milk, liquid egg and many others (Qin *et al.*, 1995). PEF treatment can also show effectiveness for inactivation of microbes for food dehydration and drying (Bajgai and Hashinaga, 2001; Taiwo *et al.,* 2002).

Principle of Pulsed Electric Field

Pulsed electric field (PEF) electroporation is a method for processing cells by means of brief pulses with duration of micro to milliseconds of a strong electric field at an intensity of order 10-80 kV/cm. In PEF processing, a substance is placed between two electrodes, then the pulsed electric field is applied. The electric field enlarges the pores of the cell membranes, which kills the cells and releases their cellular contents. After the treatment, the food is packaged asceptically and stored under refrigeration. The electric field can be applied exponentially decaying, square wave, bipolar, or oscillatory pulses and it holds potential as a type of low-temperature alternative pasteurization process for sterilizing food products. Food is capable of transferring electricity because of the presence of several ions, giving the product a certain degree of electrical conductivity. Applying electric field, the electric current flows into the liquid food and is transferred to each point in the liquid because of the charged molecule (Zhang *et al.,* 1995). The processing time is calculated by multiplying the number of pulses times with effective pulse duration. The process is based on pulsed electrical currents delivered to a product placed between a set of electrodes; the distance between electrodes is termed as the treatment gap of the PEF chamber.

Design of the PEF Equipment

The equipment consists of a high voltage pulse generator and a treatment chamber with a suitable fluid handling system and necessary monitoring and controlling devices (Figure 18.1) Food product is placed in the treatment chamber, either in a static or continuous design, where two electrodes are connected together with a nonconductive material to avoid electrical flow from one to the other. Generated high voltage electrical pulses are applied to the electrodes, which then conduct the high intensity electrical pulse to the product placed between the two electrodes. The food product experiences a force per unit charge, the so-called electric field, which is responsible for the irreversible cell membrane breakdown in microorganisms. This leads to dielectric breakdown of the microbial cell membranes and to interaction with the charged molecules of food (Fernandez-Díaz *et al.,* 2000; Zimmermann, 1986). Hence, PEF technology has been suggested for the pasteurization of foods such as juices, milk, yogurt, soups, and liquid eggs (Bendicho *et al.,* 2003).

Application of PEF in Food Processing and Preservation

Pulsed electric fields technology has been successfully applied for the pasteurization of foods such as juices, milk, yogurt, soups, and liquid eggs. Application of PEF processing is restricted to food products with no air bubbles and with low electrical conductivity. The maximum particle size in the liquid must be smaller than the gap of the treatment region in the chamber in order to ensure proper treatment. PEF is a continuous processing method, which is not suitable for solid food products that are not pumpable. PEF is also applied to enhance extraction of sugars and other cellular content from plant cells, such as sugar beets. PEF also found applicable in reducing the solid volume (sludge) of wastewater. PEF is especially promising for the

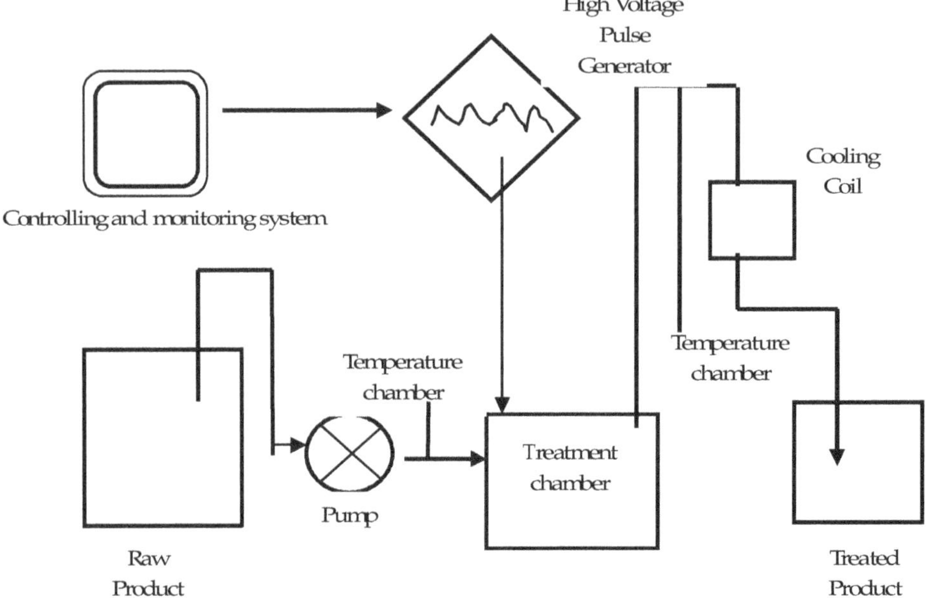

Figure 18.1: Flow Chart of PEF in Food Processing System with Basic Component.

citrus industry, which is concerned with the spoilage microorganisms and resultant production of off-flavor compounds such as lactic acid bacteria (Hendrix and Red, 1995). Jemai and Vorobiev (2007) of Berlin University stated the enhancing effect of a PEF treatment on the diffusion coefficients of soluble substances in apple slices using a peak voltage of 20 kV, an average power of 7 kW and a production capacity of 2 ton/h has been developed for the treatment of fruit mashes. Among all liquid products, PEF technology has been most widely applied to apple juice, orange juice, milk, liquid egg, and brine solutions (Qin *et al., 1995*). Studies conducted on the effects of PEF on dairy products such as skim milk, whole milk, and yogurt compromised a major section of PEF applications (Alvarez and Ji, 2003).

PEF treated *E. coli* and *L. innocua* suspended in pasteurised skim milk and in phosphate buffer (with similar pH and conductivities) with inlet and outlet temperatures of 17°C and 37°C, a flow rate of 0.5 L/min, frequency of 3 Hz and field intensity of 41 kV/cm was used by (Dutreux *et al.,* 2000). The number of surviving organisms was determined after the application of 0, 3, 10, 20, 35 and 60 pulses. Changes in the cytoplasm were observed and the cell surface appeared rough with outer membranes being partially destroyed, showing leakage of the cytoplasm. Combination of PEF treatment with antimicrobial compound nisin was used to inactivate *Listeria innocua* in skim milk by (Calderon-Miranda *et al.,* 1999). The selected field intensities (and temperatures) were 30 (22°C), 40 (28°C) and 50 (34°C) kV/cm and the number of pulses applied were 10.6, 21.3 and 32 pulses respectively. The sensitization exhibited by PEF treated *L. innocua* to nisin was assessed for 10 or 100 IU nisin/mL. *Listeria innocua* count was reduced to 2, 2.7 and 3.4 logs after exposure to the field intensities of 30, 40 and 50 kV/cm in presence of 10 IU nisin/mL while at 100 IU nisin/mL under the same PEF treatment conditions the reduction increased to 2.5, 3 and 3.8 logs. The increase in microbial reduction was attributed to the additive effect of nisin on PEF treatment.

Wüst *et al.* (2004) assessed the effect of PEF on functionality of protein and reported that the physical attributes of cottage cheese made from PEF-treated skim milk. The treatment was conducted by applying bipolar square pulses of 2 μs at field intensities of 25 and 28 kV/cm with pulse frequencies of 200 and 400 Hz and flow rate of 120 mL/min at a treatment temperature of <45°C. It was found that increasing the field strength decreased the strength of the cottage cheese gel and marginally increased the yield of cottage cheese compared to cheeses made from raw or pasteurised skim milk.

High Pressure Processing

Pascalization, bridgmanization, or high pressure (HHP), is a method of preserving and sterilizing food, in which a product is processed under very high pressure, leading to the inactivation of certain microorganisms and enzymes in the food. The technique was named after Blaise Pascal, a French scientist of the 17th century whose work included detailing the effects of pressure on fluids. During pascalization, more than 50,000 pounds per square inch (340 MPa) may be applied for around fifteen minutes, leading to the inactivation of yeast, molds and bacteria. Chemical activity caused by microorganisms that play a role in the deterioration of foods can be stopped by pascalization. The treatment occurs at low temperatures and does not include the use of food additives. From 1990, some juices, jellies, and jams have been preserved using pascalization in Japan. The technique is now used there to preserve fish and meats, salad dressing, rice cakes, and yogurts. An early use of pascalization in the United States was to treat guacamole. It did not change the guacamole's taste, texture, or color, but the shelf life of the product increased to thirty days, from three days. However, some treated foods still require cold storage because pascalization does not stop all enzyme activity caused by proteins, some of which affects shelf-life of the products.

Principle of HHP

Food processing by high hydrostatic pressure has been reviewed by several authors, giving particular attention to microbiological, (bio) chemical, technological, environmental and energetic aspects. The high pressure treatment of foods involves subjecting food materials to pressures that generally can range from of 100 to 1000 MPa. In agreement with the isostatic principle, during HHP pressure is applied uniformly and instantaneously through a food material (with or without packaging), independently of its mass, shape and composition. Under pressure, biomolecules maintains the Le Chatelier–Braun principle and reactions that result in reduced volume will be promoted. Such reactions affect the structure of large molecules (whose tertiary structure is important for functionality), such as proteins. HHP causes a partial unfolding of proteins that can promote covalent and non-covalent interactions during and upon release of pressure, thus triggering their denaturation. This results in the inactivation of microorganisms and enzymes (Hendrickx *et al.,* 1998) and can also promote changes in the rheological properties of the food products (Ahmed *et al.,* 2003). On the other hand, small molecules that have little secondary, tertiary and quaternary structures, such as amino acids, vitamins and flavor and aroma components contributing to the sensory and nutritional quality of food, remain unaffected (Balci and Wilbey, 1999; Cheftel, 1991). Process temperature during HHP can be specified from below 0°C (to minimize any effects of adiabatic heat) to above 100°C and exposure times can range from a millisecond pulse to a treatment time of over 20 min. In contrast to thermal processing, economic requirements for throughput may limit exposure times of treatment to less than 20 min (Food and Nutrition, 2000).

Overall, this technology offers several advantages for food preservation: (1) homogeneity of treatment due to the fact that pressure is uniformly applied around and throughout the food product; (2) minimal heat impact; (3) shelflives similar to thermal pasteurization, while maintaining the natural food quality parameters (nutrients and sensorial preservation); (4) small amount of energy needed to compress a solid or liquid to 500 MPa as compared to heating to 100°C (Tewari, 2007). HHP cold pasteurization technology is gathering applicability throughout the world in the processing of a variety of product categories. HHP has already become a commercially implemented technology, spreading from its origins in Japan, followed by USA and now in Europe, with worldwide increase almost exponentially since 2000.

Design of HHP

Main components of a high pressure system are,

1. A pressure vessel and its closure
2. A pressure generation system
3. A temperature control device
4. A materials handling system

Application of HHP

HPP finds application in food preservation in many ways and it has become a subject of major interest for both food preservation and food preparation, once it inactivates vegetative microorganisms by using pressure rather than heat to achieve pasteurization. Commercial production of pressurized foods has been reported for fruit jams, jellies, sauces, juices, rice wine, cake, avocado pulp, guacamole and cooked ham. For dairy and egg industry applications, the technique has been studied, due to changes induced the functional properties of whey protein as well as in other milk components and native constituents. For improving the shelf life of goat's cheese (Capellas *et al.,* 1996), to reduce the ripening time of cheese to 3 days at 250MPa (Yokoyama *et al.,* 1992) and to prevent over-acidification of yoghurt, increasing the shelf life to more than 2 weeks at 4°C when treated at 200MPa for 15 min at 20 °C (De Ancos *et al.,* 2000). The possibility of using HHP to reduce milk allergenicity by specific hydrolysis of B-lactoglobulin at 250 MPa was found by (Olsen *et al.,* 2003). Effect of high pressure treatment was also studied on human breast milk. The effect of

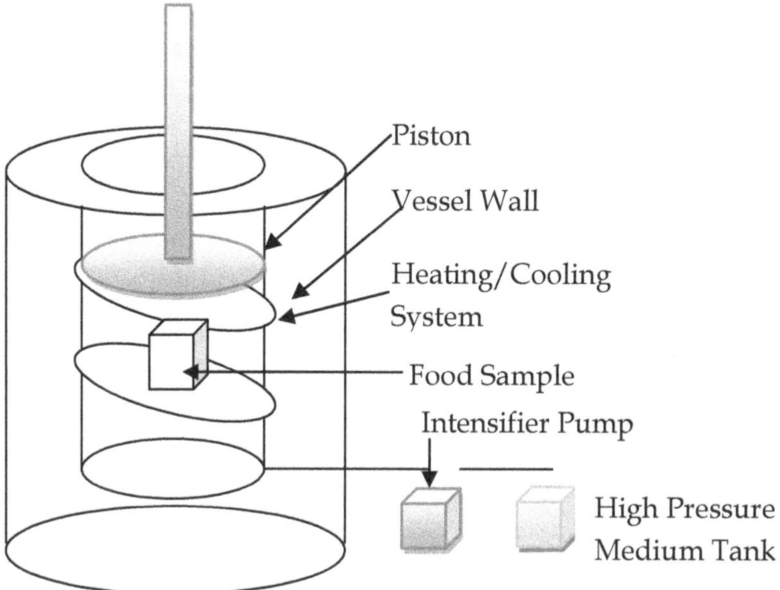

Figure 18.2: HPP Equipment.

high pressure treatment on some essential nutrients and immunological components present in breast milk and reported that treatments at 300 MPa and 50 °C maintained certain levels of Igs such as IgM (~75 per cent retention), IgA (~48 per cent) and IgG (~100 per cent), while the rest of combinations produced important decreases of their contents was studied by (Delgado *et al.,* 2013).

Advantages of HHP

HHP is suitable for products with high water content and can be modified for both batch processing and semi-continuous processing.

- ✰ Raw products can be processed without significantly altering their flavor, texture or appearance.
- ✰ Develops product packaging to withstand a changed in volume up to 15 per cent followed by a return to its original shape without losing seal integrity or barrier properties.
- ✰ It is used with hundreds of products to activate food borne packages, inactivates spoilage causing organisms, inactivates enzymes, germinates or inactivates some bacterial spores, extends shelf life, reduces the potential for food borne illness, promotes ripening of cheese and minimizes oxidative browning.
- ✰ Does not destroy the food because it is applied evenly from all side.
- ✰ Equally effective on molds, bacteria, virus.

- Reduce the processing time physical and chemical changes, retention of freshness, flavor, texture, appearance and color elimination of vitamin C loss, reduces ice crystal damage and reduces functionally alteration comparative traditional thermal processing.
- It curtails many of the diseases with compassion of raw products shown that many micro-organisms are destroyed by customary HHP operating pressure.

Limitation of HHP

- HHP is not practiced because the capital cost for a commercial scale is very high.
- It shows substantial economic losses because there is implementation of comprehensive quality assurance programmed to eliminate or reduce micro-organism in processing.
- HHP system consist of high pressure vessel, a means to close the vessel off,a system for temperature and pressure control and a material handling system so, machinery required is complex and requires extremely high precision in its construction, use and maintenance.
- HHP unit immediately becomes rate limiting steps in most processing operation.

Ultrasounds

Ultrasound technology has a wide range of current and future applications in the food industry. Ultrasound, in its most basic definition, refers to pressure waves with a frequency of 20 kHz or more (Brondum *et al.,* 1998; Butz and Tauscher, 2002). Generally, ultrasound equipment uses frequencies from 20 kHz to 10 MHz. Higher-power ultrasound at lower frequencies (20 to 100 kHz), which is referred to as "power ultrasound", has the ability to cause cavitation, which has uses in food processing to inactivate microbes. There is a great interest in ultrasound due to the fact that industries can be provided with practical and reliable ultrasound equipment. Nowadays, its emergence as green novel technology has also attracted the attention to its role in the environment sustainability (Chemat *et al.,* 2011). Ultrasound applications are based on three different methods:

- Direct application to the product.
- Coupling with the device.
- Submergence in an ultrasonic bath.

Principle of Ultrasound

Sound propagates through food materials as mechanical waves causing alternating compressions and decompressions (Blitz, 1963, 1971). These ultrasound waves have characteristic wavelength, velocity, frequency, pressure and period. The interaction of sound waves with matter alters both the velocity and attenuation of the sound waves via absorption and/or scattering mechanisms (McClements, 2005).

The velocity of sound is the product of frequency and wavelength, thereby high frequency sound waves have shorter wavelength while low frequency waves have longer wavelength. Ultrasound waves with frequencies more than 18 kHz are generated by the application of a vibration force to the surface of a material. When the vibration force is applied to the surface of a material, it is transmitted through the bonds within molecules. Further, each of the molecules transmits the motion to an adjoining molecule before returning to approximately its original position in this process. If ultrasound is applied perpendicular to the surface of the material, then a compression wave is generated within the material. Similarly, a shear wave is generated by the application of ultrasound parallel to surface. The ultrasound waves cause the layers in the material to oscillate in their original positions at the same frequency as the ultrasound waves. Thus, displacement of a fixed position in the material varies sinusoidal with time, and the time difference between two maximum positions is the period of oscillation (McClements, 1995).

The application of ultrasound to a liquid creates compressions. Thus, sound waves with sufficient high amplitude produce bubbles or cavities, and this incident is called 'cavitation'. These cavitation bubbles have a limited lifetime and break up into smaller bubbles or completely disappear. There are two types of cavitation; stable or transient. Stable cavitation occurs due to the oscillation created by ultrasound waves, which forms small bubbles in the liquid. It takes so many oscillatory cycles for the bubbles to increase their size in a stable cavitation. As the ultrasound waves pass through the liquid, they vibrate these bubbles and strong current is produced in the surrounding liquid. Further, it attracts the other small bubbles into the sonic field and microcurrents are created in the liquid. This effect is called microstreaming, which provide a subtantial force causing the cells to shear and breakdown without the collapse of bubbles. The shear force created by this process is one of the actions that lead to disruption of cells. In transient cavitation, the bubble size changes in a few oscillatory cycles and it collapses with different intensities. The larger bubbles eventually collapsed producing high pressures of up to 100 MPa and high temperatures up to 5000°K instantly. The pressure produced during bubble collapse is also sufficient to disrupt cell walls and eventually lead to cell disruption. Application of ultrasound to a liquid also leads to the formation of free radicals by sonolysis of water due to these high pressures and temperatures (Leadley and Williams, 2002; Sala *et al.*, 1995; Scherba *et al.*, 1991; Suslick, 1988).

Application of Ultrasound in Food Processing

Ultrasonic waves have a wide variety of applications in the processing and evaluation of products. From grading beef to sterilization, ultrasound has a number of applications in an increasing number of areas in the food industry. In combination with heat, it can accelerate the rate of sterilization of foods, thus reducing both the duration and intensity of thermal treatment and the resultant damage. The advantages of ultrasound over heat pasteurization include: the minimizing of flavor loss, especially in sweet juices; greater homogeneity; and significant energy savings (Crosby, 1982). Valero *et al.* (2007) studied the effect of ultrasonic treatment in treatment of orange juice. When batch ultrasonic treatment of 500 kHz/240 W for 15 min was done the microbial inactivation was upto 61.08 log CFU/ml) with no change in quality of orange juice. Homogenized skim milk treated with 20kHz at 20 and 41W under controlled temperature condition for different time interval upto 60 min and was observed that turbidity was reduced with no affect on viscosity of milk also the whey protein in milk was denatured and form soluble whey-whey/whey-casein aggregates, and these further interacted with casein micelles to form micellar aggregates during the initial 30min of sonication (Shanmugam *et al.,* 2012).

High Intensity Pulsed Light Technology

Pulsed light (PL) is an emerging nonthermal technology for decontamination of food surfaces and food packages, consisting of short time high-peak pulses of broad spectrum white light (Lopez *et al.,* 2007). It is considered an alternative to continuous ultraviolet light treatments for solid and liquid foods. High intensity light is also described as pulsed broad spectrum white light is a decontamination or sterilization technology that can be used for the rapid inactivation of micro organisms on food surfaces, equipments and food packaging materials (Mathavi *et al.,* 2007). Surface decontamination of food products using pulsed high intensity light has many potential benefits to the food industry. In 1999, pulsed light treatment of food has been approved by FDA (Federal Register, 1999). Basically, a pulsed UV-light lamp produces a continual broadband spectrum from the deep UV to infrared. This continual spectrum is especially rich and efficient in the UV range below 400 nm, which is germicidal. The flashes of pulsed-light are created by compressing electrical energy into short pulses and by using these pulses to energize a xenon lamp (XENON, 2005). The lamp then emits an intense flash of light for a few hundred microseconds. Pulsed UV-light is more cost-efficient compared to the continuous UV-light. Conventional UV-light systems produce continuous UV light with a power output in the range of 100 to 1000 Watts. Generating these power levels of high intensity can be costly to the user. For this reason, systems are designed to efficiently maximize the conversion and collection of UV radiation. However, pulsed-UV systems can generate many megawatts of electrical power inside the light source (MacGregor *et al.,* 1997). Therefore, a modest energy input can yield high peak power levels (~35 MW/cm^2).

Principle of High Intensity Pulsed Light Technology

The principle involved in generating high intensity light is that a gradual increase of low to moderate power energy can be released in highly concentrated bursts of more powerful energy. The key component of a Pulsed Light unit is a flash lamp is filled with an inert gas, such as Xenon, which emits broadband radiation that ranges from the UV cut off of the envelope material (about 180 nm) to NIR (around 1100 nm). A high-voltage, high-current electrical pulse is applied to the inert gas in the lamp, and the strong collision between electrons and gas molecules cause excitation of the latter, which then emit an intense, very short light pulse (1 μs to 0.1 s). UV plays a critical role in microbial inactivation. The antimicrobial effects of UV light on bacteria are attributed to structural changes in the DNA, as well as abnormal ion flow, increased cell membrane permeability and depolarization of the cell membrane. Some studies also indicates observable injurious effects on yeast cells and mold spores

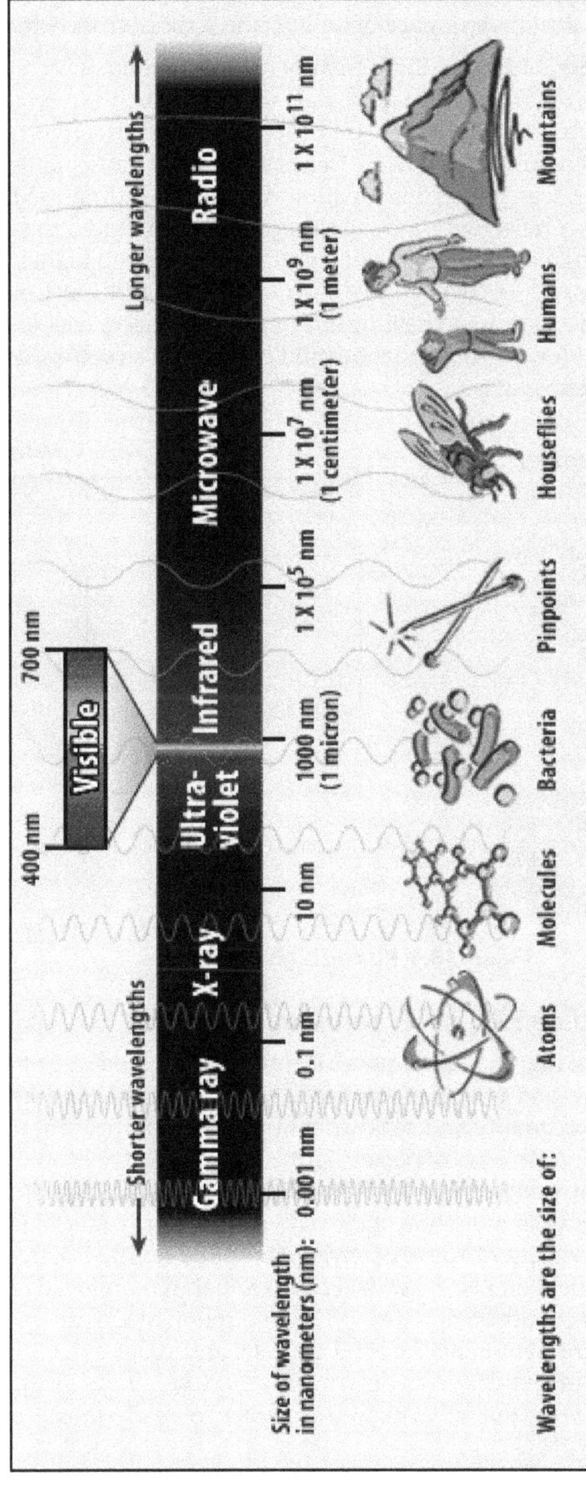

Figure 18.3: Electromagnetic Spectrum Adapted from http://amazingspace.stsci.edu/resources/explorations/groundup/lesson/basics/g1a/.

following exposure to pulsed light. The survival curves for the PL treatment display an obvious nonlinear decline, evidence of tailing, and a concave upward shape.

The main limitation of Pulsed Light treatment is its limited penetration depth. Since the effectiveness of Pulsed Light is strongly influenced by the interaction of the substrate with the incident light, the treatment is most effective on smooth, nonreflecting surfaces or in liquids that are free of suspended particulates. In surface treatments, rough surfaces hinder inactivation due to cell hiding, while for very smooth surfaces, surface reflectivity and cell clumping caused by hydrophobic effects are also limiting the degree of microbial reduction. For any Pulsed Light treatment to be fully effective, uniform, 360° exposure of the treated food is critical. Depending on the product characteristics, limited heating effects can be noticed. Heating is usually too modest to account for microbial inactivation or to cause structural and sensory changes in the treated products.

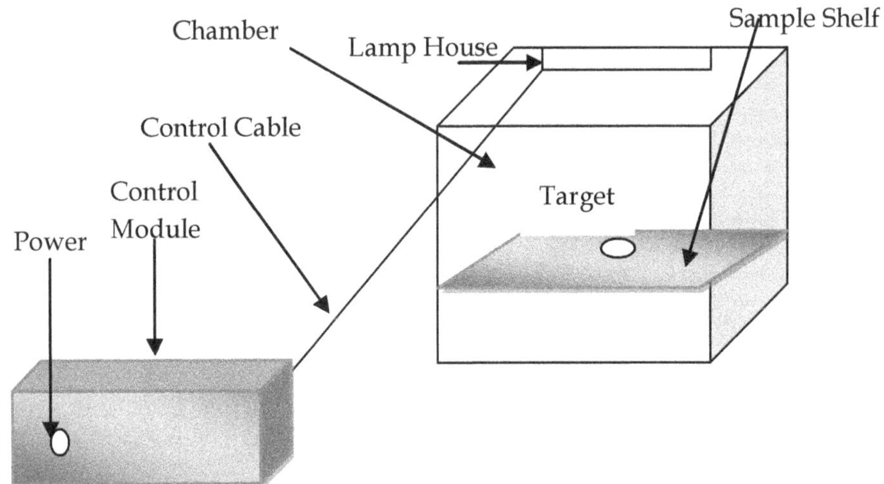

Figure 18.4: Pulsed Light Setup.

Application of Pulsed Light Treatment

Light pulses are a high potential novel technology for food decontamination. This technology is fast and environment friendly. Gemma *et al.,* 2010 studied the effect of pulsed light on quality and antioxidant properties of fresh cut mushrooms and observed that the application of pulsed light at doses of 4.8Jcm^2 could extend the shelf life of fresh-cut mushrooms without dramatically affecting texture and antioxidant properties with reduction of the native microflora of sliced mushrooms ranged from 0.6 to 2.2 log after 15 days of refrigerated storage. The inactivation effect of intense pulse light field on *Listeria monocytogens* and the commercial feasibility of this sterilization method, approximately 4-5 log reduction of the cell was achieved with intense pulse light treatments for 5000, 600, 300 and 100µs at 10, 15, 20 and 25 kV of voltage pulse respectively was determined by (Choi *et al.,* 2010). Testing of pulsed light to improve the microbial quality and safety of ready-to-eat cooked meat products and observed that when vacuum packaged ham and bologna slices were superficially inoculated with *Listeria monocytogens* and treated with 0.7, 2.1, 4.2 and 8.4 J/cm^2 was done by (Hierro *et al.*, 2011). The effect of 8.4 J/cm^2 reduced *L. monocytogenes* by 1.78 cfu/cm^2 in cooked ham and by 1.11 cfu/cm^2 in bologna also no effect on sensory attributes of cooked ham was observed while treatment above 2.1 J/cm^2 negatively influenced the sensory properties of bologna. Effect of high intensity light treatment pulses for control of *E. coli* and *Listeria inocua* in apple juice, orange juice and milk was studied and it was observed that *E.coli* and *Listeria inocua* was inactivated at pre inoculated sample and no effect on sensory qualities was seen (Palgan *et al.*, 2011). The effect of Pulsed light treatment was studied on β lactoglobulin and concluded that this treatment implementation can be worthwile for making value added ingredients because using this treatment the surface activity increased and also increase in viscoelasticity in adsorbed layer and higher foaming capacity was seen at air water interface. Pulsed light can also be used for decontamination of vegetables, dairy products, microbial inactivation of water, sanitation of packaging materials and disinfection of equipment surfaces.

Irradiation

Food irradiation is the process of treating food with a specific dosage of ionizing Radiation (WHO, 1991). The major effect of irradiation is to generate short-lived and transient radicals (*e.g.* the hydroxy radical, the hydrogen atom and solvated electrons) that in turn damage DNA and intercellular structures. The target organism ceases all processes related to maturation or reproduction. At high enough doses the target organism does not survive. Application of ionizing radiation treatment of foods on an industrial scale was started at the beginning of the 1980s after the joint FAO/IAEA/WHO expert committee accepted the application of a 10 kGy overall average dose for foods (WHO, 1981). In 1997 the FDA approved the use of low-dose ionizing radiation to eliminate pathogens in red meat. This food processing technology can improve the safety of food and extend the shelf life of certain foods by eliminating pathogenic bacteria, parasites and other microorganisms that cause food-borne disease. Currently, forty-two countries practice some form of food irradiation. In the past four decades, a vast knowledge has been accumulated on the chemical and biological effects of ionizing irradiation, which has contributed to promote its utilization (Diehl and Josephson, 1994; Josephson *et al.*, 1978; Olsen, 1998; Radomyski *et al.*, 1994). Food irradiation is currently permitted by over 50 countries, and the volume of food treated is estimated to exceed 5×10^5 metric tons annually worldwide. However, the extent of clearances is varying significantly, from a single food category (dried herbs, spices and vegetable seasonings) in Austria, Germany, and many other countries of the European Union to any food in Brazil. The recommended dose levels are: low level at 1 kGy to inhibit insect infestation and delay ripening; medium at 1 to 10 kGy to reduce bacterial load (particularly of pathogens); and high at 10 to 50 kGy for commercial sterilization and elimination of viruses (WHO, 1981, 1999). Irradiation is technique is still controversial till today due to its confusion with radioactive contamination and an imputed association with nuclear industry also due to thought of consumer about chemical changes in irradiated food will be different than that of normal heating of food (as ionizing radiation produces a higher energy transfer per collision than conventional radiant heat.

Principle of Irradiation

According to the Codex General Standard for Irradiated Foods (CAC, 2003), ionising radiations foreseen for food processing are limited to high energy photons (gamma rays of radionuclides ^{60}Co and, to a much smaller extent, ^{137}Cs, or, X-rays from machine sources with energies up to 5 MeV, or accelerated electrons with energies up to 10 MeV. In the USA, the Food and Drug Administration amended recently the food additive regulations by establishing a new maximum permitted energy level of X-rays for treating food of 7.5 MeV provided that the X-rays are generated from machine sources that use tantalum or gold as the target material (FDA, 2004).

The energy employed in food irradiation technology is referred to as *ionizing irradiation*. Ionizing radiation occurs when one or more electrons are removed from an atom. Electrons orbiting at minimum energy level or ground state can be raised to higher levels, becoming electronically excited (excitation). If enough energy is transferred to an orbital electron, the excited electron may be ejected from the atom (ionization). A minimum amount of absorbed energy, called *ionization potential*, exists for each electron energy level necessary to exit the atom domain. If the energy absorbed by the electron is greater than its ionization potential, the excess energy enters a kinetic state, enabling the electron to leave the atom domain. Although each electron behaves individually, electrons can be used in large numbers, called *electron beams*, to irradiate food. Electron beams are produced from commercial electron accelerators. One advantage of electron beam radiation is that the electron accelerators can be switched off when not in use, leaving no radiation hazard; however, the penetration of electron beams into foods is limited. Electrostatic forces tend to attract charged particles such as electrons (negatively charged), limiting the electron beam penetration into foods. Therefore, electron beam radiation is limited to small food items, such as grains, or to the removal of surface contamination of prepared meals.

Photons of electromagnetic radiation such as those created by x-rays and gamma rays travel without charge; thus, they do not interact with electrostatic forces while traveling through the food, and can penetrate deeper than electron beams. High energy electron beams are produced by electron accelerating machines. X-ray production starts with high-energy electrons: X-ray machines convert electron energy to electromagnetic X-rays called Bremsstrahlung. These types of radiation are chosen because:

(a) They produce the desired food preservative effects.

(b) They do not induce radioactivity in foods or packaging materials.

(c) They are available in quantities and at costs that allow commercial use of the irradiation process (Farkas, 2004).

The mechanism of microbial inactivation by ionising radiation is mainly due to the damage of nucleic acids, direct damage or indirect damage, affected by oxidative radicals originating from the radiolysis of water. Differences in radiation sensitivities among the microorganisms are related to differences in their chemical and physical structure, and in their ability to recover from radiation injury. The amount of radiation energy required to control microorganisms in food, therefore, varies according to the resistance of the particular species and according to the number of organisms present. Besides such inherent abilities, several factors such as composition of the medium, the moisture content, the temperature during irradiation, presence or absence of oxygen, the fresh or frozen state influence radiation resistance, particularly in case of vegetative cells (Farkas, 2006).

Application of Irradiation in Food

Irradiation of the main commodities such as tuber and bulb crops, stored grains, dried ingredients, meats, poultry and fish, or fruits has an enormous literature evolved during the past 60 years (Molins, 2001; Wilkinson and Gould, 1996). The chief potential values to consumers and food safety are in the area of preventing of food poisoning through elimination of non-sporeforming pathogens, particularly from some meats and seafood and they are well established. This technology is also used nowadays for drinking water pouches.The effect of gamma radiation on amino acids content of baby food proteins which was irradiated with dose levels of 0.5, 1.5, 6, 10, 30, 50kGy at room temperature with a gamma cell (Co-60) in the presence of air was studied by (Matloubi *et al.,* 2004). Destruction pattern of amino acids in this formulated food (whose ingredients were: wheat starch, skim milk powder, sugar, vegetable oils, vitamins, minerals, essences) was not very different from whole foods. Effect of electron beam irradiation on quality indicators of peanut butter over a storage period have been observed by (Rawas *et al.,* 2012). Peanut butter samples were exposed to a range of e-beam doses and examined over a 14-day period at 22 °C and changes in colour were observed also change in quality was observed as e-beam dose increased.

Table 18.1: Application of Irradiation in Food and Dose Requirement

Preservative Effects and Types of Application	*Dose Requirements (kGy)*
Inhibition of germination of potatoes and onions	0.03–0.12
Killing and sterilising insects (disinfestation of food)	0.2–0.8
Prevention of growth and multiplication of food-borne parasites	0.1–3.0
Decrease of after-ripening and delaying senescence of some fruit and vegetables; extension of shelf-life of food by reduction of microbial populations	0.5–5.0
Elimination of viable non-sporeforming pathogenic microorganisms (other than viruses) in fresh and frozen food	1.0–7.0
Reduction or elimination of microbial population in dry food ingredients	3.0–10
Reducing the number of viable microorganisms in enzyme-inactivated foods to such an extent that no microbial spoilage or food poisoning should occur (12D-reduction of botulinal spores by analogy with heat processing for shelf-stable foods)	25–6

High Pressure Homogenization

Homogenization was first invented in 1900 by Augustin and was presented at World Fair Paris. It allows the production of dairy and food emulsions with improved texture, taste, flavor and shelf-life characteristics, especially for dairy products like milk, cream and ice cream and provides also enhanced consumer acceptance of some products in comparison with the untreated product (Dickinson and Stainsby 1988). The demand of consumers for longer shelf-life and products with better stability has led to evolutions and new developments in the homogenization technology. More recently, in the early 1990s, a new generation of homogenizers, referred to as high-pressure homogenizers, has been developed. This new technology-high-pressure homogenization-has a different reaction chamber geometry which can withstand pressures 10 to 15 times higher than the classical homogenizers. Different types of equipment in this category now exist, both prototype and industrial scale equipment (Burgaud *et al.,* 1990). The development of equipment operating at increasingly higher pressure has created possibilities for new applications and product enhancements not available with lower-pressure operations. The term 'high pressure' is not precisely defined. In the early days, 34 MPa was considered as a high pressure for homogenization; but today, 150 MPa is not unusual. Some designs achieve even pressures of 300 MPa or more (Pandolf, 1998). Even though the homogenizer may commonly be thought of as a machine to process dairy products, it is now an essential part of many industrial applications. In the pharmaceutical, cosmetic, chemical and food industries high-pressure homogenization is used for the preparation or stabilization of emulsions and suspensions, or for creating physical changes, such as viscosity changes, in products (Pandolf, 1998; Paquin, 1999; Floury *et al.,* 2002). Another application is cell disruption of yeasts or bacteria in order to release intracellular products such as recombinant proteins

Principle of High Pressure Homogenization

A homogenizer consists of a positive displacement pump and a homogenizing valve. The pump is used to forcc thc fluid into the homogenizing valve where the work is done (Middelberg, 1995). Effluent from the homogenizer is normally chilled to minimize thermal damage to the product caused by friction heat generated due to the high fluid velocity that elevates the product temperature about 2.0 to 2.5°C per 10 MPa (Engler, 1990; Popper and Knorr, 1990). In the homogenizing valve, the fluid is forced under pressure through a small orifice between the valve and the valve seat. The fluid leaves the gap in the form of a radial jet that stagnates on an impact ring (Middelberg 1995). Finally, it exits the homogenizer at low velocity and essentially atmospheric pressure. The operating pressure is controlled by adjusting the distance between the valve and seat.

During the process, the treatment fluid is forced under high pressures to pass through a narrow gap. Thus, it creates a fast acceleration (200 m/s at 340 MPa) undergoing an extreme drop in pressure as the fluid exits the homogenization valve (Floury *et al.,* 2004), which leads to inactivation of microorganisms and denaturation of enzymes. HPH includes high–speed friction, cavitation collapse, strong impacts, turbulence, and heating. Such effects result in cell wall rupture and cellular death. It is an effective process to inactivate vegetative bacteria, yeasts and moulds. However, its low efficacy against bacterial spores turns the HPH into a treatment similar to thermal pasteurisation.

Application of High Pressure Homogenization

Microbial inactivation by high-pressure homogenization has been studied in buffer systems and in real food products like milk, ice cream and orange juice. High-pressure homogenization (HPH) is a promising technique, particularly suitable for continuous production of fluid foods, allowing to limit thermal damage and promote "freshness". The main industrial application of HPH is linked to the production of stable emulsions. For this reason, it is widely used in pharmaceutical, cosmetic and food industries. In particular, HPH has shown good potential for applicability mainly in the dairy sector. The applicability of high pressure homogenization for production of banana juice was studied and it was observed that pressure higher than 200Mpa was needed to obtain 4 log unit reduction of total mesophilic bacteria and pectate lyase inactivation also banana juice resulted brighter and less viscous than untreated ones (Calligaris *et al.,* 2012). The effect of pre treatment of milk with high pressure homogenization for manufacturing of "Pecorino" cheese was observed and it was found that cheese yield was increased and also reduction in enterococci, yeast and lactococci in cheese curd was seen (Vannini *et al.,* 2008). Effect of ultra high pressure homogenization on alkaline phosphatase and lactoperoxidase activity in raw skim milk and found that the pressure above 250 Mpa can be used for microbial inactivation and residual activity of both enzymes can be used as indicative of pressure level applied in milk during HPH process (Pinho *et al.,* 2011).

Oscillating Magnetic Field

Magnetic technology is one method of non thermal preservation for preserving food using the magnetic field. Food preservation based on oscillating magnetic field technology, worked by using magnetic fields generated from currents to coils. Magnetic fields can affect biological material, including microorganisms. Such field, under some conditions, damage microorganisms, but in others they show a stimulating effect or no effect at all on cell growth. Oscillating magnetic field (OMF) is applied in the form of constant amplitude or decaying sinusoidal waves. The magnetic field may be homogeneous or heterogeneous. OMF applied in the form of pulses reverses the charge for each pulse, and the intensity of each pulse decreases with time to about 10 per cent of the initial intensity (Pothakamury *et al.,* 1993). In many applications it is desirable to only preserve the food, without causing any change, except to destroy the microorganism that can cause spoilage. High intensity magnetic field can be used to destroy or inactivate the microorganisms in mainly non-electrically conductive environment. Destruction of microorganisms in food subjected to an OMF, applied in the form of pulsed can be achieved in a very short treatment time, resulting in no significant temperature rise in the food and no plasma production.

Principle of Oscillating Magnetic Field

Two theories to explain the inactivation mechanisms for cells placed in OMF was given given by (Pothakamury *et al.,* 1993). The first theory stated that a "weak" OMF could loosen the bonds between ions and proteins. Many proteins vital to the cell metabolism contain ions. In the presence of a steady background magnetic field such as that of the earth, the biological effects of OMF are more pronounced around particular frequencies, the cyclotron resonance frequency of ions (Coughlan and Hall, 1990). An ion entering a magnetic field B at velocity v experiences a force F given by:

$$F = q\vec{v} \times \vec{B} \quad (1)$$

The frequency at which the ions revolve in the magnetic field is known as the ion's gyrofrequency n, which depends on the charge/mass ratio of the ion and the magnetic field intensity:

$$n = q\,B/(2\,\pi\,m) \quad (2)$$

where, q is the charge and m is the mass of the ion. Cyclotron resonance occurs when n is equal to the frequency of the magnetic field. At 50 μ T, the resonance frequency of Na^+ and Ca^+ is 33.33 and 38.7 Hz, respectively. At cyclotron resonance, energy is transferred selectively from the magnetic field to the ions with n equivalent to frequency of the magnetic field. The interaction site of the magnetic field is the ions in the cell, and they transmit the effects of magnetic fields from the interaction site to other cells, tissues and organs.

A second theory considers the effect of OMF on calcium ions bound in calcium-binding proteins, such as calmodulin. The calcium ions continually vibrate about an equilibrium position in the binding site of calmodulin. A steady magnetic field to calmodulin causes the plane of vibration to rotate, or proceed in the direction of magnetic field at a frequency that is exactly equal to the cyclotron frequency of the bound calcium. Adding a "wobbling" magnetic field at the cyclotron frequency disturbs the precision to such an extent that it loosens the bond between the calcium ion and the calmodulin (Pothakamury *et al.,* 1993).

Hoffman (1985) suggested that the inactivation of microorganisms may be based on the theory that the OMF may couple energy into the magnetically active parts of large critical molecules such as DNA. Within 5-50 T range, the amount of energy per oscillation coupled to 1 dipole in the DNA is 10^{-2} to 10^{-3} eV. Several oscillations and collective assembly of enough local activation may result in the breakdown of covalent bonds in the DNA molecule and inhibition of the growth of microorganisms (Pothakamury *et al.,* 1993).

Application of Oscillating Magnetic Field

OMF can be applied to preserve solid or liquid foods. Preservation of solid foods with OMF involves sealing the food in a plastic bag, whereas for liquid foods, the product is pumped through a pipe in a continous flow. For OMF treatment the frequency applied is higher than 500 kHz for microbial inactivation. Magnetic field treatments are carried out at atmospheric pressure and moderate temperatures. The product is slightly heated to temperatures ranging from 2 to 5°C. OMF can be used to increase the shelf life of food products in combination with traditional processing techniques like Low pasteurization heat treatments. The effects of high density OMF pulses on the survival of certain selected pathogenic microorganisms in potatoes was studied. It was found that log reductions for bacteria of over three orders of magnitude with

OMF treatments, whereas fungi appeared to be more resistant to treatments, with less than 2 log reductions (Lipiec *et al.,* 2004).

Plasma Technology

Non thermal plasma is a new discipline in food processing. Plasma is electrically energized matter in a gaseous state that can be generated by electrical discharge. Electrical discharges in atmospheric pressure and low temperature make this process practical, inexpensive and suitable for decontamination of products where heat is not desirable (Ragni *et al.,* 2010). In 1922, the American scientist Irving Langmuir proposed that the electrons, ions and neutrals in an ionized gas could be considered as corpuscular material entrained in some kind of fluid medium and termed this entraining medium *"plasma"*, similar to the plasma, introduced by the Czech physiologist Jan Evangelista Purkinje to denote the clear fluid which remains after removal of all the corpuscular material in blood.

Plasma is ionized gas that consists of a large number of different species such as electrons, positive and negative ions, free radicals, and gas atoms, molecules in the ground or excited state and quanta of electromagnetic radiation (photons). It is considered to be the forth state of matter in the world (Tendero *et al.,* 2006). It can be generated in the large range of temperature and pressure by means of coupling energy to gaseous medium. This energy can be mechanical, thermal, nuclear, radian or carried by an electric current. These energies dissociate the gaseous molecules into collection of ions, electrons, charge – neutral gas molecules and other species (Nehra *et al.,* 2008). Depending on the type of energy supply and amount of energy transferred to the plasma, density and temperature of the electrons are changed. These lead Plasma to be distinguished into two groups, high temperature plasma and low temperature plasma (Nehra *et al.,* 2008). High temperature plasma implies that electron, ions and neutral species are in a thermal equilibrium state. Low temperature plasma is subdivided to thermal plasma, also called local thermodynamic equilibrium plasmas (LTE) and non thermal plasma (NTP), also called non-local thermodynamic equilibrium plasmas (non-LTE) (Tendero *et al.,* 2008). An equilibrium or near equality between electrons, ions and neutrals is the main characterization of thermal plasmas (TP).

Principle of Plasma Technology

Several mechanisms are considered to be responsible for microbial inactivation. During plasma treatment, killing microorganisms are result of direct contact to antimicrobial active spices. Accumulation of charged particles at the surface of the cell membrane can rupture the cell membrane. Oxidation of the lipids, amino acids and nucleic acids with reactive oxygen and nitrogen spices cause changes that lead to microbial death or injury. In addition to reactive spices, UV photons can modify DNA of microorganisms and as a result disturb cell replication. Contribution of mentioned mechanisms depends on plasma characteristics and to the type of microorganisms. The former includes voltage, working gas, water content in the gas, distance of the microorganism from the discharge flow, etc. where the latter takes account of Gram-positive, Gram-negative, spores and other types.

Applications of Plasma Technology

NTP has been applied in the food industry including decontamination of raw agricultural products (Golden Delicious apple, lettuce, almond, mangoes, and melon), egg surface and real food system (cooked meat, cheese). Gurol *et al.* (2012) studied the effect of cold plasma for decontamination of *E. coli* in milk and reported that initial pre-plasma bacterial count of 7.78 Log CFU/ml in whole milk decreased to 3.63 Log CFU/ml after 20 min of plasma application without any change in colour and pH of milk. Perni *et al.* (2008) studied decontamination effect of NTP generated by an AC voltage (variable 12 – 16 kV) on pericurp of melon and mangoes that inoculated by *Saccharomyces cerevisiae, Pantoea agglomerans, Gluconacetobacter liquefacien* and *E. coli.* It was observed that *S.cervisie* was the most resistant. *P. agglomerans* and *G. liquefaciens* were reduced below the detection limit (corresponding to 3 log) after only 2.5 s on both fruits, whereas *E. coli* required 5 s to reach the same level of inactivation. Montenegro *et al.* (2002) employed direct current corona discharges for reduction of *E. coli* O157:H7 in apple juice. After 40 s treatment at a frequency of less than 100 Hz with 4000 pulses of 9000 V peak voltage, the number of cell reduction was more than 5 log CFU/g.

Combination of Treatments

The potential of combinations of non-thermal technologies to reduce biological hazards and extend product shelf life has been investigated in a range of food matrices. The combination of novel, non-thermal technologies for preservation purposes is a recent trend in food processing research. Effect of combination of two non thermal process (manothermosonication MTS and pulsed electric field PEF) on raw milk was studied by (Halpin *et al.*, 2013). Homogenised milk was subjected to MTS (frequency; 20 kHz, amplitude; 27.9 mm, pressure; 225 kPa) at two temperatures (37 °C or 55 °C), before being immediately treated with PEF (electric field strength; 32 kV/cm, pulse width; 10 ms, frequency; 320 Hz) and found that the microbial count in milk treated with combination of MTS and PEF was lower. Palgan *et al.* (2012) studied the effect of combination of PEF and MTS on smoothie type beverage and observed that microbial inactivation can be achieved upto 4.2 log CFU/ml exceeding the minimum standards of USFDA which is of 5 log cfu/ml. Palgan *et al.*, 2011 studied the effect of combination of non thermal processes for inactivation of *Escherichia coli* and *Pichia fermentans* in apple and cranberry juice blend. Non-thermal hurdles such as ultraviolet light (UV) (5.3 J/cm2), high intensity light pulses (HILP) (3.3 J/cm2), pulsed electric fields (PEF) (34 kV/cm, 18 Hz, 93 µs) or manothermosonication (MTS) (4 bar, 43 °C, 750 W, 20 kHz) were examined. Combinations of non-thermal hurdles consisting of UV or HILP followed by either PEF or MTS resulted in comparable reductions for both microorganisms to those observed in thermally pasteurised samples (approx. 6 log cfu/ml).

Environmental Impact

Automation of production processes and rising demand of food safety during the last 30 years had led to think over the energy consumption and energy demand of food industry (Pereira and Vicente, 2010). Environment footprints has increased due to increase in cleaning cycle which lead to more consumption of hot and cold water as the consumer today are concentrating on more hygiene food. The majority of the energy input in food processing is used for thermal purposes such as heating, drying, evaporation, and frying. This trend is still applied until the present moment. The novel non-thermal technologies for food processing are being developed and evaluated continuously, many of them provide not only energy saving but also water savings, increased reliability, reduced emission, higher product quality and improved productivity and have less impact on environment.

A recent study (Lung *et al.,* 2006) provided estimated information on the potential energy savings of PEF and dielectric drying systems compared to existing technologies. Orange juice and cookies manufacturing were chosen as the representative target industrial sectors for the analysis of PEF pasteurization. Concerning PEF pasteurization, the natural gas savings were estimated at 100 per cent, since thermal processing is eliminated. The electricity savings of PEF can be up to 18 per cent, based on the assumed electricity consumption range of the base technology. Several applications of the novel technologies are still being developed in order to reduce the environmental impact of conventional processes such as peeling, blanching and drying. Conventionally, peeling of fruits and vegetables is performed both by immersion in hot caustic soda solution (lye) or by using steam, which requires high water use and provides less quality. OH, PEF, IF and RF can significantly accelerate drying processes when compared to untreated and conventionally heat pre-treated samples, allowing a precise control of the process temperature and leading to lower energy costs, reduced gas consumption and less combustion- related emissions. Also the pressure-assisted sterilization processes are the focus of numerous ongoing research projects, once HHP application is considered to be a waste-free process. In fact, the pre-sterilization of packaging by *e.g.* H_2O_2 or other chemical agents will not be required and will therefore contribute to a reduction of the amount of chemicals in the liquid effluents.

Conclusion

The application of emerging non-thermal preservation technologies holds potential application for producing high-quality and safe food products. Most novel non thermal technologies are still in their early stages of development although some emerging non thermal processes have now been implemented in industrial-scale systems for commercial and research applications. Other than irradiation, however, it is safe to say that not one non thermal process has been developed to a point where its use alone can guarantee the safety of low acid foods. Still further research and development are required to make commercialized products by maintaining all the legal safety standards laid down by the Food Safety Authority.

References

Ahmed, J., Ramaswamy, H. S., Alli, I., Ngadi, M. (2003). Effect of high-pressure on rheological characteristics of liquid egg. *Food Sci. Tech.*, 36, 517–524.

Alvarez, I., Pagan, R., Condon, S., Raso, J. (2003b). The influence of process parameters for the inactivation of *Listeria monocytogenes* by pulsed electric fields, *Int. J. Food Microbiol.* 87: 87-95.

Alvarez, I., Raso, I., Sala, F., Condon, S. (2003c). Inactivation of *Yersinia enterocolitica* by pulsed electric fields, *Food Microbiol.* 20, 691-700.

Bajgai, T. R., Hashinaga, E. (2001). High electric field drying of Japanese radish, *Drying Technol.* 19, 2291-2302.

Balci, A., and Wilbey, R. (1999). High pressure processing of milk the first 100 years in the development of a new technology. *Int J. Dairy Tech.* 52, 149–155.

Barbosa, G. V., Pierson, M. D., Zhang, Q. H., Schaffner, D. W. (2000). Pulsed electric fields, *Food Sci.* 65(8), 65-79.

Bendicho, S., Barbosa-Cánovas, G. V., Martín, O. (2003). Reduction of protease activity in simulated milk ultrafiltrate by continuous flow high intensity pulsed electric field treatments. *J Food Sci.* 68(3), 952–957.

Blitz, J. (1963). Fundamentals of ultrasonics, London: Butterworths and Co., 214.

Blitz, J. (1971). Ultrasonics: Methods and applications. New York: Van Nostrand Reinhold Co.

Burgaud, I., Dickinson, E., Nelson, R. V. (1990). An improved high-pressure homogenizer for making fine emulsions on a small scale. *Int. J. Food Sd. Technol.* 25, 39-46.

Butz, P., Tauscher, B. (2002). Emerging technologies: Chemical aspects. *Food Res. Int.*, 35, 279–284.

CAC (Codex Alimentarius Commission). (2003). *Codex general standard for irradiated foods.* CODEX STAN 106-1983. Rev. 1-2003.

Calderon, M. L., Barbosa-Canovas, G. V., Swanson, B.G. (1999). Transmission electron microscopy of *Listeria innocua* treated by pulsed electric fields and nisin in skimmed milk, *Int. J. Food Microbiol.*, 51, 31-38.

Calligaris, S., Foschia, M., Bartolomeoli, I., Maifreni, M., Manzocco, L. (2012). Study on the applicability of high-pressure homogenization for the production of banana juices, *LWT - Food Sci. Technol.*, 45, 117-121.

Capellas, M., Mor-Mur, M., Sendra, E., Pla, R., Guamis, B. (1996). Populations of aerobic mesophiles and inoculated *E. coli* during storage of fresh goat's milk cheese treated with high pressure. *J. Food Prot.*, 59(6), 582–587.

Cheftel, J. C. (1991). Applications des hautes pressions en technologie alimentaire. *Industries Agro-Alimentaires*, 108, 141–153.

Chemat, F., Huma, Z., Khan, M. K. (2011). Applications of ultrasound in food technology: Processing, preservation and extraction. *Ultrasonics Sonochemistry.* 18, 813–835.

Choi, M. S., Chegh, C. I., Jeong, E. A., Shin, J. K., Chung, M. S. (2010). Non thermal sterilization of *Listeria monocytogens* in infant foods by intense pulsed-light treatment. *J. Food Eng.*, 97, 504-509.

Coughlan, A., Hall, N. (1990). How magnetic field can influence your ions? *New Scientist.* 8(4):30

Crosby, L. (1982). Juices pasteurized ultrasonically. *Food Production/Management*, 16.

De Ancos, B., Pilar Cano, M., G!omez, R. (2000). Characteristics of stirred low-fat yoghurt as affected by high pressure. *Int. Dairy J.* 10, 105–111.

Delgado, F. J, Contador, R., Barrientos, A. A., Cava, R., Adamez J. D., Ramirez, R. (2013). Effect of high pressure thermal processing on some essential nutrients and immunological components present in breast milk. *Innovative Food Science and Emerging Technologies*. 19, 50–56.

Dickinson, E., Stainby, G. (1988). Emulsion Stability. In E. Dickinson and G. Stainby, (eds.), Advances in Food Emulsions and Foams, Elsevier Applied Sciences, London, Chp 1.

Diehl, J. F., Josephson, E. S. (1994). Assessment of the wholesomeness of irradiated food (a review). *Acta Alimentaria*. 23 (2), 195–214.

Dutreux, N., Notermans, S., Wijtzes, T., Gongora-Nieto, M. M., Barbosa-Canovas, G. V. Swanson, B. G. (2000). Pulsed electric fields inactivation of attached and free-living *Escherichia coli* and *Listeria innocua* under several conditions, *Int. J. Food Microbiol.* 54, 91-98.

Engler, C. R. (1990). Cell disruption by homogenizer. In J.A. Asenjo (ed.). Separation processes in biotechnology. Marcel Dekker, New York, 95-105.

Farkas, J. (2006). Irradiation for better foods, *Trends in Food Sci. Technol.* 17(4), 148-152.

FDA. (2004). Irradiation in the production, processing and handling of food Federal Register, 69 (246). 76844–76847.

Federal Register. 1999. Pulsed light treatment of food. *Fed. Regist.* 66, 338829-338830.

Fernandez-Díaz, M. D., Barsotti, L., Dumacy, E., Chefter, J. C. (2000). Effects of pulsed electric fields on ovalbumin solutions and dialyzed egg white. *J. Agric. Food Chem.* 48, 2332–2339.

Floury, J., Bellettre, J., Legrand, J., Desrumaux, A. (2004). Analysis of a new type of high pressure homogeniser. A study of the flow pattern. *Chem. Eng. Sci.* 59(4), 843-853.

Floury, J., Desrumaux, A., Axelos, M. A. V., Legrand, J. (2002). Degradation of methylcellulose during ultra-high pressure homogenization. *Food Hydrocolloid.* 16, 47-53.

Food Irradiation – A technique for preserving and improving the safety of food, WHO, Geneva, 1991.

Food, for Food Safety and Nutrition, A. (2000). Kinetics of microbial inactivation for alternative food processing technologies: High pressure processing. <http://www.foodsafety.gov/comm/ift-hpp.html>.

Gemma, O. O., Aguayo, A., Belloso, I. M., Fortuny, O. S. (2010). Effects of pulsed light treatments on quality and antioxidant properties of fresh-cut mushrooms (Agaricus bisporus). *Postharvest Boil. Tech.* 56(3), 35-50.

Gurol, C., Ekinci, F. Y., Aslan, N., Korachi, M. (2012). Low Temperature Plasma for decontamination of E. coli in milk. *Int. J. Food Microbiol.* 157, 1–5.

Halpin, R. M., Alberti, O. C., Whyte, P., Lyng, J. G., Noci, F. (2013).Combined treatment with mild heat, manothermosonication and pulsed electric fields reduces microbial growth in milk. *Food Control*, 34, 364-371.

Hendrickx, M., Ludikhuyze, L., Van den Broeck, I., Weemaes, C. (1998). Effects of high pressure on enzymes related to food quality. *Trends in Food Sci. Technol.* 9(5), 197–203.

Hendrix, C. M., Red, J. B. (1995), Chemistry and technology of citrus juices and byproducts, In: *Production and Packaging of Non-Carbonated Fruit Juices and Fruit Beverages*(P.R. Ashurst, ed.), Blackie Academic and Professional, Glasgow, UK, 53-87.

Hofmann, G.A. 1985. Deactivation of microorganisms by an oscillating magnetic field. U.S. Patent 4, 524, 079.

http://allnaturalfreshness.com/hpp/

http://amazing-pace.stsci.edu/resources/explorations/groundup/lesson/basics/g1a/

http://www.foodinnovationnetwork.co.nz/foodbowl/what-we-do/high-pressure-pasteurisation/

Jemai, A. B., Vorobiev, E. (2007). Pulsed Electric Field Assisted Pressing of Sugar Beet Slices: towards a Novel Process of Cold Juice Extraction. *Biosystems Eng.*, 93 (1), 57–68.

Josephson, E.S., Thomas, M.H., Calhoun, W.K. (1978). Nutritional aspects of food irradiation: an overview. *J. Food Proces Preserve.* 2, 299–313.

Leadley, C., Williams, A. (2002). Power ultrasound - current and potential applications for food processing, UK: Campden and Chorleywood Food Research Association Group.

Lipiec, J., Jana, P., Barabasz, W. (2004). Effect of oscillating magnetic field pulses on the survival of selected microorganisms. *International Agrophysics*, 18(4), 325-328.

Lopez, V. M., Ragaert, P., Debevere, J., devlieghere, F. (2007). Pulsed light for food decontamination: a review, *Trends in Food Science and Technol.* 18, 464-473.

Lung, R., Masanet, E., and McKane, A. (2006). The role of emerging technologies in improving energy efficiency: Examples from the food processing industry. In Proceedings of the Industrial Energy Technologies Conference, New Orleans, Louisiana.

MacGregor, S. J., Turnbull, S.M., Tuema, F.A., Farish, O. (1997). Factors affecting and methods of improving the pulse repetition frequency of pulse-charged and de-charged high pressure gas switches. *IEEE Trans. Plasma Sci.* 25, 110-117.

Matavi, V., Sujatha, G., Ramya S. B., Devi, B. K. (2013). New trends in food processing. *Int. J. Advances in Eng. Technol.* 5(2), 176-187.

Matloubi, H., Aflaki, F., Hadjiezadegan, M. (2004). Effect of gamma-irradiation on amino acid content of baby food proteins. *J. Food Comp. Anal.* 17, 133-139.

McClements, D. J. (1995). Advances in the application of ultrasound in food analyses and processing, *Trends Food Sci. Technol.* 6, 293-299.

McClements, D. J. (2005). Food emulsions: Principles, practices, and techniques. CRC.

Middelberg, A. P. J. (1995). Process-scale disruption of microorganisms. *Biotechnol. Adv.* 13, 491-551.

Miguel, N. G., Fortuny, R. S., Canovas, G. V. B., Belloso, O. M. (2011). Use of Oscillating magnetic fields in food preservation, Wiley Blackwell, Chp. 16, 222-236.

Molins R. (Ed.) (2001). Food irradiation: Principles and applications, Wiley/ Interscience, New York.

Montenegro, J., Ruan, R., Ma, H., Chen, P. (2002). Inactivation of *E. coli* O157:H7 Using a Pulsed Nonthermal Plasma System. *J. Food Sci.* 67, 646-648.

Mozumder, A., Hatano, Y. (2004). Charged particle and photon interactions with matter, Marcel Dekker, New York, 785–812.

Nehra, V., Kumar, A., Dwivedi, H. (2008). Atmospheric non-thermal plasma sources. *Int. J. Eng.* 2, 53.

Olsen, D., 1998. Irradiation of food. Scientific status summary. *Food Technol.* 52 (1), 56–62.

Olsen, K., Kristiansen, K. R., Skibsted, L. H. (2003). Effect of high hydrostatic pressure on the steady-state kinetics of tryptic hydrolysis of b-lactoglobulin. *Food Chem.* 80, 255–260.

Palgan, I., Caminiti, I. M., Munoz, A., Noci, F., Whyte, P., Morgan, D. J., Cronin, D. A., Lyng, J. G. (2011). Effectiveness of High Intensity Light Pulses (HILP) treatments for the control of *Escherichia coli* and *Listeria innocua* in apple juice, orange juice and milk. *Food Microbiol.* 28, 14-20.

Palgan, I., Caminiti, I. M., Noci, M. F., Whyte, P., morgan, D. J., Cronin, D. A., Lyng, J. G. (2011). Combined effect of selected non-thermal technologies on *Escherichia coli* and *Pichia fermentans* inactivation in an apple and cranberry juice blend and on product shelf life. *Int. J. Food Microbiol.* 151, 1–6.

Palgan, I., Noci, M. F., Whyte, P., Morgan, D. J., Cronin, D. A., Lyng, J. G. (2012). Effectiveness of combined Pulsed Electric Field (PEF) and Manothermosonication (MTS) for the control of Listeria innocua in a smoothie type beverage. *Food Control.* 25, 621-625.

Pandolf, W.D. (1998). High-pressure homogenization: Latest technology expands performance and product possibilities. *Chem. Process.* 61, 39-43.

Paquin, P. (1999). Technological properties of high-pressure homogenizers: The effect of fat globules, milk proteins and polysaccharides. *Int. Dairy J.* 9, 329-335.

Pereira, R. N., Vicente, A. A. (2010). Environmental impact of novel thermal and non-thermal technologies in food processing, *Food Res. Int.* 43, 1936–1943.

Perni, S., Liu, D. W., Shama, G., Kong, M. G. (2008). Cold Atmospheric Plasma Decontamination of the Pericarps of Fruit. *J. Food Prot.* 71, 302-308.

Pinho, C. R. G., Franchi, M. A., Tribst, A. A. L., Cristianini, M. (2011). Effect of ultra high pressure homogenization on alkaline phosphatase and lactoperoxidase activity in raw skim milk, *Procedia Food Science*, 1, 874 – 878.

Popper, L., Knorr, D. (1990). Applications of high-pressure homogenization for food preservation. *Food Technol.* 44, 84-89.

Pothakamury, U. R., Barbosa-Canovas, G. V., Swanson, B. G. (1993). Magnetic field inactivation of microorganisms and generation of biological changes. *Food Technol.* 47(12), 85-93.

Qin, B. L., Chang, F., Barbosa-Cfinovas, G. V., Swanson, B. G. (1995). Nonthermal inactivation of Saccharomyces cerevisiae in apple juice using pulsed electric fields. *Lebensm.-Wiss. Technol.* 28, 564-568.

Qin, B. L., Zhang, Q., Barbosa-Cánovas, G., Swanson, B., and Pedrow, P. D. (1995) Pulsed electric field treatment chamber design for liquid food pasteurization using finite element method, *Transactions of the ASAE*, 38 (2), 557-565.

Quass, D. W. (1997). Pulsed electric field processing in the food industry. A status report on pulsed electric field. Palo Alto, CA. Electric Power Research Institute. CR- 1097. 42, 23-35.

Radomyski, T., Murano, E. A., Olson, D. G., Murano, P. S. (1994). Elimination of pathogens of significance in food by low-dose irradiation: a review. *J. Food Prot.* 57 (1), 73–86.

Ragni, L., Berardinelli, A., Vannini, L., Montanari, C., Sirri, F., Guerzoni, M. E., Guarnieri, A. (2010). Non-thermal atmospheric gas plasma device for surface decontamination of shell eggs. *J. Food Eng.* 100, 125-132.

Rawas, A. E., Hvizdzak, A., Davenport, M., Beamer, S., Jaczynski, J., Matak, K. (2012). Effect of electron beam irradiation on quality indicators of peanut butter over a storage period. *Food Chem.* 133, 212–219.

Sala, F. J., Burgos, J., Condon, S., Lopez, P., Raso, J. (1995). Effect of heat and ultrasound on microorganisms and enzymes. In New methods of food preservation, G. W. Gould (Ed.), Glasgow: Blackie Academic and Professional, 117-204.

Scherba, G., Weigel, R. M., O'Brien, W. D. (1991). Quantitative assessment of the germicidal efficacy of ultrasonic energy, *Appl. Environ. Microbiol.* 57(7), 2079-2084.

Shanmugam, A., Chandrapala, J., Ashokkumar, M. (2012). The effect of ultrasound on the physical and functional properties of skim milk. *Inno. Food Sci. Emerg. Technol.* 16, 251–258.

Suslick, K. S. (1988). Ultrasound: its chemical, physical and biological effects, New York: VCH Publishers.

Taiwo, K. A., Angersbach, A., Knorr, D. (2002). Influenceof high intensity electric field pulses and osmotic dehydration on the rehydration characteristics of apple slices at different temperatures. *J. Food Eng.* 52, 185-192.

Tendero, C., Tixier, C., Tristant, P., Desmaison, J., Leprince, P. (2006). Atmospheric pressure plasmas: A review. *Spectrochimica Acta Part B: Atomic Spectroscopy*, 61, 2-30.

Tewari, G. (2007). High-pressure processing of foods. In G. Tewari and V. Juneja (Eds.), Advances in thermal and non-thermal food preservation. Oxford, UK: Blackwell Publishing.

Valero, M., Recrosio, N., Saura, D., Munoz, N., Marti, N., Lizama, V. (1982). Effects of ultrasonic treatments in orange juice processing. *Journal of Food Engineering* 80 (2007) 509–516.

Van Loey, A., Verachtert, B., Hendrickx, M. (2002). Effects of high electric fields pulses on enzymes, *Trends Food Sci. Technol.* 12, 94-102.

Vannini, L., Patrignani, F., Lucci, L., Ndagijimana, M., Vallicelli, M., Lanciotti, R., Guerzoni, M. E. (2008). Effect of a pre-treatment of milk with high pressure homogenization on yield as well as on microbiological, lipolytic and proteolytic patterns of "Pecorino" cheese, *Int. J. Food Microbiol.* 128, 329–335.

WHO, 1981. Wholesomeness of irradiated food, Report of a joint FAO/IAEA/WHO Expert Committee. WHO Tech.Rep. Ser. 659. Geneva.

WHO, 1999. High-dose irradiation: wholesomeness of food irradiated with doses above 10 kGy, Report of a joint FAO/IAEA/WHO study group. WHO technical Report Series 890, World Health Organization, Geneva.

Wilkinson, V. M., G.W. Gould G. W. (1996). Food irradiation. A reference guide Butterworth/Heinemann, Oxford.

Wüst, R., Pearce, R., Ortega-Rivas, E., Sherkat, F. (2004). Pulsed electric fields treatment of milk affects the properties of the cottage cheese gel. *9th International Congress on Engineering and Food*, Abst. No 896: 87. 7-11 March, Montpellier, France.

XENON. (2007). Pulsed UV-Light Benefits. Available at: http://www.xenoncorp.com/perf_cure.html.

Yokoyama, H., Sawamura, N., Motobayashi, N. (1992). Method for accelerating cheese ripening, European Patent Application EP 0 469 857 A1.

Zhang, Q. H., Barbosa-Cánovas, G. V., Swanson, B. G. (1995). Engineering aspects of pulsed electric field pasteurization. *J. Food Eng.* 25, 261–281.

Zimmermann, U. (1986). Electric breakdown, electropermeab ilization and electrofusion, Rev. Phys. Biochem. *Pharmacol.* 105, 196-256, 176-257.

Index

G

H

I

K